Signaling Pathways and Targeting Transcription in Cancer

Signaling Pathways and Targeting Transcription in Cancer

Editor: Finn McFarland

FOSTER ACADEMICS
www.fosteracademics.com

www.fosteracademics.com

FOSTER ACADEMICS

Cataloging-in-Publication Data

Signaling pathways and targeting transcription in cancer / edited by Finn McFarland.
p. cm.
Includes bibliographical references and index.
ISBN 978-1-64646-555-2
1. Cancer. 2. Cellular signal transduction. 3. Cell interaction. I. McFarland, Finn.
RC254.6 .S54 2023
616.994--dc23

Foster Academics,
118-35 Queens Blvd., Suite 400,
Forest Hills, NY 11375, USA

ISBN 978-1-64646-555-2 (Hardback)

Contents

Preface

The purpose of the book is to provide a glimpse into the dynamics and to present opinions and studies of some of the scientists engaged in the development of new ideas in the field from very different standpoints. This book will prove useful to students and researchers owing to its high content quality.

A series of chemical reactions, wherein a group of molecules in a cell work together to control a cell function is called a signaling pathway. In cancer, signaling pathways are frequently altered. For instance, signaling pathways that regulate cell cycle progression (e.g., RTK-CDK4/6) and genomic stability (BRCA1, p53 and p21) are altered. In cases of glioblastoma, melanoma, and breast adenocarcinoma, chromosomal rearrangements occur due to genomic instability. The signals in cancer have their own special characteristics during the cancer progression and these can be targeted for anti-cancer therapy. In the usual course of cell production, two processes are involved, namely, transcription and translation. However, in cancer, these processes are up regulated to match with the proliferative state of these cells. Many signaling pathways are involved in cancer progression responsible for carcinogenesis, tumor growth, metastasis, etc. and multiple factors are involved in the regulation of signaling pathways. Targeted therapies are being developed for targeting different signaling pathways. Certain therapies involve targeting transcription factors, which inhibit transcription mechanisms such as factor-cofactor protein-protein interactions. Additionally, modulation of auto-inhibition, proteolysis targeting chimaeras (PROTACs), use of cysteine reactive inhibitors, are novice approaches for targeting transcription factors. This book explores all the important aspects of signaling pathways and targeting transcription in cancer in the present day scenario. It will serve as a reference to a broad spectrum of readers.

At the end, I would like to appreciate all the efforts made by the authors in completing their chapters professionally. I express my deepest gratitude to all of them for contributing to this book by sharing their valuable works. A special thanks to my family and friends for their constant support in this journey.

Editor

Therapeutic Targeting of Stat3 Using Lipopolyplex Nanoparticle-Formulated siRNA in a Syngeneic Orthotopic Mouse Glioma Model

Benedikt Linder [1,*], Ulrike Weirauch [2], Alexander Ewe [2], Anja Uhmann [3], Volker Seifert [4], Michel Mittelbronn [5,6,7,8,9], Patrick N. Harter [5,10], Achim Aigner [2,†] and Donat Kögel [1,†]

1 Experimental Neurosurgery, Department of Neurosurgery, Neuroscience Center, Goethe University Hospital, 60528 Frankfurt am Main, Germany; koegel@em.uni-frankfurt.de
2 Rudolf-Boehm-Institute for Pharmacology and Toxicology, Clinical Pharmacology, University of Leipzig, 04107 Leipzig, Germany; Ulrike.Weirauch@medizin.uni-leipzig.de (U.W.); Alexander.Ewe@medizin.uni-leipzig.de (A.E.); Achim.Aigner@medizin.uni-leipzig.de (A.A.)
3 Institute of Human Genetics, Tumor Genetics Group, University of Göttingen, 37073 Göttingen, Germany; auhmann@gwdg.de
4 Department of Neurosurgery, Goethe University Hospital, 60528 Frankfurt am Main, Germany; V.seifert@em.uni-frankfurt.de
5 Institute of Neurology (Edinger-Institute), University Hospital Frankfurt, Goethe University, 60528 Frankfurt am Main, Germany; michel.mittelbronn@lns.etat.lu (M.M.); patrick.harter@kgu.de (P.N.H.)
6 Luxembourg Centre for Systems Biomedicine (LCSB), University of Luxembourg, 4362 Esch-sur-Alzette, Luxembourg
7 Laboratoire national de santé (LNS), 3555 Dudelange, Luxembourg
8 Luxembourg Centre of Neuropathology (LCNP), 3555 Dudelange, Luxembourg
9 NORLUX Neuro-Oncology Laboratory, Luxembourg Institute of Health (LIH), 1526 Luxembourg, Luxembourg
10 German Cancer Consortium DKTK Partner site Frankfurt/Main, Frankfurt am Main, Germany and German Cancer Research Center DKFZ, 69120 Heidelberg, Germany
* Correspondence: linder@med.uni-frankfurt.de
† These authors contributed equally to this work.

Abstract: Glioblastoma (GBM), WHO grade IV, is the most aggressive primary brain tumor in adults. The median survival time using standard therapy is only 12–15 months with a 5-year survival rate of around 5%. Thus, new and effective treatment modalities are of significant importance. Signal transducer and activator of transcription 3 (Stat3) is a key signaling protein driving major hallmarks of cancer and represents a promising target for the development of targeted glioblastoma therapies. Here we present data showing that the therapeutic application of siRNAs, formulated in nanoscale lipopolyplexes (LPP) based on polyethylenimine (PEI) and the phospholipid 1,2-dipalmitoyl-sn-glycero-3-phosphocholine (DPPC), represents a promising new approach to target Stat3 in glioma. We demonstrate that the LPP-mediated delivery of siRNA mediates efficient knockdown of Stat3, suppresses Stat3 activity and limits cell growth in murine (Tu2449) and human (U87, Mz18) glioma cells in vitro. In a therapeutic setting, intracranial application of the siRNA-containing LPP leads to knockdown of STAT3 target gene expression, decreased tumor growth and significantly prolonged survival in Tu2449 glioma-bearing mice compared to negative control-treated animals. This is a proof-of-concept study introducing PEI-based lipopolyplexes as an efficient strategy for therapeutically targeting oncoproteins with otherwise limited druggability.

Keywords: STAT3; siRNA/RNAi; polyethylenimine; PEI; lipopolyplex; siRNA delivery; glioma; glioblastoma

1. Introduction

Glioblastoma (grade IV glioma, WHO) is the most aggressive primary brain tumor in adults, with a median survival of ~15 months following standard radiochemotherapy with temozolomide (TMZ) and a 5-year survival rate below 5% [1]. Due to diffusely infiltrating tumor cells that cannot be surgically resected, most tumors will quickly recur and afflict the patients with even more aggressive disease [2]. GBM are further subdivided into distinct subtypes based on their genetic profile. Different groups reported different subtypes, but it is generally agreed that a proneural subtype with the best prognosis and a mesenchymal subtype with the worst prognosis exist [3,4]. Another subtype is the classical one [4], which is sometimes subdivided into a proneural and a neural subtype [3]. One key molecule that is frequently highly expressed and overactivated in GBM and associated with the most aggressive and treatment-resistant mesenchymal subtype [5,6] is the oncogenic transcription factor signal transducer and activator of transcription 3 (Stat3) that acts as a signaling hub promoting most hallmarks of cancer, including proliferation, cell survival, angiogenesis and immune evasion (reviewed in [7,8]). STAT3 activation is facilitated by phosphorylation of multiple upstream kinases [9]. Stat3 can be phosphorylated on two different sites (reviewed in [10]), Tyrosine705 (p^{Y705}Stat3) and Serine727 (p^{S727}Stat3), which can exert different functions in different cell types. In glioma it was shown that S727-phosphorylation is dependent on Y705-phoshorylation, that is necessary for maximal activation of Stat3 [11]. Activated Stat3 dimerizes and induces gene expression of a variety of genes, many of which are known to be important for hallmarks of cancers like migration/invasion (Mmp2, Mmp9; [12,13]) or EMT-like features and immune evasion/suppression (e.g., Snail; [14,15]). Stat3 is activated in many other human cancers (reviewed in [16]) in addition to gliomas. In particular Stat3 expression increases with tumor grade in these tumors (Figure S1 and reviewed in [9,17]), and is negatively correlated with patient survival (reviewed in [9,17]). Additionally, Stat3 is known to be necessary for activation of microglia [18], the macrophage-like population of immune cells residing in the brain. In glioma, these cells usually acquire a pro-tumorigenic phenotype that promotes glioma cell migration and invasion [18].

Consequently, we and others previously demonstrated that targeting Stat3 by upstream signaling inhibitors of the Stat3 pathway decreases glioma cell proliferation and migration in vitro and prolongs overall survival of tumor-bearing mice in vivo [19–24].

One strategy to target aberrantly activated and/or overexpressed oncoproteins in tumor therapy is to ablate their expression by RNA interference (RNAi). This knockdown strategy relies on small interfering RNA (siRNA) that, due to high sequence specificity, generally exhibits fewer off-target effects in comparison to the application of pharmacological agents (reviewed in [25,26]). The second major advantage of RNAi is that it allows to selectively interfere with the activity of otherwise un-druggable or hard-to-drug targets like Stat3 and other oncogenic transcription factors. Using a pre-transplantational genetic knockdown approach, we could previously demonstrate that ablation of Stat3 function with lentiviral shRNA limits tumor growth of Tu2449 gliomas in an orthotopic, syngeneic mouse model [21], providing support for the concept of therapeutic intervention with Stat3 activity in vivo. Therapeutic intervention in already established glioma, however, requires the development of siRNA formulations for their direct application. We have previously established siRNA complexation with nanoscale polyethlyenimine (PEI)-based complexes for therapeutic siRNA delivery in vivo. This confers protection, cellular delivery and intracellular release of formulated small RNA molecules (siRNAs, miRNAs) in vitro and in vivo, and thus allows for their therapeutic application ([27–30]; see [31] for review). More recently, this was extended towards combinations with liposomes, leading to lipopolyplexes (LPP) that combine properties of both components [32].

For gene delivery, LPP comprising phospholipids had been previously shown to display strongly reduced surface charges, enhanced transfection efficiencies, decreased cytotoxicity and high colloidal stability as compared to their unmodified complex counterparts [33,34]. Likewise, we were able to demonstrate for siRNA delivery that DPPC-based LPP without co-lipids displayed very good transfection efficiency and further enhanced biocompatibility in cell culture. Intracellular siRNA release was not impaired by the liposomal component. Due to strongly reduced surface charges, storage stabilities under various conditions were markedly increased by protecting the LPP from aggregation [35]. Importantly, these favorable properties also translated into therapeutic in vivo efficacy upon systemic injection into tumor xenograft-bearing mice [32], thus providing the basis for now switching to another model and another mode of administration.

Here we investigated the therapeutic potential of intracranially applied LPP containing Stat3-specific siRNA in an orthotopic Tu2449 glioma model. We demonstrate that specific targeting of Stat3 using LPP significantly reduces tumor growth and improves overall survival. This proof-of-concept study thus supports the notion that highly therapy-resistant cancers such as malignant gliomas and hard-to-drug genes like Stat3 can be efficiently targeted using siRNA in vivo.

2. Results

2.1. LPP Nanoparticles Provide a Stable and Efficient siRNA Formulation

As shown previously, the complexation of siRNAs with the ~4–10 kDa low molecular weight polyethylenimine PEI F25-LMW [36] leads to the formation of positively charged nanoscale complexes ('polyplexes') in a size range of ~100–350 nm, dependent on the buffer conditions used for complexation [27]. This was confirmed here by NanoSight measurements, yielding PEI/siRNA complexes of ~130 nm diameter with a zeta potential of 27 mV (Figure 1a,c). The combination of these polyplexes with DPPC liposomes of about the same size (Figure 1a,b) resulted in the formation of slightly larger lipopolyplexes (Figure 1a,b,d; [37,38]). The liposomal contribution also affected the zeta potential of the LPP, which was almost neutral as opposed to the polyplexes (Figure 1a). Size determinations by NanoSight measurements were confirmed by dynamic light scattering (DLS) which, however, also indicated the tendency of PEI/siRNA complexes to aggregate, leading to larger particle sizes (Figure 1a, Supplementary Figure S2). Notably, this was inhibited by lipopolyplex (LPP) formation (Figure 1a; note PEI/siRNA complex size in DLS measurements). The colloidal stability of LPP provides an advantage over polyplexes and prompted us to prefer LPP for intracranial siRNA delivery. In an LDH release assay in Tu2449 cells, the LPP showed excellent biocompatibility in vitro (Figure 1e). The absence of appreciable cell damage upon transfection suggests the applicability of LPP also in a sensitive environment in vivo (CNS), and similar results in the case of the parent polyplexes also indicates that even an LPP decomposition with possible polyplex release would not lead to toxic effects. Since highly positive surface charges of nanoparticles may lead to enhanced cellular uptake, we tested next whether the considerably reduced zeta potentials of LPP would negatively affect biological efficacies. Notably, upon transfection of stable EGFP (Figure 1f) or luciferase (Luc3) reporter cell lines (Figure 1g) with siRNAs formulated in polyplexes or in LPP, no differences were observed. More specifically, the comparison of cells transfected with negative control siRNA (siLuc2) vs specific siRNA (siEGFP or siLuc3, respectively) revealed a profound 50–65% knockdown after 72 h, indicating biological efficacy of both, polyplexes and LPP. For the reasons stated above, LPP were selected for further studies.

a

Nanoparticle	Nanosight Mean Diameter (nm)	SD	DLS Mean Diameter (nm)	SD	Zeta Potential (mV)	SD
DPPC liposomes	164	66	126	4.1	0.4	4.0
PEI/siRNA	129	80	486	11	27	4.2
LPP (in vitro)	254	183	293	6	1.6	1.4

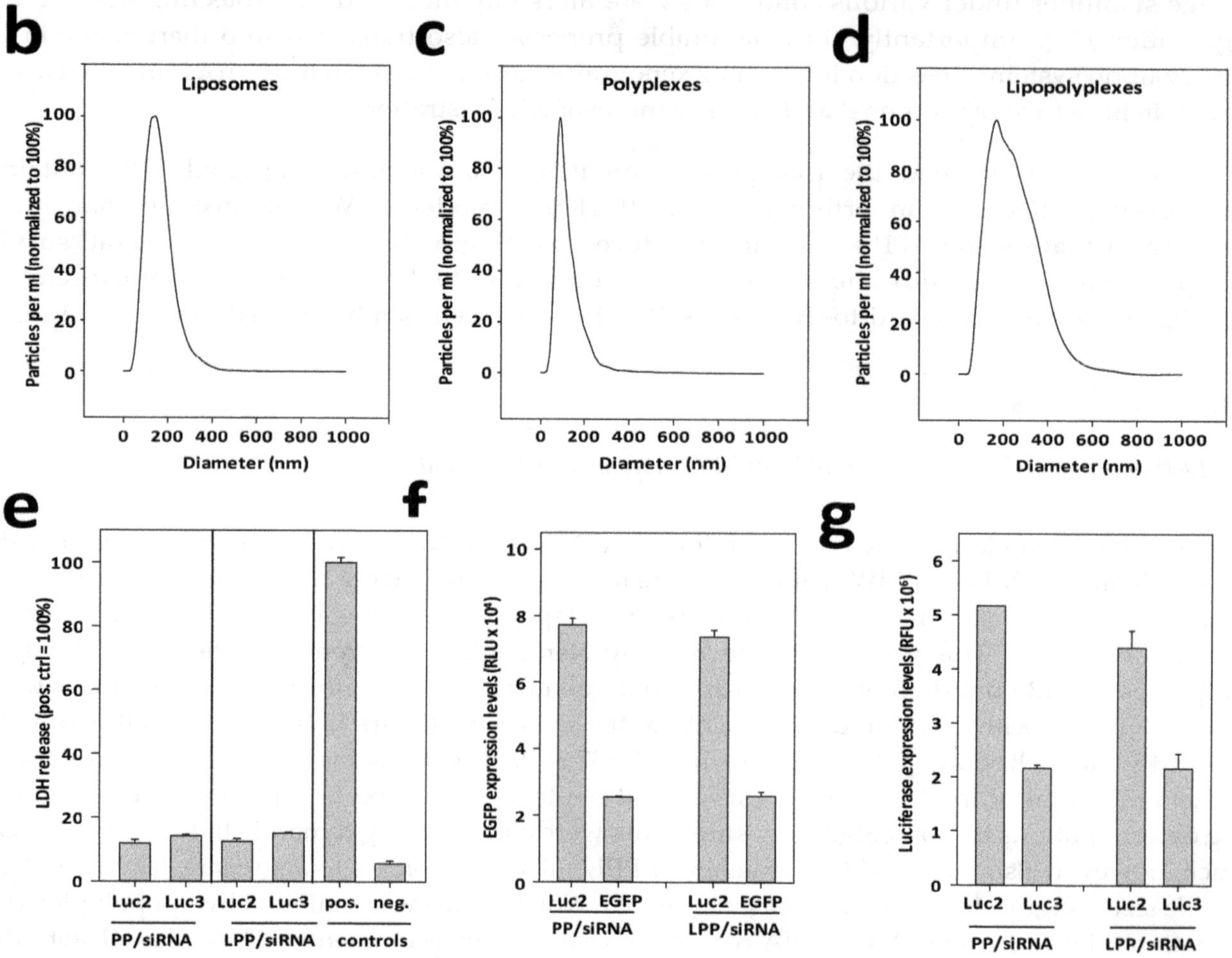

Figure 1. (**a**) Compilation of nanoparticle sizes and zeta potentials, as determined by Zetasizer and Nanosight. (**b–d**) Original diagrams of size measurements by Nanosight. (**e–g**) Biological properties of nanoparticles. (**e**) Determination of cytotoxicity by LDH release assay. (**f,g**) Determination of knockdown efficacies in EGFP/Luciferase (Luc3) Tu2449 reporter cells. Knockdown of (**f**) EGFP and (**g**) luciferase (Luc3) is shown. For (**a–d**) two independent samples were measured 10 times; (**e**) was performed twice in 4 biological replicates and (**f,g**) were performed twice in biological duplicate; shown are the summaries of all experiments.

2.2. *Targeting STAT3 Reduces Cancer Cell Proliferation In Vitro*

To test the therapeutic potential of LPP-formulated siRNA in vitro and in vivo, we selected Stat3 as a target because it is frequently overactivated in glioma and also hard-to-drug. Consistent with a number of previous reports, analysis of the TCGA data-set indicated that high expression of STAT3 is associated with an especially poor survival of GBM patients (Figure 2a). To validate these clinical observations in vitro, we knocked down STAT3 in two human GBM cell lines, U87 (Figure 2b) and Mz18 (Figure 2c). Using two different siRNAs we could significantly reduce *STAT3* mRNA expression in both cell lines, with siSTAT3-2 being more effective than siSTAT3-1. Consistently, STAT3 suppression was also achieved on the protein level in both cell lines (Figure 2d). Notably, we frequently observed a second band below the STAT3 signal in U87, but since both siRNAs target all three protein coding sequences of STAT3 (NM_213662.1, NM_003150.3 and NM_139276.2) this is likely an unspecific signal. To assess the antitumor effects of STAT3 depletion, we evaluated the growth kinetics of U87 (Figure 2e) and Mz18 cells (Figure 2f) after siSTAT3 treatment. Both cell lines showed significantly reduced proliferation 192 h after siSTAT3-treatment, with siSTAT3-2 again being more effective than siSTAT3-1. Of note, U87 cells were more sensitive to STAT3 depletion than Mz18 cells, indicating that this line may be particularly addicted to STAT3 activity, in line with findings described earlier [39]. Mz18 cells also express STAT3 and we could previously show that this line exhibits moderate levels of tyrosine-phosphorylated STAT3, which could be inhibited by upstream JAK2-inhibition [22]. We also tested the murine GBM cell line Tu2449, which we previously had used for in vivo experiments with pre-transplantational depletion of Stat3 with shRNA [21]. First, we sought out to test if siRNA-mediated Stat3-knockdown also inhibits proliferation and indeed we observed that siRNA delivery using conventional in vitro reagents like INTERFERin^{TM} also achieved a reduction in proliferation (Figure 2g). Next, we applied siRNA complexed as polyplexes, in order to verify that the delivery method does not affect knockdown efficiency. Accordingly, LPP mediated siStat3 delivery strongly inhibited proliferation (Figure 2h) and was able to efficiently reduce Stat3 and phospho-Stat3 protein levels (Figure 2i), whereas polyplexes without liposomal content were accompanied by increased nonspecific toxicities although a knockdown could also be achieved (data not shown). Thus, in these experiments LPP were found to be superior over polyplexes.

Cell cycle analysis of Tu2449 cells showed a significant increase in G1 phase and concomitant decrease in G2 phase upon siStat3 transfection, suggesting that the observed antiproliferative effect is at least in part due to a G1 arrest upon Stat3 knockdown (Figure 3a). Decreased cell cycle progression was also confirmed in the human cell lines U87 and Mz18 (Supplementary Figure S3a,b). To further verify the dependency of Tu2449 cells on Stat3 in a more complex cell culture system, we generated Tu2449 tumor spheroids, which resemble an in vivo situation more closely with regard to gradient access to oxygen, nutrients, as well as therapeutics. siRNA-mediated knockdown of Stat3 lead to distinctly smaller spheroids than control treatment (Figure 3b,c), also demonstrating that LPP are efficient in transfecting cells in spheroids.

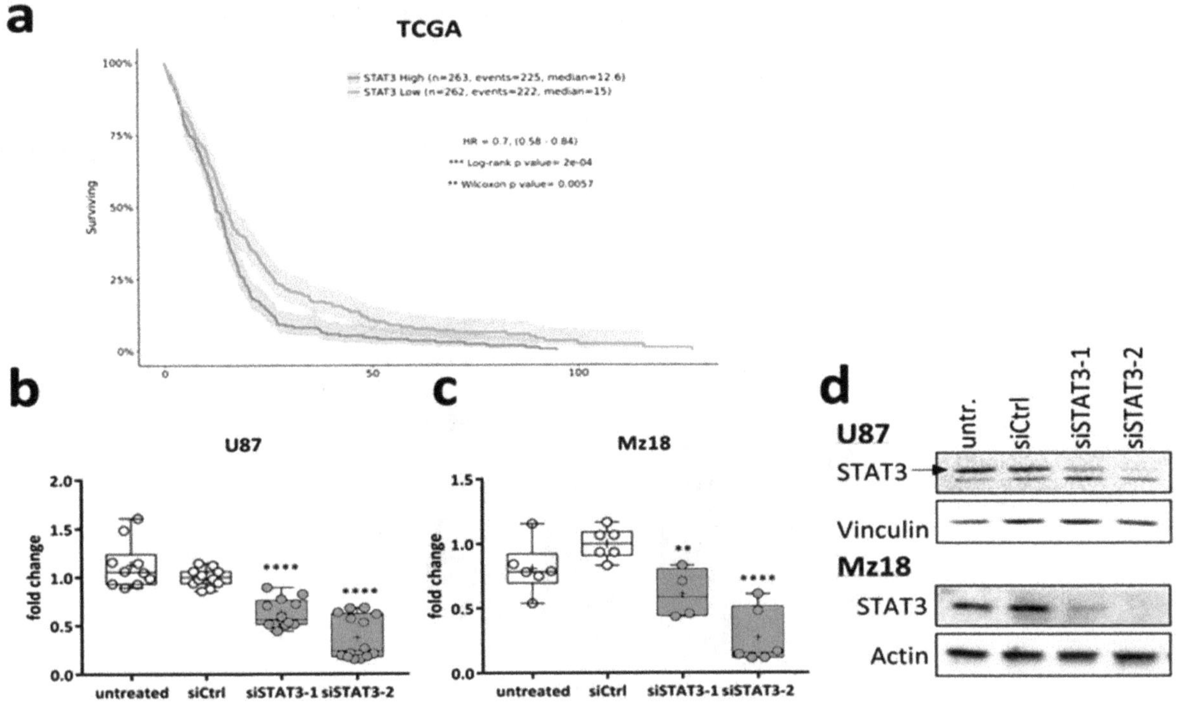

Figure 2. *Cont.*

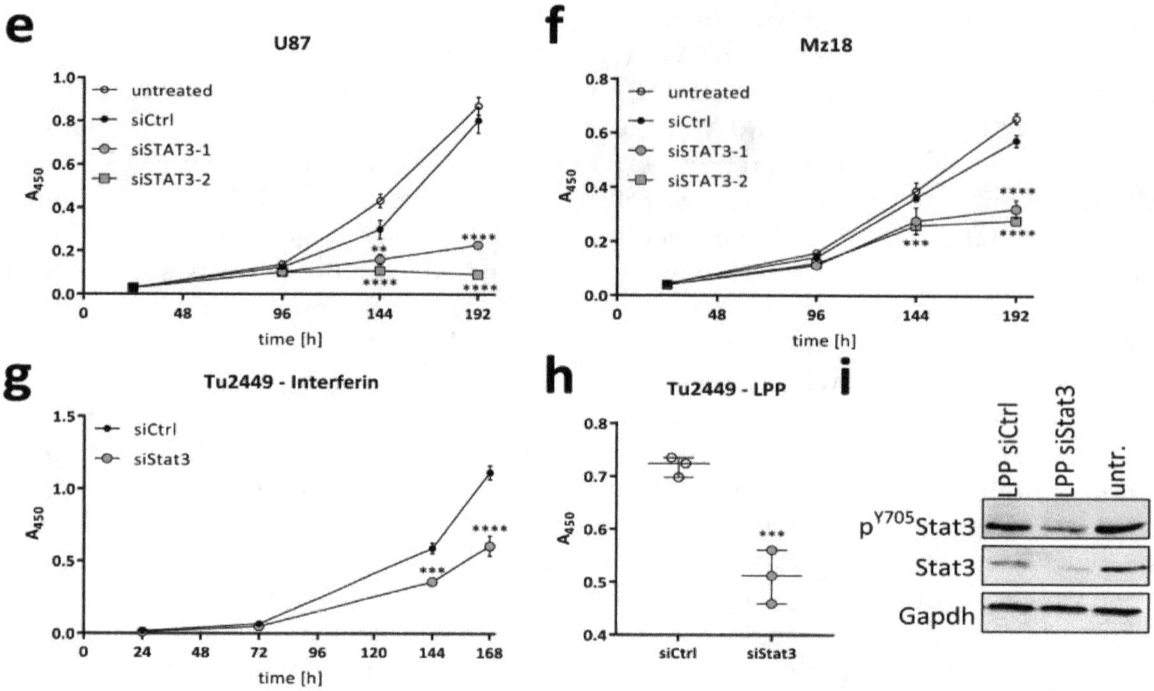

Figure 2. (**a**) Kaplan-Meier-Survival Plot from TCGA dataset GBM [40] showing that high STAT3 expression is associated with shorter survival; (**b**,**c**) qRT-PCR from (**b**) U87 and (**c**) Mz18 human glioma cell lines after transfection with control siRNA (siCtrl) or two siRNAs against STAT3 (siSTAT3-1 and siSTAT3-2). STAT3-expression was normalized to Actin as housekeeper and siCtrl-transfected cells as control sample using the ΔΔCt-method. The data are presented as box-plots (min-to-max) with all samples displayed as circles; the horizontal line in the box depicts the median value, the plus-symbol the mean. (**d**) Western Blot of U87 and Mz18 after transfection as in (**b**,**c**) after transfection of siCtrl, siSTAT3-1 or siSTAT3-2. (**e**–**h**) Proliferation (WST-1) assays of the human glioma cell lines (**e**) U87 and (**f**) Mz18, using INTERFERin and the two different siSTAT3 for comparison, and in the murine glioma cell line Tu2449 after transfection with (**g**) INTERFERin^{TM} or (**h**) LPP. The data in (**e**–**g**) are presented as mean +/− SEM; the data in (**h**) are presented as Box-Plots (min-to-max) with all samples displayed. (**i**) Western Blot of Tu2449 cells 96 h after transfection with 150 pmol LPP siCtrl or LPP siStat3. (**b**,**c**) shows the summary of at least three independent experiments performed in biological duplicates; (**d**) was performed twice; (**e**,**f**,**h**) were performed three (**g**) two times in biological triplicates; (**i**) was performed three times. **: $p < 0.01$; ***: $p < 0.001$ and ****: $p < 0.0001$ compared to siCtrl treatment.

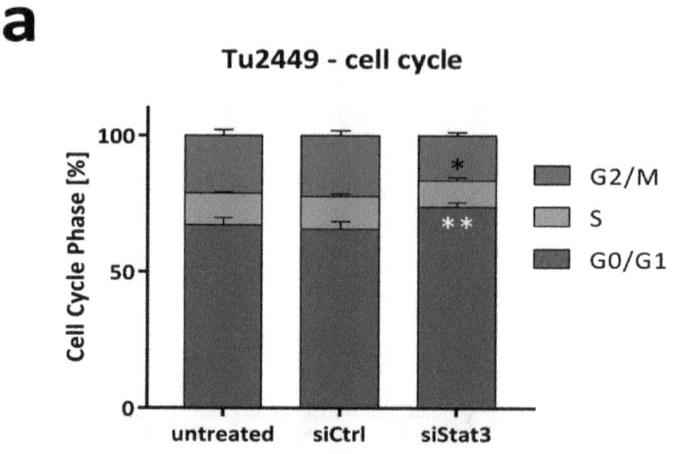

Figure 3. *Cont.*

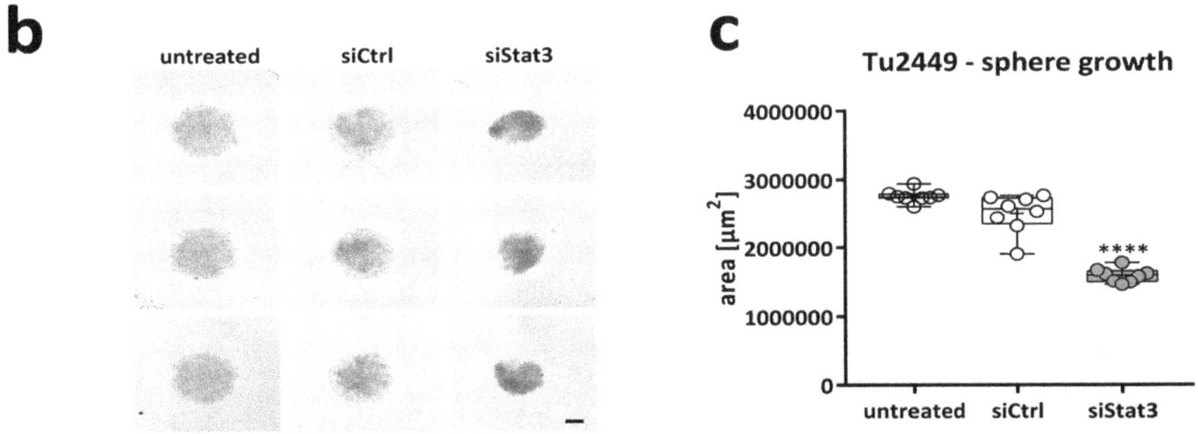

Figure 3. (**a**) Cell cycle analysis of Tu2449 cells after transfection of siCtrl or siStat3. A significant increase in the percentage of cells in the G1 phase suggests antiproliferative properties of Stat3-depletion. (**b**,**c**) Sphere growth assay of Tu2449 after transfection of siCtrl or siStat3. Three representative spheres for each condition are presented in (**b**) the quantification after 10 days of sphere growth is shown. Scale bar: 500 µm. (**c**) as box-plots (min-to-max) with all samples displayed as circles; the horizontal line in the box depicts the median value, the plus-symbol the mean. *: $p < 0.0.05$; **: $p < 0.01$ and ****: $p < 0.0001$ compared to siCtrl treatment. A representative experiment of at least three independent repetitions is shown.

In conclusion, we could successfully establish siRNA-mediated STAT3/Stat3 depletion and suppression of tumor cell proliferation by LPP siSTAT3 in vitro and verify Tu2449 as suitable cell line to explore a Stat3 targeting therapy.

2.3. LPP siStat3 Prolongs Overall Survival and Reduces Tumor Size in Glioma-Bearing Mice

Based on these findings, we pushed our system further towards therapeutic intervention, using a complex and pathophysiological relevant system. To this end, we employed our syngenic, orthotopic transplantation model [41] to assess LPP applicability and efficacy in vivo. The treatment scheme is outlined in Figure 4a. Briefly, after implantation of 10,000 Tu2449 cells into the striatum of 42 mice, the tumors were allowed to grow for one week. Hereafter, the mice were randomly divided in two groups of 21 animals each and the intracranial (i.e., intratumoral) treatment with LPPs was performed every third day. After three weeks of tumor growth, 8 mice were sacrificed for histological and molecular analyses. 1 tumor in the LPP siCtrl and LPP siStat3 cohort planned for histological assessment did not grow and were removed from the analysis. The remaining mice were monitored until they succumbed to the disease. Of those, 2 and 4 tumors of the LPP siCtrl and LPP siStat3 cohort did not grow well, respectively, and these animals were also removed from the analysis. The survival analysis (Figure 4b) showed that after LPP siStat3 none of the mice died during the treatment period, whereas six mice treated with LPP siCtrl had to be euthanized. The median survival was significantly improved after treatment with LPP siStat3 in comparison to LPP siCtrl from 26 to 33 days, respectively. Histological analyses (Figure 4c,d) confirmed that the tumors of mice treated with LPP siStat3 were significantly smaller compared to LPP siCtrl treated mice.

2.4. LPP siStat3 Slightly Reduces Stat3 Activation, But Does Not Affect Gross Tumor Proliferation Rates In Vivo

Next, we analyzed LPP siStat3 effects on the cellular and molecular level in vivo. When performing analyses of randomly chosen tumor areas (excluding the tumor border and necrotic areas), we found phospho-Stat3 (Tyr705) slightly reduced (Figure 5b). This, however, was not accompanied by alterations in proliferation rates, as determined by Ki67 staining for proliferating cells (Figure 5a). Decreased phospho-Stat3 levels were also confirmed in western blot analyses of whole tumor lysates (Figure 5c, upper panel), while Stat3 protein levels remained stable. Since Stat3 was proposed to modulate the host immune response including activation of microglia [18], we also analyzed the presence of T-cells and microglia in the tumor. Staining against CD4 and CD8a for T-cells and Iba1 for microglia showed a slight reduction of CD4-positive cells after LPP siStat3 treatment (Figure 6a) while no major differences between LPP siCtrl and LPP siStat3 treated animals were seen with regard to Cd8a and Iba1 (Figure 6b,c).

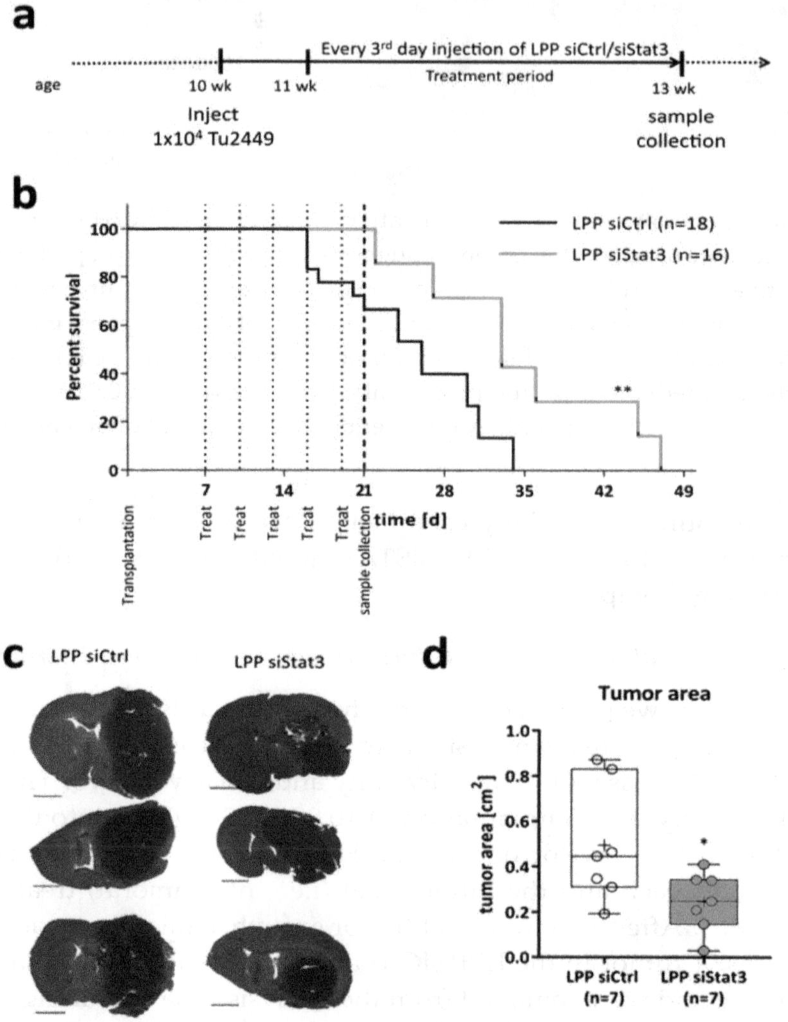

Figure 4. (**a**) Schematic presentation of the in vivo experiments. Briefly, 42 mice were injected with 10,000 Tu2449 cells into the right striatum. After 7 days the mice were randomly divided into two groups of 21 animals each and intracranial treatment was performed every third day. After 21 days 10 mice were euthanized for histological and molecular analysis. The remaining mice were monitored for long-term survival. Mice without established tumors were excluded from the analysis (3 (1 for histology; 2 for survival) for LPP siCtrl; 5 (1 for histology; 4 for survival) for LPP siStat3) (**b**) Kaplan-Meier-Survival Plot of B6C3F1 mice after implantation of Tu2449 and treatment with LPP siCtrl or LPP siStat3. The vertical dotted lines depict the treatment days, the vertical dashed line the day of sample collection. LPP siStat3 treatment significantly increases median survival from 26 days to 33 after LPP siCtrl and LPP siStat3 treatment respectively (**c**) Three representative brain sections with tumors 21 days after tumor cell implantation and treatment with LPP siCtrl (left side) or LPP siStat3 (right side); scale bar: 2 mm. (**d**) Box-Plots (min-to-max) with all samples displayed as circles; the horizontal line in the box depicts the median value, the plus-symbol the mean of tumor areas 21 days after tumor cell implantation after treatment with LPP siCtrl or LPP siStat3. *: $p < 0.05$; **: $p < 0.01$ compared to LPP siCtrl.

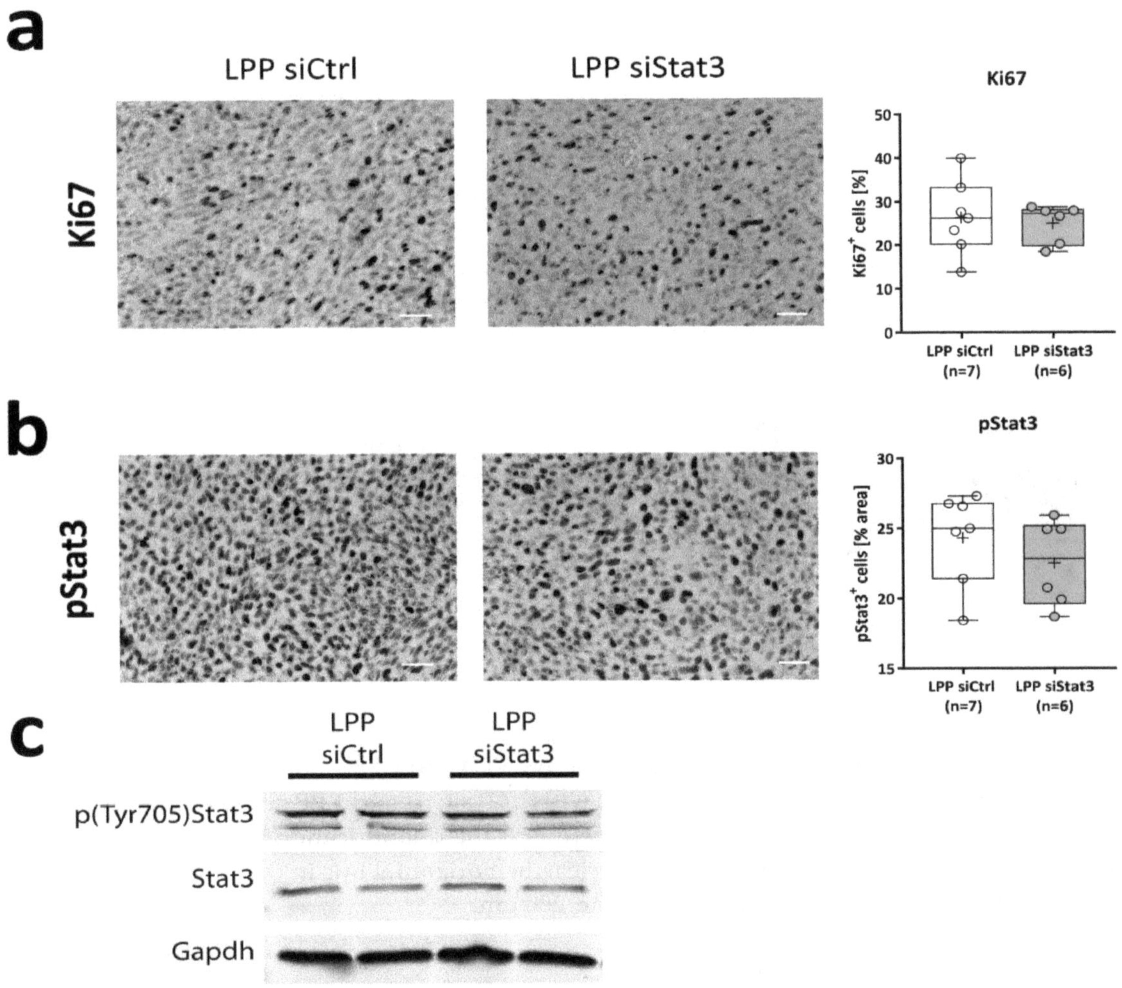

Figure 5. (**a**,**b**) Immunohistochemical analysis of (**a**) Ki67 and (**b**) tyrosine-phosphorylated Stat3 (pStat3) expressing tumor cells 21 days after implantation and treatment with LPP siCtrl and LPP siStat3. One representative picture is shown for each staining and treatment and the quantification is presented as box-plots (min-to-max) with all samples displayed as circles; the horizontal line in the box depicts the median value, the plus-symbol the mean of tumor areas 21 days after tumor cell implantation after treatment with LPP siCtrl or LPP siStat3. Scale bar: 200 µm. (**c**) Western Blot of whole-tumor lysates from two mice each after treatment with LPP siCtrl or LPP siStat3; Gapdh was used as the loading control.

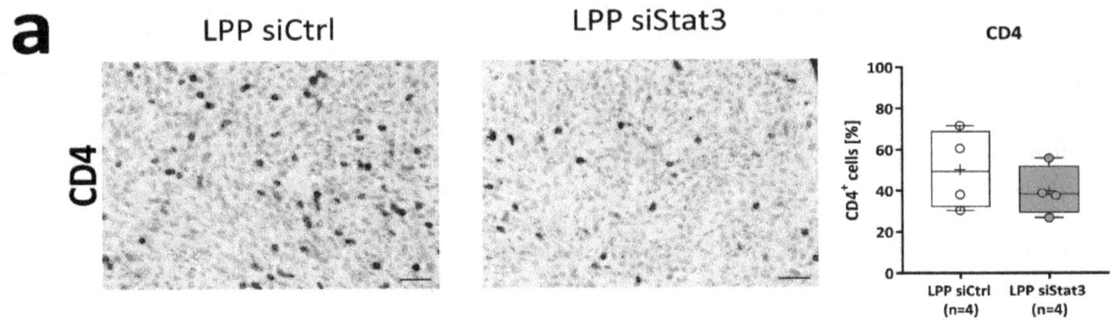

Figure 6. *Cont.*

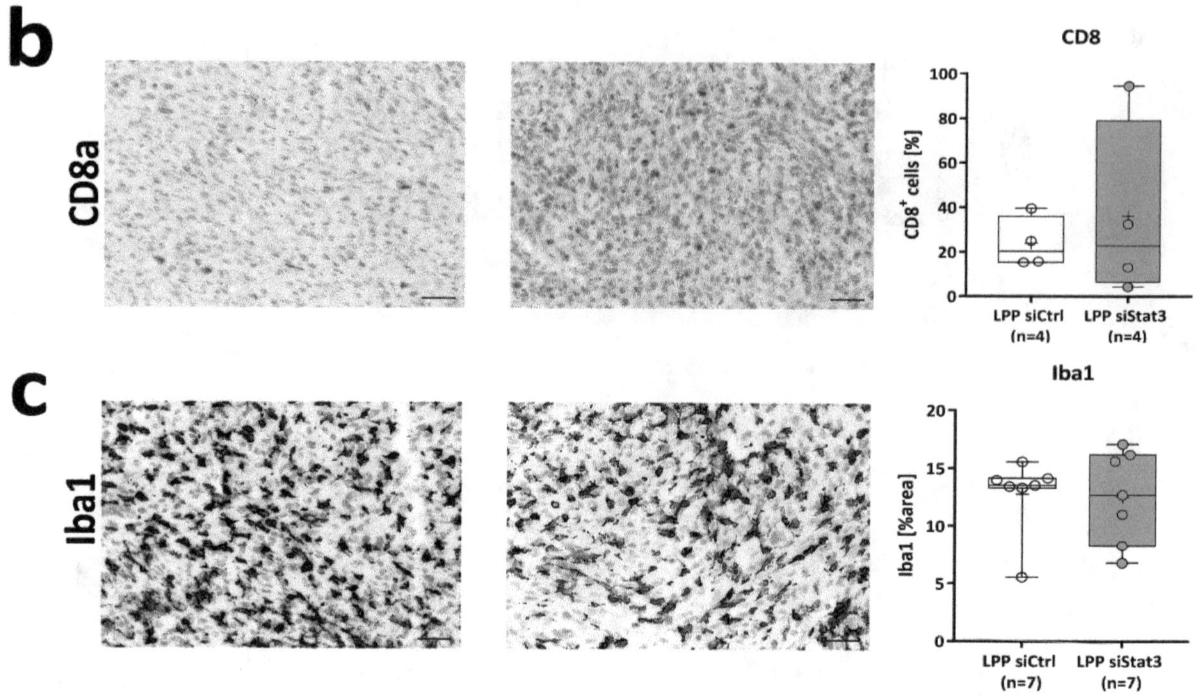

Figure 6. (**a–c**) Immunohistochemical analysis of (**a**) CD4, (**b**) CD8a and (**c**) Iba1 expressing tumor cells 21 days after implantation and treatment with LPP siCtrl and LPP siStat3. One representative picture is shown for each staining and treatment and the quantification is presented as box-plots (min-to-max) with all samples displayed as circles; the horizontal line in the box depicts the median value, the plus-symbol the mean value. scale bar: 200 μm.

2.5. LPP siStat3 Reduces Stat3 Expression in the Core Region of the Tumor

The fact that we observed significantly improved survival and reduced tumor sizes following treatment with LPP/siStat3, but could not observe a significant Stat3-repression in whole tumor samples led us to hypothesize that Stat3 depletion might be locally restricted to the area surrounding the injection site. To further investigate this hypothesis, we performed laser-capture microdissection to excise the central area of the tumor (Figure 7a) that equals the site of LPP injection. In case a sample showed necrotic areas in this region, these were first excised and discarded. Next, we isolated RNA from excised tumor core regions and performed qRT-PCR. We used 6 samples per cohort for the analyses, and obtained quantifiable data from 4 samples for *Stat3* and the housekeeping gene *Tbp*, but only 3 for the additional housekeeping *Hprt* (Figure 7b). These analyses showed that mRNA expression of *Stat3* and the known Stat3 target gene *Mmp9* was significantly decreased following LPP siStat3 treatment. Additionally, *Snai1* showed a tendency for reduced expression. The expression of *Mmp2* remained unchanged. Due to the small sample sizes, this can be interpreted as a trend towards expressional suppression.

These results support our previously proposed hypothesis that LPP siStat3 can specifically target cancer cells to limit tumor growth in vivo. Despite the robust effects of LPP-formulated siStat3 on tumor growth, our results also indicate that only a small fraction of tumor cells can be reached using the current formulations and mode of application. These observations suggest that improvement of the bioavailability may allow further enhancement of therapeutic efficacies of this approach in vivo.

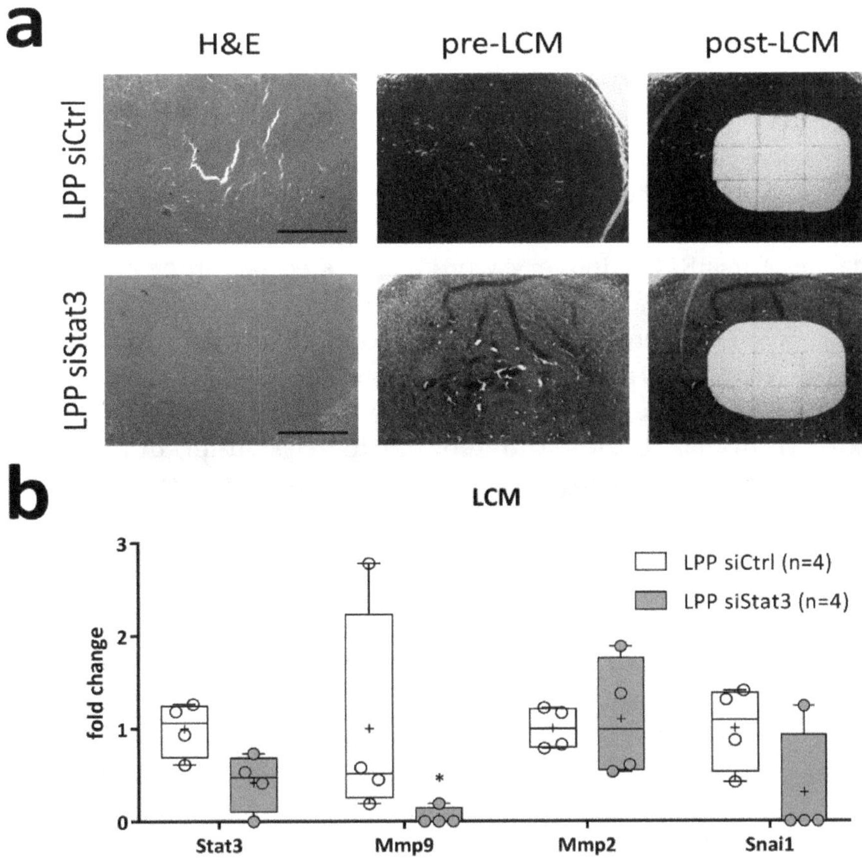

Figure 7. (**a**) Representative histology of tumor centers stained with H&E (left panel) and of unlabeled sections before (middle panel, pre-LCM) and after (right panel, post-LCM) microdissection of the central part of the tumor. Scale bar: 400 µm. (**b**) Taqman-based gene expression analysis normalized using the housekeepers *Hprt* and *Tbp* depicted as box-plots (min-to-max) with all samples displayed as circles; the horizontal line in the box depicts the median value, the plus-symbol the mean value. *: $p < 0.05$ compared to LPP siCtrl.

3. Discussion

Even after decades of research GBM still remains an incurable disease with one of the most dismal prognosis for cancer patients. The oncogenic transcription factor STAT3 represents a key signaling hub regulating many tumor-related processes including proliferation, migration, apoptosis-resistance, angiogenesis and immune evasion (reviewed in [12]). Furthermore, STAT3 expression is correlated with the highly aggressive mesenchymal subtype of glioma that is particularly resistant to conventional therapy [5,6] and is known to regulate stemness in glioma cells [6,23]. Previous findings from our and other groups had demonstrated that upstream pharmacological inhibition [22,42–44] and stable lentiviral depletion [17,21,23] of Stat3 provokes great antitumoral responses in vitro and improves survival of tumor-bearing mice in vivo.

To further advance the concept of targeting STAT3 in GBM, we decided to use a gene therapeutic approach aimed at interfering with STAT3 expression. To this end, we employed a novel approach by using siRNAs formulated in lipopolyplexes [32,37,45] that was investigated for its effects on cultured glioma cells and locally administered to syngeneic tumors of orthotopically transplanted mice that faithfully mimic all the hallmarks of human gliomas including high mitotic activity, focal necrosis surrounded by pseudopalisading cells and diffuse infiltration into the brain parenchyma [41]. Using this approach we could show that LPPs loaded with Stat3-siRNA inhibit proliferation of glioma cells in the absence of unspecific toxicity in vitro, significantly limit tumor growth in vivo and improve the overall survival of glioma-bearing mice. Our results also show that these antitumor effects were not associated with detectable differences in the numbers of infiltrating T-cells and microglia, suggesting that tumor cell-derived STAT3 represents the primary target of LPP-PEI-siRNA in vivo. The analyses of microglia were of particular interest because it was proposed that STAT3 is required for the activation of microglia [18,46] which usually acquire a pro-tumorigenic phenotype in glioma [18]. Since we did neither observe detectable differences in the number of microglia nor in their morphology, we concluded that Stat3 siRNA had no major influence on these brain-resident immune cells if supplied as LPP complexes, however.

Various approaches for siRNA delivery were also tested by other groups, e.g., by targeting Beclin1 using intranasal delivery of PEIs [47], by targeting Survivin upon stereotactic injection [28], by targeting Eg5 using siRNA delivered via an viral envelope [48], by using multifunctional surfactants as packaging reagents to target Hif-1α [49] or by dual targeting of EGFR and Akt2 using peptides [50]. This exemplifies that specific targeting of a molecular vulnerability of cancers can be exploited by various means. Accordingly, all these different approaches suffer, in principle, from the same drawbacks. These are, limited tissue distribution, securing the release in the target cells and avoidance of siRNA-degradation and/or capture by unintended cells (e.g., immune cells) or components of the extracellular matrix. Another recent approach using a compound drug consisting of an aptamer targeting PDGFbeta and an siRNA targeting Stat3 in glioma showed similar growth retardation in vitro and in vivo [19], although this study only used subcutaneous xenografts for the in vivo experiments. Interestingly, this study also indicated that a stable fraction of Stat3 protein remained present in the tumor cells that cannot be depleted using their approach, whereas reduction of Stat3 phosphorylation was very well possible. In line with these findings, we could observe depletion of phospho-Stat3 in vitro and in vivo, but did not observe depletion of total Stat3 following administration of LPP-siStat3 in vivo.

The major limitation of this study lies in the comparatively low perfusion range of siRNA achieved, providing an explanation why reduction of Stat3 expression (and its target genes) in the bulk tumor can be considered only as minor. Our data suggest that the LPP PEIs only reach a sub-fraction of the tumor cells, as seen by a clear tendency towards reduced *Stat3*-mRNA expression in the tumor core region. Due to the very low sample size (4 samples per group) a definitive answer cannot be inferred from this data. Based on the fact that (1) most samples after LPP siStat3 treatment have *Stat3* expression values below the mean of LPP siCtrl-treated animals and (2) the mRNA of the known Stat3 target gene *Mmp9* was significantly reduced after LPP siStat3 treatment, it can be deduced that LPP-mediated delivery of siRNA was successfully achieved. Additionally, based on the in vitro data we deduce that the siRNA can specifically and effectively deplete Stat3 expression, because we could successfully inhibit proliferation of Tu2449 cells and spheroids after LPP-mediated delivery of siStat3. Therefore we conclude that delivery to Tu2449 is possible even in more complex settings (i.e., spheroids). Hence, it is more likely that the limiting factor in our current approach is tissue dispersion of LPP complexes rather than siRNA specificity. Indeed, when exploring organotypic tissue slice cultures of intact tumor (xenograft) material with regard to nanoparticle tissue penetration using fluorophore-labeled siRNAs for microscopic evaluation of tissue penetrance, we did observe LPP to diffuse into the tissue only to a certain extent. LPP tissue penetration was found to be better than for polyplexes, reflecting lesser impairment of nanoparticle diffusion with reduced surface charge (zeta potential; [51]). However, in another in vivo system (non-tumorous mouse brain in an alpha-synuclein mouse model), even our polyplexes were found to distribute across the CNS down to the lumbar spinal cord after a single intracerebroventricular infusion, thus emphasizing a distribution sufficient for exerting biological effects [52].

The changes of Stat3-target genes like *Mmp9* and *Snai1* also provide mechanistic hints on how Stat3-depletion increases survival. In particular, the repression of *Mmp9* expression might reduce the migratory capacity of tumor cells due to their inability to modify the ECM. This could be further potentiated by repression of *Snai1*, which is a master regulator of EMT [15]. Furthermore, Snai1 has been shown to enhance an immunosuppressive phenotype in cancers [15]. LPP-siStat3-mediated

Snail depletion might therefore alleviate this phenotype and in turn make the tumors more accessible or amenable for immunotherapy. Future research should therefore be directed towards increasing intratumoral dispersion of siRNA in order to increase the knockdown efficiency. This could be achieved by further improving the existing complexation formulation for LPPs for example by reducing size and/or surface charges, or by enhancing the LPP concentration for being able to inject larger siRNA amounts in the maximum possible injection volume and/or development of alternative, increasingly sophisticated nanoparticle systems for delivery [26].

A different explanation for the limited Stat3-depletion could be changes in Stat3 turnover or stability. Accordingly, the half-life of Stat3 has been previously determined to be between 30 h in primary murine neurons to 90 h in hepatocytes [53] and around 50 h in Epstein-Barr-Virus-infected PBMCs [54] indicating pronounced variations between cell types. Considering that HeLa cells have a 50% turnover rate of ~60% of their entire proteome within 5 h [55] it might also be possible that the in vivo microenvironment stimulates the tumor cells to promote faster protein turnover, because of increased cell divisions. In this case, the treatment frequency every third day might be too low. However, the fact that we already observe strong increases in survival by 7 days (~25%) indicates that by continuous delivery of siRNA this issue can likely be resolved. One possibility to achieve continuous delivery could be to employ convection-enhanced delivery (CED) using implanted catheters. Accordingly, Chen et al. [56] could show that irinotecan-treatment using CED had the best survival responses in orthotopic, murine xenografts compared to conventional treatment approaches. In fact, the use of CED for treating glioma patients has also been tested in several clinical trials for the delivery of small molecules [57], antisense oligonucleotides [58] or siRNAs [59]. More recent reports also show the potential for multiplexed targeting using various siRNAs in established murine tumors [60]. Another study also reported the use of the so-called Cleveland Multiport Catheter (CMC) [61] which has already been tested in clinical trials (NCT02278510). The main advantage of the CMC is that it consists of four independent microcatheters and could therefore be employed to deliver multiple siRNAs and/or chemotherapeutics to more specifically target the vulnerabilities of a specific tumor. Despite these considerations, even the somewhat limited perfusion of Stat3-siRNA observed in our in vivo experiments with LPP therapy allowed to achieve impressive effects on tumor growth and overall survival, further supporting the suitability of STAT3 as an excellent target for GBM therapy. However, bearing in mind the enhanced colloidal stability of lipopolyplexes over their polyplex counterparts [35], the above mentioned approach of continuous delivery via direct infusion of the LPP into the tumor, using convection-enhanced delivery (CED) via an intratumoral catheter connected to a portable pump, also becomes feasible, especially because it is to be expected that siRNA stability or release kinetics would not be negatively affected by this approach.

4. Materials and Methods

4.1. Analysis of TCGA Data Sets

The publicly available TCGA datasets GBMLGG [62] and GBM [40] were accessed, analyzed and the plots were exported via the GlioVis portal (gliovis.bioinfo.cnio.es [63]). Comparison of Stat3 expression in different glioma grades was plotted based on the histology. The survival curves were derived from the GBM dataset of all GBM subtypes combined with a median cut-off, including confidence intervals.

4.2. Cells and Compounds

The murine GBM cell line Tu2449 [64] and double reporter cell line Tu2449-EGFP/Luc stably expressing EGFP and luciferase (Luc3) (kindly provided by Dr Alexander Wurm, Universitätsklinikum Leipzig, Leipzig, Germany), the human GBM cell line U-87 MG (U87, obtained from ATCC, Manassas, VA, USA) and Mz18 [65] were cultivated in DMEM Glutamax (high Glucose, Gibco/Thermo Fisher Scientific, Karlsruhe, Germany) with 10% heat-inactivated fetal calf serum and 1%

penicillin/streptomycin (all from Gibco). Cultures were maintained in a humidified incubator at 37 °C and 5% CO_2 and checked for mycoplasma contamination monthly by PCR Mycoplasma Test Kit II (AppliChem, Darmstadt, Germany) according to the manufacturer's instructions.

4.3. siRNAs + Sequences

Chemically synthesized siRNAs were purchased from MWG (Ebersberg, Germany), Dharmacon/GE Healthcare (Lafayette, CO, USA) or Eurogentec (Seraing, Belgium), with sequences as follows: siLuc2/siCtrl (human cell lines): 5′-CGU ACG CGG AAU ACU UCG AdTdT-3′ (sense), 5′-UCG AAG UAU UCC GCG UAC GdTdT-3′ (antisense); siLuc3: 5′-CUU ACG CUG AGU ACU UCG AdTdT-3′ (sense), 5′-UCG AAG UAC UCA GCG UAA GdTdT-3′ (antisense); siEGFP: 5′-GCA GCA CGA CUU CUU CAA G-dTdT-3′ (sense), 5′-CUU GAA GAA GUC GUG CUG CdTdT-3′ (antisense), human STAT3 (siSTAT3-1: CGU UAU AUA GGA ACC GUA AdTdT (sense) and 5′-UUA CGG UUC CUA UAU AAC GdTdT (antisense), siSTAT3-2: GCC UCU CUG CAG AAU UCA AdTdT (sense) and UUG AAU UCU GCA GAG AGG CdTdG (antisense) or murine Stat3 (siStat3: 5′-GGC AUA UCG AGC CAG CAA AdTdT-3′ (sense) and 5′-UUU GCU GGC UGC AUA UGC CdTdT-3′ (antisense)), or negative control siRNAs (mouse: 5′-AAU CCG CUG UCG GCU GGA AdTdT-3′ (sense) and 5′-UUC CAG CCG ACA GCG GAU UdTdT-3′ (antisense)).

4.4. Polyplex and Lipopolyplex Preparation

Liposomes from 1,2-dipalmitoyl-sn-glycero-3-phosphocholine (DPPC; Avanti Polar Lipids, Alabaster, AL, USA) were prepared by the hydration/extrusion method as described previously [32]. For polyplex formation, siRNAs were complexed in HN buffer (150 mM NaCl, 10 mM HEPES, pH 7.4) with branched 4–12 kDa PEI F25-LMW [36] at PEI/siRNA mass ratio = 7.5:1 (N/P ratio 57) as previously described [27]. Lipopolyplexes (LPP) for in vitro experiments were prepared by incubating equal volumes of PEI/siRNA complexes with DPPC liposomes at a PEI/lipid mass ratio of 2.6 [32]. For this, 25 μL polyplex containing 0.8 μg siRNA, 6 μg PEI F25 and 25 μL DPPC liposomes comprising 16 μg lipid were properly mixed by pipetting and vortexing, and incubated for at least 1 h at room temperature. For in vivo application, lipopolyplexes were prepared by complexing 0.5 μg siRNA (200 μM), 3.75 μg PEI F25 (4.2 mg/mL) in 3 μL and mixed with 10 μg DPPC (5 mg/mL; 2 μL) to a total volume of 5 μL per injection.

4.5. qRT-PCR

For preparation of total cellular RNA from cell lines, phenol/chloroform extraction using 250 μL TRI-Reagent (Sigma-Aldrich, Munich, Germany) according to the manufacturer's protocol was used. Employing the RevertAidTM H Minus First Strand cDNA Synthesis Kit (Fermentas, St. Leon-Roth, Germany), cDNA was transcribed from 800 ng RNA according to the manufacturer's protocol. Quantitative PCR (qPCR) was performed in a StepOnePlus Real-Time PCR System (Applied Biosystems, Darmstadt, Germany) using the PerfeCTa SYBR® Green FastMix ROX (Quantabio, Beverly, MA, USA). All procedures were conducted according to the manufacturers' protocols with 4 μL cDNA (diluted 1:10 with nuclease free water), 1 μL primers (5 μM) and 5 μL SYBR Green master mix. After pre-incubation for 15 s at 95 °C, 40 amplification cycles followed: 10 s at 95 °C, 10 s at 55 °C and 10 s at 72 °C. A melting curve for PCR product analysis was recorded by rapid cooling down from 95 °C to 65 °C followed by incubation at 65 °C for 15 s prior to heating to 95 °C. Actin-specific primer sets were always run in parallel for each sample, to normalize for equal mRNA/cDNA amounts. Target levels were determined by the ΔΔCt method [66]. Following primers were used: human STAT3 fwd: GAG GAC TGA GCA TCG AGC A and rev: CAT GTG ATC TGA CAC CCT GAA, human Actin fwd: CCA ACC GCG AGA AGA TGA and rev: CCA GAG GCG TAC AGG GAT AG.

Pooled samples from laser-capture-microdissected formalin-fixed paraffin embedded (FFPE) tissues were subjected to RNA-Isolation using the Arcturus® Paradise® PLUS FFPE RNA Isolation Kit (Thermo Fisher) and cDNA-synthesis using SuperScript IV Vilo (life technologies/Thermo

Fisher) per the manufacturer's instructions. Briefly, the RNA was eluted in 12 µL DEPC-treated H_2O and used entirely for the cDNA-synthesis without prior RNA quantification. The qPCR was performed using 20× Taqman Probes (Applied Biosystems, Darmstadt, Germany) and 2× Fast-Start Universal Probe Master Mix (Roche) and 2 µL of cDNA on a StepOne Plus System (Applied Biosystems) using the standard setting. The gene expression values were normalized to TATA-Box binding protein (Tbp) or Hypoxanthine-guanine phosphoribosyltransferase (Hprt) by calculating the mean Ct value of both housekeeping gene and using this value as reference for further normalization using the ΔΔCt method [66]. The following Probes were used: Hprt (Mm00446966_m1); Mmp2 (Mm00439506_m1); Mmp9 (Mm00442991_m1); Snai1 (Snail, Mm01249564_g1); Stat3 (Mm01219775_m1); Tbp (Mm01219775_m1).

4.6. Western Blotting

Preparation of protein lysates from Tu2449 cells [67] and mouse brain tumors [68] were prepared as described. SDS-PAGE and Western Blotting was performed as described [67]. Membranes were blocked in 5% BSA/TBS-Tween20 (TBS-T) for 1 h at room temperature. Antibodies were incubated at 4 °C overnight in 5% BSA/TBS-T, secondary goat anti-mouse or goat anti-rabbit (dilution 1:10,000, Li-Cor Biosciences, Bad Homburg, Germany) were incubated at room temperature for 1 h and detection was achieved using a LI-COR Odyssey reader (LI-COR Biosciences, Bad Homburg, Germany).

For protein analysis of human cell cultures, cells were seeded and transfected in 24-well plates as described above. After 72 h, medium was removed and cells were washed once with PBS. 80 µL RIPA lysis buffer (50 mM Tris (pH 7.4), 150 mM NaCl, 1% Triton X-100, 0.5% sodium deoxycholate, 0.1% SDS, 2.5 mM sodium pyrophosphate, 1 mM EDTA, 10 mM NaF, Protease Inhibitor Cocktail Set III (EDTA-free, Merck, Darmstadt, Germany) was added per well and plates were incubated on ice for 10 min. The suspension was transferred to Eppendorf tubes. After centrifugation (10,000 rpm, 4 °C, 10 min), the supernatant was transferred to a new Eppendorf tube.

Using the Bio-Rad DCTM Protein-Assay (Bio-Rad, Munich, Germany), protein concentration was determined according to manufacturer's protocol. 4× loading buffer was added (0.25 mM Tris-HCl, pH 6.8, 20% glycerol, 10% beta-mercaptoethanol, 8% SDS, 0.08% bromophenol blue) to yield a 1× concentration and 20 µg protein was loaded onto 10% polyacrylamide gels. Proteins were separated by SDS-PAGE and transferred to a 0.45 µm ImmobilonTM-P Transfer PVDF Membrane (Millipore, Burlington, MA, USA). Membranes were blocked with blocking buffer (5% (*w*/*v*) milk powder in TBST (10 mM Tris-HCl, pH 7.6, 150 mM NaCl, 0.1% Tween 20) for 30 min. After washing in TBST, membranes were incubated with primary antibodies diluted in 3% milk powder (*w*/*v*) in TBST: anti-human STAT3 (Thermo Fisher Scientific), anti-Actin (Cell signaling Technologies (CST (Frankfurt am Main, Germany)), or anti-Vinculin (Sigma-Aldrich) overnight at 4 °C. Then, blots were washed in TBST and incubated for 1 h with horseradish peroxidase-coupled goat anti-rabbit IgG (CST) or horseradish peroxidase-coupled goat anti-mouse IgG (Thermo Fisher Scientific) in 3% milk powder (*w*/*v*) in TBST before washing again. Bound antibodies were visualized by enhanced chemiluminescence (ECL kits: SignalFireTM (CST) or SuperSignal® West Femto (Thermo Fisher Scientific)).

4.7. Antibodies

The following antibodies and dilutions were used: Actin (CST, 13E5, 1:1000 in 3% milk); Cluster of Differentiation 4 (CD4); Affymetrix eBioscience/Thermo Fisher; clone 4SM95; 14-97664; 1:100 (IHC); CD8a (Synaptic Systems, Göttingen, Germany) 361003; 1:500 (IHC); Gapdh (Calbiochem, #CB1001, Darmstadt, Germany) 1:20,000; Iba1 (Wako, 019-19741); 1:1000 (IHC); Ki67 (SP6, Thermo Fisher, MA5-14520); 1:100 (IHC); Stat3 (Santa Cruz Biotechnology (Heidelberg, Germany)), c-20; sc-482; 1:200 (WB); Stat3 (phospho Tyr705; CST Technologies (CST (Frankfurt am Main, Germany)); D3A7; 9145S; 1:1000 for Western Blot (WB) and 1:500 for Immunohistochemistry (IHC)); STAT3 (Thermo Fisher, 9D8, 1:5000 in 3% milk); Vinculin (Sigma Aldrich, hVIN-1, 1:2000 in 3% milk).

4.8. *Nanoparticle Characterization*

Particle sizes and zeta potentials were determined by photon correlation spectroscopy (PCS) and phase analysis light scattering (PALS), using a Brookhaven ZetaPALS system (Brookhaven Instruments, Holtsville, NY, USA). The data were analyzed using the manufacturer's software and applying a viscosity and refractive index of pure water at 25 °C. For size determination, the complexes were analyzed in five runs with a run duration of 1 min. Results are expressed as intensity weighted mean diameter from different experiments. Zeta potentials were measured in ten runs, with each run containing ten cycles, and applying the Smoluchowski model. Additionally, hydrodynamic diameters of the nanoparticles were determined by nanoparticle tracking analysis (NTA), using a NanoSight LM 10 HS apparatus (Malvern) equipped with a 640 nm sCMOS camera and a temperature controlled sample chamber, as described previously [45,69].

4.9. *Cell Transfection*

EGFP and luciferase knockdown experiments were performed in 24-well plates. The day before, Tu2449-EGFP/Luc cells were seeded at a density of 30,000 cells per well in 0.5 mL fully supplemented medium. Polyplexes and lipopolyplexes comprising 0.8 µg siRNA per well were prepared as described above, containing siRNAs against the targets EGFP or luciferase, or the negative control siRNA (see above Section 4.3).

For transfection using INTERFERin^{TM} (Polyplus, Illkirch, France), 2×10^2 cells were seeded in a 96-well plate (proliferation assay; U87 or Mz18 cells) or 2×10^5 cells in a 24-well plate (cytotoxicity assay; Tu2449 cells), respectively. Cells were incubated under standard conditions unless stated otherwise. 10 nM siRNA were transfected according to the manufacturer's protocol and incubated for the indicated periods of time.

4.10. *Determination of Knockdown Efficacies*

Luciferase activities were determined in the double reporter cell line Tu2449-EGFP/Luc 96 h post transfection using the Beetle-Juice Kit (PJK, Kleinblittersdorf, Germany). The medium was aspirated and 300 µL lysis buffer (Promega, Mannheim, Germany) was added, prior to incubation for 30 min at RT. In a test tube, 25 µL luciferin substrate was mixed with 10 µL cell lysate and luminescence was immediately measured in a luminometer (Berthold, Bad Wildbad, Germany).

The EGFP knockdown was quantitated after 96 h by flow cytometry. Cells were trypsinized and centrifuged for 3 min at 3000 rpm followed by a washing step with 1% BSA in PBS (1 mL), and finally resuspended in 0.5 mL of the same buffer. The cells were measured in an Attune® Acoustic Focusing Cytometer (Life Technologies, Darmstadt, Germany) with appropriate instrument settings. 20,000 events were gated and analyzed using the Attune® software (V2.1.0). The data are presented as mean fluorescent intensities.

4.11. *Cell Proliferation and Viability Assays*

For proliferation assays, U87 or Mz18 cells were seeded and transfected in 96-well plates as described above. The number of viable cells was determined at the time points indicated using a colorimetric assay. Briefly, 50 µL of a 1:10 dilution of Cell proliferation Reagent WST-1 (Roche Molecular Biochemicals, Mannheim, Germany) in serum-free medium was added to the cells after aspirating the media. Cells were incubated for 1 h at 37 °C and absorbance at 450 nm was measured in an ELISA reader.

Acute cell damage upon transfection was analyzed in Tu2449 cells by the lactate dehydrogenase (LDH) release assay, using the Cytotoxicity Detection Kit (Roche, Mannheim, Germany) according to the manufacturer's protocol. Briefly, conditioned medium from knockdown experiments was collected after 24 h. Conditioned medium from untreated cells served as negative control, and for the determination of maximum LDH release (100% value), cells were lysed by adding Triton X-100 to final concentration of 2% into the medium. In a 96 well plate, 50 μL sample medium was mixed with 50 μL reagent mix and incubated for 30 min in the dark. The reaction was stopped with 50 μL 1 M acetic acid and the absorption at 490 nm and 620 nm as reference filter was measured in an ELISA reader. Fresh fully supplemented medium and reagent mix served as background and was subtracted from all values. Cytotoxicity values are shown in percent of the maximum value.

4.12. Cell Cycle Analysis

Cell cycle distribution was assessed using flow cytometry. Cells were seeded in 24-well plates and transfected as described above. 48 h after transfection, cells were trypsinized, transferred to Eppendorf tubes and centrifuged (1500 rpm, room temperature, 5 min). The supernatant was aspirated and the cell pellet was resuspended in ice cold 70% ethanol and incubated for at least 1 h at 4 °C for fixation. After centrifugation (1500 rpm, room temperature, 5 min) and discarding the supernatant, cells were resuspended in an RNase A solution (50 μg/mL RNase A in PBS) and incubated for 30 min at 37 °C. Propidium iodide solution was added to yield a 50 μg/mL concentration. Cell cycle distribution was analyzed in the BL2 channel using an Attune™ Flow Cytometer.

4.13. Spheroid Assay

To generate spheroids, 3×10^3 Tu2449 cells were seeded in a 96-well U-shaped ultra-low attachment plate and incubated under standard conditions. Using SAINT-sRNA transfection reagent (Synvolux, Leiden, The Netherlands), 90 nM siRNA were transfected according to the manufacturer's protocol.After 10 days, spheroid size was documented and measured using a Celigo Imaging Cytometer (Nexcelom Bioscience, Lawrence, MA, USA) and the corresponding software.

4.14. Animal Experiments

Intracranial implantation of tumor cells into the right striatum of B6C3F1 mice (Envigo, Huntingdon, UK; average weight 25 g) was performed as described [21]. Briefly, 10,000 viable Tu2449 in 1 μL were injected with a flow rate of 0.5 μL/min at a depth of 3 mm 1.5 mm posterior and 2 mm lateral to the bregma. 42 mice were implanted with tumor cells and one week after implantation the animals were randomly divided into two groups; one received LPP siCtrl treatment, the other LPP siStat3 treatment. Application of LPPs (equivalent to 0.5 μg siRNA in 5 μL which, due to limitations regarding the total lipopolyplex volume, was the maximum possible amount in this experimental setting) was performed using the same drill hole in the skull as for tumor cell implantation every third day one week after tumor cell implantation. 21 days after tumor implantation 10 mice per group were sacrificed; of those, 8 mice were used for histological analysis and two for molecular analyses. Those tumors were divided and one half was immediately frozen on dry ice and used for protein isolation, the other half was stored in RNAlater (Qiagen, Hilden, Germany) and used for RNA isolation. All intracranial injections were performed on anesthetized animals and treatment was done unblinded. Histoligical analysis showed that in the LPP siCtrl and LPP siStat3 group three and five tumors could not be established. Of those, one and two tumors were planned for histological assessment after 21 days. These mice were removed from the survival-analysis. Animals at the collection time point or those that showed neurological symptoms related to tumor burden were euthanized by cervical dislocation after they received a lethal injection of anesthetic. Hereafter the brains were immediately removed, rinsed in PBS and fixed for 2–7 days in 4% PFA (Chemcruz/Santa Cruz). Prior to paraffin embedding the fixed brains were cut coronally in 2–3 mm thick section and the resulting sections were aligned horizontally. Embedding in paraffin was performed according to standard paraffin embedding procedures known from routine pathology. Sildes were first stained with hematoxylin and eosin. Immunohistochemistry was performed on an automated staining system Bond III (Leica Microsystems, Wetzlar, Germany) using standard protocols. All animal experiments were approved by the local administrative council (Regierungspräsidium Darmstadt, FK/1011).

4.15. Determination of Tumor Size

The area of tumors was determined by measuring the longest distance (a) between tumor borders and the respective orthogonal line (b) from all tumor sections visible from H&E stained tumor sections. Using the formulas for ellipses the area (A) was calculated:

$$A = \pi \times 0.5 \times a \times b$$

If a brain contained multiple tumor sections those values were summed to account for the depth of the tumor. If multiple sections from one mouse were available the area, indicative for the tumor center, was chosen.

4.16. Laser-Capture Microdissection

Laser capture microdissection (LCM) was performed of 10 μm thick, serial sections mounted on UV-irradiated MMI membrane slides (MMI, Eching, Germany) and collected on MMI Isolation caps with diffuser caps (MMI). Prior to microdissection, the sections were dried at 37 °C for 1 h and rehydrated (2 times 10′ xylene, 99%, 96%, 70% EtOH, 30″ each) and the first and last section of the series were stained with hematoxylin and eosin, dehydrated and mounted using Pertex (Leica, Wetzlar, Germany). LCM was performed using a laser capture microscope (Oylmpus IX71 equipped with MMI CellCamera und MMI CellCut) operated by the software MMI CellTools (v. 4.0).

4.17. Statistics

All statistical analysis were performed using GraphPad Prism 7 (GraphPad Software, La Jolla, CA, USA). Kaplan-Meier-Survival curves were analyzed using log-rank test, all other experiments using two-tailed T-Test, with assuming equal SD and without corrections for multiple comparisons. Cell cycle measurement was analyzed by two-way-ANOVA of matched samples with Dunnett's multiple comparison test against siCtrl treated cells. Unless stated otherwise, all samples are presented as Box-Plots (min-to-max), where the horizontal line depicts the median value and the "+" symbol depticts the calculated mean.

5. Conclusions

In conclusion, we provide first proof-of-concept evidence that intratumoral targeting of oncogenes like Stat3 using lipopolyplexsosomal siRNA for the treatment of brain tumors is feasible and leads to improved overall survival in a murine, syngeneic, orthotopic transplantation model. Future development of nanoparticle systems and the modes of their application will hopefully enhance the restricted tissue distribution achieved in this study, thereby further improving the tumor targeting capacity of siRNA. These efforts may lead to the development of new and highly specific treatments selectively targeting the expression of oncogenes such as STAT3, an avenue recently opened in a clinical trial focusing on another gene target in brain tumors (NCT03020017).

Author Contributions: Conceptualization, B.L., A.A. and D.K.; Formal analysis, B.L. and U.W.; Funding acquisition, V.S., A.A. and D.K.; Investigation, B.L., U.W., A.E., A.U., M.M. and P.N.H.; Project administration, A.A. and D.K.; Resources, A.U., V.S., M.M., P.N.H., A.A. and D.K.; Supervision, V.S., A.A. and D.K.; Visualization, B.L. and U.W.; Writing—original draft, B.L., U.W., A.A. and D.K.; Writing—review & editing, B.L., A.A. and D.K.

Acknowledgments: We would like to thank Gabriele Köpf and Hildegard König for excellent technical assistance and Tatjana Starzetz and Maika Dunst (Institute of Neurology (Edinger-Institute), University Hospital Frankfurt, Goethe University, Frankfurt am Main, Germany) for sectioning and staining of FFPE tissue. We are also grateful to Hanibal Bohnenberger (Institute of Pathology, University Medical Centre, Göttingen, Germany) for access to and help with the LCM microscope. M.M. would like to thank the Luxembourg National Research Fund (FNR) for the support (FNR PEARL P16/BM/11192868 grant).

References

1. Louis, D.N.; Perry, A.; Reifenberger, G.; von Deimling, A.; Figarella-Branger, D.; Cavenee, W.K.; Ohgaki, H.; Wiestler, O.D.; Kleihues, P.; Ellison, D.W. The 2016 World Health Organization Classification of Tumors of the Central Nervous System: A summary. *Acta Neuropathol.* **2016**, *131*, 803–820. [CrossRef]
2. Becker, K.P.; Yu, J. Status quo–standard-of-care medical and radiation therapy for glioblastoma. *Cancer J.* **2012**, *18*, 12–19. [CrossRef]
3. Verhaak, R.G.; Hoadley, K.A.; Purdom, E.; Wang, V.; Qi, Y.; Wilkerson, M.D.; Miller, C.R.; Ding, L.; Golub, T.; Mesirov, J.P.; et al. Integrated genomic analysis identifies clinically relevant subtypes of glioblastoma characterized by abnormalities in PDGFRA, IDH1, EGFR, and NF1. *Cancer Cell* **2010**, *17*, 98–110. [CrossRef]
4. Olar, A.; Aldape, K.D. Using the molecular classification of glioblastoma to inform personalized treatment. *J. Pathol.* **2014**, *232*, 165–177. [CrossRef]
5. Carro, M.S.; Lim, W.K.; Alvarez, M.J.; Bollo, R.J.; Zhao, X.; Snyder, E.Y.; Sulman, E.P.; Anne, S.L.; Doetsch, F.; Colman, H.; et al. The transcriptional network for mesenchymal transformation of brain tumours. *Nature* **2010**, *463*, 318–325. [CrossRef]
6. Hossain, A.; Gumin, J.; Gao, F.; Figueroa, J.; Shinojima, N.; Takezaki, T.; Priebe, W.; Villarreal, D.; Kang, S.G.; Joyce, C.; et al. Mesenchymal Stem Cells Isolated From Human Gliomas Increase Proliferation and Maintain Stemness of Glioma Stem Cells Through the IL-6/gp130/STAT3 Pathway. *Stem. Cells* **2015**, *33*, 2400–2415. [CrossRef]
7. Yu, H.; Lee, H.; Herrmann, A.; Buettner, R.; Jove, R. Revisiting STAT3 signalling in cancer: New and unexpected biological functions. *Nat. Rev. Cancer* **2014**, *14*, 736–746. [CrossRef]
8. Galoczova, M.; Coates, P.; Vojtesek, B. STAT3, stem cells, cancer stem cells and p63. *Cell. Mol. Biol. Lett.* **2018**, *23*, 12. [CrossRef]
9. Kim, J.E.; Patel, M.; Ruzevick, J.; Jackson, C.M.; Lim, M. STAT3 Activation in Glioblastoma: Biochemical and Therapeutic Implications. *Cancers (Basel)* **2014**, *6*, 376–395. [CrossRef]
10. Brantley, E.C.; Benveniste, E.N. Signal transducer and activator of transcription-3: A molecular hub for signaling pathways in gliomas. *Mol. Cancer Res.* **2008**, *6*, 675–684. [CrossRef]
11. Ganguly, D.; Fan, M.; Yang, C.H.; Zbytek, B.; Finkelstein, D.; Roussel, M.F.; Pfeffer, L.M. The critical role that STAT3 plays in glioma-initiating cells: STAT3 addiction in glioma. *Oncotarget* **2018**, *9*, 22095–22112. [CrossRef]
12. Luwor, R.B.; Baradaran, B.; Taylor, L.E.; Iaria, J.; Nheu, T.V.; Amiry, N.; Hovens, C.M.; Wang, B.; Kaye, A.H.; Zhu, H.J. Targeting Stat3 and Smad7 to restore TGF-beta cytostatic regulation of tumor cells in vitro and in vivo. *Oncogene* **2013**, *32*, 2433–2441. [CrossRef]
13. Westermarck, J.; Kahari, V.M. Regulation of matrix metalloproteinase expression in tumor invasion. *FASEB J.* **1999**, *13*, 781–792. [CrossRef]
14. Wendt, M.K.; Balanis, N.; Carlin, C.R.; Schiemann, W.P. STAT3 and epithelial-mesenchymal transitions in carcinomas. *JAKSTAT* **2014**, *3*, e28975. [CrossRef]
15. Kudo-Saito, C.; Shirako, H.; Takeuchi, T.; Kawakami, Y. Cancer metastasis is accelerated through immunosuppression during Snail-induced EMT of cancer cells. *Cancer Cell* **2009**, *15*, 195–206. [CrossRef]
16. Avalle, L.; Camporeale, A.; Camperi, A.; Poli, V. STAT3 in cancer: A double edged sword. *Cytokine* **2017**, *98*, 42–50. [CrossRef]
17. Liu, Y.; Li, C.; Lin, J. STAT3 as a Therapeutic Target for Glioblastoma. *Anticancer Agents Med. Chem* **2010**, *10*, 512–519. [CrossRef]
18. Li, W.; Graeber, M.B. The molecular profile of microglia under the influence of glioma. *Neuro Oncol.* **2012**, *14*, 958–978. [CrossRef]
19. Esposito, C.L.; Nuzzo, S.; Catuogno, S.; Romano, S.; de Nigris, F.; de Franciscis, V. STAT3 Gene Silencing by Aptamer-siRNA Chimera as Selective Therapeutic for Glioblastoma. *Mol. Nucleic Acids* **2018**, *10*, 398–411. [CrossRef]

20. Kiprianova, I.; Remy, J.; Milosch, N.; Mohrenz, I.V.; Seifert, V.; Aigner, A.; Kogel, D. Sorafenib Sensitizes Glioma Cells to the BH3 Mimetic ABT-737 by Targeting MCL1 in a STAT3-Dependent Manner. *Neoplasia* **2015**, *17*, 564–573. [CrossRef]
21. Priester, M.; Copanaki, E.; Vafaizadeh, V.; Hensel, S.; Bernreuther, C.; Glatzel, M.; Seifert, V.; Groner, B.; Kogel, D.; Weissenberger, J. STAT3 silencing inhibits glioma single cell infiltration and tumor growth. *Neuro Oncol.* **2013**, *15*, 840–852. [CrossRef]
22. Senft, C.; Priester, M.; Polacin, M.; Schroder, K.; Seifert, V.; Kogel, D.; Weissenberger, J. Inhibition of the JAK-2/STAT3 signaling pathway impedes the migratory and invasive potential of human glioblastoma cells. *J. Neurooncol.* **2011**, *101*, 393–403. [CrossRef]
23. Sherry, M.M.; Reeves, A.; Wu, J.L.K.; Cochran, B.H. STAT3 Is Required for Proliferation and Maintenance of Multipotency in Glioblastoma Stem Cells. *Stem. Cells* **2009**, *27*, 2383–2392. [CrossRef]
24. Takeda, K.; Kaisho, T.; Yoshida, N.; Takeda, J.; Kishimoto, T.; Akira, S. Stat3 activation is responsible for IL-6-dependent T cell proliferation through preventing apoptosis: Generation and characterization of T cell-specific Stat3-deficient mice. *J. Immunol.* **1998**, *161*, 4652–4660. [CrossRef]
25. Aigner, A. Applications of RNA interference: Current state and prospects for siRNA-based strategies in vivo. *Appl. Microbiol. Biotechnol.* **2007**, *76*, 9–21. [CrossRef]
26. Aigner, A.; Kogel, D. Nanoparticle/siRNA-based therapy strategies in glioma: Which nanoparticles, which siRNAs? *Nanomedicine (London)* **2018**, *13*, 89–103. [CrossRef]
27. Hobel, S.; Koburger, I.; John, M.; Czubayko, F.; Hadwiger, P.; Vornlocher, H.P.; Aigner, A. Polyethylenimine/small interfering RNA-mediated knockdown of vascular endothelial growth factor in vivo exerts anti-tumor effects synergistically with Bevacizumab. *J. Gene Med.* **2010**, *12*, 287–300.
28. Hendruschk, S.; Wiedemuth, R.; Aigner, A.; Topfer, K.; Cartellieri, M.; Martin, D.; Kirsch, M.; Ikonomidou, C.; Schackert, G.; Temme, A. RNA interference targeting survivin exerts antitumoral effects in vitro and in established glioma xenografts in vivo. *Neuro Oncol.* **2011**, *13*, 1074–1089. [CrossRef]
29. Ibrahim, A.F.; Weirauch, U.; Thomas, M.; Grunweller, A.; Hartmann, R.K.; Aigner, A. MicroRNA replacement therapy for miR-145 and miR-33a is efficacious in a model of colon carcinoma. *Cancer Res.* **2011**, *71*, 5214–5224. [CrossRef]
30. Hampl, V.; Martin, C.; Aigner, A.; Hoebel, S.; Singer, S.; Frank, N.; Sarikas, A.; Ebert, O.; Prywes, R.; Gudermann, T.; et al. Depletion of the transcriptional coactivators megakaryoblastic leukaemia 1 and 2 abolishes hepatocellular carcinoma xenograft growth by inducing oncogene-induced senescence. *EMBO Mol. Med.* **2013**, *5*, 1367–1382. [CrossRef]
31. Hobel, S.; Aigner, A. Polyethylenimines for siRNA and miRNA delivery in vivo. *Wiley Interdiscip. Rev. Nanomed. Nanobiotechnol.* **2013**, *5*, 484–501. [CrossRef]
32. Ewe, A.; Panchal, O.; Pinnapireddy, S.R.; Bakowsky, U.; Przybylski, S.; Temme, A.; Aigner, A. Liposome-polyethylenimine complexes (DPPC-PEI lipopolyplexes) for therapeutic siRNA delivery in vivo. *Nanomedicine* **2017**, *13*, 209–218. [CrossRef]
33. Heyes, J.; Palmer, L.; Chan, K.; Giesbrecht, C.; Jeffs, L.; MacLachlan, I. Lipid encapsulation enables the effective systemic delivery of polyplex plasmid DNA. *Molcure* **2007**, *15*, 713–720. [CrossRef]
34. Ko, Y.T.; Kale, A.; Hartner, W.C.; Papahadjopoulos-Sternberg, B.; Torchilin, V.P. Self-assembling micelle-like nanoparticles based on phospholipid-polyethyleneimine conjugates for systemic gene delivery. *J. Control. Release* **2009**, *133*, 132–138. [CrossRef]
35. Ewe, A.; Schaper, A.; Barnert, S.; Schubert, R.; Temme, A.; Bakowsky, U.; Aigner, A. Storage stability of optimal liposome-polyethylenimine complexes (lipopolyplexes) for DNA or siRNA delivery. *Acta Biomater.* **2014**, *10*, 2663–2673. [CrossRef]
36. Werth, S.; Urban-Klein, B.; Dai, L.; Hobel, S.; Grzelinski, M.; Bakowsky, U.; Czubayko, F.; Aigner, A. A low molecular weight fraction of polyethylenimine (PEI) displays increased transfection efficiency of DNA and siRNA in fresh or lyophilized complexes. *J. Control. Release* **2006**, *112*, 257–270. [CrossRef]
37. Schafer, J.; Hobel, S.; Bakowsky, U.; Aigner, A. Liposome-polyethylenimine complexes for enhanced DNA and siRNA delivery. *Biomaterials* **2010**, *31*, 6892–6900. [CrossRef]
38. Ewe, A.; Aigner, A. Cationic Lipid-Coated Polyplexes (Lipopolyplexes) for DNA and Small RNA Delivery. *Methods Mol. Biol.* **2016**, *1445*, 187–200.

39. Dasgupta, A.; Raychaudhuri, B.; Haqqi, T.; Prayson, R.; Van Meir, E.G.; Vogelbaum, M.; Haque, S.J. Stat3 activation is required for the growth of U87 cell-derived tumours in mice. *Eur. J. Cancer* **2009**, *45*, 677–684. [CrossRef]
40. The Cancer Genome Atlas Research Network. Comprehensive genomic characterization defines human glioblastoma genes and core pathways. *Nature* **2008**, *455*, 1061–1068.
41. Smilowitz, H.M.; Weissenberger, J.; Weis, J.; Brown, J.D.; O'Neill, R.J.; Laissue, J.A. Orthotopic transplantation of v-src-expressing glioma cell lines into immunocompetent mice: Establishment of a new transplantable in vivo model for malignant glioma. *J. Neurosurg.* **2007**, *106*, 652–659. [CrossRef] [PubMed]
42. Stechishin, O.D.; Luchman, H.A.; Ruan, Y.; Blough, M.D.; Nguyen, S.A.; Kelly, J.J.; Cairncross, J.G.; Weiss, S. On-target JAK2/STAT3 inhibition slows disease progression in orthotopic xenografts of human glioblastoma brain tumor stem cells. *Neuro Oncol.* **2013**, *15*, 198–207. [CrossRef]
43. Doucette, T.A.; Kong, L.Y.; Yang, Y.; Ferguson, S.D.; Yang, J.; Wei, J.; Qiao, W.; Fuller, G.N.; Bhat, K.P.; Aldape, K.; et al. Signal transducer and activator of transcription 3 promotes angiogenesis and drives malignant progression in glioma. *Neuro Oncol.* **2012**, *14*, 1136–1145. [CrossRef] [PubMed]
44. Weissenberger, J.; Priester, M.; Bernreuther, C.; Rakel, S.; Glatzel, M.; Seifert, V.; Kogel, D. Dietary curcumin attenuates glioma growth in a syngeneic mouse model by inhibition of the JAK1,2/STAT3 signaling pathway. *Clin. Cancer Res.* **2010**, *16*, 5781–5795. [CrossRef] [PubMed]
45. Ewe, A.; Aigner, A. Nebulization of liposome–polyethylenimine complexes (lipopolyplexes) for DNA or siRNA delivery: Physicochemical properties and biological activity. *Eur. J. Lipid Sci. Technol.* **2014**, *116*, 1195–1204. [CrossRef]
46. Planas, A.M.; Soriano, M.A.; Berruezo, M.; Justicia, C.; Estrada, A.; Pitarch, S.; Ferrer, I. Induction of Stat3, a signal transducer and transcription factor, in reactive microglia following transient focal cerebral ischaemia. *Eur. J. Neurosci.* **1996**, *8*, 2612–2618. [CrossRef]
47. Rodriguez, M.; Lapierre, J.; Ojha, C.R.; Kaushik, A.; Batrakova, E.; Kashanchi, F.; Dever, S.M.; Nair, M.; El-Hage, N. Intranasal drug delivery of small interfering RNA targeting Beclin1 encapsulated with polyethylenimine (PEI) in mouse brain to achieve HIV attenuation. *Sci. Rep.* **2017**, *7*, 1862. [CrossRef]
48. Matsuda, M.; Yamamoto, T.; Matsumura, A.; Kaneda, Y. Highly efficient eradication of intracranial glioblastoma using Eg5 siRNA combined with HVJ envelope. *Gene* **2009**, *16*, 1465–1476. [CrossRef]
49. Gillespie, D.L.; Aguirre, M.T.; Ravichandran, S.; Leishman, L.L.; Berrondo, C.; Gamboa, J.T.; Wang, L.; King, R.; Wang, X.; Tan, M.; et al. RNA interference targeting hypoxia-inducible factor 1alpha via a novel multifunctional surfactant attenuates glioma growth in an intracranial mouse model. *J. Neurosurg.* **2015**, *122*, 331–341. [CrossRef]
50. Michiue, H.; Eguchi, A.; Scadeng, M.; Dowdy, S.F. Induction of in vivo synthetic lethal RNAi responses to treat glioblastoma. *Cancer Biol.* **2009**, *8*, 2306–2313. [CrossRef]
51. Merz, L.; Hobel, S.; Kallendrusch, S.; Ewe, A.; Bechmann, I.; Franke, H.; Merz, F.; Aigner, A. Tumor tissue slice cultures as a platform for analyzing tissue-penetration and biological activities of nanoparticles. *Eur. J. Pharm. Biopharm.* **2017**, *112*, 45–50. [CrossRef]
52. Helmschrodt, C.; Hobel, S.; Schoniger, S.; Bauer, A.; Bonicelli, J.; Gringmuth, M.; Fietz, S.A.; Aigner, A.; Richter, A.; Richter, F. Polyethylenimine Nanoparticle-Mediated siRNA Delivery to Reduce alpha-Synuclein Expression in a Model of Parkinson's Disease. *Mol. Nucleic Acids* **2017**, *9*, 57–68. [CrossRef] [PubMed]
53. Mathieson, T.; Franken, H.; Kosinski, J.; Kurzawa, N.; Zinn, N.; Sweetman, G.; Poeckel, D.; Ratnu, V.S.; Schramm, M.; Becher, I.; et al. Systematic analysis of protein turnover in primary cells. *Nat. Commun.* **2018**, *9*, 689. [CrossRef] [PubMed]
54. Bocchini, C.E.; Nahmod, K.; Katsonis, P.; Kim, S.; Kasembeli, M.M.; Freeman, A.; Lichtarge, O.; Makedonas, G.; Tweardy, D.J. Protein stabilization improves STAT3 function in autosomal dominant hyper-IgE syndrome. *Blood* **2016**, *128*, 3061–3072. [CrossRef] [PubMed]
55. Boisvert, F.M.; Ahmad, Y.; Gierlinski, M.; Charriere, F.; Lamont, D.; Scott, M.; Barton, G.; Lamond, A.I. A quantitative spatial proteomics analysis of proteome turnover in human cells. *Mol. Cell. Proteom.* **2012**, *11*, M111.011429. [CrossRef] [PubMed]
56. Chen, P.Y.; Ozawa, T.; Drummond, D.C.; Kalra, A.; Fitzgerald, J.B.; Kirpotin, D.B.; Wei, K.C.; Butowski, N.; Prados, M.D.; Berger, M.S.; et al. Comparing routes of delivery for nanoliposomal irinotecan shows superior anti-tumor activity of local administration in treating intracranial glioblastoma xenografts. *Neuro Oncol.* **2013**, *15*, 189–197. [CrossRef] [PubMed]

57. Bruce, J.N.; Fine, R.L.; Canoll, P.; Yun, J.; Kennedy, B.C.; Rosenfeld, S.S.; Sands, S.A.; Surapaneni, K.; Lai, R.; Yanes, C.L.; et al. Regression of recurrent malignant gliomas with convection-enhanced delivery of topotecan. *Neurosurgery* **2011**, *69*, 1272–1279, discussion 1279–1280. [CrossRef] [PubMed]
58. Jaschinski, F.; Rothhammer, T.; Jachimczak, P.; Seitz, C.; Schneider, A.; Schlingensiepen, K.H. The antisense oligonucleotide trabedersen (AP 12009) for the targeted inhibition of TGF-beta2. *Curr. Pharm. Biotechnol.* **2011**, *12*, 2203–2213. [CrossRef] [PubMed]
59. Voges, J.; Reszka, R.; Gossmann, A.; Dittmar, C.; Richter, R.; Garlip, G.; Kracht, L.; Coenen, H.H.; Sturm, V.; Wienhard, K.; et al. Imaging-guided convection-enhanced delivery and gene therapy of glioblastoma. *Ann. Neurol.* **2003**, *54*, 479–487. [CrossRef] [PubMed]
60. Yu, D.; Khan, O.F.; Suva, M.L.; Dong, B.; Panek, W.K.; Xiao, T.; Wu, M.; Han, Y.; Ahmed, A.U.; Balyasnikova, I.V.; et al. Multiplexed RNAi therapy against brain tumor-initiating cells via lipopolymeric nanoparticle infusion delays glioblastoma progression. *Proc. Natl. Acad. Sci. USA* **2017**, *114*, E6147–E6156. [CrossRef] [PubMed]
61. Vogelbaum, M.A.; Brewer, C.; Barnett, G.H.; Mohammadi, A.M.; Peereboom, D.M.; Ahluwalia, M.S.; Gao, S. First-in-human evaluation of the Cleveland Multiport Catheter for convection-enhanced delivery of topotecan in recurrent high-grade glioma: Results of pilot trial 1. *J. Neurosurg.* **2018**. [CrossRef] [PubMed]
62. Ceccarelli, M.; Barthel, F.P.; Malta, T.M.; Sabedot, T.S.; Salama, S.R.; Murray, B.A.; Morozova, O.; Newton, Y.; Radenbaugh, A.; Pagnotta, S.M.; et al. Molecular Profiling Reveals Biologically Discrete Subsets and Pathways of Progression in Diffuse Glioma. *Cell* **2016**, *164*, 550–563. [CrossRef] [PubMed]
63. Bowman, R.L.; Wang, Q.; Carro, A.; Verhaak, R.G.; Squatrito, M. GlioVis data portal for visualization and analysis of brain tumor expression datasets. *Neuro Oncol.* **2017**, *19*, 139–141. [CrossRef] [PubMed]
64. Pohl, U.; Wick, W.; Weissenberger, J.; Steinbach, J.P.; Dichgans, J.; Aguzzi, A.; Weller, M. Characterization of Tu-2449, a glioma cell line derived from a spontaneous tumor in GFAP-v-src-transgenic mice: Comparison with established murine glioma cell lines. *Int. J. Oncol.* **1999**, *15*, 829–834. [CrossRef] [PubMed]
65. Hetschko, H.; Voss, V.; Horn, S.; Seifert, V.; Prehn, J.H.; Kogel, D. Pharmacological inhibition of Bcl-2 family members reactivates TRAIL-induced apoptosis in malignant glioma. *J. Neurooncol.* **2008**, *86*, 265–272. [CrossRef] [PubMed]
66. Livak, K.J.; Schmittgen, T.D. Analysis of relative gene expression data using real-time quantitative PCR and the 2(-Delta Delta C(T)) Method. *Methods* **2001**, *25*, 402–408. [CrossRef] [PubMed]
67. Antonietti, P.; Linder, B.; Hehlgans, S.; Mildenberger, I.C.; Burger, M.C.; Fulda, S.; Steinbach, J.P.; Gessler, F.; Rodel, F.; Mittelbronn, M.; et al. Interference with the HSF1/HSP70/BAG3 Pathway Primes Glioma Cells to Matrix Detachment and BH3 Mimetic-Induced Apoptosis. *Mol. Cancer* **2017**, *16*, 156–168. [CrossRef]
68. Weissenberger, J.; Loeffler, S.; Kappeler, A.; Kopf, M.; Lukes, A.; Afanasieva, T.A.; Aguzzi, A.; Weis, J. IL-6 is required for glioma development in a mouse model. *Oncogene* **2004**, *23*, 3308–3316. [CrossRef]
69. Schulze, J.; Hendrikx, S.; Schulz-Siegmund, M.; Aigner, A. Microparticulate poly(vinyl alcohol) hydrogel formulations for embedding and controlled release of polyethylenimine (PEI)-based nanoparticles. *Acta Biomater.* **2016**, *45*, 210–222. [CrossRef]

2

TYK2: An Upstream Kinase of STATs in Cancer

Katharina Wöss, Natalija Simonović, Birgit Strobl, Sabine Macho-Maschler and Mathias Müller *

Institute of Animal Breeding and Genetics, University of Veterinary Medicine Vienna, A-1210 Vienna, Austria; Katharina.Woess@vetmeduni.ac.at (K.W.); natalijabozovic@gmail.com (N.S.); birgit.strobl@vetmeduni.ac.at (B.S.); sabine.macho-maschler@vetmeduni.ac.at (S.M.-M.)

* Correspondence: mathias.mueller@vetmeduni.ac.at

Abstract: In this review we concentrate on the recent findings describing the oncogenic potential of the protein tyrosine kinase 2 (TYK2). The overview on the current understanding of TYK2 functions in cytokine responses and carcinogenesis focusses on the activation of the signal transducers and activators of transcription (STAT) 3 and 5. Insight gained from loss-of-function (LOF) gene-modified mice and human patients homozygous for *Tyk2/TYK2*-mutated alleles established the central role in immunological and inflammatory responses. For the description of physiological TYK2 structure/function relationships in cytokine signaling and of overarching molecular and pathologic properties in carcinogenesis, we mainly refer to the most recent reviews. Dysregulated TYK2 activation, aberrant TYK2 protein levels, and gain-of-function (GOF) TYK2 mutations are found in various cancers. We discuss the molecular consequences thereof and briefly describe the molecular means to counteract TYK2 activity under (patho-)physiological conditions by cellular effectors and by pharmacological intervention. For the role of TYK2 in tumor immune-surveillance we refer to the recent Special Issue of Cancers "JAK-STAT Signaling Pathway in Cancer".

Keywords: tyrosine kinase 2; JAK family of protein tyrosine kinases; signal transducer and activator of transcription; cytokine receptor signaling; gain-of-function mutation; tumorigenesis

1. TYK2-Mediated Cytokine Signaling and Activation of STAT3 and STAT5

TYK2 was the first identified member of a family of non-receptor kinases later termed Janus kinases (JAK), which additionally comprises JAK1-3 [1,2]. JAKs are associated with cytokine and growth factor receptors and activate STAT (STAT1-4, STAT5A, STAT5B, STAT6) family members [2,3]. JAKs share four functional domains (from N- to C-terminal): (i) a four-point-one, ezrin, radixin, moesin (FERM) homology domain; (ii) an atypical Src-homology 2 (SH2) domain, both facilitating protein-protein interactions (PPIs); (iii) a kinase-like or pseudokinase (JAK homology (JH) 2) domain negatively regulating the kinase activity; and (iv) a tyrosine kinase (JH1) domain which, upon conformational changes at ligand bound receptors, increases its catalytic activity by trans-/autophosphorylation of its activation loop [2,4].

To date, the requirement for TYK2 in signaling has been shown for numerous cytokines, including distinct interleukin (ILs) and interferons (IFNs), which comprise several subtypes (i.e., type I and III IFNs). The heterodimeric cytokine receptor complexes are composed of four distinct TYK2-associated receptor chains (IFNAR1, IL-12Rβ1, IL-10R2, and IL-13Rα1) and a respective second receptor chain associated either with JAK1 or JAK2, which serves as the signal transducing chain harboring STAT docking sites. Usually, these sites contain critical tyrosine residues that are phosphorylated by JAKs upon receptor complex activation (Figure 1). TYK2 also associates with the gp130 receptor chain, yet there is no evidence that gp130-utilizing cytokines rely on TYK2 for signal transduction [5,6]. Note that comprehensive reviews [2,7] provide lists of various other receptors utilizing TYK2-STAT signaling; however, TYK2-STAT activation/utilization is frequently only biochemically assessed by phosphorylation of critical tyrosine residues and cannot be put on a level with dissected downstream cellular activities. Here we review the cytokines which clearly transduce the TYK2 phosphorylation events into downstream physiological changes (Figure 1).

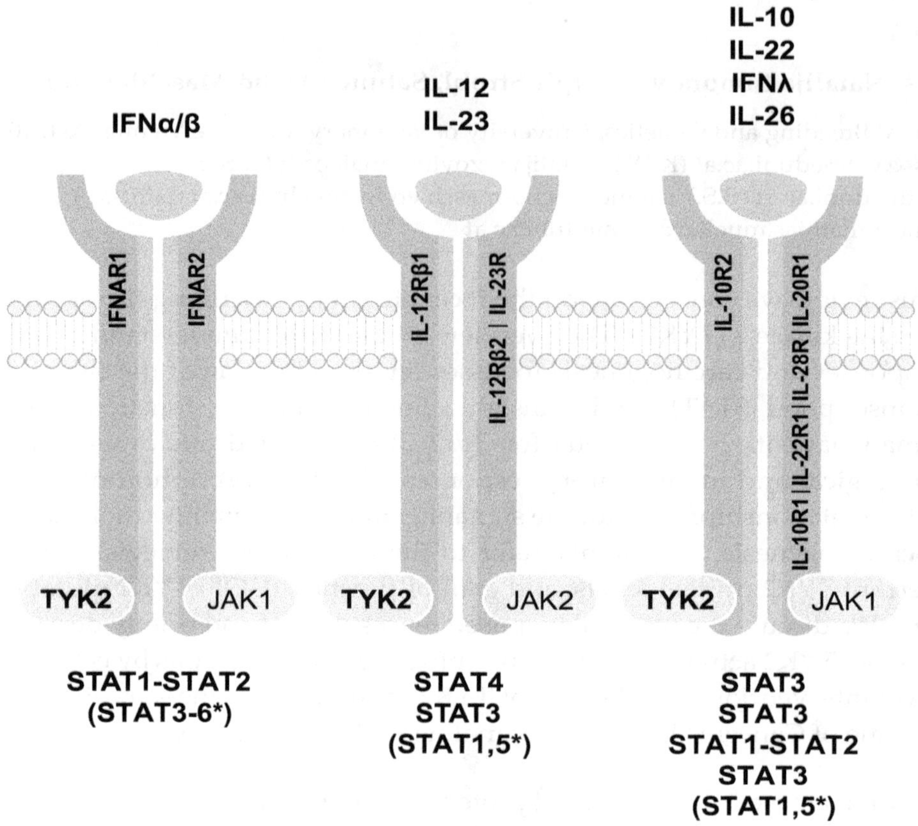

Figure 1. Cytokine receptor families signaling with the participation of TYK2 and JAK1 or JAK2. Cytokines are depicted only upon appearance in humans and mice and proof of TYK2 dependency. The color codes indicate the major STAT(s) activated by the respective cytokines. STAT1-STAT2 heterodimers combine with IFN regulatory factor (IRF) 9 and form the interferon-stimulated gene factor 3 (ISGF3) complex; * STAT activation is dependent on cell type or of less clear biological relevance.

The biological relevance for TYK2-dependent cytokines activating STAT3 is best established for the IL-10R2 utilizing IL-22 [8,9] and the IL-12Rβ1-utilizing IL-12 and IL-23 [10–12]. IL-22 is a central cytokine in tissue-barrier function, wound healing, and epithelial homeostasis and repair. Cancer promoting, as well as restraining, functions were described [13,14]. IL-23 is a key mediator of inflammation, bridges innate and adaptive immune responses, and is known to support tumorigenesis and metastasis [15,16]. IL-12 is central in promoting cell-mediated immunity to infection and cancer [12]. However, this anti-carcinogenic function can be counteracted by IL-12-STAT3-promoted production of pro-carcinogenic IL-23 [17]. While STAT3 is activated by type I and III IFN stimulation in various cell types, its biological functions in the IFN responses are less clear. Growing evidence suggests that STAT3 is a negative regulator of type I IFN activities, thereby providing a pro-viral and pro-survival cellular program [18]; there is, however, also a report on an opposite, i.e., anti-viral activity of STAT3 [19]. The role of TYK2 in IL-10 signaling through STAT3 is not entirely clear and may be cell type- or context-dependent [6]. The double-edged role of IL-10 in immunity and cancer is reviewed elsewhere [9,20]. IL-19, IL-20, IL-24, and IL-26 (absent in mice) constitute a subfamily within the IL-10 cytokine family and signal primarily through activation of STAT3 [9]. Activation of TYK2 at the respective receptors has not been formally shown but can be inferred from the receptor-chain composition. As this subfamily constitutes relatively recently discovered cytokines, cellular responses are still poorly defined, and we refer to recent publications and reviews for a potential cancer connection [21–23]. Lastly, without specification of the cytokines involved, TYK2 via STAT3 was reported to be crucial for the mediation of cell death in an auto-inflammatory context [24].

STAT5, in contrast, is not among the primarily activated STATs downstream of TYK2 (Figure 1) and occurs dependently on cell type and differentiation stage, in response to type I and III IFNs [25,26], IL-10R2-, and IL-12Rβ1-receptor family cytokines [9,12]. Neither a cytokine-TYK2-STAT5 axis nor its significance have been established under physiological conditions.

2. Aberrant Expression and/or Activity of TYK2 in Cancers

The JAK-STAT pathway is recognized as a core cancer pathway [27] and directly contributes to all hallmarks of cancer [28]. Oncogenic JAK activity can originate from aberrant JAK expression, deregulated upstream signals, GOF mutations, or generation of fusion proteins, as well as loss of negative feedback regulation [2,29–31]. Initially, cancer research focused on JAK1-3, while the TYK2 impact on disease was predominantly studied in inflammatory and (auto-)immune diseases [32,33]. Table 1 summarizes the literature on constitutive or hyperactivated TYK2, as well as GOF-mutated TYK2 and the resulting activation of STATs in cancers.

2.1. Aberrant TYK2 Levels

In vitro studies with overexpressed JAKs revealed that aberrant TYK2 levels lead to cellular transformation with constitutive phosphorylation of STAT3 [34]. An unusually high expression of TYK2 associated with or causative for carcinogenesis (reviewed [35]) was described for various cancer cell lines and samples from patients suffering from prostate [36,37], ovarian [38], cervical [39], and breast cancer [40,41], as well as malignant peripheral nerve sheath tumors (MPNST) [42,43]. Conflictingly, lowered TYK2 levels in tumor samples and sections (tumor cells and stroma) are generally considered to be an unfavorable prognostic marker (e.g., [44], www.proteinatlas.org). This is supported by a recently published meta-analysis of JAKs and STATs in hepatocellular carcinoma (HCC) patients, where normal or higher TYK2 levels correlated with longer survival and were found in healthy tissue [45]. The underlying reason for these conflicting reports may be attributed to the anti-proliferative/pro-apoptotic and/or tumor surveillance properties of TYK2 [5], as well as the undetermined tumor cell intrinsic and extrinsic state of TYK2. The important role of TYK2 in immune-surveillance is also in line with findings in patients who carry mutated TYK2 alleles which lead to loss of TYK2, lowered TYK2 levels [46], or expression of kinase-inactive TYK2 [47–49], and that show primarily immunodeficiencies. Nonetheless, proteomics suggested that low TYK2 facilitates local metastasis in breast cancer [50], and a comprehensive screen for protein tyrosine kinase variants in numerous cancer cell lines identified splice variants that render TYK2 inactive [51]. On a molecular mechanistic level, the cell intrinsic tumor-promoting consequences of low TYK2 or LOF of TYK2 currently remain elusive.

2.2. Aberrant Activation of TYK2

A comprehensive list of receptors (over-)expressed in various cancer types which allows us to deduce putative upstream signals involved in hyperactivation of TYK2 was compiled recently [7]. Primary hematological neoplasm (ALCL, anaplastic large cell lymphoma; T-ALL, T cell acute lymphoblastic leukemia) patient samples and cell lines were shown to be dependent on TYK2 activated by upstream IL-10 and/or IL-22 signals and established an upregulation of anti-apoptotic BCL2 family members via STAT1 and/or STAT3 [52,53]. A similar high TYK2-STAT1/3-BCL2 axis was found in MPNST [43]. Cytotoxic T-lymphocyte-associated antigen 4 (CTLA4, CD152) is mainly expressed on T cells and is a well-established immune checkpoint. CTLA4 signaling is initiated through binding to CD80 (B7-1) or CD86 (B7-2) on the surface of antigen-presenting cells (APCs). Ectopic expression of CTLA4 was found on diverse B-cell lymphoma. Mechanistically, it was established that CD86-CTLA4 engagement resulted in recruitment/activation of TYK2, which, in turn, led to a STAT3-driven tumor-promoting transcriptional program [54]. A STAT-independent involvement of activated TYK2 in fibroblast growth factor 2 (FGF-2) mediated escape from drug-induced death was reported for a sarcoma cell line [55].

2.3. TYK2 Mutations

Oncogenic JAK2 with the prominent JAK2^{V617F} mutation found in over 50% of myeloproliferative neoplasia (MPN) patients [56] is the paradigm for the understanding of structure/function relations of JAK activity [57–59] and for the general alertness of the cancer field for mutated JAK family members as potential oncogenes. TYK2 joined the club of GOF-mutated JAKs causative for patient hematopoietic malignancies only recently: In 2013, the first TYK2 GOF point mutations were found in T-ALL cell lines and characterized to have transforming capacity via STAT1 and a BCL2 family member [53]. With respect to biochemical studies, the first GOF mutation of TYK2 was V678F, which is the homologous mutation to JAK2^{V617F} [60,61]. Until now, this mutation was not found in patients. The only mutation reported in a public cancer genome database (www.stjude.cloud) for this residue is the V678L mutation, albeit with unknown structure/function consequences. Point mutations at the TYK2 locus are distributed throughout the whole gene body, with GOF mutations—similar to the other JAKs—primarily accumulating in the JH1 and JH2 domains ([2,5] and see public databases, e.g., Genomic Data Commons of the National Cancer Institute [62,63], Catalogue of Somatic Mutations in Cancer (COSMIC [64], and cBioPortal for Cancer Genomics [65,66]).

In addition to the somatic cancer cell mutations, two GOF TYK2 germline mutations (P760L and G761V) were found in pediatric patients developing several de novo leukemias. These mutations are located in the JH2 pseudokinase domain of TYK2 and are predicted to attenuate the negative regulation on the JH1 kinase domain, leading to constitutively activated TYK2 [67].

A prominent germline TYK2 mutation is P1104A/V, which was first found to be associated with solid and hematopoietic cancers [68,69] and later with immunological and inflammatory disorders (reviewed in [5]). While analyzing MPNST tumor samples, it was proposed that TYK2^{P1104A} is an unfavorable prognostic marker for the disease [42]. Notably, this study solely genotyped the somatic cancer cells and overlooked that this mutation impairs TYK2 catalytic activity; cellular signaling, however, is not completely abrogated, and the detected induction of BCL2 expression might favor an anti-apoptotic program [69,70]. Recent studies show that TYK2^{P1104A} is a LOF mutation, because patients homozygous for this allele are either susceptible to microbial infection or protected from autoimmune disease [47,49,71]. These mechanistic and phenotypic features of TYK2^{P1104A} were confirmed in independent mouse models [48,71].

2.4. TYK2 Fusion Proteins

Chromosomal rearrangements account for a number of driver kinase fusion genes in cancer [72–74]. The first fusion kinase involving a JAK was TEL-JAK2, consisting of a 3′ portion of JAK2 and a 5′ region of TEL, a member of the ETS transcription factor family [75]. This chromosomal translocation is found in T-ALL in patients [75] and transgenic mice expressing TEL-JAK2 develop T-cell leukemia [76]. In vitro studies with a TEL-TYK2 fusion showed constitutive activation of STAT1/3/5 and transforming capacities [77], albeit respective translocations have not yet been identified in patients. As observed for GOF-mutated JAKs, JAK2 kinase fusions occur most frequently compared to the other JAKs, which suggests that the JAK2 locus is a mutation and rearrangement hotspot [56,78,79]. The first leukemia patients carrying TYK2 fusion genes described were combinations of the TYK2 kinase domain and a part of the pseudokinase domain with 5′ portions of nucleophosmin (NPM) 1, polyadenylate binding protein (PABPC) 4, or the transcription factors MYB or NFκB2 [80–82]. Structurally and mechanistically, the TYK2 fusion proteins lack the negatively regulating function of the pseudokinase (JH2) domain leading to a GOF kinase activity and hyperactivity of STAT3 and depending on the cellularity also STAT1 and 5 (reviewed in [5,58,59]).

Subsequent analysis of patient samples and cell lines [83–87] and screening of cancer data sets revealed more than 50 chromosomal *TYK2* rearrangements found mostly in hematological, but also in solid cancers [88]. For the fusions, it is currently not known if they contribute as driver oncogenes to early tumorigenesis or are rather the result of genomic instability at later tumor stages [89]. Recently, chromothripsis was identified as a new type of chromosomal rearrangement during carcinogenesis. Based on a single chromosome-shattering event and DNA repair complex, intra- and interchromosomal rearrangements, such as fusion genes, are produced within a few cell cycles. If the fusion event(s) allow for growth or survival advantages, a cancer driver gene might be generated [90,91]. Chromothripsis was assigned to genomic alterations in childhood cancer [92], and mechanistically it is caused by defects in the nuclear envelope composition or formation and failures during mitosis [93]. It is tempting to speculate that the remarkably high number of described TYK2 fusions were—at least in part—generated through chromothripsis and thus might act as driver mutations.

3. Tumor-Promoting Activities of (Hyper-)Active TYK2

The molecular contribution of TYK2 signaling and known protein–protein interactions to the hallmarks of cancer were reviewed previously [5,28]. Here, we highlight the latest findings on the consequences of TYK2 hyperactivity in cancer cells.

3.1. TYK2 Activation of (Oncogenic) STAT Signaling

As shown in Figure 1, the heterodimeric cytokine receptors with engagement of TYK2 are capable of activating all STATs. Hyperactive, GOF-mutated TYK2 or TYK2 fusions in oncogenic settings preferentially lead to aberrant activation of STAT1, STAT3, and STAT5. The oncogenic potential of STAT3 and STAT5 was recognized early on and is well documented [94,95]. STAT1 was initially considered to exert tumor suppressor functions, and its oncogenic potential emerged more recently [96–98].

STAT1/3/5 were found hyperactivated in patient-tailored cell lines with activated TYK2 [53], as well as carrying somatic or germline TYK2 GOF mutations [53,67] or TYK2-NPM1 and -NFkB2 fusions [80,82]. In other tumor samples or experimental tissue culture settings, STAT3 only, or other dual combinations of activated STAT1/3/5, are described (see Table 1).

Interestingly, TYK2 does not only phosphorylate the major phosphorylation site Y705 in STAT3, but also Y640, which represses STAT3 activation [99]. This phosphorylation site in STAT3 is often mutated in cancers [100,101]. Neither the general (patho-)physiological impact nor the contribution to malignancies of this phosphorylation event are currently known.

Table 1. (Hyper-)active TYK2, GOF-, or LOF-mutated TYK2 and STAT activation in various cancers and cancer cell lines.

TYK2 Status	Disease	Activated STAT	Ref.
	Activating somatic mutations (GOF)		
TYK2-G36D; -S47N	T-ALL	STAT1, STAT3	[53] (3) (2) (2 *)
TYK2-731I	T-ALL	STAT1, STAT3, STAT4	[53] (3) (2) (2 *)
TYK2-E957D	T-ALL	STAT1, STAT3, STAT5	[53] (3) (2) (2 *)
TYK2-R1027H	T-ALL	STAT1, STAT3	[53] (3) (2) (2 *)
TYK2-V678F	—	STAT3, STAT5	[61] (2 *)
	Inactivating germline mutations (LOF)		
TYK2-P1104A	MPNST	n.d.	[42] (4 *) (1)
TYK2-P1104A	Breast-, colon-, stomach-cancer	n.d.	[68] (1)
TYK2-P1104V	AML	n.d.	[69] (5) (1) (2 *)
	Activating germline mutations (GOF)		
TYK2-P760L	B-ALL	STAT1, STAT3, STAT5	[67] (3) (1) (2 *)
TYK2-G761V	T-ALL	STAT1, STAT3, STAT5	[67] (3) (1) (2 *)

Table 1. *Cont.*

TYK2 Status	Disease	Activated STAT	Ref.
	Oncogenic fusion proteins (GOF)		
NPM1-TYK2	CD30-positive LPDs	STAT1, STAT3, STAT5	[82] (3) (1) (2) (2 *)
NFkB2-TYK2	ALCL	STAT1, STAT3, STAT5	[80] (3) (1) (2 *)
ELAVL1-TYK2	AML	STAT3, STAT5	[84] (2)
PABPC4-TYK2	ALCL	n.d.	[80] (1)
TEL-TYK2	—	STAT1, STAT3, STAT5	[77] (2 *)
MYB-TYK2	Ph-like ALL	n.d.	[81] (1)
	High wildtype TYK2 levels		
TYK2 WT	T-ALL	STAT1, STAT3, STAT4, STAT5	[53] (1) (2) (2 *)
TYK2 WT	ALCL	STAT1, STAT3	[52] (4) (1) (2)
TYK2 WT	Hepatocarcinoma	STAT1, STAT3	[102] (3 *) (2 **)
TYK2 WT	MPNST	STAT1, STAT3	[43] (4) (?) (1) (2)
TYK2 WT	B-cell lymphoma	STAT3	[54] (3 *) (2)
TYK2 WT	Lung cancer	STAT3	[103] (3 *) (1) (2) (2 *)
TYK2 WT	Hepatocarcinoma	STAT3	[104] (3 *) (2 **)
TYK2 WT	Ovarian cancer	STAT3	[38] (3 *) (2)
TYK2 WT	Prostate cancer	n.d.	[36] (4) (1) (2)
TYK2 WT	Prostate cancer	n.d.	[37] (4) (1) (2)
TYK2 WT	Osteosarcoma	no	[55] (3 *) (2)
TYK2 WT	Breast cancer	n.d.	[40,41] (4) (1) (2)
TYK2 WT	Squamous cervical carcinoma	n.d.	[39] (4) (1)
TYK2 WT	MPNST	n.d.	[42] (4) (?) (1)
TYK2 WT	Lung cancer	STAT1	[105] (2 **)
	Low wildtype TYK2 levels		
TYK2 WT	Breast cancer (metastatic)	n.d.	[50] (6) (1) (2)

— Unrelated to disease, in vitro findings in stable cell lines, (1) found in patient samples and primary material, (2) in vitro findings endogenous TYK2 expression, (2 *) in vitro findings exogenous TYK2 expression, (2 **) in vitro findings exogenous claudin expression, (3) phosphorylated mutated TYK2 protein, (3 *) phosphorylated wildtype TYK2 protein, (4) high levels of wildtype TYK2, (4 *) high levels of mutated TYK2, (5) reduced levels of phosphorylated mutated TYK2, (6) reduced levels of wildtype TYK2, and (?) not specified if wildtype or mutated TYK2. Please note that some references did not study the activation of all STATs and that not all described STATs in the table are active in all cell systems used. Ph = Philadelphia, and n.d. not determined.

3.2. TYK2 Stimulation of Tumor Cell Invasion

The families of tight junction proteins claudins (CLDNs) and of matrix metalloproteinases (MMPs) are central for the invasion of tumor cells and, in consequence, metastasis formation [106,107]. Recent studies show that, in liver and lung carcinoma, high levels of CLDN9/12/17 caused activation of TYK2 and STAT1/3 and promoted metastasis [102,104,105]. The promoters of various MMP genes harbor STAT binding sites, and many MMPs are transcriptionally activated through TYK2-associated cytokine receptors [108,109]. Gene-targeted mice revealed that TYK2 and STAT1 are required for expression of MMP2/9/14 under inflammatory conditions [110]. Biochemical studies showed that, dependent on context and inflammatory conditions, MMP1/3 induction involves STAT1 alone [108] or also STAT3 [111]. In a hematopoietic tumor TYK2-STAT3 induced MMP9 and tumor cell invasiveness [54] and in a solid tumor TYK2-STAT3 signaling induced MMP1 expression [103].

The urokinase-type plasminogen activator (uPA)/receptor (uPAR) system is central for a cascade of proteolytic events, including activation of MMPs, which allow for tumor cell migration and metastasis [112]. Signaling via uPAR involves TYK2 and PI3K [113], and, at the post-transcriptional level, TYK2 inhibits the accumulation of plasminogen activator inhibitor (PAI) 2 [114]. In prostate cancer, high levels of TYK2 correlate with invasion and metastasis [36,37]. In an ovarian cancer cell line pY-STAT3 co-localizes with TYK2 and JAK2 at focal adhesions, and hyperactive STAT3 was shown to promote cancer cell motility [38]. Without providing molecular details, a mouse model for aggressive lymphoma showed reduced tumor cell invasiveness upon loss of TYK2 [115]. In addition, without providing molecular insights, a siRNA screen assessing the role of the tyrosine kinome in metastasis formation identified TYK2 as a promoter of invadopodia, which are cellular structures characteristic for tumor cell migration [116,117]. Connexin43 (Cx43) is the most widely expressed member of a large family of transmembrane proteins involved in gap junction formation. Cx43 can be both pro- and anti-tumorigenic, e.g., by promoting invasion and metastasis and by acting as a tumor suppressor [118,119]. TYK2 was found to play a dual role in regulation of Cx43: On the one hand, TYK2 is capable of directly phosphorylating Cx43, thereby decreasing its stability; on the other hand, angiotensin II-activated TYK2 increased Cx43 levels in a STAT3-dependent manner [120]. This regulatory loop has not yet been studied in the context of carcinogenesis. Furthermore, knockdown of TYK2 reduced migration of breast cancer cell lines [50].

3.3. TYK2 Prevention of Apoptosis

IFNs in general are capable of promoting apoptosis of cancer cells [121]; hence, provided that IFN stimulus and responsiveness in the tumor is given, TYK2 acts tumor suppressive. Tumor cells are able to resist cell death by upregulation of anti-apoptotic BCL-2 family members [122,123]. TYK2 was shown to drive either in a STAT1- and/or a STAT3-dependent manner or in a STAT-independent but ERK1/2-dependent manner high expression of BCL-2 [43,53,55] or its family members BCL-2L1 [54] and MCL1 [52,55]. In contrast, an in vitro study demonstrated that TYK2 physically interacts with SIVA-1 and promotes SIVA-1 mediated apoptosis, as well as inhibits BCL-2 [124].

3.4. TYK2 Crosstalk to Oncogenes and Proto-Oncogenic Pathways

In a mouse model of ALCL, as well as in patient cells, TYK2 showed co-operativity with the oncogenic fusion kinase NPM-ALK [52]. In contrast, no co-operation of TYK2 with mutated FLT3-ITD or $JAK2^{V617F}$ in MPN mouse models was found [125,126]. The latter is consistent with the observation that, in $JAK2^{V617F}$ MPN patients (see below) resistant to pharmacological JAK2 inhibition, only JAK1, and not TYK2, leads to heterodimeric STAT activation, despite both kinases show equal tyrosine phosphorylation at the activating loop [127]. This is to be expected, since, in contrast to the other JAKs, loss of TYK2 at heterodimeric JAK-associated cytokine receptors leads only to a partial impairment in signaling [5,6], and, as experimentally described for the IFNAR receptor, TYK2 is the subordinated JAK at cytokine receptors [128,129].

Early biochemical studies suggest that, upon type I IFN treatment, TYK2 interacts with various proto-oncogenes, including the guanine nucleotide exchange factor 1 VAV, the E3 ubiquitin-protein ligase C-CBL, and the SRC family tyrosine kinases FYN and LYN [130–134]. The importance of these PPIs for tumorigenesis is currently unknown. In cancer samples or cell lines, TYK2 was found to cooperate with other oncogenic effectors and pathways, such as the RAF/ERK [53,55,61], MAPKs [135], PIM1/2 [84], and PI3K/AKT/mTOR pathway [36,53,61]. Reported solely in the context of skin inflammation is the TYK2-STAT3 requirement for expression of IκBζ (encoded by *NFKBIZ*) [136]; however, emerging reports suggest cell-intrinsic oncogenic, as well as tumor-suppressive, functions of IκBζ [137].

The mapped and predicted PPIs of TYK2 based on proteomics [138,139] and next generation sequencing (NGS) are accessible at various open-source databases (for a review, see [140]). The TYK2 kinase domain and a STAT3-based reporter system were used to establish the first mammalian two hybrid kinase substrate sensor (KISS) screening platform [141,142]. These databases and the screening approaches should be systematically exploited to further define and fine tune the TYK2 interactome in health and disease.

4. Deactivation and Stabilization of TYK2 under (Patho-)Physiological Conditions

JAK activity is counter regulated by molecule-intrinsic events, such as post-translational modifications (PTMs) and the inhibitory function of the pseudokinase domain [143] as well as by extrinsic inhibitory regulators, such as suppressor of cytokine signaling (SOCS) proteins and protein tyrosine phosphatases (PTPs) [144].

Databases [145,146] provide curated PTMs, but with the exception of the well described activating phosphotyrosines, there is still a lack of information on the properties of JAKs that are defined by PTMs. For TYK2, ubiquitination and phosphorylation are detected at multiple residues and discussed in the context of stability/decay (PhosphoSitePlus®, [146]), albeit the (patho-) physiological relevance is unknown.

SOCS proteins are encoded by STAT target genes and are negative feedback inhibitors of JAK signaling. SOCS1 and 3 are the most potent JAK inhibitors because, in addition to recruitment of JAKs to E3 ubiquitination/degradation mediated by all SOCS family members, they also harbor a kinase inhibitory region (KIR), which efficiently shuts down JAK activity by binding to the JH1 domain [147]. Activated JAKs and cytokine receptor chains are dephosphorylated by multiple PTPs [148]. The current literature regarding deactivation of TYK2 by SOCS1/3, the PTPs PTB1B and SHP1, as well as the global impact of SOCS and PTP family members in cancer are reviewed elsewhere [5,149–151].

In vitro studies showed that in hematopoietic tumor cells the PTP SHP1 suppresses growth via accelerating the TYK2 protein degradation [152]. In lung cancer cells, overexpression of the E3 ubiquitin ligase seven-in-absentia-2 (SIAH2) accelerates the proteasomal degradation of TYK2, thereby attenuating STAT3 signaling [103].

HSP90 is a chaperone supporting folding, stability, and function of many client proteins, including JAKs and STATs [153–155]. Cancer cells frequently use HSP90 to stabilize and/or increase the function of numerous oncogenes, and HSP90 inhibitors have been studied as anticancer drugs for more than two decades [156,157]. Physical interaction of HSP90 with TYK2 was demonstrated in cancer cell lines and confirmed in a proteome-wide assessment of the HSP90 interactome [158,159]. HSP90 inhibitor treatments in various tumor settings showed beneficial effects by reducing the activity of TYK2 or its fusion proteins [158,160,161].

An emerging field is the involvement of noncoding RNAs in the regulation of the JAK-STAT pathway in carcinogenesis [162–164]. Recently, the long noncoding RNA (lncRNA) MEG3 in concert with a microRNA (miR-147) was reported to modulate JAK-STAT signaling in chronic myeloid leukemia (CML). Interestingly, the lncRNA was found to physically interact with TYK2, JAK2, and STAT3, thereby diminishing the activity level of STAT3 (and STAT5) [165].

5. Pharmaceutical TYK2 Inhibition

The first selective JAK inhibitor (JAKinib) to be tested in humans was tofacitinib, which potently inhibits JAK3 and JAK1, and, to a lesser extent JAK2, and has little effect on TYK2 [166]. Historically, JAKinibs were developed as immunosuppressive drugs for the clinical use in organ transplants and autoimmune diseases [167]. The success story of ruxolitinib, a JAK2 and JAK1 inhibitor which was the first JAKinib approved for treatment of a hematopoietic malignancy, pushed the perception of JAKinibs as anticancer drugs [168,169]. For insight in development and clinical use, as well as side effects of JAKinibs, we refer to the most recent reviews [170–173].

TYK2inibs are mainly envisaged as therapeutics for treatment of autoimmune and inflammatory diseases [33,174], in which JAKinib selectivity is currently considered not to be of utmost importance [175]. As for the other JAKinibs, the first generation TYK2inibs are directed to the JH1 domain and compete with ATP in binding to the enzymatic pocket. These inhibitors are potent in inhibiting wildtype (overexpressed) TYK2, mutated (hyperactive) TYK2, and TYK2 fusion proteins harboring the JH1 domain. Since the JAKs show high homology in the JH1 domain, it is hard to develop ATP-competing inhibitors with high selectivity for one particular JAK family member [170,172]. A next-generation inhibitor of TYK2 is directed against the JH2 domain and recently passed the phase II clinical trial for psoriasis treatment [176]. A comprehensive report on the high selectivity and the biological effects of this TYK2inib in mouse models, as well as its efficacy in human cells collected from autoimmune patients, was recently published [177]. JH2-specific TYK2inibs are currently further improved, and additional compounds are being developed [178–181]. The only TYK2inib reported and successfully tested to block TYK2 activity in an oncogenic setting is a JH1-specific TYK2inib [135].

Notably, JH2 domain inhibitors might not be working for treatment of diseases driven by TYK2 fusion genes missing parts of the JH2 domain.

6. Conclusions and Future Perspectives

Since the discovery of TYK2 and the JAK-STAT signaling paradigm in the early 1990s, enormous progress has been made in the structural and functional understanding of the linear JAK-STAT axis and the crosstalk of JAKs or STATs to other signaling hubs, as well as the cell type-specific contributions of JAKs and STATs in health and disease. The striking phenotypical similarities between mouse models deficient for TYK2 or engineered to express kinase-inactive TYK2 and human patients carrying the respective germline mutations established TYK2 as a fundamental component in both innate and adaptive immunity. The (patho-)physiological and molecular pathway similarities of TYK2 in human and mice allow for highly informative comparative biomedical studies and efficient translation of basic molecular insights into clinical applications. The use of TYK2inibs in the treatment of immunological and inflammatory diseases is within reach [182] and is also attractive for malignancies with the involvement of hyperactivated TYK2. The role of TYK2 and GOF-mutated TYK2 upstream of oncogenic STAT3—and, less frequently, STAT1—is established, while, up to now, no mechanistic evidence for an oncogenic TYK2-STAT5 axis is given. Mouse models as genetic mimics of kinase-inhibited TYK2 exist [48,71,183,184] and are currently exploited to further dissect the kinase-dependent from the scaffolding functions of TYK2.

In a short-term perspective, work should concentrate on the use of refined TYK2 mouse models that allow studying the kinase-independent and cell type-specific functions, in order to fully in vivo assess TYK2inibs with respect to their benefits and unwanted side effects. Mouse models to study the consequences of aberrant high TYK2 and GOF-mutated TYK2 are underway (K. Wöss, T. Rülicke et al., unpublished). For pharmacological intervention with oncogenic TYK2, TYK2inibs with the highest possible selectivity are required, and efforts should focus on the further development and in vivo testing of these next-generation TYK2inibs.

In a long-term perspective, the further understanding of the TYK2 function requires the in-depth elucidation of the PTMs and the interactome of TYK2 under spatiotemporal conditions. Additionally, computational modelling and structure predictions (e.g., [185]) should complement the attempts to determine the holo-crystal structure of TYK2 and to use high-resolution imaging (e.g., [186]) to gain insight into the structural features of full-length wildtype and mutated TYK2, as well as its conformation bound to various cytokine receptors.

Author Contributions: K.W., S.M.-M., and M.M. designed the draft of the review and performed the literature search. K.W. and S.M.-M. compiled the table, and N.S. designed the figure. S.M.-M., M.M., N.S., and B.S. provided the final version of the manuscript.

References

1. Krolewski, J.J.; Lee, R.; Eddy, R.; Shows, T.B.; Dalla-Favera, R. Identification and chromosomal mapping of new human tyrosine kinase genes. *Oncogene* **1990**, *5*, 277–282. [PubMed]
2. Hammaren, H.M.; Virtanen, A.T.; Raivola, J.; Silvennoinen, O. The regulation of JAKs in cytokine signaling and its breakdown in disease. *Cytokine* **2019**, *118*, 48–63. [CrossRef] [PubMed]
3. Stark, G.R.; Darnell, J.E. The JAK-STAT Pathway at Twenty. *Immunity* **2012**, *36*, 503–514. [CrossRef] [PubMed]
4. Ferrao, R.; Lupardus, P.J. The Janus Kinase (JAK) FERM and SH2 Domains: Bringing Specificity to JAK-Receptor Interactions. *Front. Endocrinol.* **2017**, *8*, 71. [CrossRef]
5. Leitner, N.R.; Witalisz-Siepracka, A.; Strobl, B.; Muller, M. Tyrosine kinase 2-Surveillant of tumours and bona fide oncogene. *Cytokine* **2017**, *89*, 209–218. [CrossRef]

6. Strobl, B.; Stoiber, D.; Sexl, V.; Mueller, M. Tyrosine kinase 2 (TYK2) in cytokine signalling and host immunity. *Front. Biosci.* **2011**, *16*, 3214–3232. [CrossRef]
7. Bousoik, E.; Montazeri Aliabadi, H. "Do We Know Jack" About JAK? A Closer Look at JAK/STAT Signaling Pathway. *Front. Oncol.* **2018**, *8*, 287. [CrossRef]
8. Dudakov, J.A.; Hanash, A.M.; van den Brink, M.R. Interleukin-22: Immunobiology and pathology. *Annu. Rev. Immunol.* **2015**, *33*, 747–785. [CrossRef]
9. Ouyang, W.; O'Garra, A. IL-10 Family Cytokines IL-10 and IL-22: From Basic Science to Clinical Translation. *Immunity* **2019**, *50*, 871–891. [CrossRef]
10. Kastelein, R.A.; Hunter, C.A.; Cua, D.J. Discovery and biology of IL-23 and IL-27: Related but functionally distinct regulators of inflammation. *Annu. Rev. Immunol.* **2007**, *25*, 221–242. [CrossRef]
11. Rutz, S.; Wang, X.; Ouyang, W. The IL-20 subfamily of cytokines–from host defence to tissue homeostasis. *Nat. Rev. Immunol.* **2014**, *14*, 783–795. [CrossRef] [PubMed]
12. Tait Wojno, E.D.; Hunter, C.A.; Stumhofer, J.S. The Immunobiology of the Interleukin-12 Family: Room for Discovery. *Immunity* **2019**, *50*, 851–870. [CrossRef] [PubMed]
13. Hernandez, P.; Gronke, K.; Diefenbach, A. A catch-22: Interleukin-22 and cancer. *Eur. J. Immunol.* **2018**, *48*, 15–31. [CrossRef] [PubMed]
14. Lim, C.; Savan, R. The role of the IL-22/IL-22R1 axis in cancer. *Cytokine Growth Factor Rev.* **2014**, *25*, 257–271. [CrossRef] [PubMed]
15. Huynh, J.; Chand, A.; Gough, D.; Ernst, M. Therapeutically exploiting STAT3 activity in cancer—Using tissue repair as a road map. *Nat. Rev. Cancer* **2019**, *19*, 82–96. [CrossRef]
16. Yan, J.M.; Smyth, M.J.; Teng, M.W.L. Interleukin (IL)-12 and IL-23 and Their Conflicting Roles in Cancer. *Cold Spring Harb. Perspect. Biol.* **2018**, *10*, a028530. [CrossRef]
17. Kortylewski, M.; Xin, H.; Kujawski, M.; Lee, H.; Liu, Y.; Harris, T.; Drake, C.; Pardoll, D.; Yu, H. Regulation of the IL-23 and IL-12 balance by Stat3 signaling in the tumor microenvironment. *Cancer Cell* **2009**, *15*, 114–123. [CrossRef]
18. Tsai, M.H.; Pai, L.M.; Lee, C.K. Fine-Tuning of Type I Interferon Response by STAT3. *Front. Immunol.* **2019**, *10*, 1448. [CrossRef]
19. Mahony, R.; Gargan, S.; Roberts, K.L.; Bourke, N.; Keating, S.E.; Bowie, A.G.; O'Farrelly, C.; Stevenson, N.J. A novel anti-viral role for STAT3 in IFN-alpha signalling responses. *Cell. Mol. Life Sci.* **2017**, *74*, 1755–1764. [CrossRef]
20. Mannino, M.H.; Zhu, Z.; Xiao, H.; Bai, Q.; Wakefield, M.R.; Fang, Y. The paradoxical role of IL-10 in immunity and cancer. *Cancer Lett.* **2015**, *367*, 103–107. [CrossRef]
21. Chen, Y.Y.; Li, C.F.; Yeh, C.H.; Chang, M.S.; Hsing, C.H. Interleukin-19 in breast cancer. *Clin. Dev. Immunol.* **2013**, *2013*, 294320. [CrossRef] [PubMed]
22. Niess, J.H.; Hruz, P.; Kaymak, T. The Interleukin-20 Cytokines in Intestinal Diseases. *Front. Immunol.* **2018**, *9*, 1373. [CrossRef] [PubMed]
23. You, W.; Tang, Q.; Zhang, C.; Wu, J.; Gu, C.; Wu, Z.; Li, X. IL-26 promotes the proliferation and survival of human gastric cancer cells by regulating the balance of STAT1 and STAT3 activation. *PLoS ONE* **2013**, *8*, e63588. [CrossRef] [PubMed]
24. Wan, J.; Fu, A.K.; Ip, F.C.; Ng, H.K.; Hugon, J.; Page, G.; Wang, J.H.; Lai, K.O.; Wu, Z.; Ip, N.Y. Tyk2/STAT3 signaling mediates beta-amyloid-induced neuronal cell death: Implications in Alzheimer's disease. *J. Neurosci.* **2010**, *30*, 6873–6881. [CrossRef] [PubMed]
25. Kotenko, S.V. IFN-lambdas. *Curr. Opin. Immunol.* **2011**, *23*, 583–590. [CrossRef] [PubMed]
26. Schreiber, G. The molecular basis for differential type I interferon signaling. *J. Biol. Chem.* **2017**, *292*, 7285–7294. [CrossRef] [PubMed]
27. Vogelstein, B.; Papadopoulos, N.; Velculescu, V.E.; Zhou, S.; Diaz, L.A., Jr.; Kinzler, K.W. Cancer genome landscapes. *Science* **2013**, *339*, 1546–1558. [CrossRef]
28. Hanahan, D.; Weinberg, R.A. Hallmarks of cancer: The next generation. *Cell* **2011**, *144*, 646–674. [CrossRef]
29. Chen, E.; Staudt, L.M.; Green, A.R. Janus kinase deregulation in leukemia and lymphoma. *Immunity* **2012**, *36*, 529–541. [CrossRef]
30. Groner, B.; von Manstein, V. Jak Stat signaling and cancer: Opportunities, benefits and side effects of targeted inhibition. *Mol. Cell. Endocrinol.* **2017**, *451*, 1–14. [CrossRef]

31. Thomas, S.J.; Snowden, J.A.; Zeidler, M.P.; Danson, S.J. The role of JAK/STAT signalling in the pathogenesis, prognosis and treatment of solid tumours. *Br. J. Cancer* **2015**, *113*, 365–371. [CrossRef] [PubMed]
32. O'Shea, J.J.; Holland, S.M.; Staudt, L.M. JAKs and STATs in immunity, immunodeficiency, and cancer. *N. Engl. J. Med.* **2013**, *368*, 161–170. [CrossRef] [PubMed]
33. Schwartz, D.M.; Kanno, Y.; Villarino, A.; Ward, M.; Gadina, M.; O'Shea, J.J. JAK inhibition as a therapeutic strategy for immune and inflammatory diseases. *Nat. Rev. Drug Discov.* **2017**. [CrossRef] [PubMed]
34. Knoops, L.; Hornakova, T.; Royer, Y.; Constantinescu, S.N.; Renauld, J.C. JAK kinases overexpression promotes in vitro cell transformation. *Oncogene* **2008**, *27*, 1511–1519. [CrossRef] [PubMed]
35. Ubel, C.; Mousset, S.; Trufa, D.; Sirbu, H.; Finotto, S. Establishing the role of tyrosine kinase 2 in cancer. *Oncoimmunology* **2013**, *2*, e22840. [CrossRef]
36. Ide, H.; Nakagawa, T.; Terado, Y.; Kamiyama, Y.; Muto, S.; Horie, S. Tyk2 expression and its signaling enhances the invasiveness of prostate cancer cells. *Biochem. Biophys. Res. Commun.* **2008**, *369*, 292–296. [CrossRef] [PubMed]
37. Santos, J.; Mesquita, D.; Barros-Silva, J.D.; Jeronimo, C.; Henrique, R.; Morais, A.; Paulo, P.; Teixeira, M.R. Uncovering potential downstream targets of oncogenic GRPR overexpression in prostate carcinomas harboring ETS rearrangements. *Oncoscience* **2015**, *2*, 497–507. [CrossRef]
38. Silver, D.L.; Naora, H.; Liu, J.; Cheng, W.; Montell, D.J. Activated signal transducer and activator of transcription (STAT) 3: Localization in focal adhesions and function in ovarian cancer cell motility. *Cancer Res.* **2004**, *64*, 3550–3558. [CrossRef] [PubMed]
39. Zhu, X.; Lv, J.; Yu, L.; Zhu, X.; Wu, J.; Zou, S.; Jiang, S. Proteomic identification of differentially-expressed proteins in squamous cervical cancer. *Gynecol. Oncol.* **2009**, *112*, 248–256. [CrossRef] [PubMed]
40. Christy, J.; Priyadharshini, L. Differential expression analysis of JAK/STAT pathway related genes in breast cancer. *Meta Gene* **2018**, *16*, 122–129. [CrossRef]
41. Song, X.C.; Fu, G.; Yang, X.; Jiang, Z.; Wang, Y.; Zhou, G.W. Protein expression profiling of breast cancer cells by dissociable antibody microarray (DAMA) staining. *Mol. Cell. Proteom.* **2008**, *7*, 163–169. [CrossRef] [PubMed]
42. Hirbe, A.C.; Kaushal, M.; Sharma, M.K.; Dahiya, S.; Pekmezci, M.; Perry, A.; Gutmann, D.H. Clinical genomic profiling identifies TYK2 mutation and overexpression in patients with neurofibromatosis type 1-associated malignant peripheral nerve sheath tumors. *Cancer* **2017**, *123*, 1194–1201. [CrossRef] [PubMed]
43. Qin, W.J.; Godec, A.; Zhang, X.C.; Zhu, C.G.; Shao, J.Y.; Tao, Y.; Bu, X.Z.; Hirlbe, A.C. TYK2 promotes malignant peripheral nerve sheath tumor progression through inhibition of cell death. *Cancer Med.* **2019**, *8*, 5232–5241. [CrossRef] [PubMed]
44. Uhlen, M.; Zhang, C.; Lee, S.; Sjostedt, E.; Fagerberg, L.; Bidkhori, G.; Benfeitas, R.; Arif, M.; Liu, Z.; Edfors, F.; et al. A pathology atlas of the human cancer transcriptome. *Science* **2017**, *357*, eaan2507. [CrossRef]
45. Wang, X.; Liao, X.; Yu, T.; Gong, Y.; Zhang, L.; Huang, J.; Yang, C.; Han, C.; Yu, L.; Zhu, G.; et al. Analysis of clinical significance and prospective molecular mechanism of main elements of the JAK/STAT pathway in hepatocellular carcinoma. *Int. J. Oncol.* **2019**, *55*, 805–822. [CrossRef]
46. Nemoto, M.; Hattori, H.; Maeda, N.; Akita, N.; Muramatsu, H.; Moritani, S.; Kawasaki, T.; Maejima, M.; Ode, H.; Hachiya, A.; et al. Compound heterozygous TYK2 mutations underlie primary immunodeficiency with T-cell lymphopenia. *Sci. Rep.* **2018**, *8*, 6956. [CrossRef]
47. Boisson-Dupuis, S.; Ramirez-Alejo, N.; Li, Z.; Patin, E.; Rao, G.; Kerner, G.; Lim, C.K.; Krementsov, D.N.; Hernandez, N.; Ma, C.S.; et al. Tuberculosis and impaired IL-23-dependent IFN-gamma immunity in humans homozygous for a common TYK2 missense variant. *Sci. Immunol.* **2018**, *3*, eaau8714. [CrossRef]
48. Dendrou, C.A.; Cortes, A.; Shipman, L.; Evans, H.G.; Attfield, K.E.; Jostins, L.; Barber, T.; Kaur, G.; Kuttikkatte, S.B.; Leach, O.A.; et al. Resolving TYK2 locus genotype-to-phenotype differences in autoimmunity. *Sci. Transl. Med.* **2016**, *8*, 363ra149. [CrossRef]
49. Kerner, G.; Ramirez-Alejo, N.; Seeleuthner, Y.; Yang, R.; Ogishi, M.; Cobat, A.; Patin, E.; Quintana-Murci, L.; Boisson-Dupuis, S.; Casanova, J.L.; et al. Homozygosity for TYK2 P1104A underlies tuberculosis in about 1% of patients in a cohort of European ancestry. *Proc. Natl. Acad. Sci. USA* **2019**, *116*, 10430–10434. [CrossRef]
50. Sang, Q.X.; Man, Y.G.; Sung, Y.M.; Khamis, Z.I.; Zhang, L.; Lee, M.H.; Byers, S.W.; Sahab, Z.J. Non-receptor tyrosine kinase 2 reaches its lowest expression levels in human breast cancer during regional nodal metastasis. *Clin. Exp. Metastasis* **2012**, *29*, 143–153. [CrossRef]

51. Ruhe, J.E.; Streit, S.; Hart, S.; Wong, C.H.; Specht, K.; Knyazev, P.; Knyazeva, T.; Tay, L.S.; Loo, H.L.; Foo, P.; et al. Genetic alterations in the tyrosine kinase transcriptome of human cancer cell lines. *Cancer Res.* **2007**, *67*, 11368–11376. [CrossRef]
52. Prutsch, N.; Gurnhofer, E.; Suske, T.; Liang, H.C.; Schlederer, M.; Roos, S.; Wu, L.C.; Simonitsch-Klupp, I.; Alvarez-Hernandez, A.; Kornauth, C.; et al. Dependency on the TYK2/STAT1/MCL1 axis in anaplastic large cell lymphoma. *Leukemia* **2019**, *33*, 696–709. [CrossRef] [PubMed]
53. Sanda, T.; Tyner, J.W.; Gutierrez, A.; Ngo, V.N.; Glover, J.; Chang, B.H.; Yost, A.; Ma, W.; Fleischman, A.G.; Zhou, W.; et al. TYK2-STAT1-BCL2 pathway dependence in T-cell acute lymphoblastic leukemia. *Cancer Discov.* **2013**, *3*, 564–577. [CrossRef] [PubMed]
54. Herrmann, A.; Lahtz, C.; Nagao, T.; Song, J.Y.; Chan, W.C.; Lee, H.; Yue, C.; Look, T.; Mulfarth, R.; Li, W.; et al. CTLA4 Promotes Tyk2-STAT3-Dependent B-cell Oncogenicity. *Cancer Res.* **2017**, *77*, 5118–5128. [CrossRef] [PubMed]
55. Carmo, C.R.; Lyons-Lewis, J.; Seckl, M.J.; Costa-Pereira, A.P. A Novel Requirement for Janus Kinases as Mediators of Drug Resistance Induced by Fibroblast Growth Factor-2 in Human Cancer Cells. *PLoS ONE* **2011**, *6*, e19861. [CrossRef]
56. Vainchenker, W.; Constantinescu, S.N. JAK/STAT signaling in hematological malignancies. *Oncogene* **2013**, *32*, 2601–2613. [CrossRef]
57. Hubbard, S.R. Mechanistic Insights into Regulation of JAK2 Tyrosine Kinase. *Front. Endocrinol.* **2017**, *8*, 361. [CrossRef]
58. Morris, R.; Kershaw, N.J.; Babon, J.J. The molecular details of cytokine signaling via the JAK/STAT pathway. *Protein Sci.* **2018**, *27*, 1984–2009. [CrossRef]
59. Silvennoinen, O.; Hubbard, S.R. Molecular insights into regulation of JAK2 in myeloproliferative neoplasms. *Blood* **2015**, *125*, 3388–3392. [CrossRef]
60. Gakovic, M.; Ragimbeau, J.; Francois, V.; Constantinescu, S.N.; Pellegrini, S. The Stat3-activating Tyk2 V678F mutant does not up-regulate signaling through the type I interferon receptor but confers ligand hypersensitivity to a homodimeric receptor. *J. Biol. Chem.* **2008**, *283*, 18522–18529. [CrossRef]
61. Staerk, J.; Kallin, A.; Demoulin, J.B.; Vainchenker, W.; Constantinescu, S.N. JAK1 and Tyk2 activation by the homologous polycythemia vera JAK2 V617F mutation: Cross-talk with IGF1 receptor. *J. Biol. Chem.* **2005**, *280*, 41893–41899. [CrossRef] [PubMed]
62. Grossman, R.L.; Heath, A.P.; Ferretti, V.; Varmus, H.E.; Lowy, D.R.; Kibbe, W.A.; Staudt, L.M. Toward a Shared Vision for Cancer Genomic Data. *N. Engl. Med.* **2016**, *375*, 1109–1112. [CrossRef] [PubMed]
63. Jensen, M.A.; Ferretti, V.; Grossman, R.L.; Staudt, L.M. The NCI Genomic Data Commons as an engine for precision medicine. *Blood* **2017**, *130*, 453–459. [CrossRef] [PubMed]
64. Tate, J.G.; Bamford, S.; Jubb, H.C.; Sondka, Z.; Beare, D.M.; Bindal, N.; Boutselakis, H.; Cole, C.G.; Creatore, C.; Dawson, E.; et al. COSMIC: The Catalogue of Somatic Mutations in Cancer. *Nucleic Acids Res.* **2019**, *47*, D941–D947. [CrossRef]
65. Cerami, E.; Gao, J.; Dogrusoz, U.; Gross, B.E.; Sumer, S.O.; Aksoy, B.A.; Jacobsen, A.; Byrne, C.J.; Heuer, M.L.; Larsson, E.; et al. The cBio cancer genomics portal: An open platform for exploring multidimensional cancer genomics data. *Cancer Discov.* **2012**, *2*, 401–404. [CrossRef]
66. Gao, J.; Aksoy, B.A.; Dogrusoz, U.; Dresdner, G.; Gross, B.; Sumer, S.O.; Sun, Y.; Jacobsen, A.; Sinha, R.; Larsson, E.; et al. Integrative analysis of complex cancer genomics and clinical profiles using the cBioPortal. *Sci. Signal* **2013**, *6*, pl1. [CrossRef]
67. Waanders, E.; Scheijen, B.; Jongmans, M.C.; Venselaar, H.; van Reijmersdal, S.V.; van Dijk, A.H.; Pastorczak, A.; Weren, R.D.; van der Schoot, C.E.; van de Vorst, M.; et al. Germline activating TYK2 mutations in pediatric patients with two primary acute lymphoblastic leukemia occurrences. *Leukemia* **2017**, *31*, 821–828. [CrossRef]
68. Kaminker, J.S.; Zhang, Y.; Waugh, A.; Haverty, P.M.; Peters, B.; Sebisanovic, D.; Stinson, J.; Forrest, W.F.; Bazan, J.F.; Seshagiri, S.; et al. Distinguishing cancer-associated missense mutations from common polymorphisms. *Cancer Res.* **2007**, *67*, 465–473. [CrossRef]
69. Tomasson, M.H.; Xiang, Z.; Walgren, R.; Zhao, Y.; Kasai, Y.; Miner, T.; Ries, R.E.; Lubman, O.; Fremont, D.H.; McLellan, M.D.; et al. Somatic mutations and germline sequence variants in the expressed tyrosine kinase genes of patients with de novo acute myeloid leukemia. *Blood* **2008**, *111*, 4797–4808. [CrossRef]

70. Li, Z.; Gakovic, M.; Ragimbeau, J.; Eloranta, M.L.; Ronnblom, L.; Michel, F.; Pellegrini, S. Two rare disease-associated Tyk2 variants are catalytically impaired but signaling competent. *J. Immunol.* **2013**, *190*, 2335–2344. [CrossRef]
71. Gorman, J.A.; Hundhausen, C.; Kinsman, M.; Arkatkar, T.; Allenspach, E.J.; Clough, C.; West, S.E.; Thomas, K.; Eken, A.; Khim, S.; et al. The TYK2-P1104A Autoimmune Protective Variant Limits Coordinate Signals Required to Generate Specialized T Cell Subsets. *Front. Immunol.* **2019**, *10*, 44. [CrossRef] [PubMed]
72. Gao, Q.S.; Liang, W.W.; Foltz, S.M.; Mutharasu, G.; Jayasinghe, R.G.; Cao, S.; Liao, W.W.; Reynolds, S.M.; Wyczalkowski, M.A.; Yao, L.J.; et al. Driver Fusions and Their Implications in the Development and Treatment of Human Cancers. *Cell Rep.* **2018**, *23*, 227–238. [CrossRef] [PubMed]
73. Medves, S.; Demoulin, J.B. Tyrosine kinase gene fusions in cancer: Translating mechanisms into targeted therapies. *J. Cell. Mol. Med.* **2012**, *16*, 237–248. [CrossRef] [PubMed]
74. Stransky, N.; Cerami, E.; Schalm, S.; Kim, J.L.; Lengauer, C. The landscape of kinase fusions in cancer. *Nat. Commun.* **2014**, *5*, 4846. [CrossRef] [PubMed]
75. Lacronique, V.; Boureux, A.; Valle, V.D.; Poirel, H.; Quang, C.T.; Mauchauffe, M.; Berthou, C.; Lessard, M.; Berger, R.; Ghysdael, J.; et al. A TEL-JAK2 fusion protein with constitutive kinase activity in human leukemia. *Science* **1997**, *278*, 1309–1312. [CrossRef]
76. Carron, C.; Cormier, F.; Janin, A.; Lacronique, V.; Giovannini, M.; Daniel, M.T.; Bernard, O.; Ghysdael, J. TEL-JAK2 transgenic mice develop T-cell leukemia. *Blood* **2000**, *95*, 3891–3899. [CrossRef]
77. Lacronique, V.; Boureux, A.; Monni, R.; Dumon, S.; Mauchauffe, M.; Mayeux, P.; Gouilleux, F.; Berger, R.; Gisselbrecht, S.; Ghysdael, J.; et al. Transforming properties of chimeric TEL-JAK proteins in Ba/F3 cells. *Blood* **2000**, *95*, 2076–2083. [CrossRef]
78. Ho, K.; Valdez, F.; Garcia, R.; Tirado, C.A. JAK2 Translocations in hematological malignancies: Review of the literature. *J. Assoc. Genet. Technol.* **2010**, *36*, 107–109.
79. Levavi, H.; Tripodi, J.; Marcellino, B.; Mascarenhas, J.; Jones, A.V.; Cross, N.C.P.; Gruenstein, D.; Najfeld, V. A Novel t(1;9)(p36;p24.1) JAK2 Translocation and Review of the Literature. *Acta Haematol.* **2019**, *142*, 105–112. [CrossRef]
80. Crescenzo, R.; Abate, F.; Lasorsa, E.; Tabbo, F.; Gaudiano, M.; Chiesa, N.; Di Giacomo, F.; Spaccarotella, E.; Barbarossa, L.; Ercole, E.; et al. Convergent mutations and kinase fusions lead to oncogenic STAT3 activation in anaplastic large cell lymphoma. *Cancer Cell* **2015**, *27*, 516–532. [CrossRef]
81. Roberts, K.G.; Li, Y.; Payne-Turner, D.; Harvey, R.C.; Yang, Y.L.; Pei, D.; McCastlain, K.; Ding, L.; Lu, C.; Song, G.; et al. Targetable kinase-activating lesions in Ph-like acute lymphoblastic leukemia. *N. Engl. J. Med.* **2014**, *371*, 1005–1015. [CrossRef] [PubMed]
82. Velusamy, T.; Kiel, M.J.; Sahasrabuddhe, A.A.; Rolland, D.; Dixon, C.A.; Bailey, N.G.; Betz, B.L.; Brown, N.A.; Hristov, A.C.; Wilcox, R.A.; et al. A novel recurrent NPM1-TYK2 gene fusion in cutaneous CD30-positive lymphoproliferative disorders. *Blood* **2014**, *124*, 3768–3771. [CrossRef] [PubMed]
83. Roberts, K.G.; Gu, Z.; Payne-Turner, D.; McCastlain, K.; Harvey, R.C.; Chen, I.M.; Pei, D.; Iacobucci, I.; Valentine, M.; Pounds, S.B.; et al. High Frequency and Poor Outcome of Philadelphia Chromosome-Like Acute Lymphoblastic Leukemia in Adults. *J. Clin. Oncol.* **2017**, *35*, 394–401. [CrossRef] [PubMed]
84. Tron, A.E.; Keeton, E.K.; Ye, M.; Casas-Selves, M.; Chen, H.; Dillman, K.S.; Gale, R.E.; Stengel, C.; Zinda, M.; Linch, D.C.; et al. Next-generation sequencing identifies a novel ELAVL1-TYK2 fusion gene in MOLM-16, an AML cell line highly sensitive to the PIM kinase inhibitor AZD1208. *Leuk. Lymphoma* **2016**, *57*, 2927–2929. [CrossRef]
85. Gu, Z.; Churchman, M.; Roberts, K.; Li, Y.; Liu, Y.; Harvey, R.C.; McCastlain, K.; Reshmi, S.C.; Payne-Turner, D.; Iacobucci, I.; et al. Genomic analyses identify recurrent MEF2D fusions in acute lymphoblastic leukaemia. *Nat. Commun.* **2016**, *7*, 13331. [CrossRef]
86. Roberts, K.G.; Yang, Y.L.; Payne-Turner, D.; Lin, W.; Files, J.K.; Dickerson, K.; Gu, Z.; Taunton, J.; Janke, L.J.; Chen, T.; et al. Oncogenic role and therapeutic targeting of ABL-class and JAK-STAT activating kinase alterations in Ph-like ALL. *Blood Adv.* **2017**, *1*, 1657–1671. [CrossRef]
87. Prasad, A.; Rabionet, R.; Espinet, B.; Zapata, L.; Puiggros, A.; Melero, C.; Puig, A.; Sarria-Trujillo, Y.; Ossowski, S.; Garcia-Muret, M.P.; et al. Identification of gene mutations and fusion genes in patients with Sezary Syndrome. *J. Investig. Dermatol.* **2016**, *136*, 1490–1499. [CrossRef]

88. Kim, P.; Zhou, X. FusionGDB: Fusion gene annotation DataBase. *Nucleic Acids Res.* **2019**, *47*, D994–D1004. [CrossRef]
89. Negrini, S.; Gorgoulis, V.G.; Halazonetis, T.D. Genomic instability—An evolving hallmark of cancer. *Nat. Rev. Mol. Cell Biol.* **2010**, *11*, 220–228. [CrossRef]
90. Ly, P.; Cleveland, D.W. Rebuilding Chromosomes After Catastrophe: Emerging Mechanisms of Chromothripsis. *Trends Cell Biol.* **2017**, *27*, 917–930. [CrossRef]
91. Rode, A.; Maass, K.K.; Willmund, K.V.; Lichter, P.; Ernst, A. Chromothripsis in cancer cells: An update. *Int. J. Cancer* **2016**, *138*, 2322–2333. [CrossRef] [PubMed]
92. Grobner, S.N.; Worst, B.C.; Weischenfeldt, J.; Buchhalter, I.; Kleinheinz, K.; Rudneva, V.A.; Johann, P.D.; Balasubramanian, G.P.; Segura-Wang, M.; Brabetz, S.; et al. The landscape of genomic alterations across childhood cancers. *Nature* **2018**, *555*, 321–327. [CrossRef] [PubMed]
93. Liu, S.; Kwon, M.; Mannino, M.; Yang, N.; Renda, F.; Khodjakov, A.; Pellman, D. Nuclear envelope assembly defects link mitotic errors to chromothripsis. *Nature* **2018**, *561*, 551–555. [CrossRef] [PubMed]
94. Bromberg, J. Stat proteins and oncogenesis. *J. Clin. Investig.* **2002**, *109*, 1139–1142. [CrossRef]
95. Wingelhofer, B.; Neubauer, H.A.; Valent, P.; Han, X.; Constantinescu, S.N.; Gunning, P.T.; Muller, M.; Moriggl, R. Implications of STAT3 and STAT5 signaling on gene regulation and chromatin remodeling in hematopoietic cancer. *Leukemia* **2018**, *32*, 1713–1726. [CrossRef]
96. Avalle, L.; Pensa, S.; Regis, G.; Novelli, F.; Poli, V. STAT1 and STAT3 in tumorigenesis: A matter of balance. *JAKSTAT* **2012**, *1*, 65–72. [CrossRef]
97. Meissl, K.; Macho-Maschler, S.; Muller, M.; Strobl, B. The good and the bad faces of STAT1 in solid tumours. *Cytokine* **2017**, *89*, 12–20. [CrossRef]
98. Zhang, Y.; Liu, Z. STAT1 in cancer: Friend or foe? *Discov. Med.* **2017**, *24*, 19–29.
99. Mori, R.; Wauman, J.; Icardi, L.; Van der Heyden, J.; De Cauwer, L.; Peelman, F.; De Bosscher, K.; Tavernier, J. TYK2-induced phosphorylation of Y640 suppresses STAT3 transcriptional activity. *Sci. Rep.* **2017**, *7*, 15919. [CrossRef]
100. Koskela, H.L.; Eldfors, S.; Ellonen, P.; van Adrichem, A.J.; Kuusanmaki, H.; Andersson, E.I.; Lagstrom, S.; Clemente, M.J.; Olson, T.; Jalkanen, S.E.; et al. Somatic STAT3 mutations in large granular lymphocytic leukemia. *N. Engl. J. Med.* **2012**, *366*, 1905–1913. [CrossRef] [PubMed]
101. Pilati, C.; Amessou, M.; Bihl, M.P.; Balabaud, C.; Nhieu, J.T.; Paradis, V.; Nault, J.C.; Izard, T.; Bioulac-Sage, P.; Couchy, G.; et al. Somatic mutations activating STAT3 in human inflammatory hepatocellular adenomas. *J. Exp. Med.* **2011**, *208*, 1359–1366. [CrossRef] [PubMed]
102. Sun, L.M.; Feng, L.S.; Cui, J.W. Increased expression of claudin-17 promotes a malignant phenotype in hepatocyte via Tyk2/Stat3 signaling and is associated with poor prognosis in patients with hepatocellular carcinoma. *Diagn. Pathol.* **2018**, *13*, 72. [CrossRef] [PubMed]
103. Muller, S.; Chen, Y.; Ginter, T.; Schafer, C.; Buchwald, M.; Schmitz, L.M.; Klitzsch, J.; Schutz, A.; Haitel, A.; Schmid, K.; et al. SIAH2 antagonizes TYK2-STAT3 signaling in lung carcinoma cells. *Oncotarget* **2014**, *5*, 3184–3196. [CrossRef]
104. Liu, H.; Wang, M.; Liang, N.; Guan, L. Claudin-9 enhances the metastatic potential of hepatocytes via Tyk2/Stat3 signaling. *Turk. J. Gastroenterol.* **2019**, *30*, 722–731. [CrossRef] [PubMed]
105. Sun, L.M.; Feng, L.S.; Cui, J.W. Increased expression of claudin-12 promotes the metastatic phenotype of human bronchial epithelial cells and is associated with poor prognosis in lung squamous cell carcinoma. *Exp. Ther. Med.* **2019**, *17*, 165–174. [CrossRef] [PubMed]
106. Cathcart, J.; Pulkoski-Gross, A.; Cao, J. Targeting matrix metalloproteinases in cancer: Bringing new life to old ideas. *Genes Dis.* **2015**, *2*, 26–34. [CrossRef] [PubMed]
107. Tabaries, S.; Siegel, P.M. The role of claudins in cancer metastasis. *Oncogene* **2017**, *36*, 1176–1190. [CrossRef]
108. Cutler, S.J.; Doecke, J.D.; Ghazawi, I.; Yang, J.; Griffiths, L.R.; Spring, K.J.; Ralph, S.J.; Mellick, A.S. Novel STAT binding elements mediate IL-6 regulation of MMP-1 and MMP-3. *Sci. Rep.* **2017**, *7*, 8526. [CrossRef]
109. Fanjul-Fernandez, M.; Folgueras, A.R.; Cabrera, S.; Lopez-Otin, C. Matrix metalloproteinases: Evolution, gene regulation and functional analysis in mouse models. *Biochim. Biophys. Acta* **2010**, *1803*, 3–19. [CrossRef]
110. Costantino, G.; Egerbacher, M.; Kolbe, T.; Karaghiosoff, M.; Strobl, B.; Vogl, C.; Helmreich, M.; Muller, M. Tyk2 and signal transducer and activator of transcription 1 contribute to intestinal I/R injury. *Shock* **2008**, *29*, 238–244. [CrossRef]

111. Araki, Y.; Tsuzuki Wada, T.; Aizaki, Y.; Sato, K.; Yokota, K.; Fujimoto, K.; Kim, Y.T.; Oda, H.; Kurokawa, R.; Mimura, T. Histone Methylation and STAT-3 Differentially Regulate Interleukin-6-Induced Matrix Metalloproteinase Gene Activation in Rheumatoid Arthritis Synovial Fibroblasts. *Arthritis Rheumatol.* **2016**, *68*, 1111–1123. [CrossRef] [PubMed]
112. Mahmood, N.; Mihalcioiu, C.; Rabbani, S.A. Multifaceted Role of the Urokinase-Type Plasminogen Activator (uPA) and Its Receptor (uPAR): Diagnostic, Prognostic, and Therapeutic Applications. *Front. Oncol.* **2018**, *8*, 24. [CrossRef] [PubMed]
113. Kusch, A.; Tkachuk, S.; Haller, H.; Dietz, R.; Gulba, D.C.; Lipp, M.; Dumler, I. Urokinase stimulates human vascular smooth muscle cell migration via a phosphatidylinositol 3-kinase-Tyk2 interaction. *J. Biol. Chem.* **2000**, *275*, 39466–39473. [CrossRef] [PubMed]
114. Radwan, M.; Miller, I.; Grunert, T.; Marchetti-Deschmann, M.; Vogl, C.; O'Donoghue, N.; Dunn, M.J.; Kolbe, T.; Allmaier, G.; Gemeiner, M.; et al. The impact of tyrosine kinase 2 (Tyk2) on the proteome of murine macrophages and their response to lipopolysaccharide (LPS). *Proteomics* **2008**, *8*, 3469–3485. [CrossRef] [PubMed]
115. Schuster, C.; Muller, M.; Freissmuth, M.; Sexl, V.; Stoiber, D. Commentary on H. Ide. Tyk2 expression and its signaling enhances the invasiveness of prostate cancer cells. *Biochem. Biophys. Res. Commun.* **2008**, *366*, 869–870. [CrossRef]
116. Murphy, D.A.; Courtneidge, S.A. The 'ins' and 'outs' of podosomes and invadopodia: Characteristics, formation and function. *Nat. Rev. Mol. Cell Biol.* **2011**, *12*, 413–426. [CrossRef]
117. Revach, O.Y.; Sandler, O.; Samuels, Y.; Geiger, B. Cross-Talk between Receptor Tyrosine Kinases AXL and ERBB3 Regulates Invadopodia Formation in Melanoma Cells. *Cancer Res.* **2019**, *79*, 2634–2648. [CrossRef]
118. Aasen, T.; Leithe, E.; Graham, S.V.; Kameritsch, P.; Mayán, M.D.; Mesnil, M.; Pogoda, K.; Tabernero, A. Connexins in cancer: Bridging the gap to the clinic. *Oncogene* **2019**, *38*, 4429–4451. [CrossRef]
119. Bonacquisti, E.E.; Nguyen, J. Connexin 43 (Cx43) in cancer: Implications for therapeutic approaches via gap junctions. *Cancer Lett.* **2019**, *442*, 439–444. [CrossRef]
120. Li, H.; Spagnol, G.; Zheng, L.; Stauch, K.L.; Sorgen, P.L. Regulation of Connexin43 Function and Expression by Tyrosine Kinase 2. *J. Biol. Chem.* **2016**, *291*, 15867–15880. [CrossRef]
121. Kotredes, K.P.; Gamero, A.M. Interferons as inducers of apoptosis in malignant cells. *J. Interferon Cytokine Res.* **2013**, *33*, 162–170. [CrossRef] [PubMed]
122. Kale, J.; Osterlund, E.J.; Andrews, D.W. BCL-2 family proteins: Changing partners in the dance towards death. *Cell Death Differ.* **2018**, *25*, 65–80. [CrossRef] [PubMed]
123. Singh, R.; Letai, A.; Sarosiek, K. Regulation of apoptosis in health and disease: The balancing act of BCL-2 family proteins. *Nat. Rev. Mol. Cell Biol.* **2019**, *20*, 175–193. [CrossRef] [PubMed]
124. Shimoda, H.K.; Shide, K.; Kameda, T.; Matsunaga, T.; Shimoda, K. Tyrosine kinase 2 interacts with the proapoptotic protein Siva-1 and augments its apoptotic functions. *Biochem. Biophys. Res. Commun.* **2010**, *400*, 252–257. [CrossRef] [PubMed]
125. Nakajima, H.; Shibata, F.; Kumagai, H.; Shimoda, K.; Kitamura, T. Tyk2 is dispensable for induction of myeloproliferative disease by mutant FLT3. *Int. J. Hematol.* **2006**, *84*, 54–59. [CrossRef]
126. Yamaji, T.; Shide, K.; Kameda, T.; Sekine, M.; Kamiunten, A.; Hidaka, T.; Kubuki, Y.; Shimoda, H.; Abe, H.; Miike, T.; et al. Loss of Tyrosine Kinase 2 Does Not Affect the Severity of Jak2V617F-induced Murine Myeloproliferative Neoplasm. *Anticancer Res.* **2017**, *37*, 3841–3847. [CrossRef]
127. Koppikar, P.; Bhagwat, N.; Kilpivaara, O.; Manshouri, T.; Adli, M.; Hricik, T.; Liu, F.; Saunders, L.M.; Mullally, A.; Abdel-Wahab, O.; et al. Heterodimeric JAK-STAT activation as a mechanism of persistence to JAK2 inhibitor therapy. *Nature* **2012**, *489*, 155–159. [CrossRef]
128. Kohlhuber, F.; Rogers, N.C.; Watling, D.; Feng, J.; Guschin, D.; Briscoe, J.; Witthuhn, B.A.; Kotenko, S.V.; Pestka, S.; Stark, G.R.; et al. A JAK1/JAK2 chimera can sustain alpha and gamma interferon responses. *Mol. Cell. Biol.* **1997**, *17*, 695–706. [CrossRef]
129. Briscoe, J.; Rogers, N.C.; Witthuhn, B.A.; Watling, D.; Harpur, A.G.; Wilks, A.; Stark, G.R.; Ihle, J.N.; Kerr, I.M. Kinase-negative mutants of JAK1 can sustain interferon-gamma-inducible gene expression but not an antiviral state. *EMBO J.* **1996**, *15*, 799–809. [CrossRef]
130. Adam, L.; Bandyopadhyay, D.; Kumar, R. Interferon-alpha signaling promotes nucleus-to-cytoplasmic redistribution of p95Vav, and formation of a multisubunit complex involving Vav, Ku80, and Tyk2. *Biochem. Biophys. Res. Commun.* **2000**, *267*, 692–696. [CrossRef]

131. Uddin, S.; Gardziola, C.; Dangat, A.; Yi, T.; Platanias, L.C. Interaction of the c-cbl proto-oncogene product with the Tyk-2 protein tyrosine kinase. *Biochem. Biophys. Res. Commun.* **1996**, *225*, 833–838. [CrossRef] [PubMed]
132. Uddin, S.; Grumbach, I.M.; Yi, T.; Colamonici, O.R.; Platanias, L.C. Interferon alpha activates the tyrosine kinase Lyn in haemopoietic cells. *Br. J. Haematol.* **1998**, *101*, 446–449. [CrossRef] [PubMed]
133. Uddin, S.; Sher, D.A.; Alsayed, Y.; Pons, S.; Colamonici, O.R.; Fish, E.N.; White, M.F.; Platanias, L.C. Interaction of p59(fyn) with interferon-activated Jak kinases. *Biochem. Biophys. Res. Commun.* **1997**, *235*, 83–88. [CrossRef] [PubMed]
134. Uddin, S.; Sweet, M.; Colamonici, O.R.; Krolewski, J.J.; Platanias, L.C. The vav proto-oncogene product (p95vav) interacts with the Tyk-2 protein tyrosine kinase. *FEBS Lett.* **1997**, *403*, 31–34. [CrossRef]
135. Akahane, K.; Li, Z.; Etchin, J.; Berezovskaya, A.; Gjini, E.; Masse, C.E.; Miao, W.; Rocnik, J.; Kapeller, R.; Greenwood, J.R.; et al. Anti-leukaemic activity of the TYK2 selective inhibitor NDI-031301 in T-cell acute lymphoblastic leukaemia. *Br. J. Haematol.* **2017**, *177*, 271–282. [CrossRef] [PubMed]
136. Muromoto, R.; Tawa, K.; Ohgakiuchi, Y.; Sato, A.; Saino, Y.; Hirashima, K.; Minoguchi, H.; Kitai, Y.; Kashiwakura, J.I.; Shimoda, K.; et al. IkappaB-zeta Expression Requires Both TYK2/STAT3 Activity and IL-17-Regulated mRNA Stabilization. *Immunohorizons* **2019**, *3*, 172–185. [CrossRef] [PubMed]
137. Willems, M.; Dubois, N.; Musumeci, L.; Bours, V.; Robe, P.A. IkappaBzeta: An emerging player in cancer. *Oncotarget* **2016**, *7*, 66310–66322. [CrossRef]
138. Luck, K.; Sheynkman, G.M.; Zhang, I.; Vidal, M. Proteome-Scale Human Interactomics. *Trends Biochem. Sci.* **2017**, *42*, 342–354. [CrossRef]
139. Rolland, T.; Tasan, M.; Charloteaux, B.; Pevzner, S.J.; Zhong, Q.; Sahni, N.; Yi, S.; Lemmens, I.; Fontanillo, C.; Mosca, R.; et al. A proteome-scale map of the human interactome network. *Cell* **2014**, *159*, 1212–1226. [CrossRef]
140. Szklarczyk, D.; Jensen, L.J. Protein-protein interaction databases. *Methods Mol. Biol.* **2015**, *1278*, 39–56. [CrossRef]
141. Lievens, S.; Gerlo, S.; Lemmens, I.; De Clercq, D.J.; Risseeuw, M.D.; Vanderroost, N.; De Smet, A.S.; Ruyssinck, E.; Chevet, E.; Van Calenbergh, S.; et al. Kinase Substrate Sensor (KISS), a mammalian in situ protein interaction sensor. *Mol. Cell. Proteom.* **2014**, *13*, 3332–3342. [CrossRef] [PubMed]
142. Masschaele, D.; Gerlo, S.; Lemmens, I.; Lievens, S.; Tavernier, J. KISS: A Mammalian Two-Hybrid Method for In Situ Analysis of Protein-Protein Interactions. *Methods Mol. Biol.* **2018**, *1794*, 269–278. [CrossRef] [PubMed]
143. Babon, J.J.; Lucet, I.S.; Murphy, J.M.; Nicola, N.A.; Varghese, L.N. The molecular regulation of Janus kinase (JAK) activation. *Biochem. J.* **2014**, *462*, 1–13. [CrossRef] [PubMed]
144. Murray, P.J. The JAK-STAT signaling pathway: Input and output integration. *J. Immunol.* **2007**, *178*, 2623–2629. [CrossRef] [PubMed]
145. Chen, C.; Huang, H.; Wu, C.H. Protein Bioinformatics Databases and Resources. *Methods Mol. Biol.* **2017**, *1558*, 3–39. [CrossRef] [PubMed]
146. Hornbeck, P.V.; Kornhauser, J.M.; Latham, V.; Murray, B.; Nandhikonda, V.; Nord, A.; Skrzypek, E.; Wheeler, T.; Zhang, B.; Gnad, F. 15 years of PhosphoSitePlus(R): Integrating post-translationally modified sites, disease variants and isoforms. *Nucleic Acids Res.* **2019**, *47*, D433–D441. [CrossRef]
147. Kershaw, N.J.; Murphy, J.M.; Lucet, I.S.; Nicola, N.A.; Babon, J.J. Regulation of Janus kinases by SOCS proteins. *Biochem. Soc. Trans.* **2013**, *41*, 1042–1047. [CrossRef]
148. Xu, D.; Qu, C.K. Protein tyrosine phosphatases in the JAK/STAT pathway. *Front. Biosci.* **2008**, *13*, 4925–4932. [CrossRef]
149. Bollu, L.R.; Mazumdar, A.; Savage, M.I.; Brown, P.H. Molecular Pathways: Targeting Protein Tyrosine Phosphatases in Cancer. *Clin. Cancer Res.* **2017**, *23*, 2136–2142. [CrossRef]
150. Inagaki-Ohara, K.; Kondo, T.; Ito, M.; Yoshimura, A. SOCS, inflammation, and cancer. *JAKSTAT* **2013**, *2*, e24053. [CrossRef]
151. Jiang, M.; Zhang, W.W.; Liu, P.; Yu, W.; Liu, T.; Yu, J. Dysregulation of SOCS-Mediated Negative Feedback of Cytokine Signaling in Carcinogenesis and Its Significance in Cancer Treatment. *Front. Immunol.* **2017**, *8*, 70. [CrossRef] [PubMed]
152. Wu, C.; Guan, Q.; Wang, Y.; Zhao, Z.J.; Zhou, G.W. SHP-1 suppresses cancer cell growth by promoting degradation of JAK kinases. *J. Cell. Biochem.* **2003**, *90*, 1026–1037. [CrossRef] [PubMed]

153. Bocchini, C.E.; Kasembeli, M.M.; Roh, S.H.; Tweardy, D.J. Contribution of chaperones to STAT pathway signaling. *JAKSTAT* **2014**, *3*, e970459. [CrossRef] [PubMed]
154. Prodromou, C. Mechanisms of Hsp90 regulation. *Biochem. J.* **2016**, *473*, 2439–2452. [CrossRef] [PubMed]
155. Taipale, M.; Jarosz, D.F.; Lindquist, S. HSP90 at the hub of protein homeostasis: Emerging mechanistic insights. *Nat. Rev. Mol. Cell Biol.* **2010**, *11*, 515–528. [CrossRef] [PubMed]
156. Jaeger, A.M.; Whitesell, L. HSP90: Enabler of Cancer Adaptation. *Annu. Rev. Cancer Biol.* **2019**, *3*, 275–297. [CrossRef]
157. Trepel, J.; Mollapour, M.; Giaccone, G.; Neckers, L. Targeting the dynamic HSP90 complex in cancer. *Nat. Rev. Cancer* **2010**, *10*, 537–549. [CrossRef]
158. Caldas-Lopes, E.; Cerchietti, L.; Ahn, J.H.; Clement, C.C.; Robles, A.I.; Rodina, A.; Moulick, K.; Taldone, T.; Gozman, A.; Guo, Y.; et al. Hsp90 inhibitor PU-H71, a multimodal inhibitor of malignancy, induces complete responses in triple-negative breast cancer models. *Proc. Natl. Acad. Sci. USA* **2009**, *106*, 8368–8373. [CrossRef]
159. Taipale, M.; Krykbaeva, I.; Koeva, M.; Kayatekin, C.; Westover, K.D.; Karras, G.I.; Lindquist, S. Quantitative analysis of HSP90-client interactions reveals principles of substrate recognition. *Cell* **2012**, *150*, 987–1001. [CrossRef]
160. Akahane, K.; Sanda, T.; Mansour, M.R.; Radimerski, T.; DeAngelo, D.J.; Weinstock, D.M.; Look, A.T. HSP90 inhibition leads to degradation of the TYK2 kinase and apoptotic cell death in T-cell acute lymphoblastic leukemia. *Leukemia* **2016**, *30*, 219–228. [CrossRef]
161. Schoof, N.; von Bonin, F.; Trumper, L.; Kube, D. HSP90 is essential for Jak-STAT signaling in classical Hodgkin Lymphoma cells. *Cell Commun. Sig.* **2009**, *7*, 17. [CrossRef] [PubMed]
162. Witte, S.; Muljo, S.A. Integrating non-coding RNAs in JAK-STAT regulatory networks. *JAKSTAT* **2014**, *3*, e28055. [CrossRef] [PubMed]
163. Mullany, L.E.; Herrick, J.S.; Sakoda, L.C.; Samowitz, W.; Stevens, J.R.; Wolff, R.K.; Slattery, M.L. MicroRNA-messenger RNA interactions involving JAK-STAT signaling genes in colorectal cancer. *Genes Cancer* **2018**, *9*, 232–246. [CrossRef] [PubMed]
164. Pencik, J.; Pham, H.T.; Schmoellerl, J.; Javaheri, T.; Schlederer, M.; Culig, Z.; Merkel, O.; Moriggl, R.; Grebien, F.; Kenner, L. JAK-STAT signaling in cancer: From cytokines to non-coding genome. *Cytokine* **2016**, *87*, 26–36. [CrossRef]
165. Li, Z.Y.; Yang, L.; Liu, X.J.; Wang, X.Z.; Pan, Y.X.; Luo, J.M. The Long Noncoding RNA MEG3 and its Target miR-147 Regulate JAK/STAT Pathway in Advanced Chronic Myeloid Leukemia. *EBioMedicine* **2018**, *34*, 61–75. [CrossRef]
166. Ghoreschi, K.; Jesson, M.I.; Li, X.; Lee, J.L.; Ghosh, S.; Alsup, J.W.; Warner, J.D.; Tanaka, M.; Steward-Tharp, S.M.; Gadina, M.; et al. Modulation of innate and adaptive immune responses by tofacitinib (CP-690,550). *J. Immunol.* **2011**, *186*, 4234–4243. [CrossRef]
167. Changelian, P.S.; Flanagan, M.E.; Ball, D.J.; Kent, C.R.; Magnuson, K.S.; Martin, W.H.; Rizzuti, B.J.; Sawyer, P.S.; Perry, B.D.; Brissette, W.H.; et al. Prevention of organ allograft rejection by a specific Janus kinase 3 inhibitor. *Science* **2003**, *302*, 875–878. [CrossRef]
168. Harrison, C.; Kiladjian, J.J.; Al-Ali, H.K.; Gisslinger, H.; Waltzman, R.; Stalbovskaya, V.; McQuitty, M.; Hunter, D.S.; Levy, R.; Knoops, L.; et al. JAK inhibition with ruxolitinib versus best available therapy for myelofibrosis. *N. Engl. J. Med.* **2012**, *366*, 787–798. [CrossRef]
169. Verstovsek, S.; Mesa, R.A.; Gotlib, J.; Levy, R.S.; Gupta, V.; DiPersio, J.F.; Catalano, J.V.; Deininger, M.; Miller, C.; Silver, R.T.; et al. A double-blind, placebo-controlled trial of ruxolitinib for myelofibrosis. *N. Engl. J. Med.* **2012**, *366*, 799–807. [CrossRef]
170. Gadina, M.; Johnson, C.; Schwartz, D.; Bonelli, M.; Hasni, S.; Kanno, Y.; Changelian, P.; Laurence, A.; O'Shea, J.J. Translational and clinical advances in JAK-STAT biology: The present and future of jakinibs. *J. Leukoc. Biol.* **2018**, *104*, 499–514. [CrossRef]
171. Gadina, M.; Le, M.T.; Schwartz, D.M.; Silvennoinen, O.; Nakayamada, S.; Yamaoka, K.; O'Shea, J.J. Janus kinases to jakinibs: From basic insights to clinical practice. *Rheumatology* **2019**, *58*, 4–16. [CrossRef] [PubMed]
172. Vainchenker, W.; Leroy, E.; Gilles, L.; Marty, C.; Plo, I.; Constantinescu, S.N. JAK inhibitors for the treatment of myeloproliferative neoplasms and other disorders. *F1000Res* **2018**, *7*, 82. [CrossRef] [PubMed]
173. Virtanen, A.T.; Haikarainen, T.; Raivola, J.; Silvennoinen, O. Selective JAKinibs: Prospects in Inflammatory and Autoimmune Diseases. *Biodrugs* **2019**, *33*, 15–32. [CrossRef] [PubMed]

174. He, X.R.; Chen, X.B.; Zhang, H.C.; Xie, T.; Ye, X.Y. Selective Tyk2 inhibitors as potential therapeutic agents: A patent review (2015-2018). *Expert. Opin. Ther. Pat.* **2019**, *29*, 137–149. [CrossRef] [PubMed]
175. Danese, S.; Argollo, M.; Le Berre, C.; Peyrin-Biroulet, L. JAK selectivity for inflammatory bowel disease treatment: Does it clinically matter? *Gut* **2019**, *68*, 1893–1899. [CrossRef] [PubMed]
176. Papp, K.; Gordon, K.; Thaci, D.; Morita, A.; Gooderham, M.; Foley, P.; Girgis, I.G.; Kundu, S.; Banerjee, S. Phase 2 Trial of Selective Tyrosine Kinase 2 Inhibition in Psoriasis. *N. Engl. J. Med.* **2018**, *379*, 1313–1321. [CrossRef]
177. Burke, J.R.; Cheng, L.; Gillooly, K.M.; Strnad, J.; Zupa-Fernandez, A.; Catlett, I.M.; Zhang, Y.; Heimrich, E.M.; McIntyre, K.W.; Cunningham, M.D.; et al. Autoimmune pathways in mice and humans are blocked by pharmacological stabilization of the TYK2 pseudokinase domain. *Sci. Transl. Med.* **2019**, *11*, eaaw1736. [CrossRef]
178. Liu, C.; Lin, J.; Moslin, R.; Tokarski, J.S.; Muckelbauer, J.; Chang, C.; Tredup, J.; Xie, D.; Park, H.; Li, P.; et al. Identification of Imidazo[1,2-b]pyridazine Derivatives as Potent, Selective, and Orally Active Tyk2 JH2 Inhibitors. *ACS Med. Chem. Lett.* **2019**, *10*, 383–388. [CrossRef]
179. Moslin, R.; Gardner, D.; Santella, J.; Zhang, Y.; Duncia, J.V.; Liu, C.; Lin, J.; Tokarski, J.S.; Strnad, J.; Pedicord, D.; et al. Identification of imidazo[1,2-b] pyridazine TYK2 pseudokinase ligands as potent and selective allosteric inhibitors of TYK2 signalling. *Medchemcomm* **2017**, *8*, 700–712. [CrossRef]
180. Moslin, R.; Zhang, Y.; Wrobleski, S.T.; Lin, S.; Mertzman, M.; Spergel, S.; Tokarski, J.S.; Strnad, J.; Gillooly, K.; McIntyre, K.W.; et al. Identification of N-Methyl Nicotinamide and N-Methyl Pyridazine-3-Carboxamide Pseudokinase Domain Ligands as Highly Selective Allosteric Inhibitors of Tyrosine Kinase 2 (TYK2). *J. Med. Chem.* **2019**, *62*, 8953–8972. [CrossRef]
181. Wrobleski, S.T.; Moslin, R.; Lin, S.; Zhang, Y.; Spergel, S.; Kempson, J.; Tokarski, J.S.; Strnad, J.; Zupa-Fernandez, A.; Cheng, L.; et al. Highly Selective Inhibition of Tyrosine Kinase 2 (TYK2) for the Treatment of Autoimmune Diseases: Discovery of the Allosteric Inhibitor BMS-986165. *J. Med. Chem.* **2019**, *62*, 8973–8995. [CrossRef] [PubMed]
182. Villanueva, M.T. TYK2 inhibition shows promise. *Nat. Rev. Drug Discov.* **2019**, *18*, 668. [CrossRef] [PubMed]
183. Prchal-Murphy, M.; Semper, C.; Lassnig, C.; Wallner, B.; Gausterer, C.; Teppner-Klymiuk, I.; Kobolak, J.; Muller, S.; Kolbe, T.; Karaghiosoff, M.; et al. TYK2 kinase activity is required for functional type I interferon responses in vivo. *PLoS ONE* **2012**, *7*, e39141. [CrossRef] [PubMed]
184. Raje, V.; Derecka, M.; Cantwell, M.; Meier, J.; Szczepanek, K.; Sisler, J.D.; Strobl, B.; Gamero, A.; Harris, T.E.; Larner, A.C. Kinase Inactive Tyrosine Kinase (Tyk2) Supports Differentiation of Brown Fat Cells. *Endocrinology* **2017**, *158*, 148–157. [CrossRef] [PubMed]
185. Lesgidou, N.; Eliopoulos, E.; Goulielmos, G.N.; Vlassi, M. Insights on the alteration of functionality of a tyrosine kinase 2 variant: A molecular dynamics study. *Bioinformatics* **2018**, *34*, i781–i786. [CrossRef] [PubMed]
186. Lupardus, P.J.; Skiniotis, G.; Rice, A.J.; Thomas, C.; Fischer, S.; Walz, T.; Garcia, K.C. Structural snapshots of full-length Jak1, a transmembrane gp130/IL-6/IL-6Ralpha cytokine receptor complex, and the receptor-Jak1 holocomplex. *Structure* **2011**, *19*, 45–55. [CrossRef] [PubMed]

3

Decoration of Anti-CD38 on Nanoparticles Carrying a STAT3 Inhibitor can Improve the Therapeutic Efficacy Against Myeloma

Yung-Hsing Huang [1], Mohammad Reza Vakili [2], Ommoleila Molavi [1,3], Yuen Morrissey [1], Chengsheng Wu [1,†], Igor Paiva [2], Amir Hasan Soleimani [2], Forugh Sanaee [2], Afsaneh Lavasanifar [2] and Raymond Lai [1,*]

1 Department of Laboratory Medicine and Pathology, University of Alberta, Edmonton, AB T6G 2R3, Canada; yunghsin@ualberta.ca (Y.-H.H.); omolavi@ualberta.ca (O.M.); yuenw@ualberta.ca (Y.M.); chengshe@ualberta.ca (C.W.)

2 Faculty of Pharmacy and Pharmaceutical Sciences, University of Alberta, Edmonton, AB T6G 2H7, Canada; vakili@ualberta.ca (M.R.V.); paiva@ualberta.ca (I.P.); asoleima@ualberta.ca (A.H.S.); fsanaee@ualberta.ca (F.S.); afsaneh@ualberta.ca (A.L.)

3 Faculty of Pharmacy, Tabriz University of Medical Science, Tabriz P.O. Box 51664-14766, East Azerbaijan Province, Iran

* Correspondence: rlai@ualberta.ca

† C.W. is now with the Department of Pathology, University of California San Diego, San Diego, CA 92093, USA.

Abstract: STAT3 is an oncoprotein which has been shown to contribute to drug resistance in multiple myeloma (MM). Nonetheless, the clinical utility of STAT3 inhibitors in treating MM has been limited, partly related to some of their pharmacologic properties. To overcome these challenges, our group had previously packaged STAT3 inhibitors using a novel formulation of nanoparticles (NP) and found encouraging results. In this study, we aimed to further improve the pharmacologic properties of these NP by decorating them with monoclonal anti-CD38 antibodies. NP loaded with S3I-1757 (a STAT3 inhibitor), labeled as S3I-NP, were generated. S3I-NP decorated with anti-CD38 (labeled as CD38-S3I-NP) were found to have a similar nanoparticular size, drug encapsulation, and loading as S3I-NP. The release of S3I-1757 at 24 h was also similar between the two formulations. Using Cy5.5 labeling of the NP, we found that the decoration of anti-CD38 on these NP significantly increased the cellular uptake by two MM cell lines ($p < 0.001$). Accordingly, CD38-S3I-NP showed a significantly lower inhibitory concentration at 50% (IC50) compared to S3I-NP in two IL6-stimulated MM cell lines ($p < 0.001$). In a xenograft mouse model, CD38-S3I-NP significantly reduced the tumor size by 4-fold compared to S3I-NP on day 12 after drug administration ($p = 0.006$). The efficacy of CD38-S3I-NP in suppressing STAT3 phosphorylation in the xenografts was confirmed by using immunocytochemistry and Western blot analysis. In conclusion, our study suggests that the decoration of anti-CD38 on NP loaded with STAT3 inhibitors can further improve their therapeutic effects against MM.

Keywords: multiple myeloma; STAT3; S3I-1757; nanoparticle; CD38

---

1. Introduction

Recent studies have shown that nanoparticles (NP) can be an effective drug delivery system to treat cancers [1]. In addition to its usefulness in delivering hydrophobic drugs, NP can promote drug accumulation at tumor sites due to the fact that NP are too large to pass through the normal capillaries but small enough to leak through the poorly-formed vasculatures frequently present in malignant tumors [2]. To further increase the targeting ability and reduce drug toxicity, researchers have conjugated NP with various tumor-specific antibodies [3]. For example, different forms of NP conjugated with anti-human epidermal receptor-2 (HER2) antibodies have been generated to treat HER2-positive breast cancer [4–7]. Based on the results of a few studies, it appears that the decoration of NP with tumor-specific antibodies can indeed result in superior cellular uptake/cytotoxicity in vitro, as well as significantly improved tumor suppression in vivo, compared to their unconjugated counterparts [7–12]. Some studies also suggest that the conjugation of antibodies on the surface of NP is useful in overcoming drug resistance in cancers which overexpress drug efflux pumps (e.g., p-glycoprotein) [8,13]. The conjugation of antibodies on NP is versatile, and a variety of antibodies have been used to achieve specific experimental objectives, such as the use of antibodies targeting the vascular endothelial growth factor receptor (i.e., to block tumor angiogenesis) [12], matrix metalloproteinases (i.e., to block tumor invasion) [14], and transferrin receptor (i.e., to facilitate the crossing of NP through the blood-brain barrier) [15].

Multiple myeloma (MM) is a hematological disease which is characterized by a high frequency of relapses and resistance to chemotherapy. A signal transducer and activator of transcription 3 (STAT3), which is found to be active in more than 50% of MM, has been shown to contribute to the resistance to bortezomib, thalidomide, and dexamethasone in MM [16–19]. In view of its significance in cancer biology, STAT3 has been postulated to be an attractive anti-cancer target, and many STAT3 inhibitors (such as Stattic, S3I-201, and S3I-1757) have been developed [20–22]. However, the clinical utility of these compounds has been limited, which is likely related to some of their pharmacologic properties, such as their small size and hydrophobicity. Consequently, STAT3 inhibitors have been found to have relatively low therapeutic efficacy and high toxicity. In this regard, a clinical trial of OPB-31121 (an orally administered STAT3 inhibitor) in a cohort of patients with various types of solid cancer has reported that >80% of patients experienced significant nausea/diarrhea without therapeutic benefits [23]. To overcome these challenges, our group has recently generated NP to package STAT3 inhibitors. Specifically, we synthesized NP based on the poly(ethylene oxide)-*block*-poly(α-benzyl carboxylate-ε-caprolactone) (PEO-*b*-PBCL) backbone, and we used these NP to package S3I-1757 (denoted as S3I-NP); we found that S3I-NP exhibited significantly better anti-tumor effects in mice xenografted with a human melanoma cell line compared to free drugs [24].

In this study, we aimed to further improve the pharmacologic properties of S3I-NP by conjugating monoclonal antibodies against human CD38, a cell-surface marker highly expressed on MM cells, on the surface of S3I-NP (denoted as CD38-S3I-NP). We hypothesized that, due to the active targeting properties of anti-CD38 for MM cells, CD38-S3I-NP will demonstrate improved cellular uptake in vitro cytotoxicity and in vivo therapeutic efficacy compared to S3I-NP. Our results have provided the proof-of-principle that anti-CD38-conjugated NP loaded with STAT3 inhibitors are useful therapeutic agents for MM patients.

2. Results

2.1. Synthesis and Characterization of CD38-S3I-NP

To generate CD38-S3I-NP, we conjugated anti-CD38 monoclonal antibodies to the hydrophilic portion of PEO-*b*-PBCL. As illustrated in Figure 1A, anti-CD38 was first thiolated at the lysine residue located in the constant region of the heavy chain of the antibody. Thiolated anti-CD38 antibodies were then combined with maleimide-functionalized PEO-*b*-PBCL, such that antibodies were attached to the surface of the polymers. Lastly, anti-CD38-conjugated polymers were mixed with NP loaded with S3I-1757 (i.e., S3I-NP) to generate CD38-S3I-NP.

We then determined if the conjugation of anti-CD38 on the surface of S3I-NP significantly altered their physical properties. As summarized in Table 1, the average size of CD38-S3I-NP was 91.4 ± 9.4 nm, which is not significantly different from that of S3I-NP (97.4 ± 5.2 nm) (p = 0.39). Similarly, there is no significant difference in drug encapsulation efficiency between CD38-S3I-NP and S3I-NP (81.6 ± 7.2% versus 87.0 ± 9.2%, p = 0.47) as well as drug loading (14.7 ± 1.3% versus 15.7 ± 1.7%,

$p = 0.47$). The polydispersity index was significantly higher in CD38-S3I-NP compared to that of S3I-NP (0.367 ± 0.016 versus 0.273 ± 0.003, $p < 0.001$), suggesting that CD38-S3I-NP is less uniform in size compared to S3I-NP, possibly due to antibody aggregation. As shown in Figure 1B, significantly more S3I-1757 was found to be released from CD38-S3I-NP than that from S3I-NP after 1, 2, and 4 hours of incubation ($p < 0.001$, $p < 0.001$, and $p = 0.002$, respectively). Nevertheless, both formulations reached a comparable amount of S3I-1757 release (~68%, $p = 0.59$) at 24 h. Taken together, the physical properties between these two formulations are not substantially different.

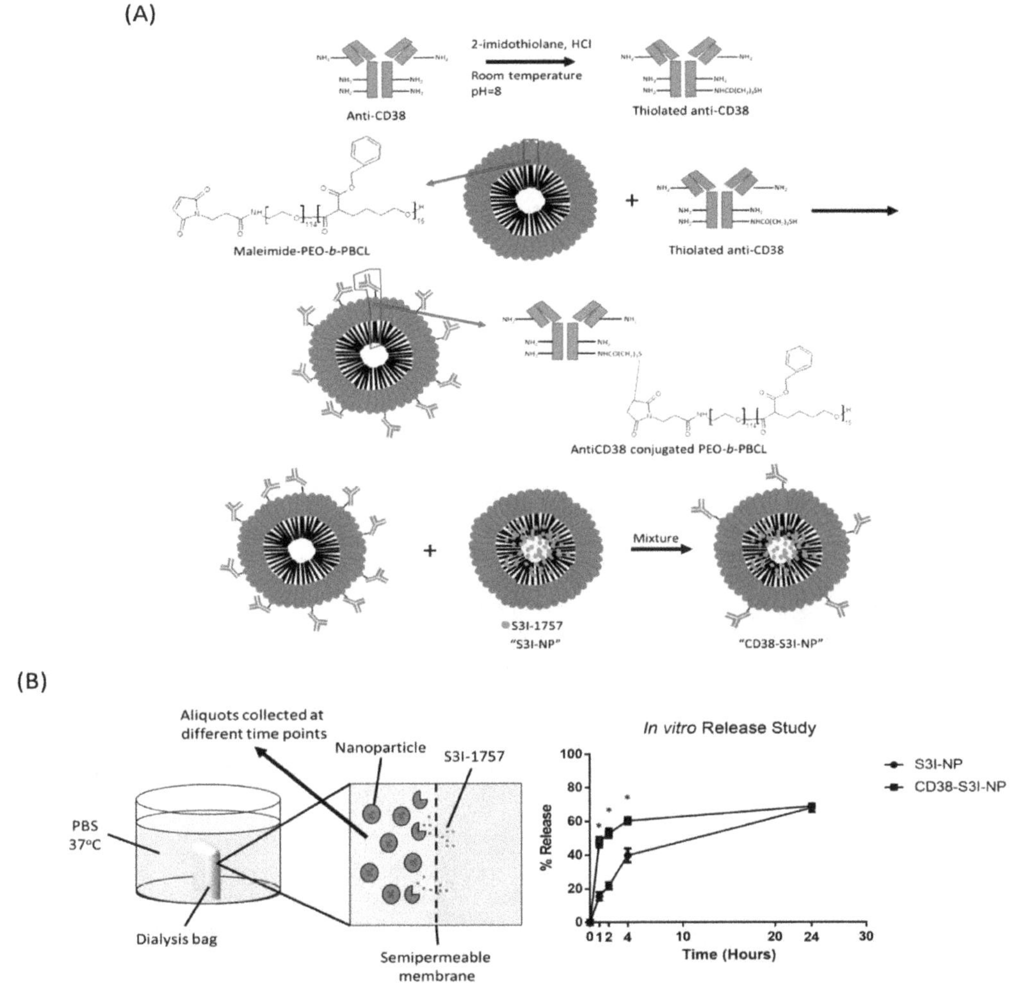

Figure 1. The synthesis of CD38-S3I-NP. (**A**) Chemical reactions of the anti-CD38 conjugation to PEO-*b*-PBCL, the building block of our nanoparticles (NP). The final product was mixed with S3I-NP to generate CD38-S3I-NP. (**B**) The release of S3I-1757 from S3I-NP or CD38-S3I-NP in vitro. The percentage of S3I-1757 released was calculated by the lost amount of S3I-1757 compared to the initial total amount of S3I-1757. The error bar represents the standard deviation from a triplicate experiment; * $p < 0.05$, via Student's *t*-test.

Table 1. Physical properties of S3I-NP and CD38-S3I-NP.

NP Formulation	Average Size (nm)	Polydispersity Index	Drug Encapsulation Efficiency (%)	Drug Loading (Weight %)
S3I-NP	97.4 ± 5.2	0.273 ± 0.003	87.0 ± 9.2%	15.7 ± 1.7%
CD38-S3I-NP	91.4 ± 9.4	0.367 ± 0.016*	81.6 ± 7.2%	14.7 ± 1.3%

* $p < 0.05$, compared to S3I-NP.

2.2. Anti-CD38 Conjugation on NP Results in More Cellular Uptake by MM Cells

We then determined if the conjugation of anti-CD38 to NP can significantly improve the uptake of NP by MM cells. To facilitate the detection and quantification of NP in vitro, we synthesized Cy5.5 (a fluorophore)-conjugated NP with or without the coating of anti-CD38 (denoted as Cy5.5-CD38-NP and Cy5.5-NP, respectively). The NP used in these experiments was not loaded with the STAT3 inhibitor to avoid drug-induced cytotoxicity, which can potentially interfere with our assays. Two MM cell lines (U266 and RPMI8226) were used. SupM2, an ALK-positive anaplastic large cell lymphoma cell line, was used as a negative control. The CD38 expression in the two MM cell lines and the absence of CD38 expression in SupM2 are illustrated in Figure S1A,B. As shown in Figure 2, both MM cell lines incubated with Cy5.5-CD38-NP for 4 h exhibited a significantly higher level of intracellular Cy5.5 compared to cells incubated with Cy5.5-NP. Specifically, in U266 cells, Cy5.5-CD38-NP treatment yielded 43.2 ± 0.1% Cy5.5-positive cells, whereas Cy5.5-NP treatment resulted in only 0.4 ± 0.1% Cy5.5-positive cells ($p < 0.001$). Similarly, in RMMI8226 cells, Cy5.5-CD38-NP yielded significantly more Cy5.5-positive cells than Cy5.5-NP treatment (76.7 ± 1.1% versus 1.2 ± 0.1%) ($p < 0.001$). Compared to the background (i.e., no treatment), Cy5.5-CD38-NP only minimally increased the proportion of Cy5.5-positive cells in SupM2 cells (9.2 ± 0.3%).

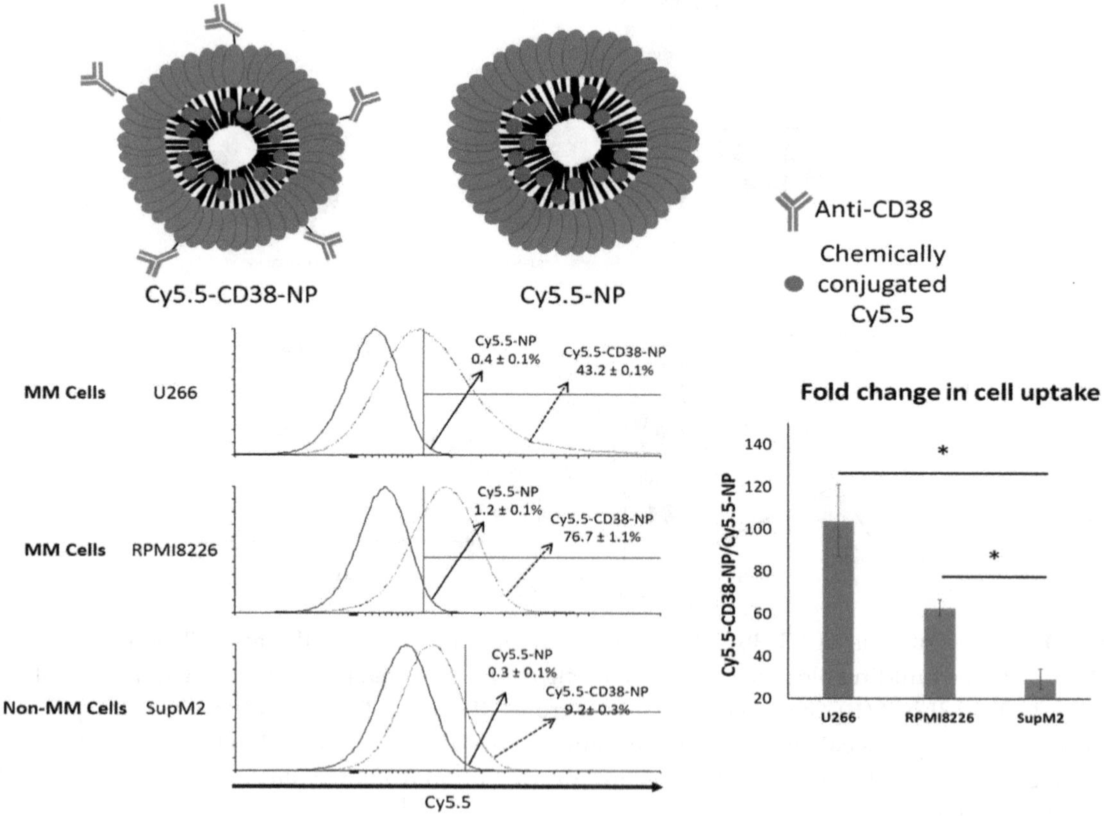

Figure 2. Flow cytometry analysis of the Cy5.5-positive cell population 4 h after treatment of Cy5.5-NP or Cy5.5-CD38-NP. Anti-CD38-conjugated NP exhibits improved cellular uptake of NP by multiple myeloma (MM) cells. Cy5.5 was chemically conjugated to the core of NP. The gated area was defined using the cells without NP treatment. The representative dot plot from a triplicate experiment is shown. The error values represent the standard deviation from the triplicate experiment. A non-MM cell line, SupM2, was included for comparison. The fold change in cell uptake was calculated by dividing the percentage of Cy5.5-positive cells with Cy5.5-CD38-NP treatment by that with Cy5.5-NP treatment. The error bar represents standard deviation from a triplicate experiment; * $p < 0.05$, via Student's t-test.

2.3. *CD38-S3I-NP is More Cytotoxic to MM Cells than S3I-NP*

We next assessed the cytotoxicity of CD38-S3I-NP compared to S3I-NP in two MM cell lines (U266 and RPMI8226) using MTS assay. In these experiments, we added exogenous IL6 (2 ng/mL) to the cell culture to enhance STAT3 activity in the cells. As shown in Figure 3A, in RPMI8226 cells, the addition of CD38-S3I-NP led to significantly lower cell viability compared to S3I-NP at 48 h (50 μM, $p = 0.001$). In U266 cells, significantly lower cell viability of CD38-S3I-NP was seen at 100 μM compared to S3I-NP ($p = 0.007$). In contrast, there was no significant difference in reducing cell viability of SupM2 between the two formulations, although these cells are known to be highly STAT3-active due to an endogenous tyrosine kinase, NPM-ALK [25]. As a comparison, we repeated the experiments using the combination of S3I-NP and free CD38 antibodies. As shown in Figure 3A, the results were similar to those of S3I-NP and significantly inferior to those CD38-S3I-NP in the two MM cell lines. The inhibitory concentration at 50% (IC_{50}) values generated from all of these experiments is summarized in Table 2. In the same experiments, we also confirmed that CD38-S3I-NP and S3I-NP were effective in suppressing STAT3 phosphorylation at residue Y705 (i.e., pSTAT3). Using Western blot analysis, we found that pSTAT3 induced by IL6 (2 ng/mL) in the two MM cell lines was substantially decreased by both formulations at 24 h (Figure 3B).

Table 2. IC_{50} values of U266, RPMI8226, and SupM2 cells treated with different NP.

Cell Line	Treatment	IC_{50} (μM)	
		24 h	48 h
U266	S3I-NP	136.7–163.6	115.4–148.6 **
	CD38-S3I-NP	127.7–151.3	106.3–114.0
	S3I-NP + Free Anti-CD38	143.5–172.4	128.6–139.4 **
RPMI8226	S3I-NP	110.9–124.5	88.2–98.1 **
	CD38-S3I-NP	100.4–109.8	64.0–73.6
	S3I-NP + Free Anti-CD38	108.9–142.2	87.2–99.5 **
SupM2	S3I-NP	88.8–105.6	110.4–144.8
	CD38-S3I-NP	86.0–98.9	83.16–119.4
	S3I-NP + Free Anti-CD38	90.1–110.0	106.9–134.9

** $p < 0.0001$, compared to the corresponding CD38-S3I-NP treatment.

2.4. *CD38-S3I-NP is More Effective in Suppressing MM Tumor Growth In Vivo Compared to S3I-NP*

We then elucidated if CD38-S3I-NP has therapeutic advantages over S3I-NP in a SCID mouse xenograft model. As detailed in Materials and Methods, we xenografted U266 cells stably expressing a luciferase expression construct (U266-luc) in SCID mice, such that the growth of tumors can be easily tracked ex vivo using bioluminescence imaging. When the tumors became detectable, S3I-NP or CD38-S3I-NP was injected intravenously. On day 15 after the injection of NP, 4/4 animals in the S3I-NP group reached the endpoints as defined in Materials and Methods, while 1/4 animals in the CD38-S3I-NP group did not. Statistical analysis reveals a trend for a longer survival for the CD38-S3I-NP group, although the difference between the two groups does not reach statistical significance ($p = 0.079$, Mantel-Cox test), most likely due to the small sample size. As a control, both mice treated with phosphate buffered saline (PBS) reached the endpoints on day 12.

Other than the time needed to reach the endpoints, we also directly assessed tumor growth in the two study groups. Specifically, we summed up the levels of detectable bioluminescence (radiance ranged between 4.00×10^5 and 1.00×10^7 p/sec/cm^2/scr) by using the IVIS Spectrum In Vivo Imaging System, as described in Materials and Methods. Images of a representative animal from each of the CD38-S3I-NP, S3I-NP, and PBS groups are shown in Figure 4A. Animals in the CD38-S3I-NP group had significantly lower tumor volume compared to the S3I-NP group at 240 and 288 h ($p = 0.018$ and $p = 0.006$, respectively). Since we had only two animals in the PBS group, the statistical significance cannot be determined for this group. Nonetheless, it is evident that the tumors grew substantially faster than those in the CD38-S3I-NP group (tumor volume at 288 h was 16.8 times higher).

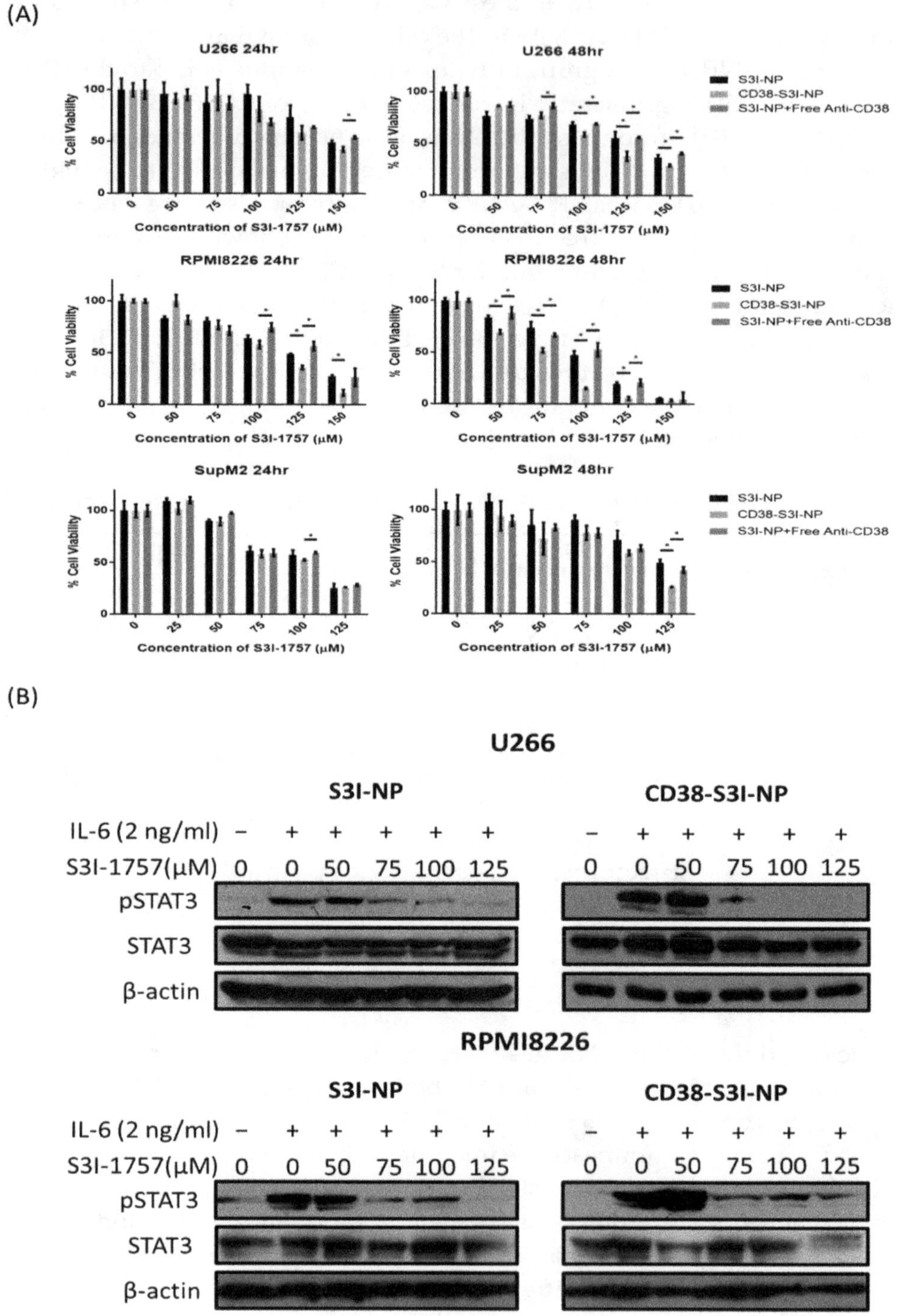

Figure 3. CD38-S3I-NP induces cytotoxicity and inhibits STAT3 activity in MM cells. (**A**) U266 and RPMI8226 cells were then treated with S3I-NP, CD38-S3I-NP, or S3I-NP with free CD38 antibody at a concentration which is equivalent to CD38-S3I-NP (1.4 mg/mL) with the presence of IL6 (2 ng/mL) for 24 and 48 h. Cell viability was measured using a 3-(4,5-dimethylthiazol-2-yl)-5-(3-carboxymethoxyphenyl)-2-(4-sulfophenyl)-2H-tetrazolium (MTS) cell viability assay in triplicate; * $p < 0.05$, via Student's t-test. (**B**) Western blot analysis of STAT3 and pSTAT3 levels in U266 and RPMI8226 cells treated with S3I-NP or CD38-S3I-NP with the presence of IL6 (2 ng/mL) for 24 h. β-actin was blotted as a loading control.

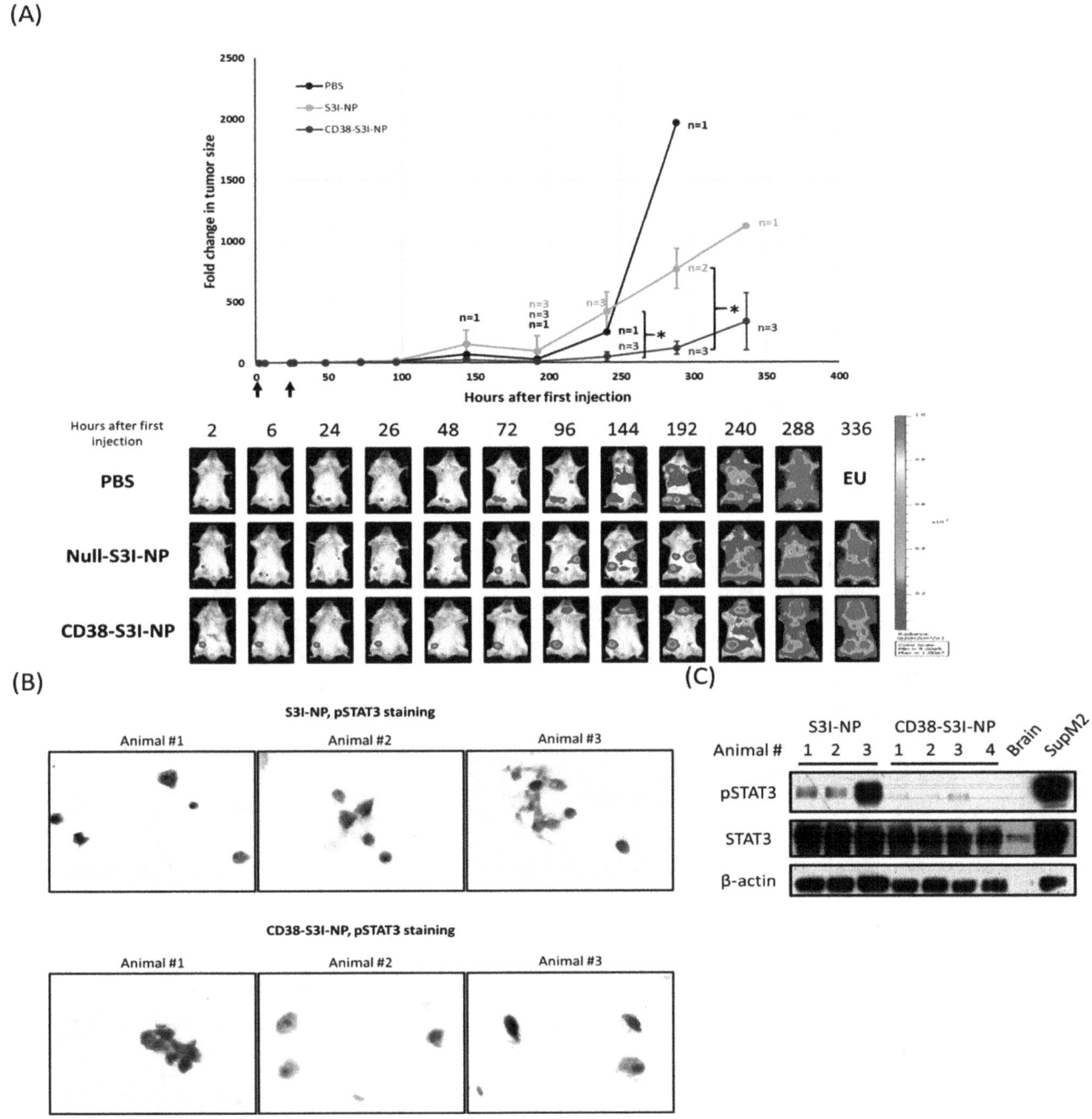

Figure 4. CD38-S3I-NP is more tumor suppressive than S3I-NP in MM xenograft. (**A**) SCID mice intravenously injected with PBS (black line, $n = 2$), 3 mg/kg S3I-NP (green line, $n = 4$) or CD38-S3I-NP (blue line, $n = 4$) every day for two days as indicated by arrows on the x-axis. Animal numbers other than the initial numbers at different time points are indicated. The tumor size was quantified by the bioluminescence intensity and normalized to the initial bioluminescence signal (i.e., 2 h post-injection). The representative bioluminescence images of animals treated with PBS, S3I-NP, or CD38-S3I-NP were shown. * $p < 0.05$, via Student's t-test; EU—euthanized. (**B**) The pSTAT3 levels in bone marrow mononuclear cells extracted from the SCID mice in (**A**) at the endpoint. Non-tumorous brain tissue from a SCID mouse was used as a negative control. SupM2 cells were used as a positive control for pSTAT3. β-actin was blotted as a loading control. (**C**) Immunocytochemical staining of pSTAT3 and in bone marrow mononuclear cells from (**B**). Each image represents bone marrow cells from one animal. The images were taken at a magnification of 400×.

We then assessed if the differences in tumor growth and survival between the CD38-S3I-NP and S3I-NP groups correlates with a difference in STAT3 down-regulation. To achieve this goal, we extracted bone marrow cells from specific bone fragments in which the involvement by MM was confirmed by the expression of bioluminescence. The expression of pSTAT3 was then detected using Western blot analysis and immunocytochemistry. The vast majority of cells extracted from the lesions (detailed in Materials and Methods) were confirmed to MM cells morphologically by using CD38 immunocytochemistry. pSTAT3 immunocytochemistry was then performed, and we found that MM cells from the CD38-S3I-NP group had no or barely detectable pSTAT3 signals, whereas MM cells from the S3I-NP group had relatively strong pSTAT3 signals in most of the cells examined (Figure 4B). This difference in pSTAT3 expression between the two groups was further confirmed by Western blot analysis (Figure 4C).

3. Discussion

It has been demonstrated that a number of novel NP drug delivery systems can significantly improve drug bioavailability in experimental models [26]. The mechanism for this improvement is believed to be attributed to the large size of the NP, which can pass through the leaky tumorous vasculature but not the normal blood vessels, resulting in the preferential accumulation of these NP in the tumors [26]. Many NP also allow the encapsulation of hydrophobic drugs which otherwise cannot be delivered to the cellular targets in their free form [27]. In this regard, our group had previously developed an NP which can effectively encapsulate and deliver cucurbitacin, a naturally occurring STAT3 inhibitor, making it more compatible for clinical use [28]. More recently, we used the same NP to encapsulate S3I-1757, a newly developed STAT3 inhibitor with higher potency and specificity, and we found that this NP (i.e., S3I-NP) demonstrated significant therapeutic efficacy against melanoma in a SCID mouse xenograft model [24].

To further improve the therapeutic potential of NP in treating cancer, various researchers have attempted to decorate NP with various targeting moieties such as monoclonal antibodies. For example, in a subcutaneous breast cancer xenograft mouse model, it was found that trastuzumab (an anti-HER2 antibody)-conjugated NP carrying doxorubicin was accumulated in the xenografts better than NP without antibodies, and this finding correlated with a >50% improvement in the reduction of tumor volume [7]. Similarly, in two other studies, cetuximab (an anti-EGFR antibody)-conjugated NP loaded with paclitaxel or gemcitabine was also found to show superior efficacy and tumor-targeting effects compared to NP without antibody conjugation [29,30]. Our group has previously found that anti-CD30 conjugated to a commercially available liposomal doxorubicin (Doxil®) was significantly more effective in treating ALK-positive anaplastic large cell lymphoma in a SCID mouse xenograft model [31]. In addition to the targeting function, antibodies conjugated on the surface of NP are also believed to prevent the uptake/removal of the drugs by the reticuloendothelial system [32].

The testing of STAT3 inhibitors in treating MM has been previously attempted. Previous studies have reported that novel STAT3 inhibitors such as atovaquone, SC99, and LLL12 can kill STAT3-active MM cells and significantly suppress subcutaneous tumors in SCID-mouse xenograft models [33–35]. However, these new anti-STAT3 agents appear to be structurally hydrophobic; without a carrier, their clinical uses will be limited as they are water-insoluble. Regarding CD38, we believed that this is a good target for MM because it is highly expressed in most cases of MM and its expression is relatively restricted in normal cells [36]. CD38 has been regarded as a therapeutic target for MM, and daratumumab is the first human anti-CD38 antibody approved by the Food and Drug Administration for treating MM [37,38]. To the best of our knowledge, our study is the second to report the development of an antibody-conjugated NP to treat MM. The first study was reported in early 2018, in which anti-CD38 conjugated NP carrying bortezomib exhibited a 2 to 3-fold increase in cell uptake by MM cells and a ~50% greater reduction of MM tumor growth compared to non-targeted NP [39]. These results are in line with our findings. Taken together, we believe that NP carrying a potent STAT3 inhibitor (such as S3I-1757) decorated by anti-CD38 is a reasonable approach to treat MM. Our data is in support of this concept.

The method we employed to conjugate anti-CD38 onto NP involved thiolation of the lysine residues of the constant region of immunoglobulins. The thiolated antibody then formed a highly stable thioester bond with the maleimide polymers. We believe that our conjugation method can provide two main advantages compared to that used in the other study of using anti-CD38 conjugated NP to treat MM, in which anti-CD38 was attached to NP via biotinylation [39]. First, the covalent thioester bond between the antibody and NP is substantially more stable than the non-covalent biotinylation bond, which likely results in a longer half-life of NP in vivo. Second, compared to biotinylation, the thiolation process of anti-CD38 is limited to its constant region, thus minimizing the risk of re-directing the antibodies in the wrong orientation (i.e., the hypervariable region of the immunoglobulin pointing inward). Thus, we believe that thiolation of the immunoglobulin can confer better stability and therapeutic efficacy to the NP.

Due to the NP barrier and the superior MM cell-targeting ability, we speculated that CD38-S3I-NP would possess a higher maximal tolerable dose and lower incidences of adverse effects compared to S3I-NP and free S3I-1757. Unlike OPB-31121, CD38-S3I-NP is not orally available since it can dissemble under conditions of adverse pH within the gastrointestinal tract. Thus, intravenous injection will be the best method to administer CD38-S3I-NP. Because STAT3 activity is linked to resistance to chemotherapy in MM, combined therapy of bortezomib, thalidomide, or dexamethasone with CD38-S3I-NP may improve response rates. One possible caveat of using CD38-S3I-NP in clinical settings is the fast clearance through the liver and kidneys. Therefore, approaches which prevent rapid NP clearance have to be developed to avoid toxicity to these organs.

4. Materials and Methods

4.1. Materials and Cell Culture

S3I-1757 (white powder with a molecular weight of 521.6 g/mol, soluble in DMSO) was obtained from Glixx Laboratories (Hopkinton, MA, USA). Methoxy-PEO (average molecular weight of 5000 g/mol), diisopropylamine (99%), benzyl chloroformate (tech 95%), sodium (in kerosin), butyllithium in hexane (2.5 M solution), palladium-coated charcoal, pyrene, and Cremophor® EL were purchased from Sigma Aldrich (St. Louis, MO, USA). α-benzyl carboxylate ε-caprolactone was prepared by Alberta Research Chemicals Incorporation (Edmonton, AB, Canada). Stannous octoate was purchased from MP Biomedicals Incorporation (Solon, OH, USA). All other chemicals were reagent grade. U266 and SupM2 cells were purchased from American Type Culture Collection (Manassas, VA, USA). RPMI8226 cells are a gift from Dr. Linda Pilarski (Department of Oncology, University of Alberta, Edmonton, AB, Canada). The cells were cultured in RPMI 1460 medium with L-glutamine (Life Technology, Carlsbad, CA, USA), 10% FBS (Life technology) and, 100U/mL penicillin/streptomycin (Sigma Aldrich). All cells were incubated at 37 °C supplied with 5% atmospheric CO_2.

4.2. Purification of Anti-CD38

The hybridoma cells (TBH-7) producing humanized anti-CD38 were cultured in RPMI1460 medium supplied with 10% fetal bovine serum. When the cells were confluent, they were transferred to RPMI1460 medium with 10% ultra-low IgG fetal bovine serum (ThermoFisher, Waltham, MA, USA). The cell supernatant was collected after 48 h. 150 mL of supernatant was concentrated to 5 mL using Amicon Ultra-15 Centrifugal Filter Unit (Millipore, Burlington, MA, USA). The concentrated supernatant was incubated in NAb™ Protein A/G Spin Column (ThermoFisher) for 10 min. The bound antibody was eluted out using the elution buffer (ThermoFisher). The concentration of purified anti-CD38 was measured by NanoDrop™ 1000 Spectrophotometer (ThermoFisher). The purified anti-CD38 was dialyzed in sterile PBS overnight prior to NP conjugation.

4.3. Preparation of NP

Poly-(ethylene oxide)-block-poly-(α-benzyl carboxylate ε-caprolactone) (PEO-b-PBCL) was synthesized using the method previously described [40]. In brief, α-benzyl carboxylate ε-caprolactone was mixed with methoxy-poly-(ethylene oxide) at 1:1.12 weight ratio with a trace amount of stannous octoate. The reaction mixture was incubated for 4 h at 140 °C in the vacuum oven and stopped by cooling the reaction at room temperature overnight. S3I-NP was prepared from PEO-b-PBCL block copolymers as previously described [24]. In brief, 20 mg of block copolymer was dissolved in tetrahydrofuran and was mixed with 2 mg S3I-1757 dissolved in DMSO. The mixture was incubated at room temperature with stirring overnight. The excess S3I-1757 was removed by centrifugation. For anti-CD38 conjugation, anti-CD38 was mixed with 2-imidothiolane at room temperature at pH 8.0 to synthesize thiolated anti-CD38. The maleimide PEO-b-PBCL was prepared by following a previously established protocol [41]. Micellized maleimide PEO-b-PBCL was mixed with thiolated anti-CD38. The anti-CD38 conjugated NP were mixed with S3I-NP in water to form CD38-S3I-NP through post-insertion method. The size and polydispersity index of S3I-NP and CD38-S3I-NP were measured by Zetasizer Nano®. The S3I-1757 concentrations in CD38-S3I-NP and S3I-NP were measured by high-performance liquid chromatography (HPLC) using a previously established protocol [24]. For the synthesis of Cy5.5-conjugated NP, a previously described protocol was followed [42].

4.4. In Vitro Release Assay

In vitro release assay was carried out as previously described [24]. In brief, 1 mL of CD38-S3I-NP or S3I-NP was put in a semi-permeable dialysis bag (molecular weight cutoff: 12,000–14,000 kDa). The bag was placed in sterile PBS and incubated in a shaking water bath at 37 °C. A 50 μL aliquot from the dialysis bag was collected at various time points for S3I-1757 concentration measurement by HPLC analysis. To maintain the total volume, 50 μL of PBS was added back to the dialysis bag after aliquot collection.

4.5. Cellular Uptake Assay

NP chemically conjugated with Cy5.5 (an amount equivalent to 0.2 ng Cy5.5) was added to 1.0×10^6 U266, RPMI8226 and SupM2 cells and cultured at 37 °C in dark for 4 h. Cells were washed with sterile PBS twice and subjected to flow cytometry (BD FACSCantoII, Franklin Lakes, NJ, USA) analysis using the Per-Cy5.5-APC fluorescence channel.

4.6. Cell Viability Assay

Cell proliferation was assessed by the CellTiter 96® AQueous one solution cell proliferation MTS assay (Promega, Madison, WI, USA). Approximately 2.5×10^4 U266 or RPMI8226 cells were seeded in each well of a 96-well plate and treated for 24 or 48 h, and 20 μL 3-(4,5-dimethylthiazol-2-yl)-5-(3-carboxymethoxyphenyl)-2-(4-sulfophenyl)-2H-tetrazolium (MTS) was added to each well. The absorbance of light at 490 nm was measured by a FLUOstar OPTIMA microplate reader (BMG Labtech, Cary, NC, USA). The IC_{50} values were calculated by Graph Prism 7 from the cell viability versus the logarithm of the concentration curve.

4.7. Western Blot Analysis

Total cell lysates were prepared and lysed with 1× RIPA buffer (10× stock solution from Cell Signaling Technology, Danvers, MA, USA) with 0.05% protease inhibitor cocktail (Millipore) and 0.05% phosphatase inhibitor cocktail (Millipore). Protein concentrations were measured using a Pierce™ BCA Protein Assay Kit (ThermoFisher). Cell lysates treated with SDS were subject to SDS-PAGE and transferred onto a nitrocellulose membrane. The membrane was probed with anti-STAT3 (1:1000, CST, #9139), anti-pSTAT3 (Y705) (1:2000, CST, #9145), and anti-β-actin (1:1000, CST, #58169) diluted in 5% BSA in TBS-Tween20 (0.05%, v/v). These antibodies were probed with anti-mouse IgG conjugated with horseradish peroxidase (1:1000, Cell Signaling). The membrane was washed three times with TBS-T after secondary antibody treatment. The bands on the membrane were visualized with Pierce™ ECL Western blotting substrate (Thermo Scientific) and exposed to X-ray films (Fuji, Tokyo, Japan).

4.8. Preparation of U266-luc Cells by Lentiviral Transduction

Lentiviral particles carrying pLenti-Puro-luc were a kind gift from Dr. Kyle Potts and Dr. Mary Hitt (Department of Oncology, University of Alberta, Edmonton, AB, Canada). Also, 2×10^6 U266 cells were transduced with 2 mL of lentiviral particles and 6 μg/mL polybrene on a 6-well plate after 30 min of spin inoculation at $1000\times g$. After 24 h, cells were washed and replenished with fresh lentiviral particles for another 24 h. Transduced cells were washed with PBS and resuspended in fresh growth medium for 2 days. The transduced cells were selected with 2 μg puromycin (ThermoFisher) in the medium. The luciferase activity of U266-luc cells was measured by bioluminescence imaging (Figure S2A). The growth rate and responsiveness to S3I-1757 of U266-luc were confirmed to be insignificantly different from parental U266 cells (Figure S2B,C).

4.9. In Vivo Studies Using MM Xenograft

The experimental protocols for all in vivo studies in this manuscript were approved by Animal Care and Use Committees, University of Alberta (#AUP00000282). Half of a million U266-luc was injected into SCID mice (Jackson, strain NOD.Cg-Prkdcscid Il2rgtm1Wjl/SzJ) intravenously. Twelve days after injection, 100 μL of S3I-NP or CD38-S3I-NP colloidal dispersions in dextrose 5 % was injected into the mice intravenously via the tail vein for two consecutive days at a dose of 3 mg/kg per day. The tumor size was measured by quantifying the total flux of bioluminescence signals detectable (radiance ranged between 4.00×10^5 and 1.00×10^7 p/sec/cm^2/scr) on the ventral side of each animal at various time points using Living Image Software (PerkinElmer, Waltham, MA, USA). At the endpoint defined by the approved protocol, the animals were euthanized and the bone marrow cells with MM bone lesion (visualized by bioluminescence) were removed from the femoral bone and split into two portions. One portion was stored at −80 °C as cell pellets for Western blot analysis, and another was stored in 4% paraformaldehyde at 4 °C for immunocytochemistry.

4.10. Immunocytochemistry

Mouse bone marrow cells in 4% paraformaldehyde (Sigma Aldrich) were pelleted and resuspended in liquid histogel (Thermo Scientific) and transferred to a 15×15 mm^2 plastic mold (Leica, Wetzlar, Germany). Upon solidification, the cell-histogel was subjected to processing and embedding. The cell blocks were sectioned into 5-μm slides. Slides were rehydrated in xylene and decreasing concentrations of ethanol. The antigens were retrieved using 1× citrate buffer (Sigma) by microwaving in a pressure cooker for 20 min. pSTAT3 (1:50, Santa Cruz, #sc-8059) and CD38 antibodies (1:100, Abcam, #ab108403) were diluted at 1:100 in antibody diluent (DAKO, Glostrup, Denmark). MACH2 mouse HRP polymer (Biocare Medical, Pacheco, CA, USA) was used as a secondary antibody. The chromogen and substrate were mixed and applied to each slide for 2 min for color development (DAKO).

4.11. Statistical Analysis

All numerical data in this study was presented as the mean from experiment replicates or independent experiments as described in the figure legends. For the comparison of IC_{50} values in Table 2, two-way ANOVA and Tukey's multiple comparison test were employed. For survival analysis in the animal study, Statistical significance between groups was analyzed using the Mantel-Cox test ($\alpha = 0.05$). For the remaining comparisons, a Student's *t*-test with $\alpha = 0.05$ was used for the statistical analysis. The analysis was done using Microsoft Excel 365, except for Table 2, Figure 4A, and the overall survival analysis in the animal study, for which GraphPad Prism 7 was used for analysis.

5. Conclusions

In summary, this study has provided the proof-of-principle that the decoration of anti-CD38 on NP loaded with STAT3 inhibitors can further improve their therapeutic effects against MM. We also predict that the use of these NP can significantly lower the unwanted side effects of STAT3 inhibitors, as they are targeted to cancer cells, and preferentially released inside of these cells. These encouraging results provide sufficient basis for a Phase I clinical trial.

Author Contributions: Y.-H.H., A.L. and R.L. designed the experiments and wrote the manuscript. Y.-H.H., Y.M. and O.M. purified the CD38 monoclonal antibody. M.R.V. synthesized the antibody-conjugated polymers. Y.-H.H., C.W. and M.V. conducted the in vitro experiments. Y.-H.H., I.P. and A.H.S. conducted the animal experiments. F.S. optimized the protocols for characterization of the synthesized nanoparticles.

Acknowledgments: The authors would like to acknowledge the technical support from Kyle Potts and his supervisor Mary Hitt on the preparation of lentiviral particles carrying pLenti-Puro-luc plasmid.

References

1. Peer, D.; Karp, J.M.; Hong, S.; Farokhzad, O.C.; Margalit, R.; Langer, R. Nanocarriers as an emerging platform for cancer therapy. *Nat. Nanotechnol.* **2007**, *2*, 751–760. [CrossRef]
2. Fang, J.; Nakamura, H.; Maeda, H. The EPR effect: Unique features of tumor blood vessels for drug delivery, factors involved, and limitations and augmentation of the effect. *Adv. Drug Deliv. Rev.* **2011**, *63*, 136–151. [CrossRef]
3. Arruebo, M.; Valladares, M.; González-Fernández, Á. Antibody-conjugated nanoparticles for biomedical applications. *J. Nanomater.* **2009**, *2009*, 439389. [CrossRef]
4. Mi, Y.; Liu, X.; Zhao, J.; Ding, J.; Feng, S.S. Multimodality treatment of cancer with herceptin conjugated, thermomagnetic iron oxides and docetaxel loaded nanoparticles of biodegradable polymers. *Biomaterials* **2012**, *33*, 7519–7529. [CrossRef]
5. Parhi, P.; Sahoo, S.K. Trastuzumab guided nanotheranostics: A lipid based multifunctional nanoformulation for targeted drug delivery and imaging in breast cancer therapy. *J. Colloid Interface Sci.* **2015**, *451*, 198–211. [CrossRef]
6. Vivek, R.; Thangam, R.; NipunBabu, V.; Rejeeth, C.; Sivasubramanian, S.; Gunasekaran, P.; Muthuchelian, K.; Kannan, S. Multifunctional HER2-Antibody Conjugated Polymeric Nanocarrier-Based Drug Delivery System for Multi-Drug-Resistant Breast Cancer Therapy. *ACS Appl. Mater. Interfaces* **2014**, *6*, 6469–6480. [CrossRef]
7. Choi, W.I.; Lee, J.H.; Kim, J.Y.; Heo, S.U.; Jeong, Y.Y.; Kim, Y.H.; Tae, G. Targeted antitumor efficacy and imaging via multifunctional nano-carrier conjugated with anti-HER2 trastuzumab. *Nanomed. Nanotechnol. Biol. Med.* **2015**, *11*, 359–368. [CrossRef]
8. Punfa, W.; Yodkeeree, S.; Pitchakarn, P.; Ampasavate, C.; Limtrakul, P. Enhancement of cellular uptake and cytotoxicity of curcumin-loaded PLGA nanoparticles by conjugation with anti-P-glycoprotein in drug resistance cancer cells. *Acta Pharmacol. Sin.* **2012**, *33*, 823–831. [CrossRef]
9. Arya, G.; Vandana, M.; Acharya, S.; Sahoo, S.K. Enhanced antiproliferative activity of Herceptin (HER2)-conjugated gemcitabine-loaded chitosan nanoparticle in pancreatic cancer therapy. *Nanomed. Nanotechnol. Biol. Med.* **2011**, *7*, 859–870. [CrossRef]
10. Sreeranganathan, M.; Uthaman, S.; Sarmento, B.; Mohan, C.G.; Park, I.K.; Jayakumar, R. In vivo evaluation of cetuximab-conjugated poly(γ-glutamic acid)-docetaxel nanomedicines in EGFR-overexpressing gastric cancer xenografts. *Int. J. Nanomed.* **2017**, *12*, 7167–7182. [CrossRef]
11. Mukerjee, A.; Ranjan, A.P.; Vishwanatha, J.K. Targeted nanocurcumin therapy using annexin A2 antibody improves tumor accumulation and therapeutic efficacy against highly metastatic breast cancer. *J. Biomed. Nanotechnol.* **2016**, *12*, 1374–1392. [CrossRef]
12. Öztürk, K.; Esendağlı, G.; Gürbüz, M.U.; Tülü, M.; Çalış, S. Effective targeting of gemcitabine to pancreatic cancer through PEG-cored Flt-1 antibody-conjugated dendrimers. *Int. J. Pharm.* **2017**, *517*, 157–167. [CrossRef]

13. Punfa, W.; Suzuki, S.; Pitchakarn, P.; Yodkeeree, S.; Naiki, T.; Takahashi, S.; Limtrakul, P. Curcumin-loaded PLGA nanoparticles conjugated with anti-P-glycoprotein antibody to overcome multidrug resistance. *Asian Pac. J. Cancer Prev.* **2014**, *15*, 9249–9258. [CrossRef]
14. Hatakeyama, H.; Akita, H.; Ishida, E.; Hashimoto, K.; Kobayashi, H.; Aoki, T.; Yasuda, J.; Obata, K.; Kikuchi, H.; Ishida, T.; et al. Tumor targeting of doxorubicin by anti-MT1-MMP antibody-modified PEG liposomes. *Int. J. Pharm.* **2007**, *342*, 194–200. [CrossRef]
15. Sun, T.; Wu, H.; Li, Y.; Huang, Y.; Yao, L.; Chen, X.; Han, X.; Zhou, Y.; Du, Z. Targeting transferrin receptor delivery of temozolomide for a potential glioma stem cell-mediated therapy. *Oncotarget* **2017**, *8*, 74451–74465. [CrossRef]
16. Bharti, A.C.; Shishodia, S.; Reuben, J.M.; Weber, D.; Alexanian, R.; Raj-Vadhan, S.; Estrov, Z.; Talpaz, M.; Aggarwal, B.B. Nuclear factor–κB and STAT3 are constitutively active in CD138 + cells derived from multiple myeloma patients, and suppression of these transcription factors leads to apoptosis. *Blood* **2004**, *103*, 3175–3184. [CrossRef]
17. Liu, T.; Fei, Z.; Gangavarapu, K.J.; Agbenowu, S.; Bhushan, A.; Lai, J.C.; Daniels, C.K.; Cao, S. Interleukin-6 and JAK2/STAT3 signaling mediate the reversion of dexamethasone resistance after dexamethasone withdrawal in 7TD1 multiple myeloma cells. *Leuk. Res.* **2013**, *37*, 1322–1328. [CrossRef]
18. Zhang, X.D.; Baladandayuthapani, V.; Lin, H.; Mulligan, G.; Li, B.-Z.; Esseltine, D.L.W.; Qi, L.; Xu, J.; Hunziker, W.; Barlogie, B.; et al. Tight jnction protein 1 modulates proteasome capacity and proteasome inhibitor sensitivity in multiple myeloma via EGFR/JAK1/STAT3 signaling. *Cancer Cell* **2016**, *29*, 639–652. [CrossRef]
19. Wu, W.; Ma, D.; Wang, P.; Cao, L.; Lu, T.; Fang, Q.; Zhao, J.; Wang, J. Potential crosstalk of the interleukin-6-heme oxygenase-1-dependent mechanism involved in resistance to lenalidomide in multiple myeloma cells. *FEBS J.* **2016**, *283*, 834–849. [CrossRef]
20. Zhang, X.; Sun, Y.; Pireddu, R.; Yang, H.; Urlam, M.K.; Lawrence, H.R.; Guida, W.C.; Lawrence, N.J.; Sebti, S.M. A novel inhibitor of STAT3 homodimerization selectively suppresses STAT3 activity and malignant transformation. *Cancer Res.* **2013**, *73*, 1922–1933. [CrossRef]
21. Zhang, X.; Yue, P.; Fletcher, S.; Zhao, W.; Gunning, P.T.; Turkson, J. A novel small-molecule disrupts Stat3 SH2 domain–phosphotyrosine interactions and Stat3-dependent tumor processes. *Biochem. Pharmacol.* **2010**, *79*, 1398–1409. [CrossRef]
22. Schust, J.; Sperl, B.; Hollis, A.; Mayer, T.U.; Berg, T. Stattic: A small-molecule inhibitor of STAT3 activation and dimerization. *Chem. Biol.* **2006**, *13*, 1235–1242. [CrossRef]
23. Oh, D.-Y.; Lee, S.-H.; Han, S.-W.; Kim, M.-J.; Kim, T.-M.; Kim, T.-Y.; Heo, D.S.; Yuasa, M.; Yanagihara, Y.; Bang, Y.J. Phase I Study of OPB-31121, an Oral STAT3 Inhibitor, in Patients with Advanced Solid Tumors. *Cancer Res. Treat.* **2015**, *47*, 607–615. [CrossRef]
24. Soleimani, A.H.; Garg, S.M.; Paiva, I.M.; Vakili, M.R.; Alshareef, A.; Huang, Y.-H.; Molavi, O.; Lai, R.; Lavasanifar, A. Micellar nano-carriers for the delivery of STAT3 dimerization inhibitors to melanoma. *Drug Deliv. Transl. Res.* **2017**, 1–11. [CrossRef]
25. Zamo, A.; Chiarle, R.; Piva, R.; Howes, J.; Fan, Y.; Chilosi, M.; Levy, D.E.; Inghirami, G. Anaplastic lymphoma kinase (ALK) activates Stat3 and protects hematopoietic cells from cell death. *Oncogene* **2002**, *21*, 1038–1047. [CrossRef]
26. Nakamura, Y.; Mochida, A.; Choyke, P.L.; Kobayashi, H. Nanodrug Delivery: Is the Enhanced Permeability and Retention Effect Sufficient for Curing Cancer? *Bioconjug. Chem.* **2016**, *27*, 2225–2238. [CrossRef]
27. Martin, C.; Aibani, N.; Callan, J.F.; Callan, B. Recent advances in amphiphilic polymers for simultaneous delivery of hydrophobic and hydrophilic drugs. *Ther. Deliv.* **2016**, *7*, 15–31. [CrossRef]
28. Molavi, O.; Ma, Z.; Mahmud, A.; Alshamsan, A.; Samuel, J.; Lai, R.; Kwon, G.S.; Lavasanifar, A. Polymeric micelles for the solubilization and delivery of STAT3 inhibitor cucurbitacins in solid tumors. *Int. J. Pharm.* **2008**, *347*, 118–127. [CrossRef]
29. Wang, X.B.; Zhou, H.Y. Molecularly targeted gemcitabine-loaded nanoparticulate system towards the treatment of EGFR overexpressing lung cancer. *Biomed. Pharmacother.* **2015**, *70*, 123–128. [CrossRef]
30. Karra, N.; Nassar, T.; Ripin, A.N.; Schwob, O.; Borlak, J.; Benita, S. Antibody conjugated PLGA nanoparticles for targeted delivery of paclitaxel palmitate: Efficacy and biofate in a lung cancer mouse model. *Small* **2013**, *9*, 4221–4236. [CrossRef]

31. Molavi, O.; Xiong, X.B.; Douglas, D.; Kneteman, N.; Nagata, S.; Pastan, I.; Chu, Q.; Lavasanifar, A.; Lai, R. Anti-CD30 antibody conjugated liposomal doxorubicin with significantly improved therapeutic efficacy against anaplastic large cell lymphoma. *Biomaterials* **2013**, *34*, 8718–8725. [CrossRef]
32. Badkas, A.; Frank, E.; Zhou, Z.; Jafari, M.; Chandra, H.; Sriram, V.; Lee, J.Y.; Yadav, J.S. Modulation of in vitro phagocytic uptake and immunogenicity potential of modified Herceptin®-conjugated PLGA-PEG nanoparticles for drug delivery. *Colloids Surf. B Biointerfaces* **2018**, *162*, 271–278. [CrossRef]
33. Lin, L.; Benson, D.M.; Deangelis, S.; Bakan, C.E.; Li, P.K.; Li, C.; Lin, J. A small molecule, LLL12 inhibits constitutive STAT3 and IL-6-induced STAT3 signaling and exhibits potent growth suppressive activity in human multiple myeloma cells. *Int. J. Cancer* **2012**, *130*, 1459–1469. [CrossRef]
34. Zhang, Z.; Mao, H.; Du, X.; Zhu, J.; Xu, Y.; Wang, S. A novel small molecule agent displays potent anti-myeloma activity by inhibiting the JAK2-STAT3 signaling pathway. *Oncotarget* **2016**, *7*, 9296–9308. [CrossRef]
35. Xiang, M.; Kim, H.; Ho, V.T.; Walker, S.R.; Bar-Natan, M.; Anahtar, M.; Liu, S.; Toniolo, P.A.; Kroll, Y.; Jones, N.; et al. Gene expression-based discovery of atovaquone as a STAT3 inhibitor and anticancer agent. *Blood* **2016**, *128*, 1845–1853. [CrossRef]
36. Malavasi, F.; Funaro, A.; Alessio, M.; DeMonte, L.B.; Ausiello, C.M.; Dianzani, U.; Lanza, F.; Magrini, E.; Momo, M.; Roggero, S. CD38: A multi-lineage cell activation molecule with a split personality. *Int. J. Clin. Lab. Res.* **1992**, *22*, 73–80. [CrossRef]
37. De Weers, M.; Tai, Y.-T.; van derVeer, M.S.; Bakker, J.M.; Vink, T.; Jacobs, D.C.H.; Oomen, L.A.; Peipp, M.; Valerius, T.; Slootstra, J.W.; et al. Daratumumab, a Novel Therapeutic Human CD38 Monoclonal Antibody, Induces Killing of Multiple Myeloma and Other Hematological Tumors. *J. Immunol.* **2011**, *186*, 1840–1848. [CrossRef]
38. Beum, P.V.; Lindorfer, M.A.; Peek, E.M.; Stukenberg, P.T.; deWeers, M.; Beurskens, F.J.; Parren, P.W.; van de Winkel, J.G.; Taylor, R.P. Penetration of antibody-opsonized cells by the membrane attack complex of complement promotes Ca^{2+} influx and induces streamers. *Eur. J. Immunol.* **2011**, *41*, 2436–2446. [CrossRef]
39. Dela Puente, P.; Luderer, M.J.; Federico, C.; Jin, A.; Gilson, R.C.; Egbulefu, C.; Alhallak, K.; Shah, S.; Muz, B.; Sun, J.; et al. Enhancing proteasome-inhibitory activity and specificity of bortezomib by CD38 targeted nanoparticles in multiple myeloma. *J. Control. Release* **2018**, *270*, 158–176. [CrossRef]
40. Mahmud, A.; Xiong, X.B.; Lavasanifar, A. Novel self-associating poly(ethylene oxide)-block-poly(ε-caprolactone) block copolymers with functional side groups on the polyester block for drug delivery. *Macromolecules* **2006**, *39*, 9419–9428. [CrossRef]
41. Kogan, T. The synthesis of substituted methoxy-poly (ethyleneglycol) derivatives suitable for selective protein modification. *Synth. Commun.* **1992**, *22*, 2417–2424. [CrossRef]
42. Garg, S.M.; Paiva, I.M.; Vakili, M.R.; Soudy, R.; Agopsowicz, K.; Soleimani, A.H.; Hitt, M.; Kaur, K.; Lavasanifar, A. Traceable PEO-poly(ester) micelles for breast cancer targeting: The effect of core structure and targeting peptide on micellar tumor accumulation. *Biomaterials* **2017**, *144*, 17–29. [CrossRef] [PubMed]

4

STAT5a/b Deficiency Delays, but does not Prevent, Prolactin-Driven Prostate Tumorigenesis in Mice

Florence Boutillon [1,2], Natascha Pigat [1,2], Lucila Sackmann Sala [1,2], Edouard Reyes-Gomez [3,4,5], Richard Moriggl [6,7], Jacques-Emmanuel Guidotti [1,2,†] and Vincent Goffin [1,2,*,†]

1 Institut Necker Enfants Malades, Inserm U1151, 75014 Paris, France
2 Faculté de Médecine, Université Paris Descartes, 75014 Paris, France
3 Unité d'Histologie et d'Anatomie Pathologique, Ecole Nationale Vétérinaire d'Alfort, 94704 Maisons-Alfort, France
4 Laboratoire d'Anatomo-Cytopathologie, BioPôle Alfort, Ecole Nationale Vétérinaire d'Alfort, 94704 Maisons-Alfort, France
5 U955—IMRB, Inserm, Ecole Nationale Vétérinaire d'Alfort, UPEC, 94704 Maisons-Alfort, France
6 Institute of Animal Breeding and Genetics, University of Veterinary Medicine Vienna, 1210 Vienna, Austria
7 Medical University of Vienna, 1090 Vienna, Austria
* Correspondence: Vincent.goffin@inserm.fr
† These authors contributed equally to this paper.

Abstract: The canonical prolactin (PRL) Signal Transducer and Activator of Transcription (STAT) 5 pathway has been suggested to contribute to human prostate tumorigenesis via an autocrine/paracrine mechanism. The probasin (Pb)-PRL transgenic mouse models this mechanism by overexpressing PRL specifically in the prostate epithelium leading to strong STAT5 activation in luminal cells. These mice exhibit hypertrophic prostates harboring various pre-neoplastic lesions that aggravate with age and accumulation of castration-resistant stem/progenitor cells. As STAT5 signaling is largely predominant over other classical PRL-triggered pathways in Pb-PRL prostates, we reasoned that Pb-Cre recombinase-driven genetic deletion of a floxed *Stat5a/b* locus should prevent prostate tumorigenesis in so-called Pb-PRL$^{\Delta STAT5}$ mice. Anterior and dorsal prostate lobes displayed the highest *Stat5a/b* deletion efficiency with no overt compensatory activation of other PRLR signaling cascade at 6 months of age; hence the development of tumor hallmarks was markedly reduced. *Stat5a/b* deletion also reversed the accumulation of stem/progenitor cells, indicating that STAT5 signaling regulates prostate epithelial cell hierarchy. Interestingly, ERK1/2 and AKT, but not STAT3 and androgen signaling, emerged as escape mechanisms leading to delayed tumor development in aged Pb-PRL$^{\Delta STAT5}$ mice. Unexpectedly, we found that Pb-PRL prostates spontaneously exhibited age-dependent decline of STAT5 signaling, also to the benefit of AKT and ERK1/2 signaling. As a consequence, both Pb-PRL and Pb-PRL$^{\Delta STAT5}$ mice ultimately displayed similar pathological prostate phenotypes at 18 months of age. This preclinical study provides insight on STAT5-dependent mechanisms of PRL-induced prostate tumorigenesis and alternative pathways bypassing STAT5 signaling down-regulation upon prostate neoplasia progression.

Keywords: STAT5; AKT; ERK1/2; prolactin; androgens; prostate cancer; knockout; escape mechanisms; stem/progenitor cells; cell hierarchy

---

1. Introduction

Studies of human prostate cancer specimens support a role for prolactin (PRL) signaling in disease progression and recurrence [1]. Indeed, PRL is expressed in more than half of prostate tumors (including local, locally advanced, and hormone refractory) and over 60% metastases [2,3], and its level of expression in primary tumors is positively associated with high Gleason score (i.e., high disease severity) [2]. Since circulating PRL levels are not correlated to prostate cancer risk [4,5], this suggests that the contribution of PRL signaling to prostate tumorigenesis mainly occurs through an autocrine/paracrine mechanism.

The canonical signaling pathways activated by the PRL receptor (PRLR) involve Janus kinase 2 (JAK2)/Signal Transducer and Activator of Transcription (STAT) pathway, the extracellular regulated kinase (ERK) 1/2 pathway and the phosphoinositide 3-kinase (PI3K)–AKT pathway [6,7]. In human and rodent prostate, the PRLR preferentially signals via the STAT5 pathway; in fact, activation of the other pathways is not detected [2,8,9]. STAT5 involves two highly homologous proteins referred to as STAT5A (94 kDa) and STAT5B (92 kDa) that are encoded by two distinct genes [10]. In PRL signaling STAT5A and STAT5B are activated by JAK2-mediated phosphorylation of a conserved C-terminal tyrosine upon which they form a parallel dimer that allows more efficient translocation into the nucleus where they activate the transcription of target genes involving many protooncogenes. In various preclinical models of prostate cancer, STAT5A/B (hereafter referred to as STAT5 unless specifically discriminated for gene product) has been shown to be critical for cell survival and proliferation through both androgen-dependent and androgen-independent mechanisms [9,11–14]. Recently, STAT5 was shown to promote epithelial-to-mesenchymal transition (EMT) and stem-like features in human prostate cancer cells [15], supporting the earlier finding that this cascade promotes metastatic properties of prostate cancer cells [16]. Consistent with these observations, inhibition of STAT5 signaling using a dominant-negative STAT5B mutant inhibited the in vitro growth and invasiveness of cell lines derived from the TRAMP (transgenic adenocarcinoma of the mouse prostate) mouse model [17]. Additionally, pharmacological inhibition of the JAK2/STAT5 cascade using the JAK2 inhibitor AZD1480 blocked the growth of primary androgen-dependent as well as the growth of recurrent castrate-resistant prostate cancer (CRPC) xenografts [18]. However, so far no transgenic mouse model studies exist for complete STAT5 deletion to ultimately test its role in prostate cancer initiation and progression, which we investigated here.

In clinical specimens of human prostate cancer, the *STAT5a/b* gene locus was shown to undergo amplification during prostate cancer progression towards metastatic CRPC [19]. Accordingly, the *STAT5a* gene locus was found to be amplified in up to 20% of metastatic CRPC with the neuroendocrine phenotype [20]. Furthermore, STAT5 was shown to be overexpressed in prostate cancer compared to healthy prostate samples, to be positively correlated with Gleason score and to predict recurrence after prostatectomy [2,19,21,22]. These findings corroborated the results obtained in preclinical models and further supported the relevance to develop therapeutic strategies aimed to inhibit STAT5 signaling in prostate cancer. Small molecule inhibitors were designed to block the docking of the SH2 domain of STAT5 to the critical tyrosine of the receptor-JAK2 complex and these have shown potency to inhibit STAT5 tyrosine phosphorylation, nuclear translocation and transcriptional activity [23]. The lead compound (IST5) induced massive apoptosis of prostate cancer cell lines and explant culture of patient-derived prostate cancer [23]. More recently this inhibitor was shown to sensitize prostate cancer to radiation by inhibiting STAT5-mediated DNA repair via the homologous recombination mechanism [24]. Additional observations supporting the role of PRL/STAT5 signaling in prostate tumorigenesis and cancer progression can be found in various review articles published within the last decade [1,14,25–27].

In contrast to the human prostate [28], expression of the *Prl* gene is not detected in the mouse prostate. Hence, to decipher the molecular and cellular mechanisms linking the autocrine PRL/STAT5 loop to prostate tumorogenesis, we use here a prostate-specific PRL transgenic mouse model. This model involves expression of rat PRL under the control of the prostate-specific, androgen-regulated probasin (Pb) minimal promoter [29]. At six months of age, Pb-PRL mice exhibit dramatically hypertrophied prostates (all lobes) harboring various pre-neoplastic lesions including prostate-intraepithelial neoplasia (PINs), increased stroma density, and inflammation [1,30,31]. They also displayed distended ducts filled with abundant secretions. All these phenotypes aggravate with age. While our seminal report indicated occasional occurrence of invasive cancer in 20 month-old mice [30], these findings were not consistently found in subsequent studies, suggesting possible interference with the genetic background and/or health/microbiota/immune cell status of transgenics that could drift particularly if animal facilities were relocated, as was the case.

According to earlier findings mentioned above, STAT5 was the only typical PRLR signaling pathway that was found to be activated in Pb-PRL prostates [30]. STAT5 was highly activated in dorsal and lateral lobe, while the ventral lobe displayed much lower level of activation presumably due to PRLR down-regulation specifically in that lobe [31]. Of interest, we discovered that Pb-PRL prostates displayed altered cell hierarchy. First, the basal/stem cell compartment was found to be markedly amplified in the prostate epithelium [30]. Second, these prostates showed amplification of another primitive cell population that had never been described before as it is very rare in the healthy prostate. This newly-identified epithelial cell pool, called LSC^{med} according to its cell sorting profile (Lin−/Sca-1+/CD49f^{med}), combines luminal (Cytokeratin [CK] 8 positive) and stem (Stem-cell antigen-1 positive) phenotypic features and exhibits stem/progenitor properties in functional assays [31,32]. Notably, while the basal/stem cell compartment has been long described to be castrate-resistant in rats [33], we recently reported that the LSC^{med} cell compartment survived castration even better than basal/stem cells in mice [32].

The amplification of castrate-resistant, stem/progenitor cells in Pb-PRL prostates further supports a role for PRL signaling in the progression of prostate tumors including escape to androgen deprivation. Of interest, careful analysis of immunostaining data suggested that the emergence of these two primitive cell populations in Pb-PRL tumors may be zonally correlated to elevated STAT5 activation [31]. However, as neither basal/stem nor LSC^{med} cell population exhibit detectable levels of PRLR expression [32], the role of STAT5 signaling in their amplification remains elusive and may involve paracrine mechanisms. In order to elucidate the actual contribution of STAT5 signaling in the various hallmarks of PRL-induced prostate tumorigenesis, we took advantage of previously developed floxed *Stat5a/b* mice [34] to abolish STAT5 expression in the epithelial cells of Pb-PRL mice. As reported below, in the anterior and dorsal prostate lobes in which *Stat5a/b* locus deletion was highly efficient, STAT5 deficiency delayed the occurrence of PRL-induced pathological phenotypes in young mice, but could not prevent prostate tumorigenesis in older animals due to the emergence of alternative signaling pathways.

2. Results

2.1. STAT5 Deletion has no Detectable Effect on the Prostate Tissue

The analyses of STAT5 expression in the prostate were performed on 6 month-old animals. As shown by RT-qPCR (Figure S1A) *Stat5a* expression was largely prevalent (~5 fold) over *Stat5b* expression in control mice (STAT5$^{f/f}$), but both genes displayed similar lobe-specific expression patterns with the highest expression in anterior and lateral lobes. While the lobe differences were less marked at the protein level, the same profile was observed by immunoblot using the G2 antibody that cross-reacts with both STAT5 isoforms (Figure S1B,C). In ΔSTAT5 mice the fold-reduction in STAT5 expression was the highest in lateral lobe and the lowest in ventral lobe (Figure S1B,C). The residual expression of STAT5 detected at the mRNA and protein levels in tissue extracts of the different lobes of ΔSTAT5 mouse prostates accounted for the unaltered STAT5 expression in non-epithelial cells and possibly to incomplete *Stat5* deletion in the epithelium especially in the ventral lobe (mosaicism). To address this further, we monitored STAT5 expression by immunohistochemistry (IHC) using the G2 antibody (Figure S1D). In ΔSTAT5 prostates a homogenous reduction of STAT5 immunostaining was observed in the epithelium of all but the ventral lobes compared to STAT5$^{f/f}$ prostates (Figure S1D). Together these analyses indicated down-regulation of STAT5A/B expression in the epithelium of the lateral, dorsal and anterior lobes of ΔSTAT5 mouse prostates.

Prostates harvested from 6, 12, and 18 months old ΔSTAT5 animals failed to display any macroscopic alterations (e.g., organ atrophy or hypertrophy). Accordingly, the weights of the different lobes (normalized to mouse weight) were unaffected by STAT5 deletion (Figure S2A). Histopathological analyses of prostate sections from these animals were performed blinded by a trained veterinarian pathologist (E.R.-G.). This analysis failed to identify detectable alterations in ΔSTAT5 compared to $STAT5^{F/F}$ prostates regarding prostate architecture and histology (Figure S2B). These findings are consistent with an older study that failed to observe histological defects in dorsal and lateral lobes of *Stat5a*-null mice [35]. In fact, systemic STAT5A deficiency induced local disorganization within acinar epithelium of the ventral lobe only, which could not be confirmed in our study as Pb-Cre4-driven *Stat5* gene ablation was inefficient in that lobe.

Taken together, these results indicate that Pb-Cre4 recombinase-driven *Stat5* deletion has no major impact on prostate tissue integrity.

2.2. Lobe-Specific Pattern of STAT5 Deletion in Pb-PRL$^{\Delta STAT5}$ Mice

We next generated Pb-PRL mice harboring *Stat5a/b* gene deletion in the epithelium (hereafter called Pb-PRL$^{\Delta STAT5}$). The analyses of *Stat5a/b* expression were performed on 6 month-old animals. As observed in $STAT5^{f/f}$ mice, *Stat5a* mRNA was predominant over *Stat5b* in Pb-PRL$^{STAT5f/f}$ prostates, with highest levels in anterior and lateral lobes (Figure 1A). In Pb-PRL$^{\Delta STAT5}$ the reduction in *Stat5a/b* mRNA expression was significant in all but the ventral lobes, and more marked in anterior and dorsal (>3-fold) than lateral (<2 fold) lobes (Figure 1A). This pattern was globally confirmed at the protein level using immunoblot and IHC analyses. As shown in Figure 1B (immunoblot) and 1C (quantification), STAT5 protein was efficiently deleted in anterior and dorsal lobes of Pb-PRL$^{\Delta STAT5}$ mice, and the deletion was homogeneous in the epithelium (Figure 1E). In contrast, no significant reduction in STAT5 protein levels were observed in ventral prostate, while deletion in the lateral lobe was intermediate (Figure 1B,C and Figure S3A).

As earlier reported, STAT5 is massively activated by transgenic PRL overexpressed in Pb-PRL prostates. Therefore tyrosine phosphorylated (p) STAT5 was also monitored by immunoblot (Figure 1B,D) and IHC (Figure 1F,G and Figure S3B) using a validated anti-pSTAT5 antibody [30,31]. Thanks to the contrasted nuclear staining obtained in IHC using the latter antibody (Figure 1F), we could quantify the actual level of activated (nuclear) STAT5 in the luminal epithelium of the four lobes. As expected from above-mentioned studies, the level of STAT5 tyrosine phosphorylation in Pb-PRL$^{STAT5f/f}$ mice was the highest in dorsal lobe and the lowest in ventral lobe (Figure 1G). Also, the pattern of nuclear STAT5 was zonal in the anterior and ventral prostates and more uniformly spread in dorsal and lateral prostates (Figures 1F and S3B). In Pb-PRL$^{\Delta STAT5}$ mice, the significant reduction in STAT5 phosphorylation in anterior and dorsal lobes was assessed by immunoblot (Figure 1B,D) confirming our IHC results of STAT5 down-regulation in epithelial cells (Figure 1F,G). In agreement with the low efficiency of STAT5 deletion in the ventral lobe (Figure 1A,C), no significant reduction of pSTAT5 could be detected in that lobe (Figure 1D,G). Finally, in the lateral lobe, despite of the reduction of STAT5 expression (Figure 1C) there was no significant reduction of STAT5 tyrosine phosphorylation, as determined by blot (Figure 1B) or by IHC (Figure 1G). This is consistent with the fact that the degree of STAT5 phosphorylation in that the lobe is intrinsically moderate in Pb-PRL$^{STAT5f/f}$ mice (Figure 1G), which limits the impact of mild decreased STAT5 expression.

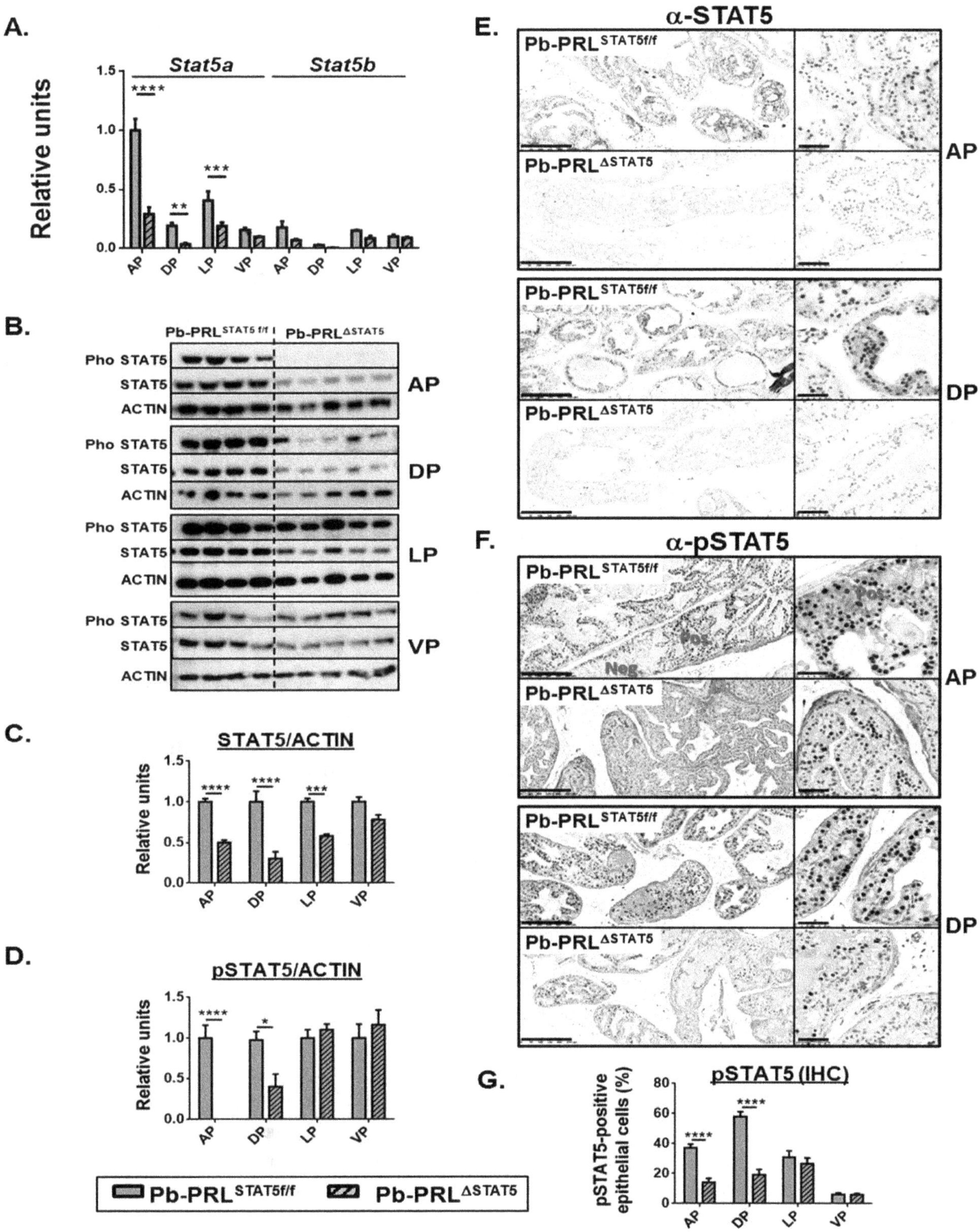

Figure 1. Lobe-specific pattern of STAT5 deletion in 6 month-old Pb-PRL$^{\Delta STAT5}$ mice. (**A**). Lobe-specific expression of *Stat5a* and *Stat5b* mRNA in Pb-PRL$^{STAT5f/f}$ and Pb-PRL$^{\Delta STAT5}$ mice as determined by RT-qPCR is shown. (**B–D**). Lobe-specific expression and phosphorylation of STAT5 protein in Pb-PRL$^{STAT5f/f}$ and Pb-PRL$^{\Delta STAT5}$ mice as determined by immunoblot is shown. Quantification of STAT5/ACTIN (C) and pSTAT5/ACTIN (D) was performed by densitometry and is shown as fold

change versus Pb-PRL$^{STAT5f/f}$ mice for each lobe. (**E**,**F**). Immunohistochemical analysis of STAT5 protein expression (**E**) and phosphorylation (**F**) in anterior (AP) and dorsal (DP) prostates of Pb-PRL$^{STAT5f/f}$ and Pb-PRL$^{\Delta STAT5}$ mice, as indicated. In panel F, negatively (neg) next to positively (pos) immunostained areas are highlighted. See Figure S3 for the other lobes. (**G**). The percentage of pSTAT5-positive epithelial cells as determined using Calopix software is shown for each lobe. *Statistics*: Stars (* $p < 0.05$, ** $p < 0.01$, *** $p < 0.001$, **** $p < 0.0001$; idem in all figures below) denote significant differences in a repeated-measures two-way ANOVA with Sidak's multiple comparisons. Size bars: 250 μm in large images and 50 μm in insets.

To confirm that the down-regulation of STAT5 phosphorylation resulted in lower STAT5 signaling, we monitored the levels of expression of Suppressor of Cytokine Signaling (SOCS)/Cytokine-Inducible SH2 Containing Protein (CISH) genes as typical negative regulators and direct targets of JAK/STAT signaling. Based on former reports, SOCS1/2/3 and CISH can be efficiently induced by PRLR/JAK/STAT signaling [7]. SOCS2 was predominantly expressed in the mouse prostate, while SOCS1 was hardly detected (see Figure 6D below for relative expression levels). SOCS2, SOCS3 and CISH were all up-regulated in Pb-PRL$^{STAT5f/f}$ dorsal prostates and significantly down-regulated in Pb-PRL$^{\Delta STAT5}$ confirming lower STAT5 signaling in the latter (Figure S3C). In addition to SOCS genes, we earlier reported that PRLR expression (all isoforms) was down-regulated in Pb-PRL prostates compared to wild-type (WT) littermates [31]; this was confirmed in this study using the dorsal prostate of Pb-PRL$^{STAT5f/f}$ mice (Figure S3D). Notably, normal levels of PRLR expression were restored in Pb-PRL$^{\Delta STAT5}$ prostates suggesting that PRLR expression is negatively regulated by STAT5 signaling (Figure S3D).

In summary, Pb-Cre4-driven *Stat5a/b* deletion was highly efficient to down-regulate STAT5 expression and signaling in the anterior and dorsal prostates but not in the ventral prostate; lateral prostate *Stat5a/b* deletion was intermediate. As described below, these lobe-specific effects were used to delineate the actual involvement of epithelial STAT5 pathway in the various tumor-related phenotypes analyzed.

2.3. STAT5 Deletion Reduces Hallmarks of PRL-Induced Prostate Tumorigenesis at 6 Month of Age

To evaluate the role of STAT5 pathway in PRL-driven prostate phenotypes, we compared the prostates of Pb-PRL$^{STAT5f/f}$ and Pb-PRL$^{\Delta STAT5}$ mice regarding the major hallmarks of early tumorigenesis previously characterized in Pb-PRL prostates [30–32,36].

2.3.1. Prostate Growth

According to former reports involving Pb-PRL mice [30], this study confirmed that prostate hypertrophy was detectable from 3 months of age in Pb-PRL$^{STAT5f/f}$ mice compared to control mice. At 6 months of age prostate weight was more than doubled compared to controls (Figure 2A). All lobes significantly contributed to prostate hypertrophy (Figures 2A and S4A). This phenotype was primarily due to cell hyperplasia, as reflected by the higher number of cells obtained in cell sorting (see below Section 2.4) and was further magnified by the increase in prostatic secretions, especially in the anterior prostate. Prostate weight of Pb-PRL$^{\Delta STAT5}$ was significantly reduced compared to Pb-PRL$^{STAT5f/f}$ mice, but remained higher than in control STAT5$^{f/f}$ mice (Figure 2A, top panel). In fact, anterior and dorsal lobes of Pb-PRL$^{\Delta STAT5}$ mice exhibited significant weight loss compared to their counterparts in Pb-PRL$^{STAT5f/f}$ mice (Figure 2A, bottom panels), but this was not the case for ventral and lateral prostates (Figure S4A). These results are in agreement with the lobe-specific pattern of STAT5 signaling down-regulation described above.

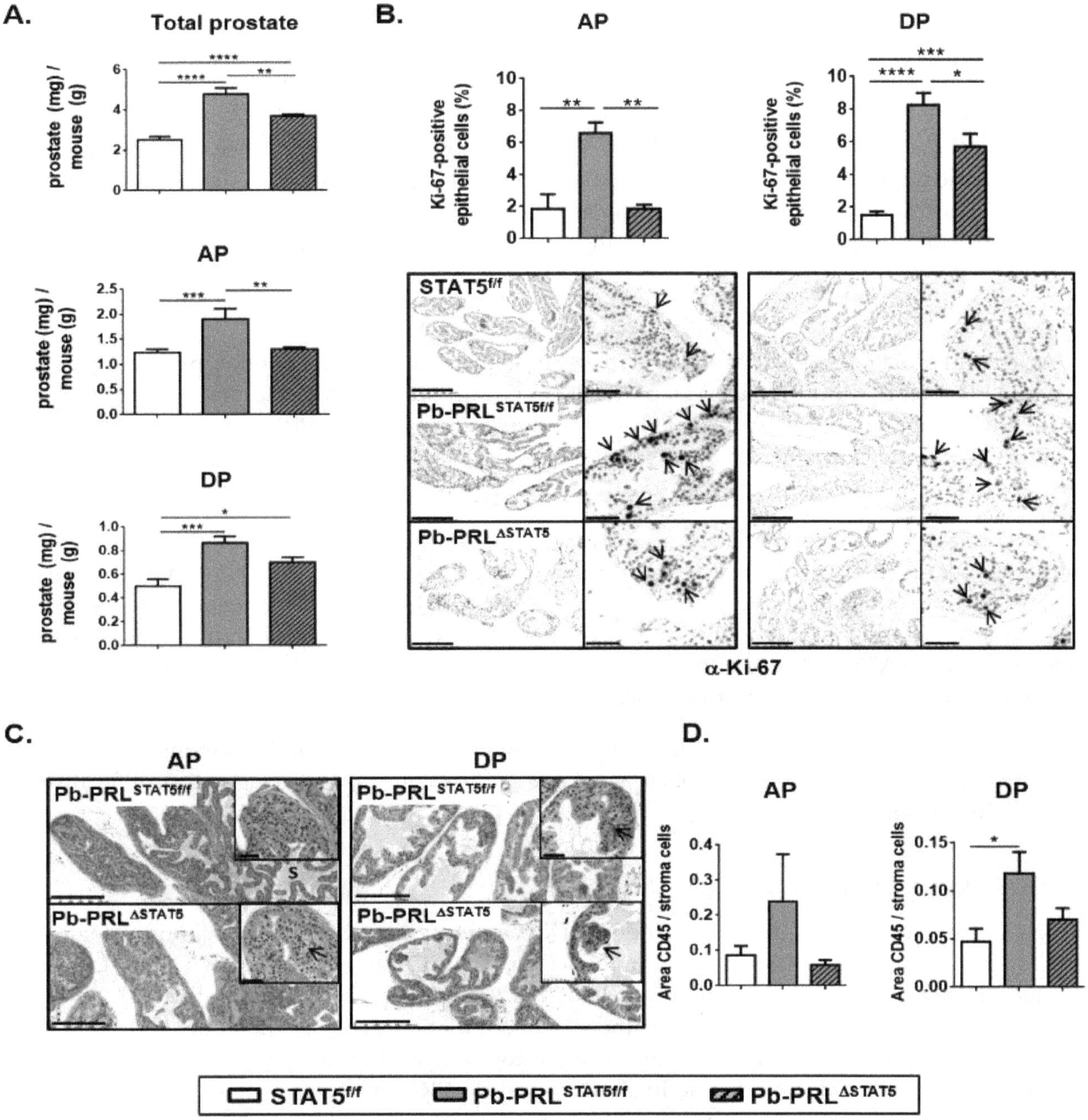

Figure 2. STAT5 deletion reduces hallmarks of early prostate tumorigenesis in 6 month-old Pb-PRL$^{\Delta STAT5}$ mice. (**A**). The weight of total prostate and of anterior (AP) and dorsal (DP) prostate lobes of STAT5$^{f/f}$, Pb-PRL$^{STAT5f/f}$ and Pb-PRL$^{\Delta STAT5}$ mice is expressed as the ratio of prostate tissue weight normalized to the weight of corresponding animal (see Figure S4A for other lobes). (**B**). Cell proliferation in anterior and dorsal lobes was assessed by immunostaining using anti-Ki-67 antibody (hematoxylin counterstaining). Arrows in insets show representative Ki-67 nuclear immunostaining in epithelial cells. The proliferation index (ratio of Ki-67-positive versus total epithelial cells) was quantified using Calopix software (see Figure S4 for other lobes). (**C**). Histological analysis (hematoxylin counterstaining) of anterior and dorsal lobes of Pb-PRL$^{STAT5f/f}$ and Pb-PRL$^{\Delta STAT5}$ mice showing similar hyperplasia in both genotypes (see Figure S4 for other lobes). (**D**). Inflammation was identified using CD45 immunostaining. The degree of inflammation was quantified using Calopix software and is represented as the ratio of CD45+ area versus stroma area. *Statistics*: Stars denote significant differences in a repeated-measures one-way ANOVA with Tukey's multiple comparisons. Size bars: 250 μm in large images and 50 μm in insets.

2.3.2. Cell Proliferation

According to their hypertrophic phenotype, all prostate lobes of Pb-PRL mice were previously shown to exhibit increased cell proliferation index (Ki-67 IHC) compared to control mice [36]. Using a computer-assisted image analysis methodology for the quantification of Ki-67-positive cells, this was confirmed in this study involving Pb-PRL$^{STAT5f/f}$ compared to STAT5$^{f/f}$ mice. In agreement with lobe weight reduction and STAT5 signaling down-regulation (see above), a significant reduction in cell proliferation index was observed in the anterior, and to a lesser extent, in the dorsal lobes of Pb-PRL$^{\Delta STAT5}$ mice (Figure 2B), while no change was observed in ventral prostate (Figure S4B).

2.3.3. Histopathology and Inflammation

At 6 months of age, Pb-PRL displays relatively mild histopathological phenotypes mainly including low grade PINs (early tumorigenesis step) [37,38]. In this study, we failed to find obvious differences between the histopathological features of Pb-PRL$^{STAT5f/f}$ compared to Pb-PRL$^{\Delta STAT5}$ prostates (Figure 2C and Figure S4C). The main phenotypical hallmark was in fact the reduction of the secretory phenotype in the latter, in agreement with the role of PRL signaling in prostate secretory function [39].

Inflammatory cell infiltrates are scarce in WT prostates. In contrast, according to previous reports [38,40], Pb-PRL$^{STAT5f/f}$ mouse prostates displayed several foci of CD45+ inflammatory cells especially in the ventral lobe (Figure 2D and Figure S4D). Pb-PRL$^{\Delta STAT5}$ mouse prostates displayed a less marked inflammatory phenotype in the anterior and dorsal lobes (Figure 2D), suggesting a role for luminal STAT5 in this phenotype. In contrast, there was a trend for an increased number of inflammatory cell foci in the two other lobes compared to Pb-PRL$^{STAT5f/f}$ prostates (Figure S4D).

Taken together, our analyses support a protective role of *Stat5* ablation against the early steps of PRL-triggered prostate tumorigenesis, weakening hyperplasia, lowering cell proliferation, and reducing chronic inflammation.

2.4. STAT5 Deletion Alters Epithelial Cell Hierarchy

The composition of the prostate epithelium can be determined using cell sorting strategies based on cell surface expression of Stem cell antigen-1 (Sca-1) and Integrin alpha 6 (CD49f). Using these validated markers, luminal and basal (also called LSC^{high}) cells can be discriminated [41]. We recently discovered in the mouse prostate a third epithelial cell population that we called LSC^{med} [31]. The basal and LSC^{med} cell populations include stem/progenitor cells and exhibit tumor-initiating capacities when transformed [32,42]. Of interest, in pre-tumoral Pb-PRL prostates both basal and LSC^{med} cell populations are amplified at the expense of the luminal compartment [30–32].

Due to the lobe-specific efficacy of *Stat5* deletion (see Figure 1), we investigated the distribution of these three epithelial cell populations individually in each lobe of the three genotypes of interest using cell sorting. Data in Figure 3A were obtained by pooling 3–6 animals to ensure enough material for analysis.

In STAT5$^{f/f}$ prostates luminal cells were predominant in all lobes (60% to 90%) and the analysis revealed very low prevalence of basal/LSC^{med} cells in ventral prostate compared to other lobes (Figure 3A,B). In Pb-PRL$^{STAT5f/f}$ prostate, basal and LSC^{med} cells were dramatically amplified in all lobes at the expense of luminal cells that dropped to <20% of the hyperplastic epithelium (Figure 3B). As shown in Figure 3C, the contents in basal and LSC^{med} cells in the various lobes were inversely related (the more basal, the less LSC^{med}). Plotting these data against the level of STAT5 phosphorylation in the epithelium of the various lobes of Pb-PRL$^{STAT5f/f}$ mouse prostates, as determined by IHC (Figure 1G) revealed that higher levels of STAT5 activation were associated with higher content in basal cells (Figure 3C).

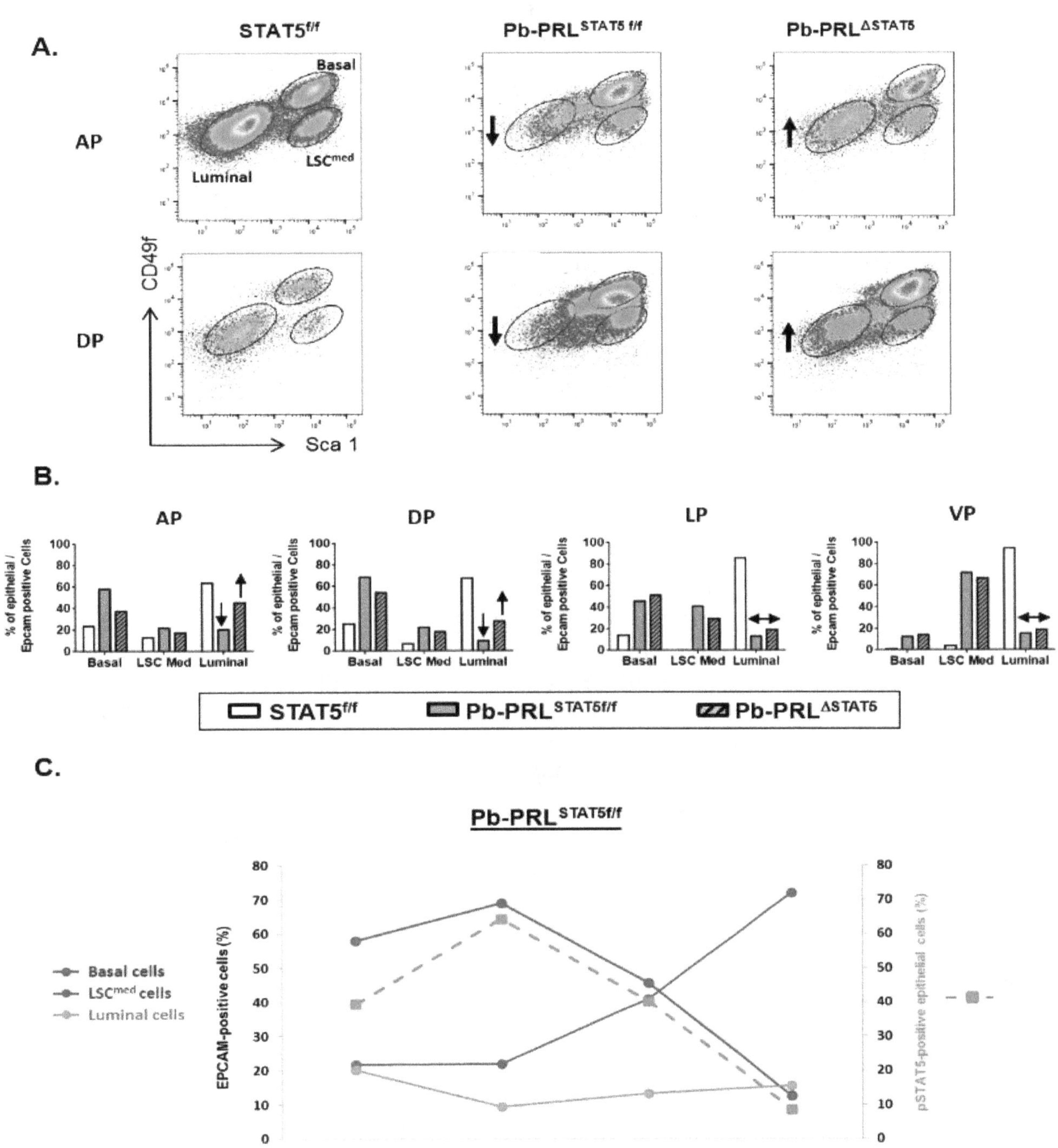

Figure 3. STAT5 deletion alters epithelial cell hierarchy in 6 month-old Pb-PRL$^{\Delta STAT5}$ mice. (**A**). Representative FACS profiles of anterior (AP) and dorsal (DP) prostate lobes from STAT5$^{f/f}$, Pb-PRL$^{STAT5f/f}$ and Pb-PRL$^{\Delta STAT5}$ mice (3–6 animals per genotype). Graphs depict epithelial cells only (gated as Lin-/Epithelial cell adhesion molecule (EPCAM)+), with large gates for percentage analysis. Each FACS profile shows gated epithelial populations: basal/stem (LSC^{high}), LSC^{med} and luminal cells. (**B**). Quantification of the three cell populations in the four prostate lobes from the three genotypes is shown. In panels A and B, the arrows indicate the loss of luminal cells in Pb-PRL$^{STAT5f/f}$ versus STAT5$^{f/f}$ mice, and their partial rescue in anterior and dorsal prostate of Pb-PRL$^{\Delta STAT5}$ mice (no similar effect was observed in the two other lobes). (**C**). For each prostate lobe of Pb-PRL$^{STAT5f/f}$ mice, the percentage of the three epithelial cell populations was plotted versus the level of STAT5 phosphorylation, as determined in Figure 1G.

Analysis of Pb-PRL$^{\Delta STAT5}$ mouse prostates showed that this enrichment in basal/ LSC^{med} cells was partly reversed upon STAT5 signaling down-regulation. Compared to Pb-PRL$^{STAT5f/f}$ mice the content in luminal cells was partly rescued in the anterior and dorsal lobes of Pb-PRL$^{\Delta STAT5}$ mice at the expense of basal/LSC^{med} cells (Figure 3A,B). No similar effect was observed in lateral and ventral lobes of Pb-PRL$^{\Delta STAT5}$ mice (Figure 3B) in agreement with the unaltered levels of epithelial pSTAT5 compared to Pb-PRL$^{STAT5f/f}$ mice (Figure 1G).

Taken together, these observations revealed the central role of epithelial STAT5 in the control of epithelial cell hierarchy by local PRL in the mouse prostate. In particular, up-regulation of STAT5 activation led to the accumulation of stem/progenitor cells previously shown to exhibit tumor-initiating capacities when transformed.

2.5. STAT5 Deletion does not Promote Alternative PRLR Signaling Cascades

In order to investigate whether *Stat5* gene deletion on the Pb-PRL background could promote compensatory PRLR-triggered signaling pathways that may be responsible for the effects reported above in Pb-PRL$^{\Delta STAT5}$ animals, we analyzed the phosphorylation status of STAT5, STAT3, ERK1/2, and AKT in prostates of 6 month old mice of the three genotypes. The immunoblots are shown in Figure 4A (anterior), B (dorsal) and C (ventral) and quantification for each lobe is shown in Figure 4D.

Virtually no basal phosphorylation could be detected for STAT5 in any lobe of STAT5$^{f/f}$ prostates, which contrasted with STAT3, ERK1/2, and AKT that all showed some background activation. Based on densitometric quantification, STAT5 was strongly activated in dorsal and anterior prostates of Pb-PRL$^{STAT5f/f}$ mice, reaching >30-fold induction compared to control mice. Mild activation (<2 fold) of STAT3 (anterior lobe) and AKT (dorsal lobe) was also observed. In ventral prostate, STAT5 was also activated, but to a much lower degree (~5 fold) than in the other lobes, in agreement with IHC data (Figure S3B). In contrast to the other lobes, the degree of activation of STAT3 in ventral lobe (~7 fold) was in the same range as STAT5 activation. Finally, ERK1/2 was not significantly triggered by PRL in any lobe as shown by unchanged pho/total ERK1/2 ratio (Figure 4D). However, due to slightly enhanced expression of ERK1/2 protein in the anterior lobe of Pb-PRL$^{STAT5f/f}$ prostates compared to STAT5$^{f/f}$ mice (Figure 4A), the tissue content in pERK1/2 in that lobe was significantly increased ($p < 0.01$), and the same trend (not significant) was observed in the other lobes (Figure 4B,C).

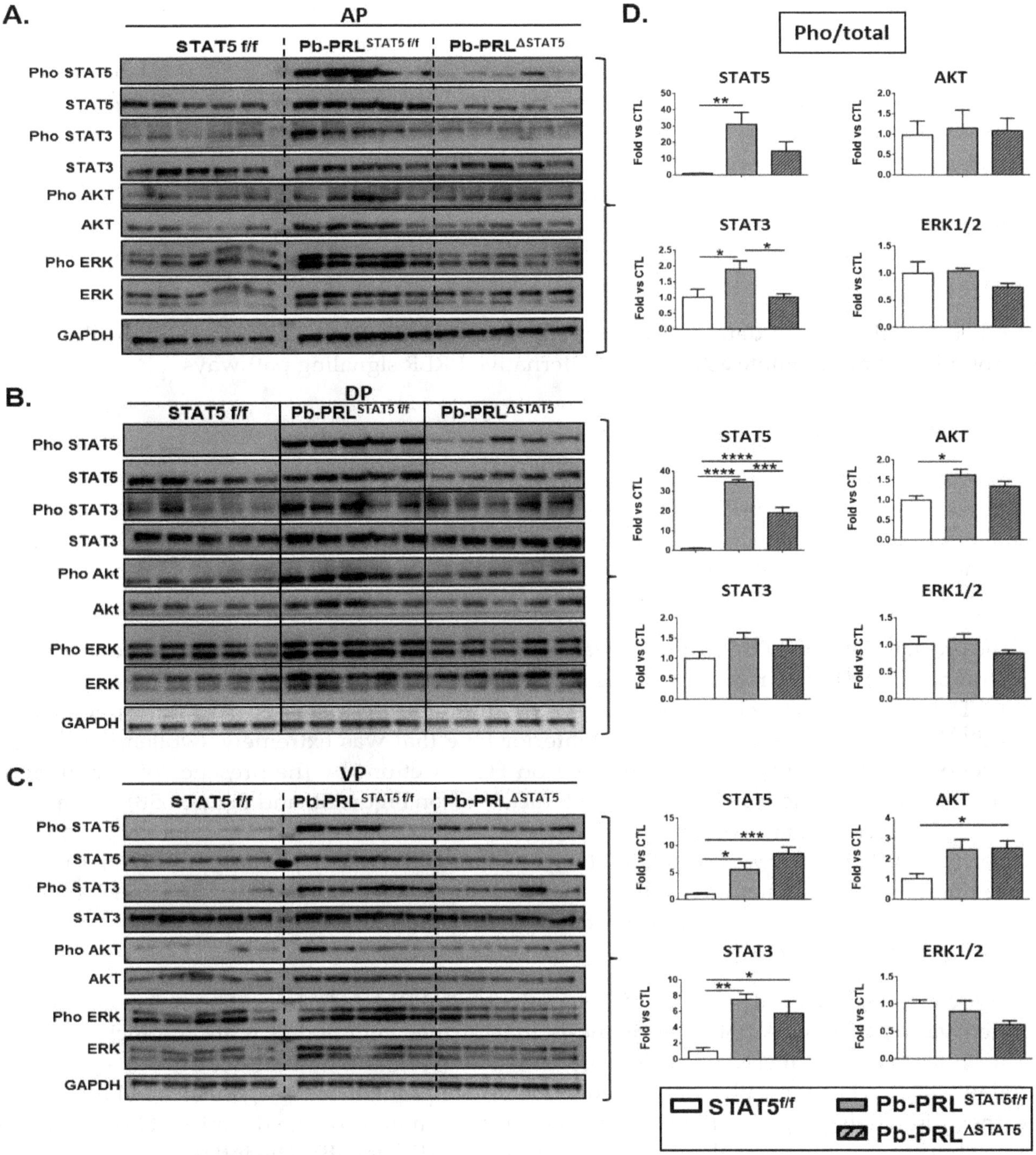

Figure 4. STAT5 deletion does not promote alternative signaling in 6 month-old Pb-PRL$^{\Delta STAT5}$ mice. (**A**–**C**). Canonical PRLR signaling pathways were analyzed by immunoblot in anterior (**A**), dorsal (**B**) and ventral (**C**) prostate lobes from STAT5$^{f/f}$, Pb-PRL$^{STAT5f/f}$ and Pb-PRL$^{\Delta STAT5}$ mice (each lane corresponds to a different mouse). (**D**). The activation of each pathway determined by the ratio of phosphorylated versus total protein (densitometry) is shown as fold-induction versus STAT5$^{f/f}$ samples. *Statistics*: Stars denote significant differences in a repeated-measures one-way ANOVA with Tukey's multiple comparisons.

The down-regulation of STAT5 in the anterior (Figure 4A) and dorsal (Figure 4B) lobes of Pb-PRL$^{\Delta STAT5}$ mice did not result in compensatory activation of alternative PRLR signaling pathways. In fact, the pathways that were mildly activated in these lobes in Pb-PRL$^{STAT5f/f}$ prostates, i.e., STAT3 in the anterior lobe and AKT in the dorsal lobe, were back to control levels in Pb-PRL$^{\Delta STAT5}$ prostates (Figure 4D). While this effect suggests that activation of these pathways could be STAT5-dependent, it may also reflect the down-regulation of transgenic PRL protein that was observed in these lobes (Figure S4E). Since the level of PRL mRNA was unaltered (Figure S4F), this suggests that STAT5 signaling may contribute to PRL protein stability. Finally, in agreement with the poor efficiency of *Stat5* deletion in the ventral lobe of Pb-PRL$^{\Delta STAT5}$ mice, PRL expression (Figure S4E) and PRLR signaling pathways (Figure 4C,D) were unchanged in this lobe compared to Pb-PRL$^{STAT5f/f}$ mice.

Taken together, these data indicate that at 6 months of age, STAT5 deletion on the Pb-PRL background does not promote activation of alternative PRLR signaling pathways.

2.6. STAT5 Deletion does not Prevent Prostate Tumor Progression in Aged Pb-PRL Mice

In order to address the long-term effects of STAT5 signaling deletion on PRL-induced tumor progression, we analyzed Pb-PRL$^{STAT5f/f}$ and Pb-PRL$^{\Delta STAT5}$ mice at 12 and 18 months of age. In the latter, the ventral, lateral, and dorsal lobes were often undistinguishable (hereafter called half prostate) due to massive tissue hypertrophy so that lobe-specific interpretation could not be performed for all investigations.

The prostate weight of Pb-PRL$^{STAT5f/f}$ was significantly increased at 12 and even more at 18 months of age compared to STAT5$^{f/f}$ mouse prostates (Figure 5A for total prostate and Figure S5A for individual lobes). This weight gain was largely due to the dramatically increased secretory phenotype of Pb-PRL$^{STAT5f/f}$ prostates, especially in the anterior lobe that was extremely swollen/hypertrophic at dissection. This phenotype was confirmed on H&E sections by the presence of abundant and dense eosinophilic secretions in hypertrophied ducts ("S" on Figure 5B and Figure S5B). Compared to Pb-PRL$^{STAT5f/f}$ mice, prostate weight was significantly reduced in Pb-PRL$^{\Delta STAT5}$ mice at 12 months of age (Figure 5A). This effect was correlated to a less marked secretory phenotype and a reduction in cell proliferation that was significant in anterior and dorsal lobes only (Figure 5C). In 18 month-old mice, there was still a trend for reduced prostate weight (Figure 5A) and reduced proliferation index in Pb-PRL$^{\Delta STAT5}$ mice (Figure 5C).

At the histological level, prostates of 12 month old Pb-PRL$^{STAT5f/f}$ mice (Figure 5 and Figure S5B) exhibited various hallmarks of tumorigenesis that further aggravated with age including PINs, cribriform lesions ("c"), increased stromal density (stars) and inflammation (arrows), especially in ventral prostate. While these phenotypes tended to be less pronounced in age-matched Pb-PRL$^{\Delta STAT5}$ mice, they ultimately progressed in the latter so that at 18 months of age the mice of both genotypes became almost undistinguishable irrespective of the lobe (Figure 5B). The inflammatory phenotype was highly heterogeneous at 12 months of age precluding any clear genotype-related difference in any lobe (Figure 5D). There was nevertheless a global trend for lower inflammation in Pb-PRL$^{\Delta STAT5}$ mice, which persisted at 18 months of age (Figure 5D).

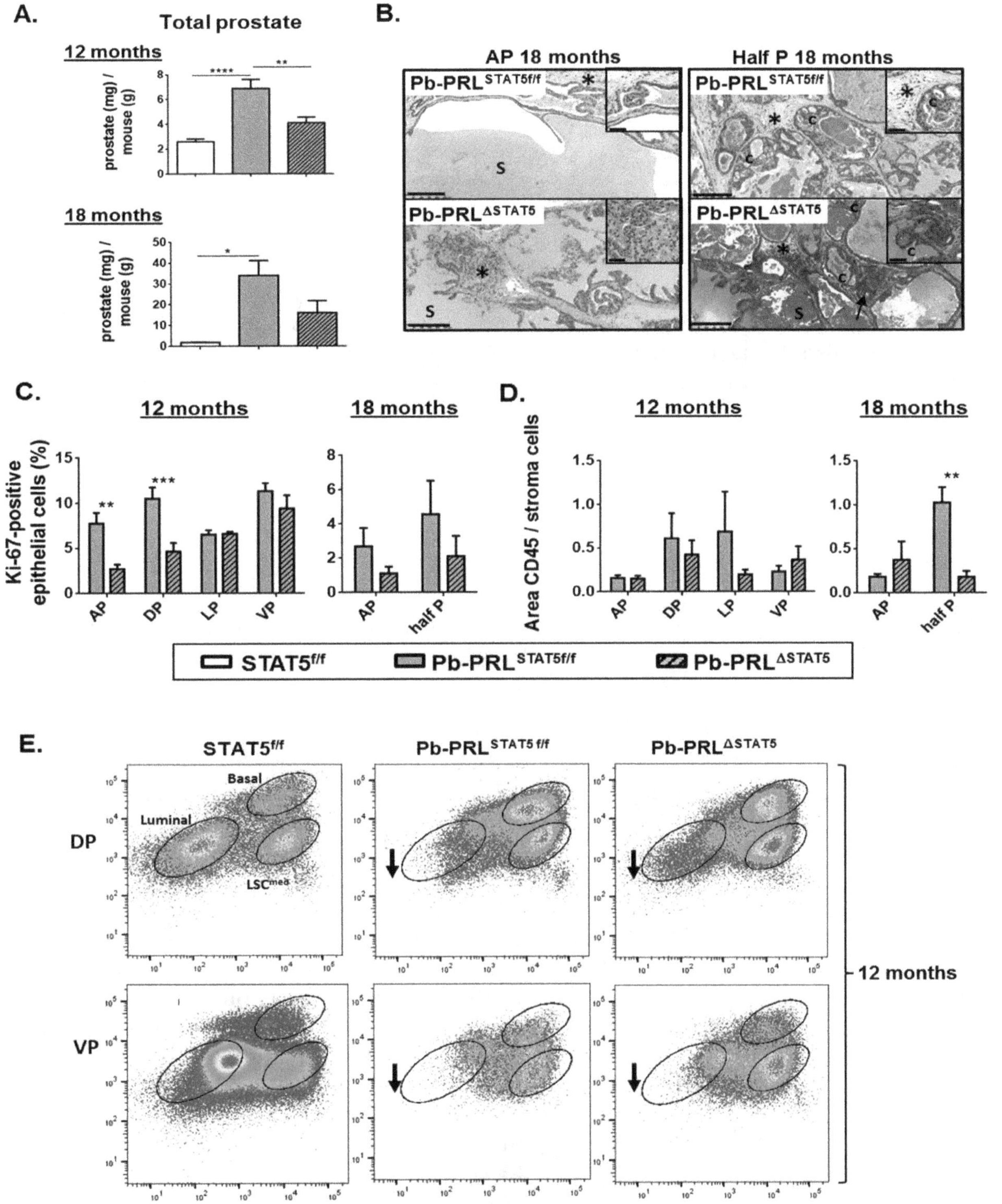

Figure 5. STAT5 deletion does not prevent prostate tumor progression in aged Pb-PRL mice. (**A**). The prostate weight in 12 and 18 month-old STAT5$^{f/f}$, Pb-PRL$^{STAT5f/f}$ and Pb-PRL$^{\Delta STAT5}$ mice is expressed as the ratio of total prostate weight normalized to the weight of corresponding animal (see Figure S5A for data per lobe). (**B**). Histological analysis (hematoxylin counterstaining) of prostate tumors of 18 month-old Pb-PRL$^{STAT5f/f}$ and Pb-PRL$^{\Delta STAT5}$ mice showing similar abnormal histology including PINs, cribriform lesions ("c"), increased stromal density (stars), inflammation (arrows), and dense eosinophilic secretions (S). Sections from the anterior lobe (AP) and from fused dorsal/lateral/ventral lobes are shown. See Figure S5B for 12 month-old animals. (**C**). The proliferation index in prostates from 12 (all lobes) and 18 month old (anterior lobe and total prostate) mice is shown (see Figure 2 for

details). (**D**). Inflammation was identified using CD45 immunostaining. The degree of inflammation was quantified using Calopix software and is represented as the ratio of CD45+ area versus stroma area. (**E**). Representative FACS profiles of dorsal (DP) and ventral (VP) prostate lobes from 12 month-old $STAT5^{f/f}$, $Pb\text{-}PRL^{STAT5f/f}$ and $Pb\text{-}PRL^{\Delta STAT5}$ mice (3–4 animals per genotype; see Figure 3A for details). Arrows show that the loss of luminal cells observed in $Pb\text{-}PRL^{STAT5f/f}$ versus $STAT5^{f/f}$ mice was not rescued in $Pb\text{-}PRL^{\Delta STAT5}$ mice. *Statistics*: Stars denote significant differences in a repeated-measures one-way ANOVA with Tukey's multiple comparisons (A) or two-way ANOVA with Sidak's multiple comparisons (C). Size bars: 250 μm in large images and 50 μm in insets.

According to the low efficiency of *Stat5* deletion in the ventral lobe, 12 month-old $Pb\text{-}PRL^{\Delta STAT5}$ and $Pb\text{-}PRL^{STAT5f/f}$ mice displayed very similar cell sorting profiles for this lobe, with accumulation of basal and LSC^{med} cells and low luminal cell content (Figure 5E). Importantly, the dorsal prostates of these animals also displayed very similar profiles enriched in stem/progenitor cells (Figure 5E), indicating that the partial rescue of the luminal cells observed in younger animals upon *Stat5* deletion (Figure 3A,B) was no longer maintained in aged mice.

Taken together these data indicate that the deletion of STAT5 delayed, but could not prevent the appearance of histopathological hallmarks of PRL-driven prostate tumorigenesis.

2.7. *Spontaneous STAT5 Signaling Shutdown and Emergence of AKT and ERK1/2 Signaling in Aged Pb-PRL Mice*

To understand further why the protective effect of STAT5 signaling inhibition observed at 6 months of age was progressively lost in old $Pb\text{-}PRL^{\Delta STAT5}$ mice, we compared the canonical PRLR signaling pathways in the dorsal prostates of the three genotypes at 6, 12, and 18 months of age (Figure 6A).

We discovered that in $Pb\text{-}PRL^{STAT5f/f}$ mice (lanes 1–3, 7–9 and 13–15 on Figure 6A), the degree of STAT5 activation (pSTAT5/STAT5 ratio) as well as the tissue content in activated STAT5 (pSTAT5/ACTIN) markedly declined between 6 and 12 months of age (Figure 6A,B). We investigated this phenotype further by monitoring the mRNA expression of key players of the PRLR/STAT5 pathway in $Pb\text{-}PRL^{STAT5f/f}$ prostate (Figure 6D). There was neither alteration of transgenic *Prl* and *Prlr* expression, nor up-regulation of short PRLR isoforms that have been shown to act as dominant-negative on STAT5 signaling [43]. The lower level of SOCS2 expression at 12- versus 6-months of age assessed that STAT5 signaling down-regulation in the latter was not due to unexpected SOCS overexpression; rather, it was associated with lower STAT5 signaling activity. SOCS7 [44] and nuclear receptor co-repressor 2 (NCOR2, also known as silencing mediator for retinoid or thyroid-hormone receptors, SMRT) [45] are two other negative regulators of STAT5 signaling. None of them were affected at the transcriptional level (Figure 6D and S6A). Otherwise, there was a significant age-related decrease in *Stat5a* mRNA expression (Figure 6D) that was also observed at the protein level in prostate lysates (STAT5/ACTIN ratio), although it did not reach significance, presumably due to unaltered *Stat5* expression in non-epithelial cells (Figure 6B). Comparison of pSTAT5 immunostaining in the epithelium of dorsal prostate of 6- versus 12-month old $Pb\text{-}PRL^{STAT5f/f}$ mice revealed that the reduced tissue content in activated STAT5 in older mice was associated with both lower number of pSTAT5-positive cells and, in the latter, reduced intensity of pSTAT5 staining (Figure 6E). Indeed, pSTAT5-positive cells in 12 month old $Pb\text{-}PRL^{STAT5f/f}$ mice exhibited faint nuclear staining similar to that observed in the few pSTAT5-positive epithelial cells of healthy prostates (Figure 6E, panels a and c), i.e., markedly lower compared to younger $Pb\text{-}PRL^{STAT5f/f}$ animals (Figure 6E, panel b). As a consequence, the level of STAT5 signaling in old $Pb\text{-}PRL^{STAT5f/f}$ mice became undistinguishable from that of age-matched $Pb\text{-}PRL^{\Delta STAT5}$ mice (see quantifications reported in Figure 6E). Accordingly, most of the targets that were regulated by STAT5 signaling in young animals were not significantly affected by *Stat5* deletion in older mice (Figure S6A).

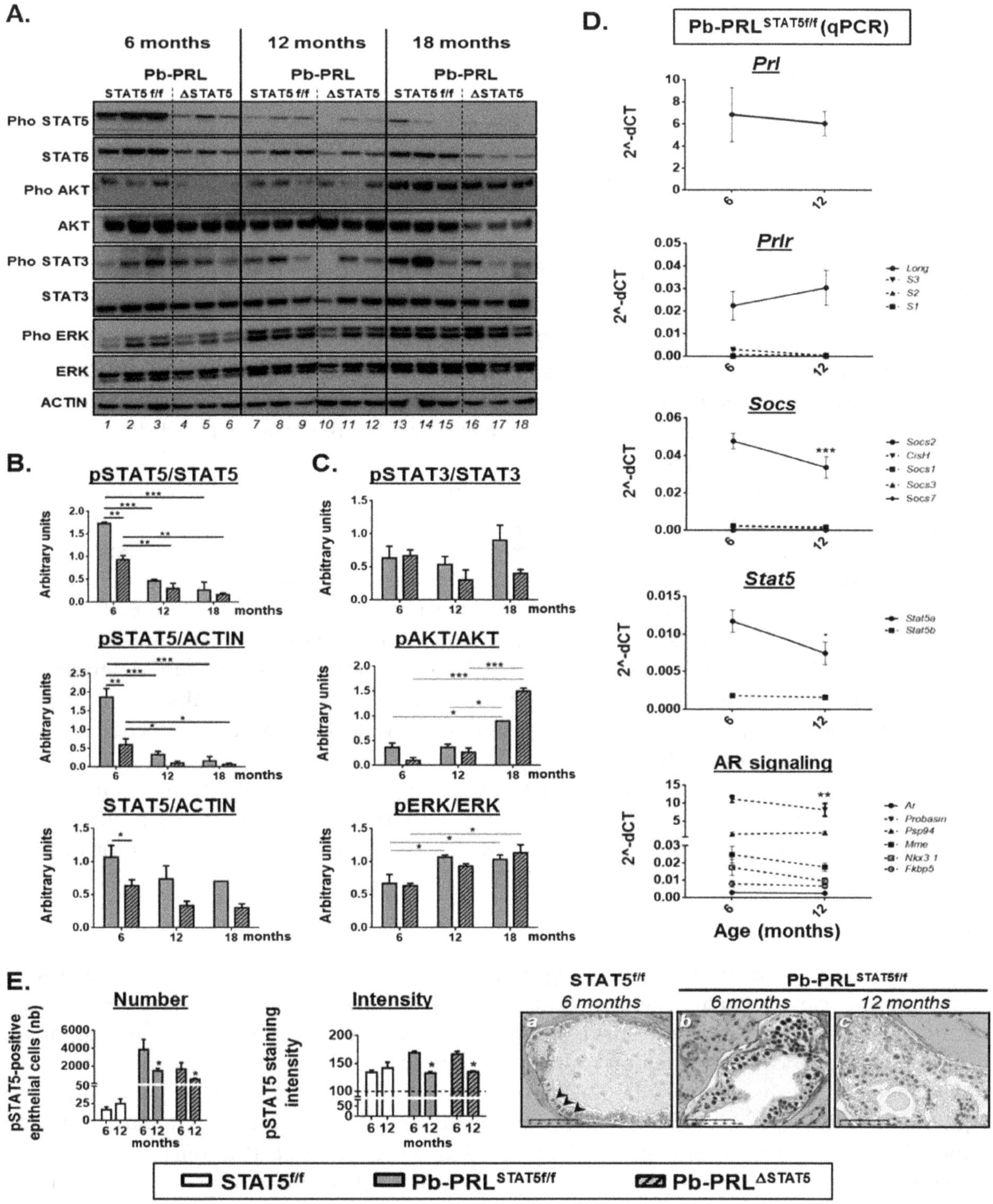

Figure 6. Spontaneous STAT5 signaling shutdown and emergence of AKT and ERK1/2 signaling in aged Pb-PRL mice. (**A**). Canonical PRLR signaling pathways activated in dorsal prostates of Pb-PRL$^{STAT5f/f}$ (lanes 1–3, 7–9, 13–15) and Pb-PRL$^{\Delta STAT5}$ (lanes 4–6, 10–12, 16–18) mice were compared at 6, 12 and 18 months of age as indicated. (**B**). Quantifications of the STAT5 pathway regarding STAT5 activation (pSTAT5/STAT5 ratio), tissue content in activated (pSTAT5/ACTIN) and total (STAT5/ACTIN) STAT5 as determined in panel A. (**C**). The activation of STAT3, AKT and ERK1/2 pathways as shown in panel A was quantified by densitometry as the ratio of phosphorylated versus total protein. (**D**). Age-dependent expression of various actors of the PRLR and AR pathways as determined by RT-qPCR in 6 and

12 month-old Pb-PRL$^{STAT5f/f}$ mice. The dotted line represents low expressed genes of the PRLR pathway (upper panels) and target genes of the AR pathway (bottom panel). See Figure S6 for genotype-dependent expression in 12 month-old mice. (**E**). Comparison of the number of pSTAT5-positive cells and of the intensity of pSTAT5 immunostaining (the horizontal dotted line is the background threshold) in 6 versus 12 month-old STAT5$^{f/f}$, Pb-PRL$^{STAT5f/f}$ and Pb-PRL$^{\Delta STAT5}$ mice. Representative images of low number (arrows)/low intensity (STAT5$^{f/f}$ mice), high number/high intensity (6 month-old Pb-PRL$^{STAT5f/f}$), and high number/low intensity (12 month-old Pb-PRL$^{STAT5f/f}$) are shown (positive cells are red circled). *Statistics*: Stars denote significant differences in a repeated-measures two-way ANOVA with Sidak's multiple comparisons. Size bars: 50 μm.

Concomitant to the age-dependent decline of STAT5 signaling in Pb-PRL$^{STAT5f/f}$ tumors, we observed a significant increase in ERK1/2 and AKT phosphorylation at 12 and 18 months of age, respectively (Figure 6A,C). Of interest, this phenomenon was also observed in Pb-PRL$^{\Delta STAT5}$ mice, indicating it occurred irrespective of *Stat5* status. There was no clear age-related alteration of STAT3 activation as it remained low in Pb-PRL$^{\Delta STAT5}$ mice and was more variable, but did not significantly increase in PRL$^{STAT5f/f}$ mice (Figure 6A,C).

Finally, we addressed potential contribution of androgen receptor (AR) signaling to the mechanisms favoring STAT5-independent tumor progression in aged mice of the Pb-PRL background. To that end we investigated the mRNA levels of *Ar* and of several AR target genes [32] in the dorsal prostate of 6 and 12 month old mice. Irrespective of the age, AR signaling was globally lower in Pb-PRL$^{STAT5f/f}$ mice compared to STAT5$^{f/f}$ mice based on the down-regulation of several genes of the AR signature (*Mme, Probasin, Nkx3.1*; Figure S6B). Additionally, there was also no evidence for age-related increase of AR signaling in Pb-PRL$^{STAT5f/f}$ mice (Figure 6D, bottom panel). Finally, *Stat5* deletion failed to rescue normal AR signaling levels in Pb-PRL$^{\Delta STAT5}$ mice, especially at 12 months of age (Figure S6B). These data clearly indicate that AR signaling up-regulation is not a compensatory mechanism of STAT5 signaling down-regulation in Pb-PRL mice.

Taken together, our analyses show that STAT5 signaling spontaneously decreases in PRL-driven prostate tumors with aging and our results suggest that alternative compensatory pathways such as AKT and ERK1/2 signaling, but not STAT3 and AR signaling, promote prostate tumor progression upon hyper PRL signaling in the absence of STAT5.

3. Discussion

Reports have provided strong evidence regarding the role of PRL/STAT5 pathway in the progression of human prostate cancer via an autocrine/paracrine mechanism (for reviews, [1,14,25–27]. Indeed, this pathway was shown to promote survival, proliferation, EMT, stemness, invasiveness, and DNA repair of/in prostate cancer cells, thereby contributing to prostate cancer progression and resistance to treatment. Several underlying molecular mechanisms have been described, non-exhaustively including up-regulation of PRL expression, STAT5 gene amplification and cooperation with androgen signaling pathway. While some of these biological effects are also observed in Pb-PRL mice (e.g., cell survival, proliferation, and stemness), the latter mice do not develop prostate cancer despite strong STAT5 action [30,31]. This suggested that hyper-activation of STAT5 signaling per se may not act as an oncogene in the prostate, but it could tweak prostate cancer progression in a more proto-oncogenic role driving e.g., enhanced Myc, Bcl2 or D type cyclin family member expressions. However, the promotion of prostate cancer progression with other oncogenic driver pathways identified in the prostate cancer genomic landscape could be complex and STAT5 signaling could be more important in other mutational context [46]. We here present the first genetically-modified mouse model allowing delineation of the actual role of STAT5 signaling in PRL-driven mouse prostate tumorigenesis.

The lobe-specific pattern of *Stat5* gene deficiency—high in dorsal and anterior lobes, low in ventral lobe of Pb-PRL$^{\Delta STAT5}$ mice—complicated the phenotype/genotype analyses, but at the same time provided the opportunity to assess the functions that are directly controlled by STAT5 signaling by comparing the different lobes within the same prostate sample. These analyses highlighted that epithelial cell survival/proliferation, immune cell infiltration, *Prlr* mRNA expression, and PRL protein stability are tightly dependent on epithelial STAT5 as their regulation (up or down) in Pb-PRL$^{STAT5f/f}$ mice (compared to control mice) was partly or totally reversed in the dorsal and anterior lobes, but not in the ventral lobe of Pb-PRL$^{\Delta STAT5}$ mice. While the STAT5-dependency of prostate epithelial cell survival/proliferation confirms many earlier reports (for a review, see Reference [26]), the molecular mechanisms mediating the epithelial STAT5-dependence of immune cell infiltration remains unknown. While high cell proliferation rate can in some instances be accompanied by increased apoptosis leading to the release of damage-associated molecular pattern (DAMPs) promoting inflammation. However, this is unlikely as Dillner and colleagues provided molecular and immunostaining evidence for reduced apoptosis in Pb-PRL prostates compared to healthy controls [47]. Alternatively, the increased secretory pattern of Pb-PRL prostates may feed the microenvironment with STAT5-induced cytokines/chemokines acting as immune cell chemoattractant. The locally high concentration of PRL may also contribute to inflammation as this hormone has been suggested to mediate inflammation in various pathological contexts, including association with prostate cancer inflammation [48,49]. Finally, the down-regulation of *Prlr* expression by PRL signaling confirmed our previous observation [31] and the reversibility of this phenotype upon *Stat5* ablation (Pb-PRL$^{\Delta STAT5}$ mice) suggested that this negative feedback effect is mediated by STAT5 signaling.

One of the striking phenotypes harbored by Pb-PRL mice is the amplification of the basal and LSC^{med} cell compartments. This was assessed both by cell sorting and by immunochemistry using cell population-specific phenotypic markers (p63/CK5 for basal and CK4 for LSC^{med} cells) [30–32]. The stem/progenitor cells contained in these two cell compartments are highly relevant to cancer progression as they are castrate-resistant and they exhibit tumor-initiating properties when transformed (e.g., by AKT/ERG overexpression or loss of key tumor suppressors in prostate cancer such as PTEN or p53) [32,42,50]. In xenograft experiments, the number of tumor-initiating cells present in the engrafted cell pool is directly correlated to the tumor incidence observed in host mice. In CWR22Rv xenografts, adenovirus-mediated overexpression of a dominant-negative STAT5A/B variant was shown to reduce tumor incidence [11], indicating a reduction in stem-like properties following STAT5 inhibition. This is in agreement with the capacity of STAT5 signaling to increase stem-like phenotypic features and functional properties of human prostate cancer cells in vitro [15]. Thus, the capacity of PRL/STAT5 signaling to induce the amplification of stem/progenitor cells may directly contribute to cancer progression by feeding the prostate tissue with cells able to resist treatments and to regrow a tumor (cancer recurrence).

In this study, the lobe-specific analyses of pSTAT5 by immunohistochemistry on the one hand, and of the three epithelial cell populations (luminal, basal, LSC^{med}) by cell sorting on the other hand, revealed that the prevalence of basal cells was positively correlated to the level of STAT5 activation in the luminal epithelium. The outcome of these analyses is schematized in Figure 7. This finding is in agreement with our former quantitative immunostaining investigations showing that basal cell clusters within the prostates epithelium of Pb-PRL mice are frequently surrounded by luminal cells displaying elevated levels of pSTAT5 [31]. Furthermore, the enrichment in basal/stem cells was partly reversed upon *Stat5* ablation in anterior and dorsal lobes of Pb-PRL$^{\Delta STAT5}$ mice. Another finding of these analyses is that the lobe-specific prevalence of basal/stem cells was inversely correlated to that of LSC^{med} cells. This supports that the latter emanate from the former in the prostate cell hierarchy, as proposed in our detailed study reporting the discovery of LSC^{med} progenitors [31]. Altogether, these observations demonstrate that the prostate cell hierarchy is tightly controlled by epithelial STAT5, which appears to limit luminal differentiation and to favor amplification of the basal/stem compartment (Figure 7). This effect is opposite to what happens in the mammary gland where STAT5A is mandatory for luminal cell differentiation [51]. The mechanism underlying the regulation of prostate cell hierarchy by STAT5 signaling is yet to be elucidated. Transcriptomic profiling of basal and LSC^{med} cells showed that they express only minimal levels of *Prlr* (Ref. [32] and our unpublished qPCR data) therefore their direct regulation by local PRL is unlikely. Paracrine regulation of basal cells by secreted factors downstream of PRL/STAT5 signaling in luminal cells is a possibility that is currently under investigation. As basal and LSC^{med} cells also express detectable levels of *Stat5* (Ref. [32] and our unpublished qPCR data), a cell-autonomous effect may also contribute to the partial reversion of basal cell enrichment in Pb-PRL$^{\Delta STAT5}$ mice. Indeed, although the androgen-regulated *Pb* promoter driving Cre recombinase expression is preferentially active in luminal cells, some reports suggest that it might also be active in basal cells [52].

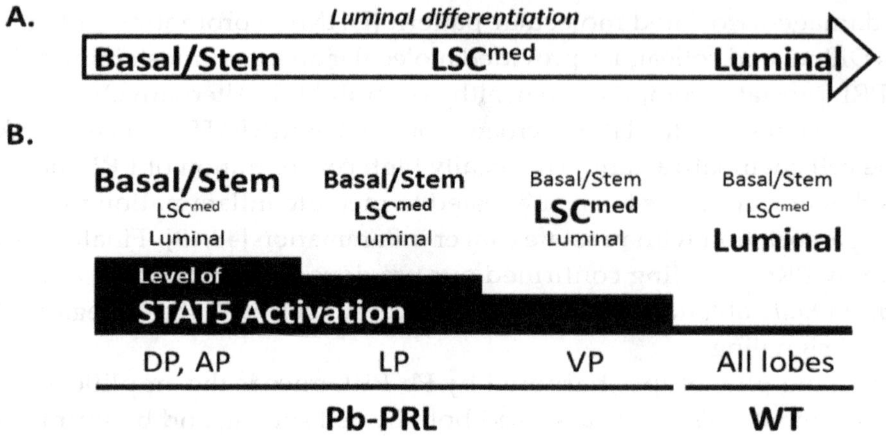

Figure 7. STAT5 signaling prevents luminal differentiation of the prostate epithelium. (**A**). Schematic representation of luminal differentiation placing LSC^{med} cells in-between basal and luminal cells. (**B**). The relative content in the three epithelial cell populations (basal/stem, LSC^{med}, luminal) in the various prostate lobes of 6-month old Pb-PRL mice or in WT mice (any lobe), as determined in Figure 3 is symbolized by the size of each population name. Below, the level of STAT5 activation in each lobe as determined in Figure 1 is also schematically represented by the black stair-bar. Based on these measurements, the highest the level of STAT5 activation, the lowest the level of epithelium differentiation was observed (see text for discussion).

While we earlier identified STAT5 as the main PRLR signaling pathway activated in Pb-PRL mouse prostates [30,31], the fact that pre-neoplastic prostate lesions observed in these animals never (or if at all, only exceptionally) evolve towards malignancy has always seemed in contradiction with the strong experimental and clinical evidence supporting the promoting role of this pathway in human prostate cancer (see Introduction). Human prostate specific PRL signaling could be more important than that of rodent prostates to drive tumorigenesis. Additionally, human prostate cancer is predominant in elder men and the various genetic and environmental risk factors (e.g., smoking, alcohol abuse, etc.) could insufficiently be recapitulated in inbred *C57Bl/6J* transgenics as used here as a study subject. Irrespective, the present work provides a possible explanation for the absence of prostate cancer development in Pb-PRL mice as we show that the prostates of these animals progressively lack STAT5 signaling with age. To address whether continuously elevated STAT5 signaling is actually required for the transformation of pre-neoplastic prostate lesions, it would be interesting to generate mice with enforced expression of a constitutively-active STAT5 variant in the prostate [53]. In conclusion, we show, for the first time, that aging can promote alternative compensatory pathway activations to bypass STAT5 function, but inflammation, cell proliferation and survival as well as prostate epithelium differentiation are significantly affected by PRL-STAT5 signaling.

The mechanism driving age-related decline of STAT5 signaling in the prostates of ageing Pb-PRL$^{STAT5f/f}$ mice remains unclear. No parallel up-regulation of the classical negative regulators of STAT5 signaling (SOCS proteins, short PRLR isoforms) could explain this observation. Although we noticed a decrease in STAT5 expression (mRNA and protein) in the dorsal prostates of 12 month old Pb-PRL$^{STAT5f/f}$ mice, this alone was insufficient to account for the pronounced loss of STAT5 activation (Figure 6B), suggesting that other mechanisms may be involved. In the breast cancer context, it has been shown that increased stiffness/density of the extracellular matrix around the tumor as a result of altered collagen deposition during cancer progression shifted the balance of PRLR signaling from JAK2/STAT5 to other pathways including ERK1/2 and AKT pathways through a Focal adhesion kinase-dependent mechanism [54,55]. Of interest, the pre-neoplastic lesions of Pb-PRL mouse prostates also exhibit strong remodeling of the extracellular matrix compared to controls, including >10 fold-increased expression of various Collagens [47]. This phenotype aggravated with age, concomitant to the decline in STAT5 signaling and up-regulation of AKT and ERK1/2 signaling, therefore it is possible that a similar mechanism as described in breast cancer contributes to the signaling shift observed in the prostates of ageing Pb-PRL$^{STAT5f/f}$ mice. No such up-regulation of AKT or ERK1/2 signaling was observed when STAT5 signaling was genetically impaired in younger Pb-PRL$^{\Delta STAT5}$ animals, further supporting that the progressive up-regulation of alternative signaling pathways in Pb-PRL$^{STAT5f/f}$ mice involves age-related mechanisms that require time to establish. These alternative signaling pathways allowed pre-neoplastic lesions to develop further in a STAT5-independent manner, as reflected by the elevated levels of cell proliferation and inflammation, and by the appearance of various histopathological phenotypes including cribriform lesions in old mice. In fact, these phenotypes were observed irrespective of *Stat5* status. Thus, although *Stat5* deficiency delayed the occurrence of various pre-neoplastic hallmarks, its protective effect was progressively diluted with age to become almost undetectable in 18 month-old animals. Of note, neither androgen nor STAT3 signaling could be identified as a compensatory pathway of age- or genetically-induced down-regulation of STAT5 signaling. There was even a trend for lower STAT3 signaling in *Stat5*-deficient prostates, suggesting some level of cooperation between both pathways.

Preventing the progression of established prostate cancers is where the therapeutic challenge stands. In agreement, there is currently strong research to develop STAT5 inhibitors that could be used for the treatment of various malignancies, including prostate cancer and myeloid leukemia [23,24,56]. Whatever the mechanisms driving the intrinsic decline of prostatic STAT5 signaling in Pb-PRL mice, the various genotype- and age-related phenomena observed in our study may shed light on possible outcomes of long-term pharmacological inhibition of STAT5 in prostate cancer context. On the good side, as low STAT5 signaling parallels luminal differentiation, pharmacological STAT5 inhibitors are not expected to result in tumor enrichment in stem/progenitor cells, but they can also act on the stromal cells such as immune cells. This effect may distinguish anti-STAT5 treatments from current therapeutic strategies (castration, chemotherapy) which generally kill luminal cells and let stem/progenitor cells unaffected, which is believed to favor tumor recurrence based on the tumor-initiating capacities of the latter. In addition, STAT5 inhibition may also decrease autocrine PRL stability as suggested from our results, which may further contribute to drug efficacy as PRL is presumably the major upstream factor triggering STAT5 pathway in the prostate. On the negative side, our study identified potential escape mechanisms to STAT5-targeted therapeutic strategies. In young Pb-PRL$^{\Delta STAT5}$ mice, which model early stages of STAT5 inhibition, no activation of alternative PRLR signaling pathway could be identified, and cell proliferation was actually reduced. With time, which is modeled by old Pb-PRL$^{\Delta STAT5}$ mice, we observed up-regulation of AKT and ERK1/2 signaling. One might speculate that in a context of pharmacological STAT5 inhibition these pathways may contribute to STAT5-independent prostate cancer progression. It is currently unknown whether AKT and ERK1/2 are triggered by PRL only or whether other paracrine growth factors will contribute. With the aim to design appropriate treatment combination limiting escape mechanisms to STAT5 inhibition, it may be relevant to address this mechanism in future studies.

4. Materials and Methods

4.1. Transgenic Mouse Strains

The Pb-PRL mouse colony carrying the rat prolactin transgene driven by the short *probasin* promoter was established on the *C57BL/6J* background (>20 backcrosses) in the Paris laboratory as previously described [30]. Mice carrying floxed *Stat5a/b* alleles (STAT5$^{f/f}$ mice) allowing targeted deletion of the whole *Stat5* locus (encompassing both *Stat5a* and *Stat5b* genes) were originally developed by L. Hennighausen [34]. Animals were obtained from the in-house colony of R. Moriggl (Vienna, Austria) established on a pure *C57Bl/6* genetic background. Pb-Cre mice (C57Bl/6 background) expressing the Cre recombinase under the control of the *short probasin* promoter [57] were purchased from the Frederick National Laboratory for Cancer Research (Frederick, MD, USA). Figure S7 displays the breeding scheme used to generate mice carrying or lacking prostate epithelium-specific deletion of *Stat5a/b* genes on the WT (STAT5$^{f/f}$ and ΔSTAT5 mice) and Pb-PRL (Pb-PRL$^{STAT5f/f}$ and Pb-PRL$^{\Delta STAT5}$ mice) backgrounds. Note that only males were used to transmit the *Pb-Cre* allele.

Colonies were housed in controlled conditions, on a 12/12-hour light/dark cycle with normal food and water provided *ad libitum*. Mice were analyzed at 6, 12 and 18 months of age, and prostate were harvested immediately after sacrifice by cervical dislocation. To isolate the prostate, dissection of the urinary tract was performed and left lobes were separately dissected, weighed then rapidly snap frozen while the right-sided half prostate was fixed in paraformaldehyde (PFA) without being further dissected, so that tissue organization was preserved for histological analysis [36]. For all genotypes, we analyzed the four prostate lobes (anterior, dorsal, lateral, ventral) at all ages. Note that at 18 months of age, dorsal, lateral, and ventral lobes were often fused due to tissue hypertrophy.

Animal experiments were approved by the local ethical committee—Comité d'Ethique en matière d'Expérimentation Animale Paris Descartes (CEEA 34) for animal experimentation (authorization CEEA34.VG.095.12) and performed according to the European guidelines for animal experimentation.

4.2. Prostate Subpopulation Sorting by FACS

The procedures for sorting the three epithelial cell populations based on their Lin/Sca-1/CD49f antigenic profile was previously described [31,32,41]. Cell sorting was performed on a BD FACSAria III (BD Biosciences, San Jose, CA, USA). Sorted cells were collected in DMEM medium, supplemented with 50% FBS, glutamine, and penicillin-streptomycin, or in RA1 Lysis Buffer (Macherey-Nagel, Düren, Germany) to perform RNA extraction as earlier described [32].

4.3. Quantitative PCR

Total RNA was isolated from separate prostate lobes using the NucleoSpin® RNA XS (Macherey Nagel, Hoerd, France) according to manufacturer's instructions. RNA integrity was assessed on Agilent BioAnalyzer (all RNAs scored 7–10). RNA (250 ng) was reverse transcribed using SuperScript™II Reverse transcriptase with the SuperScript™II First-Strand Synthesis System for RT-PCR kit (Invitrogen, CA, USA). The cDNAs were then subjected to real-time PCR amplification using gene-specific primers (0.5 μM final concentration) purchased from IDT DNA (Integrated DNA Technologies, BVBA, Leuven, Belgium; HPLC purification; referred to as Mm.PT in Table S3) or Eurogentec (Liège, Belgium; Oligold quality, sequence given in Table S1). PPIA (Peptidyl Prolyl Isomerase A) that encodes Cyclophilin A that was used as the housekeeping gene in each reaction. Real-time PCR was performed using a Qtower 2.0 (Analytik Jena, Germany). The qPCR reaction contained 2 μL cDNA sample (12.5 ng) and 8 μL mastermix with 1× GoTaq® qPCR Master Mix (Promega, Charbonnières-les-Bains, France) and 0.5 μM primer. The Qtower 2.0 Instrument was used with the following program: Enzyme activation: 95 °C for 2 min; amplification (40 cycles): 95 °C for 15 s, 60 °C for 60 s. Results were generated with the Qtower 2.0 software and were analyzed by the comparative cycle threshold method and presented as fold change in gene expression relative to internal calibrators as mentioned in figures.

4.4. Western Blotting

Freshly dissected prostate lobes were snap frozen in liquid nitrogen and stored at −80 °C until processing as previously described [31]. Protein samples (20–50 μg) were resolved in 4–12% gradient SDS-PAGE in NuPAGE Bis-Tris Precast Gels (Life Technologies, Saint-Aubin, Fance) or 10% SDS-PAGE gels. Proteins were then transferred onto nitrocellulose membranes (BioRad, Marnes-la-coquette,

France). Membranes were cut horizontally according to the size of the proteins of interest and stained with primary antibodies as described in Table S1. For band detection, Horseradish Peroxidase (HRP)-coupled secondary anti-rabbit (7074, Cell Signaling, Saint-Quentin-en-Yvelines, France) or anti-mouse (NA931, GE Healthcare Europe, Freiburg, Germany) antibodies were added before Enhanced Chemiluminescent (ECL) substrate (Immobilon Western Chemiluminescent HRP Substrate, Millipore or LumiGLO Reagent, Cell Signaling). Representative parts of blots are shown in the Figures, whole blot membranes are displayed in Figure S8.

4.5. Prostate Histopathology

Histopathological diagnosis of prostate sections from all mice was performed in blind by an independent veterinary pathologist (E.R.-G.) trained for mouse prostate analysis [36] following the recommendations of the Mouse Models of Human Cancer Consortium Prostate Pathology Committee and the reference classification of prostate intraepithelial neoplasia (PIN) lesions in genetically-modified animals [58,59]. This qualitative analysis was complemented by quantitative IHC analyses, as described below.

4.6. Immunohistochemistry (IHC)

All prostate samples were fixed in 4% PFA, paraffin embedded, and underwent heat-induced antigen retrieval in citrate buffer at pH 6. IHC and IF were performed as previously described [31] using antibodies listed in Table S2. Vector Elite ABC HRP kit with DAB substrate (Vector Laboratories, Burlingame, CA, USA) was used for detection of IHC slides, with hematoxylin as counterstain.

4.7. Image Acquisition and Histology Quantifications

Prostate tissue sections (H&E or IHC) were digitally scanned using a NanoZoomer-2.0 RT scanner (Hamamatsu, Photonics, France) coupled to NDP.view2 software analysis beta version U12388-01 (Hamamatsu, Photonics, France). For quantification of immunostaining, computer-assisted analysis of digital (scanned) images was performed using Calopix software (http://www.tribvn.com/). To quantify nuclear immunostaining (pSTAT5, Ki-67) in the epithelium, first the Random Forest Tree for tissue recognition was used to delineate and only include the glandular areas in the analysis. Then, the "morphometry" software was applied to each prostate lobe to discriminate staining-positive (DAB+) versus staining-negative (DAB−) cells. A value of marking intensity (optical density, <100 considered as background) was also provided by the algorithm for both classes of cells. The index of proliferation (Ki-67+) and of STAT5 activation was calculated as the ratio of the number of positive versus total (positive + negative) cells counted. The whole procedure was validated by comparing the results obtained by Calopix-assisted versus manual counting of digital images as previously reported [36,38,60]. For the quantification of CD45 staining, results are expressed as the ratio between the surface of CD45+ cell foci versus the total surface of stroma [36,38,60].

4.8. Statistics

The specific statistical tests performed are described in the legends for all results reported. In summary, Analysis of Variance (ANOVA) tests were used to evaluate differences among three or more groups. Depending on the number of factors tested, One or Two-way ANOVAs were used. When prostate lobe samples issued from the same mice were compared in the analysis, a repeated-measures test was carried out. Post hoc multiple comparisons were performed with Sidak's or Tukey's multiple comparisons. One, two, three or four symbols illustrating significance represent p values <0.05; <0.01; <0.001 and <0.0001, respectively. A value of $p < 0.05$ was used as significance cutoff for all tests. Error bars represent S.E.M. All analyses were performed using GraphPad Prism version 6.00 for Windows (GraphPad Software, San Diego, CA, USA).

5. Conclusions

This work confirmed STAT5 signaling as a major driver of autocrine PRL-mediated prostate tumorigenesis as genetically-induced STAT5 down-regulation delayed the onset of various PRL-induced tumor hallmarks. The age-related decline in STAT5 signaling observed in Pb-PRL mice harboring intact *Stat5* genes was unexpected, and may explain why malignancies failed to develop despite alternative signaling pathways were switched on and presumably contributed to the progression of pre-neoplastic lesions. *Stat5* genotype-related differences of prostate phenotypes observed in young animals were progressively lost in ageing mice. Together, these findings highlight an important role of PRL/STAT5 in the development of prostate tumors.

Supplementary Materials:
Figure S1. Lobe-specific pattern of STAT5 deletion in 6 month-old STAT5 mice, Figure S2. STAT5 deletion has no detectable effect on the prostate tissue, Figure S3. Lobe-specific pattern of STAT5 deletion of 6 month-old Pb-PRL STAT5 mice (complementary to Figure 1), Figure S4. Lobe-specific effects of STAT5 deletion of 6 month-old Pb-PRL STAT5 mice (complementary to Figure 4), Figure S5. STAT5 deletion does not prevent prostate tumor progression in aging Pb-PRL mice (complementary to Figure 5), Figure S6. Characterization of PRLR and AR signaling pathways in dorsal prostate of 12 month-old STAT5f/f, Pb-PRLSTAT5f/f and Pb-PRL STAT5 mice (complementary to Figure 6), Figure S7. Breeding scheme, Figure S8. Whole Western blot membranes corresponding to representative images shown in main and supplemental Figures, as indicated. Table S1: Antibody references and conditions of use, Table S2: Primers used for qPCR.

Author Contributions: Conceptualization, L.S.S., J.-E.G. and V.G.; Formal analysis, F.B. and N.P.; Funding acquisition, J.-E.G. and V.G.; Investigation, F.B., N.P. and L.S.S.; Methodology, F.B., N.P., L.S.S., J.-E.G. and V.G.; Project administration, J.-E.G.; Resources, R.M. and V.G.; Validation, E.R.-G., J.-E.G. and V.G.; Visualization, F.B. and N.P.; Writing—original draft, F.B., N.P. and V.G.; Writing—review & editing, L.S.S., R.M., J.-E.G. and V.G.

Acknowledgments: We thank Lothar Hennighausen for sharing STAT5-floxed mice originally developed in his laboratory. We are grateful to the personnel of the histology (Noémie Gadessaud, Sara Fontoura, Sophie Berissi), cytometry (Corinne Cordier, Jérome Mégret), and animal (Nadia Elganfoud and all zootechnicians) core facilities at the Structure Fédérative de Recherche Necker. We thank the Austrian Science Fund (FWF) for supporting the fees of this publication.

References

1. Goffin, V.; Hoang, D.T.; Bogorad, R.L.; Nevalainen, M.T. Prolactin regulation of the prostate gland: A female player in a male game. *Nat. Rev. Urol.* **2011**, *8*, 597–607. [CrossRef] [PubMed]
2. Li, H.; Ahonen, T.J.; Alanen, K.; Xie, J.; LeBaron, M.J.; Pretlow, T.G.; Ealley, E.L.; Zhang, Y.; Nurmi, M.; Singh, B.; et al. Activation of signal transducer and activator of transcription 5 in human prostate cancer is associated with high histological grade. *Cancer Res.* **2004**, *64*, 4774–4782. [CrossRef] [PubMed]
3. Dagvadorj, A.; collins, S.; Jomain, J.B.; Abdulghani, J.; Karras, J.; Zellweger, T.; Li, H.; Nurmi, M.; Alanen, K.; Mirtti, T.; et al. Autocrine prolactin promotes prostate cancer cell growth via Janus kinase-2-signal transducer and activator of transcription-5a/b signaling pathway. *Endocrinology* **2007**, *148*, 3089–3101. [CrossRef]
4. Stattin, P.; Rinaldi, S.; Stenman, U.H.; Riboli, E.; Hallmans, G.; Bergh, A.; Kaaks, R. Plasma prolactin and prostate cancer risk: A prospective study. *Int. J. Cancer* **2001**, *92*, 463–465. [CrossRef]
5. Berinder, K.; Akre, O.; Granath, F.; Hulting, A.L. Cancer risk in hyperprolactinemia patients: A population-based cohort study. *Eur. J. Endocrinol.* **2011**, *165*, 209–215. [CrossRef] [PubMed]
6. Aksamitiene, E.; Achanta, S.; Kolch, W.; Kholodenko, B.N.; Hoek, J.B.; Kiyatkin, A. Prolactin-stimulated activation of ERK1/2 mitogen-activated protein kinases is controlled by PI3-kinase/Rac/PAK signaling pathway in breast cancer cells. *Cell. Signal.* **2011**, *23*, 1794–1805. [CrossRef]
7. Bole-Feysot, C.; Goffin, V.; Edery, M.; Binart, N.; Kelly, P.A. Prolactin and its receptor: Actions, signal transduction pathways and phenotypes observed in prolactin receptor knockout mice. *Endocr. Rev.* **1998**, *19*, 225–268. [CrossRef]

8. Ahonen, T.J.; Harkonen, P.L.; Rui, H.; Nevalainen, M.T. PRL signal transduction in the epithelial compartment of rat prostate maintained as long-term organ cultures in vitro. *Endocrinology* **2002**, *143*, 228–238. [CrossRef]
9. Ahonen, T.J.; Xie, J.; LeBaron, M.J.; Zhu, J.; Nurmi, M.; Alanen, K.; Rui, H.; Nevalainen, M.T. Inhibition of transcription factor Stat5 induces cell death of human prostate cancer cells. *J. Biol. Chem.* **2003**, *278*, 27287–27292. [CrossRef]
10. Levy, D.E.; Darnell, J.E., Jr. Stats: Transcriptional control and biological impact. *Nat. Rev. Mol. Cell Biol.* **2002**, *3*, 651–662. [CrossRef]
11. Dagvadorj, A.; Kirken, R.A.; Leiby, B.; Karras, J.; Nevalainen, M.T. Transcription factor signal transducer and activator of transcription 5 promotes growth of human prostate cancer cells in vivo. *Clin. Cancer Res.* **2008**, *14*, 1317–1324. [CrossRef] [PubMed]
12. Thomas, C.; Zoubeidi, A.; Kuruma, H.; Fazli, L.; Lamoureux, F.; Beraldi, E.; Monia, B.P.; MacLeod, A.R.; Thuroff, J.W.; Gleave, M.E. Transcription factor Stat5 knockdown enhances androgen receptor degradation and delays castration-resistant prostate cancer progression in vivo. *Mol. Cancer Ther.* **2011**, *10*, 347–359. [CrossRef] [PubMed]
13. Tan, S.H.; Dagvadorj, A.; Shen, F.; Gu, L.; Liao, Z.; Abdulghani, J.; Zhang, Y.; Gelmann, E.P.; Zellweger, T.; Culig, Z.; et al. Transcription factor Stat5 synergizes with androgen receptor in prostate cancer cells. *Cancer Res.* **2008**, *68*, 236–248. [CrossRef] [PubMed]
14. Hoang, D.T.; Iczkowski, K.A.; Kilari, D.; See, W.; Nevalainen, M.T. Androgen receptor-dependent and -independent mechanisms driving prostate cancer progression: Opportunities for therapeutic targeting from multiple angles. *Oncotarget* **2017**, *8*, 3724–3745. [CrossRef] [PubMed]
15. Talati, P.G.; Gu, L.; Ellsworth, E.M.; Girondo, M.A.; Trerotola, M.; Hoang, D.T.; Leiby, B.; Dagvadorj, A.; McCue, P.A.; Lallas, C.D.; et al. Jak2-Stat5a/b Signaling Induces Epithelial-to-Mesenchymal Transition and Stem-Like Cell Properties in Prostate Cancer. *Am. J. Pathol.* **2015**, *185*, 2505–2522. [CrossRef] [PubMed]
16. Gu, L.; Vogiatzi, P.; Puhr, M.; Dagvadorj, A.; Lutz, J.; Ryder, A.; Addya, S.; Fortina, P.; Cooper, C.; Leiby, B.; et al. Stat5 promotes metastatic behavior of human prostate cancer cells in vitro and in vivo. *Endocr. Relat. Cancer* **2010**, *17*, 481–493. [CrossRef]
17. Kazansky, A.V.; Spencer, D.M.; Greenberg, N.M. Activation of signal transducer and activator of transcription 5 is required for progression of autochthonous prostate cancer: Evidence from the transgenic adenocarcinoma of the mouse prostate system. *Cancer Res.* **2003**, *63*, 8757–8762. [PubMed]
18. Gu, L.; Liao, Z.; Hoang, D.T.; Dagvadorj, A.; Gupta, S.; Blackmon, S.; Ellsworth, E.; Talati, P.; Leiby, B.; Zinda, M.; et al. Pharmacologic inhibition of Jak2-Stat5 signaling By Jak2 inhibitor AZD1480 potently suppresses growth of both primary and castrate-resistant prostate cancer. *Clin. Cancer Res.* **2013**, *19*, 5658–5674. [CrossRef] [PubMed]
19. Haddad, B.R.; Gu, L.; Mirtti, T.; Dagvadorj, A.; Vogiatzi, P.; Hoang, D.T.; Bajaj, R.; Leiby, B.; Ellsworth, E.; Blackmon, S.; et al. STAT5A/B gene locus undergoes amplification during human prostate cancer progression. *Am. J. Pathol.* **2013**, *182*, 2264–2275. [CrossRef]
20. Mohanty, S.K.; Yagiz, K.; Pradhan, D.; Luthringer, D.J.; Amin, M.B.; Alkan, S.; Cinar, B. STAT3 and STAT5A are potential therapeutic targets in castration-resistant prostate cancer. *Oncotarget* **2017**, *8*, 85997–86010. [CrossRef]
21. Mirtti, T.; Leiby, B.E.; Abdulghani, J.; Aaltonen, E.; Pavela, M.; Mamtani, A.; Alanen, K.; Egevad, L.; Granfors, T.; Josefsson, A.; et al. Nuclear Stat5a/b predicts early recurrence and prostate cancer-specific death in patients treated by radical prostatectomy. *Hum. Pathol.* **2013**, *44*, 310–319. [CrossRef] [PubMed]
22. Li, H.Z.; Zhang, Y.; Glass, A.; Zellweger, T.; Gehan, E.; Bubendorf, L.; Gelmann, E.P.; Nevalainen, M.T. Activation of signal transducer and activator of transcription-5 in prostate cancer predicts early recurrence. *Clin. Cancer Res.* **2005**, *11*, 5863–5868. [CrossRef] [PubMed]
23. Liao, Z.; Gu, L.; Vergalli, J.; Mariani, S.A.; De Dominici, M.; Lokareddy, R.K.; Dagvadorj, A.; Purushottamachar, P.; McCue, P.A.; Trabulsi, E.; et al. Structure-Based Screen Identifies a Potent Small Molecule Inhibitor of Stat5a/b with Therapeutic Potential for Prostate Cancer and Chronic Myeloid Leukemia. *Mol. Cancer Ther.* **2015**, *14*, 1777–1793. [CrossRef] [PubMed]
24. Maranto, C.; Udhane, V.; Hoang, D.T.; Gu, L.; Alexeev, V.; Malas, K.; Cardenas, K.; Brody, J.R.; Rodeck, U.; Bergom, C.; et al. STAT5A/B Blockade Sensitizes Prostate Cancer to Radiation through Inhibition of RAD51 and DNA Repair. *Clin. Cancer Res.* **2018**, *24*, 1917–1931. [CrossRef] [PubMed]

25. Liao, Z.; Nevalainen, M.T. Targeting transcription factor Stat5a/b as a therapeutic strategy for prostate cancer. *Am. J. Transl. Res.* **2011**, *3*, 133–138. [PubMed]
26. Tan, S.H.; Nevalainen, M.T. Signal transducer and activator of transcription 5A/B in prostate and breast cancers. *Endocr. Relat. Cancer* **2008**, *15*, 367–390. [CrossRef] [PubMed]
27. Goffin, V. Prolactin receptor targeting in breast and prostate cancers: New insights into an old challenge. *Pharmacol. Ther.* **2017**, *179*, 111–126. [CrossRef]
28. Nevalainen, M.T.; Valve, E.M.; Ingleton, P.M.; Nurmi, M.; Martikainen, P.M.; Harkonen, P.L. Prolactin and prolactin receptors are expressed and functioning in human prostate. *J. Clin. Investig.* **1997**, *99*, 618–627. [CrossRef]
29. Kindblom, J.; Dillner, K.; Sahlin, L.; Robertson, F.; Ormandy, C.; Tornell, J.; Wennbo, H. Prostate hyperplasia in a transgenic mouse with prostate-specific expression of prolactin. *Endocrinology* **2003**, *144*, 2269–2278. [CrossRef]
30. Rouet, V.; Bogorad, R.L.; Kayser, C.; Kessal, K.; Genestie, C.; Bardier, A.; Grattan, D.R.; Kelder, B.; Kopchick, J.J.; Kelly, P.A.; et al. Local prolactin is a target to prevent expansion of basal/stem cells in prostate tumors. *Proc. Natl. Acad. Sci. USA* **2010**, *107*, 15199–15204. [CrossRef]
31. Sackmann-Sala, L.; Chiche, A.; Mosquera-Garrote, N.; Boutillon, F.; Cordier, C.; Pourmir, I.; Pascual-Mathey, L.; Kessal, K.; Pigat, N.; Camparo, P.; et al. Prolactin-Induced Prostate Tumorigenesis Links Sustained Stat5 Signaling with the Amplification of Basal/Stem Cells and Emergence of Putative Luminal Progenitors. *Am. J. Pathol.* **2014**, *184*, 3105–3119. [CrossRef] [PubMed]
32. Sackmann Sala, L.; Boutillon, F.; Menara, G.; De Goyon-Pelard, A.; Leprevost, M.; Codzamanian, J.; Lister, N.; Pencik, J.; Clark, A.; Cagnard, N.; et al. A rare castration-resistant progenitor cell population is highly enriched in Pten-null prostate tumors. *J. Pathol.* **2017**, *243*, 54–64. [CrossRef] [PubMed]
33. English, H.F.; Santen, R.J.; Isaacs, J.T. Response of glandular versus basal rat ventral prostatic epithelial cells to androgen withdrawal and replacement. *Prostate* **1987**, *11*, 229–242. [CrossRef] [PubMed]
34. Cui, Y.; Riedlinger, G.; Miyoshi, K.; Tang, W.; Li, C.; Deng, C.X.; Robinson, G.W.; Hennighausen, L. Inactivation of Stat5 in mouse mammary epithelium during pregnancy reveals distinct functions in cell proliferation, survival, and differentiation. *Mol. Cell Biol.* **2004**, *24*, 8037–8047. [CrossRef] [PubMed]
35. Nevalainen, M.T.; Ahonen, T.J.; Yamashita, H.; Chandrashekar, V.; Bartke, A.; Grimley, P.M.; Robinson, G.W.; Hennighausen, L.; Rui, H. Epithelial defect in prostates of Stat5a-null mice. *Lab. Invest.* **2000**, *80*, 993–1006. [CrossRef] [PubMed]
36. Bernichtein, S.; Pigat, N.; Barry Delongchamps, N.; Boutillon, F.; Verkarre, V.; Camparo, P.; Reyes-Gomez, E.; Mejean, A.; Oudard, S.M.; Lepicard, E.M.; et al. Vitamin D3 prevents calcium-induced progression of early-stage prostate tumors by counteracting TRPC6 and calcium sensing receptor upregulation. *Cancer Res.* **2017**, *77*, 355–365. [CrossRef] [PubMed]
37. Capiod, T.; Barry Delongchamps, N.; Pigat, N.; Souberbielle, J.C.; Goffin, V. Do dietary calcium and vitamin D matter in men with prostate cancer? *Nat. Rev. Urol.* **2018**, *15*, 453–461. [CrossRef]
38. Bernichtein, S.; Pigat, N.; Camparo, P.; Latil, A.; Viltard, M.; Friedlander, G.; Goffin, V. Anti-inflammatory properties of Lipidosterolic extract of Serenoa repens (Permixon(R)) in a mouse model of prostate hyperplasia. *Prostate* **2015**, *75*, 706–722. [CrossRef]
39. Costello, L.C.; Franklin, R.B. Effect of prolactin on the prostate. *Prostate* **1994**, *24*, 162–166. [CrossRef]
40. Pigat, N.; Reyes-Gomez, E.; Boutillon, F.; Palea, S.; Barry Delongchamps, N.; Koch, E.; Goffin, V. Combined Sabal and Urtica Extracts (WS((R)) 1541) Exert Anti-proliferative and Anti-inflammatory Effects in a Mouse Model of Benign Prostate Hyperplasia. *Front. Pharmacol.* **2019**, *10*, 311. [CrossRef]
41. Lukacs, R.U.; Goldstein, A.S.; Lawson, D.A.; Cheng, D.; Witte, O.N. Isolation, cultivation and characterization of adult murine prostate stem cells. *Nat. Protoc.* **2010**, *5*, 702–713. [CrossRef] [PubMed]
42. Lawson, D.A.; Zong, Y.; Memarzadeh, S.; Xin, L.; Huang, J.; Witte, O.N. Basal epithelial stem cells are efficient targets for prostate cancer initiation. *Proc. Natl. Acad. Sci. USA* **2010**, *107*, 2610–2615. [CrossRef] [PubMed]
43. Berlanga, J.J.; Garcia-Ruiz, J.P.; Perrot-Applanat, M.; Kelly, P.A.; Edery, M. The short form of the prolactin receptor silences prolactin induction of the b-casein gene promoter. *Mol. Endocrinol.* **1997**, *11*, 1449–1457. [CrossRef]
44. Martens, N.; Uzan, G.; Wery, M.; Hooghe, R.; Hooghe-Peters, E.L.; Gertler, A. Suppressor of cytokine signaling 7 inhibits prolactin, growth hormone, and leptin signaling by interacting with STAT5 or STAT3 and attenuating their nuclear translocation. *J Biol. Chem.* **2005**, *280*, 13817–13823. [CrossRef] [PubMed]

45. Nakajima, H.; Brindle, P.K.; Handa, M.; Ihle, J.N. Functional interaction of STAT5 and nuclear receptor co-repressor SMRT: Implications in negative regulation of STAT5-dependent transcription. *EMBO J.* **2001**, *20*, 6836–6844. [CrossRef] [PubMed]
46. Taylor, B.S.; Schultz, N.; Hieronymus, H.; Gopalan, A.; Xiao, Y.; Carver, B.S.; Arora, V.K.; Kaushik, P.; Cerami, E.; Reva, B.; et al. Integrative genomic profiling of human prostate cancer. *Cancer Cell* **2010**, *18*, 11–22. [CrossRef]
47. Dillner, K.; Kindblom, J.; Flores-Morales, A.; Shao, R.; Tornell, J.; Norstedt, G.; Wennbo, H. Gene expression analysis of prostate hyperplasia in mice overexpressing the prolactin gene specifically in the prostate. *Endocrinology* **2003**, *144*, 4955–4966. [CrossRef]
48. McPherson, S.J.; Wang, H.; Jones, M.E.; Pedersen, J.; Iismaa, T.P.; Wreford, N.; Simpson, E.R.; Risbridger, G.P. Elevated androgens and prolactin in aromatase-deficient mice cause enlargement, but not malignancy, of the prostate gland. *Endocrinology* **2001**, *142*, 2458–2467. [CrossRef]
49. Tang, M.W.; Reedquist, K.A.; Garcia, S.; Fernandez, B.M.; Codullo, V.; Vieira-Sousa, E.; Goffin, V.; Reuwer, A.Q.; Twickler, M.T.; Gerlag, D.M.; et al. The prolactin receptor is expressed in rheumatoid arthritis and psoriatic arthritis synovial tissue and contributes to macrophage activation. *Rheumatology* **2016**, *55*, 2248–2259. [CrossRef]
50. Goldstein, A.S.; Huang, J.; Guo, C.; Garraway, I.P.; Witte, O.N. Identification of a cell of origin for human prostate cancer. *Science* **2010**, *329*, 568–571. [CrossRef]
51. Liu, X.; Robinson, G.W.; Wagner, K.U.; Garrett, L.; Wynshaw-Boris, A.; Hennighausen, L. Stat5a is mandatory for adult mammary gland development and lactogenesis. *Genes Dev.* **1997**, *11*, 179–186. [CrossRef] [PubMed]
52. Parisotto, M.; Metzger, D. Genetically engineered mouse models of prostate cancer. *Mol. Oncol.* **2013**, *7*, 190–205. [CrossRef] [PubMed]
53. Farrar, M.A. Design and use of constitutively active STAT5 constructs. *Methods Enzymol.* **2010**, *485*, 583–596. [CrossRef] [PubMed]
54. Barcus, C.E.; Keely, P.J.; Eliceiri, K.W.; Schuler, L.A. Prolactin signaling through focal adhesion complexes is amplified by stiff extracellular matrices in breast cancer cells. *Oncotarget* **2016**, *7*, 48093–48106. [CrossRef] [PubMed]
55. Barcus, C.E.; Keely, P.J.; Eliceiri, K.W.; Schuler, L.A. Stiff collagen matrices increase tumorigenic prolactin signaling in breast cancer cells. *J. Biol. Chem.* **2013**, *288*, 12722–12732. [CrossRef] [PubMed]
56. Wingelhofer, B.; Maurer, B.; Heyes, E.C.; Cumaraswamy, A.A.; Berger-Becvar, A.; de Araujo, E.D.; Orlova, A.; Freund, P.; Ruge, F.; Park, J.; et al. Pharmacologic inhibition of STAT5 in acute myeloid leukemia. *Leukemia* **2018**, *32*, 1135–1146. [CrossRef] [PubMed]
57. Wu, X.; Wu, J.; Huang, J.; Powell, W.C.; Zhang, J.; Matusik, R.J.; Sangiorgi, F.O.; Maxson, R.E.; Sucov, H.M.; Roy-Burman, P. Generation of a prostate epithelial cell-specific Cre transgenic mouse model for tissue-specific gene ablation. *Mech. Dev.* **2001**, *101*, 61–69. [CrossRef]
58. Park, J.H.; Walls, J.E.; Galvez, J.J.; Kim, M.J.; Abate-Shen, C.; Shen, M.M.; Cardiff, R.D. Prostatic Intraepithelial neoplasia in genetically engineered mice. *Am. J. Pathol.* **2002**, *161*, 727–735. [CrossRef]
59. Shappell, S.B.; Thomas, G.V.; Roberts, R.L.; Herbert, R.; Ittmann, M.M.; Rubin, M.A.; Humphrey, P.A.; Sundberg, J.P.; Rozengurt, N.; Barrios, R.; et al. Prostate pathology of genetically engineered mice: Definitions and classification. The consensus report from the Bar Harbor meeting of the Mouse Models of Human Cancer Consortium Prostate Pathology Committee. *Cancer Res.* **2004**, *64*, 2270–2305. [CrossRef]
60. Bernichtein, S.; Pigat, N.; Capiod, T.; Boutillon, F.; Verkarre, V.; Camparo, P.; Viltard, M.; Mejean, A.; Oudard, S.; Souberbielle, J.C.; et al. High milk consumption does not affect prostate tumor progression in two mouse models of benign and neoplastic lesions. *PLoS ONE* **2015**, *10*, e0125423. [CrossRef]

5

Chaperoning STAT3/5 by Heat Shock Proteins: Interest of their Targeting in Cancer Therapy

Gaëtan Jego [1,2,*,†], François Hermetet [1,2,†], François Girodon [1,2,3] and Carmen Garrido [1,2,4,*]

1 INSERM, LNC UMR1231, team HSP-Pathies, University of Bourgogne Franche-Comté, F-21000 Dijon, France; francois.hermetet@u-bourgogne.fr (F.H.); francois.girodon@chu-dijon.fr (F.G.)
2 UFR des Sciences de Santé, University of Burgundy and Franche-Comté, F-21000 Dijon, France
3 Haematology laboratory, Dijon University Hospital, F-21000 Dijon, France
4 Centre Georges François Leclerc, 21000 Dijon, France
* Correspondence: cgarrido@u-bourgogne.fr (C.G.); gaetan.jego@u-bourgogne.fr (G.J.);
† These authors contributed equally to this paper.

Abstract: While cells from multicellular organisms are dependent upon exogenous signals for their survival, growth, and proliferation, commitment to a specific cell fate requires the correct folding and maturation of proteins, as well as the degradation of misfolded or aggregated proteins within the cell. This general control of protein quality involves the expression and the activity of molecular chaperones such as heat shock proteins (HSPs). HSPs, through their interaction with the STAT3/STAT5 transcription factor pathway, can be crucial both for the tumorigenic properties of cancer cells (cell proliferation, survival) and for the microenvironmental immune cell compartment (differentiation, activation, cytokine secretion) that contributes to immunosuppression, which, in turn, potentially promotes tumor progression. Understanding the contribution of chaperones such as HSP27, HSP70, HSP90, and HSP110 to the STAT3/5 signaling pathway has raised the possibility of targeting such HSPs to specifically restrain STAT3/5 oncogenic functions. In this review, we present how HSPs control STAT3 and STAT5 activation, and vice versa, how the STAT signaling pathways modulate HSP expression. We also discuss whether targeting HSPs is a valid therapeutic option and which HSP would be the best candidate for such a strategy.

Keywords: heat shock proteins; chaperones; stabilization; targeted therapy

---

1. Introduction to Heat Shock Proteins/Chaperones

Heat shock proteins (HSPs), also called stress proteins, are highly conserved molecular chaperones induced by a broad variety of exogenous or intracellular stresses, including chemotherapy. Based on their molecular weight, HSPs have been classified into five conserved families: HSP110 (also called HSPH), HSP90 (HSPC), HSP70 (HSPA), HSP60 (HSPD/E), and the small HSPs (HSPB). The expression of HSPs is mostly regulated by heat shock factor 1 (HSF1), which is able to translocate from the cytoplasm to the nucleus following stress to bind to a short, highly conserved DNA sequence known as a heat shock element (HSE) [1]. As cytoprotective proteins, HSPs participate in the correct folding, activity, transport, and stability of proteins [2], which are essential processes for cell survival. In physiological conditions, these proteins support neosynthesized proteins, favoring post-translational modification processes and protein folding. Otherwise, the functions attributed to them include the subcellular transport of their "client" proteins, or participation in certain signaling pathways. In response to stress, many partially denatured proteins accumulate and cluster together forming protein aggregates via the exposure of their hydrophobic residues. Some HSP proteins are then able to bind to these partially denatured proteins, thus preventing protein aggregation and favoring their correct folding. However, HSPs can also promote the elimination of these proteins by orienting them towards different degradation pathways, notably the ubiquitin–proteasome system, when correct folding is no longer possible. It has also been reported that HSP proteins are able to inhibit the intrinsic apoptotic (inhibition of apoptosome formation) and extrinsic (inhibition of signal transduction of death receptors) processes [3,4]. Interestingly, HSPs can also be secreted and act extracellularly via membrane receptors or within extracellular vesicles as a damage-associated molecular pattern and exhibit immune-cell dysregulation properties.

2. HSP Chaperones and Cancer

As cancer cells accumulate mutations and generate physiologically stressful conditions, they require a constitutively high level of HSPs for their survival and maintenance. In 2011, in order to simplify the complexity of this disease, researchers suggested that tumor development was organized around six essential alterations [5]. These major modifications include (i) self-sufficiency in growth signals; (ii) insensitivity to growth inhibition; (iii) tissue invasion and capacity to develop metastases; (iv) unlimited replication potential; (v) de novo angiogenesis; and (vi) inhibition of programmed cell death. Although the appearance of these changes is mainly linked to instability genomics, many studies have demonstrated the involvement of HSPs in these processes, indicating that these molecular chaperones have an oncogenic role. Comprehensive discussion about the oncogene-like functions of these different molecular chaperones and their participation in the progression of resistance to cancer treatment can be found in excellent recent reviews elsewhere [6,7]. Given the role of HSPs in cancer biology, these chaperones have also been suggested as potential therapeutic targets [3,8]. A number of these proteins have been correlated to cancer aggressiveness and/or cancer resistance to radiotherapy and adjuvant chemotherapy [9].

Targeting HSPs has emerged as a promising sensitization strategy in cancer therapy since HSPs have oncogene-like functions and mediate "non-oncogene addiction" of stressed tumor cells that must adapt to a hostile microenvironment. Except for one inhibitor of HSP27 (an antisense oligonucleotide in phase I/II) [10], all the HSP inhibitors used in clinical trials target HSP90 [11,12]. In this review, we mainly focused on the chaperones HSP90, HSP70, HSP110, and HSP27 and their regulation of protein misfolding and signaling in TYK2-STAT3/5 core cancer pathways, as well as the possibility of targeting such HSPs to specifically restrain STAT3/5 oncogenic functions. We also discuss the machinery behind the chaperones, which is becoming a major therapeutic target in cancer, and the emergence of promising HSP inhibitor-based drugs, which are currently being clinically tested or developed for cancer treatment (Table 1).

Table 1. Summary of the main strategies for HSP inhibition.

Inhibitor		Study Type	Cancer Model	Ref.
Name	**Nature/Structure**			
Target: HSP27				
Apatorsen (OGX-427)	2nd generation 2′-methoxyethyl-modified ASOs	in vitro/preclinical	Prostate, Ovary	[13,14]
		clinical trial (phase I)	CRPC, Breast, Ovary, Lung, Bladder	[10]
		in vitro/preclinical	Pancreatic, NSCLC	[15,16]
		clinical trial (phase II)	Stage IV non-squamous NSCLC	[17]
3-arylethynyltriazolyl ribonucleoside	ASOs	in vitro	Pancreatic	[18]
ASOs-Hsp27	ASOs	in vitro	Lymphoma	[19]
RP101 (Brivudine)	Uridine derivative and nucleoside analog	in vitro/preclinical/clinical	Pancreatic	[20]

Table 1. *Cont.*

Inhibitor		Study Type	Cancer Model	Ref.
Name	Nature/Structure			
		Target: HSP70		
Pifithrin-μ (PFTμ, PES)	Drug-like small molecule	in vitro	AML, ALL, Primary AML blasts	[21,22]
VER-155008	ATP-derivative inhibitor	in vitro	Breast, Colon, Prostatic, Myeloma	[23–25]
A17/A8	Peptide aptamer	in vitro/preclinical	Cervix (HeLa cells), Melanoma	[26]
ADD70	Peptide aptamer	in vitro/preclinical	Rat colon carcinoma, Mouse melanoma	[27]
cmHsp70.1	Antibody	preclinical	Colorectal	[28]
Hsp70-peptide targeted NK based adoptive immunotherapy	A specific amino acid sequence (TKD) of Hsp70	clinical trials (phase I/II)	NSCLC (and colon cancer) patients with ex vivo Hsp70 peptide activated, autologous NK	[29]
		Target: HSP90		
Radicicol	natural product isolated from the fungus *Monosporium bonorden*	in vitro	CML	[30]
17-AAG; 17-DMAG	Derivative of the antibiotic geldanamycin	in vitro/preclinical	Breast, Brain, Medulloblastoma	[31–33]
17-DMAG		in vitro	CLL	[34]
IPI-504 (retaspimycin)	Water-soluble derivate of 17-AAG	in vitro/preclinical	Breast, Pancreatic, Metastatic gastrointestinal stromal tumor	[35–39]
		in vitro/preclinical	NSCLC	[40]
IPI-504, AUY922 Ganetespib, Onalespib	-	clinical trials (phase I–III)	NSCLC Breast, Ovary, Colon	[41]
Novobiocin	Aminocoumarin antibiotic, produced by the actinomycete *Streptomyces nivens*	in vitro/preclinical	Leukemia, Prostate	[42–44]
Panaxynol	Natural pesticide and fatty alcohol	in vitro/preclinical	Lung	[45]
Ganetespib (STA-9090)	Synthetic, non-geldanamycin, small molecule inhibitor	preclinical	Thyroid	[46]
		in vitro	Breast	[47]
BIIB021 (CNF2024)	Orally available, fully synthetic purine scaffold, small molecule inhibitor	in vitro/preclinical	Blood malignancies, Solid tumors	[48]
PU-H71	Non-ansamycin, purine scaffold inhibitor	preclinical	mouse models of the MPN PV and ET	[49]
			MPN	[50]

Table 1. *Cont.*

Inhibitor		Study Type	Cancer Model	Ref.
Name	Nature/Structure			
NVP_AUY922 (AUY922)	Esorcinylic isoxazole amide, second-generation non-geldanamycin inhibitor	in vitro/preclinical	Gastric, Small cell lung, Thyroid	[51–55]
		in vitro	32D mouse hematopoietic cells expressing wild-type BCR-ABL (b3a2, 32Dp210) and mutant BCR-ABL imatinib-resistant cell lines	[56]
		in vitro/preclinical	Drug-resistant chronic myelogenous leukemia	[57]
		clinical trial (phase II)	Myeloproliferative neoplasms	[58]
		clinical trials (phase I/II)	EGFR-mutant lung cancer with acquired resistance to epidermal growth factor receptor tyrosine kinase inhibitors	[59]
AUY922, HSP990, PU-H71	-	in vitro/preclinical	Leukemia	[60]
Onalespib (AT13387)	second-generation, non-ansamycin inhibitor	in vitro	Transformed kidney cells, primary lung adenocarcinoma	[61]
		in vitro/preclinical	Melanoma	[62]
		in vitro/preclinical	NSCLC	[63]
		in vitro/preclinical	NSCLC	[64]
XL888	Orally available inhibitor with high selectivity for HSP90α and HSP90β	clinical trial (phase I)	Melanoma	[65]
SNX2112 SNX5422	Orally bioavailable, synthetic, small molecule inhibitors that competitively bind to HSP90α, HSP90β, Grp94 and Trap-1	in vitro/preclinical	Head and neck squamous cell carcinoma	[66]
			NSCLC	[67]
CUDC-305, Ganetespib CH5164840, WK88-1 17-DMAG	-	preclinical	NSCLC	[68–71]
Target: HSP110				
Foldamers 33 and 52	Protein–protein interaction inhibitors, based on pyridyl scaffolds mimicking α-helix	in vitro/preclinical	Colorectal	[72]

ALL: acute lymphoblastic leukemia; AML: acute myeloid leukemia; ASOs: antisense oligonucleotides; CLL: chronic lymphocytic leukemia; CML: chronic myelogenous leukemia; CRPC: castration-resistant prostate cancer; EGFR: epidermal growth factor receptor; ET: essential thrombocytosis; MPN: myeloproliferative neoplasm; NK: natural killer cells; NSCLC: non-small cell lung cancer; PES: 2-phenylethynesulfonamide; PV: polycythemia vera.

3. HSP90

3.1. HSP90 Structure and Functions

HSP90 (also known as HSPC) is one of the most abundant chaperones in eukaryotic cells in the absence of stress. HSP90 is critical for the operation of cellular machinery under physiological conditions through interactions with so-called "client" proteins. This is only achieved through the formation of a multimeric protein complex of cochaperones that binds to all three domains of HSP90. Hundreds of client proteins for HSP90 have been identified so far [73]. Many of these proteins are involved in essential cellular functions that promote cell growth, proliferation, cell survival, and immune responses. Most of these processes are also involved in cancer development. Three main groups of "client proteins" can be described for HSP90: first, the group of kinases represents the main group because HSP90 interacts with 60% of them [74]; second, the group of multiprotein complexes for which HSP90 promotes assembly [75]; and third, the group of ligands that HSP90 stabilizes with their receptors. It is difficult to identify new client proteins because HSP90 does not bind particular sequences. In contrast, the interaction seems to be based on the overall structural instability of the client proteins [74,76]. Among the client proteins, here we focus on the kinases and receptor tyrosine kinases involved in the STAT3/5 signaling pathway.

3.2. HSP90 and Nonfusion Protein Kinases

3.2.1. Jak Kinases

The mammalian family of Janus kinases (JAKs) is composed of 4 members: JAK1, JAK2, JAK3, and Tyrosine kinase 2 (Tyk2). This family is the main activator of STAT proteins. JAK regulation by HSP90 was discovered by studying the effect of HSP90 inhibitors on the type I and II Interferon (IFN) signaling. In several cell lines, this treatment suppressed the expression of multiple IFN-γ-induced genes and decreased IFN-γ-induced STAT1 phosphorylation on Tyr-701, required for dimerization, and on Ser-727, required for transcription factor activation. As JAK1/2 were known to be the protein kinases responsible for STAT1 phosphorylation, Shang et al. investigated the effect of HSP90 inhibitors on JAK1/2. They showed that HSP90 inhibition led to the proteasome-mediated degradation of JAK1/2. Further they showed that JAK1 interacted with HSP90 (and the CDC37 cochaperone [77]), and that both interactions were destabilized by HSP90 inhibitors [78]. As overactivation or constitutive JAK1/2 signaling promotes cell proliferation and survival in a variety of solid tumors and leukemia [79,80], this discovery paved the way for the identification of the critical role of HSP90 in the aberrant JAK/STAT signaling pathway. In particular, an activating point mutation in JAK2 (*JAK2*^{V617F}) was described as being highly frequent in chronic myeloproliferative neoplasms (MPN) that promote disease progression [81–84].

Despite this activating mutation, HSP90 inhibition in cell lines homozygous for *JAK2*^{V617F} reduced total and phospho-JAK2, and subsequently cell viability [85]. In vivo experiments in a mouse model of MPN confirmed the efficacy of HSP90 targeting because treatment with the HSP90 inhibitor PU-H71 resulted in significant reductions in disease parameters and better chances of survival [49]. Furthermore, combined treatment that included HSP90 inhibitors and JAK2 inhibitors induced a greater depletion of the signaling proteins than a single inhibitor alone, and synergistically induced apoptosis in human primary CD34(+) MPN cells harboring JAK2^{V617F} [50]. Therefore, HSP90 interaction with JAK2 is not altered by activated mutations, but instead could be used as a therapeutic target. This point is of great value, as mutations within the JAK2 kinase domain that confer resistance across a panel of JAK inhibitors have been described (G935R, Y931C, and E864K). Fortunately, genetic resistance to JAK2 enzymatic inhibitors can be overcome by HSP90 inhibitors, which still promote the degradation of both wild-type and mutant JAK2 [60]. Recently, results from a phase II clinical trial with the HSP90 inhibitor AUY922 (Novartis, transferred to Vernalis) have been published and have demonstrated a clinical response in five out of seven patients with MPN [58]. This response correlated with a reduction in overall levels of JAK2, pYSTAT3, and pYSTAT5. Unfortunately, most patients experienced severe adverse effects due to the toxicity of the inhibitor, a phenomenon that has already been observed with other HSP90 inhibitors.

3.2.2. Src Kinases

The members of the Src family of nonreceptor tyrosine kinases (Src, Fyn, Yes, Blk, Yrk, Rak, Fgr, Hck, Lck, Srm, and Lyn) are implicated in numerous important functions in eukaryotic cells. They control proliferation, survival, and differentiation, therefore playing a critical role in many cancer types [86]. Src members can activate STAT3 directly and synergize with JAK family tyrosine kinase action [87]. Among the family members, c-Src has been linked to cancer development [88]. The viral homolog of c-Src kinase, v-Src (from the *Rous sarcoma* virus), has a constitutive kinase activity and was the first discovered oncogene [89,90]. Both homologs bind to the HSP90/CDC37 complex but with striking differences. HSP90, which binds weakly and transiently to c-Src, binds strongly to v-Src, which appears to be its strongest client protein [91–93]. Accordingly, v-Src kinase activity depends strongly on HSP90 [74,94]. Recently, Boczek et al. provided more insight by determining the influence of HSP90 isoforms α and β on purified c-Src and v-Src activity. They have shown that HSP90 does not affect c-Src activity in vitro, whereas v-Src activity was increased two-fold when human HSP90β (but not HSP90α) was added to the experimental setting. HSP90β also stabilized v-Src at high temperatures when it would be inactive otherwise. Until recently, the mechanism behind this striking difference was unknown [76,95]. To solve this issue, Bolcek et al. generated an Src mutant that mimics the oncogenic v-Src kinase activity (c-src3MΔC). This mutant exhibited a more extended activation loop (A-loop) (usually present in an open form during wild-type Src active state to allow substrate binding). The A-loop from c-Src3MΔC is also less stable in comparison with the wild-type Src. Consequently, the c-Src3MΔC is conformationally uncontrolled, which enhances its interaction with HSP90 and suggests this could be a more general mechanism for the interaction between HSP90 and oncogenic kinases than the presence of a general client sequence motif. Indeed, HSP90 potentially interacts more strongly with structurally extended kinases, a frequent state observed upon activating mutations. Interestingly, a very similar mechanism needed to aid the initial folding of immature kinases such as c-Src, which is furtive is this case, governs the binding of HSP90 to conformationally unstable but mature kinases like v-Src. In this context, CDC37 appears to bind to parts of the unfolded kinase first (which might be considered as an independent kinase binding unit), partly unfolding it further before HSP90 clamps around the CDC37/kinase complex [96]. Other Src family members, like Lck^{Y505F} and HCK^{499F}, are probably stabilized by the same mechanism [97,98].

3.2.3. ACK1

Another nonreceptor tyrosine kinase, activated CDC42-associated kinase-1 (ACK1), catalyzes the phosphorylation of STAT1, STAT3, and STAT5. HSP90 interacts with ACK1 [99] and is necessary for the phosphorylation of STAT1 in transformed kidney cells and STAT3 in primary lung adenocarcinoma by ACK1 [61].

3.2.4. BRAF

The activated serine/threonine kinase BRAF mutant is a main driver of melanoma growth and progression [100] and is a HSP90 client protein [74,101]. Inhibition of HSP90 by AT13387 delays the emergence of resistance to BRAF inhibitors [62]. A recent phase I dose escalation clinical trial in melanoma has shown that another HSP90 inhibitor (XL888) in combination with a specific anti BRAF inhibitor (vemurafenib) has clinical activity in patients with advanced $BRAF^{V600}$-mutant melanoma, with a tolerable side effect profile [65].

3.3. HSP90 and Fusion Protein Kinases

3.3.1. BCR-ABL

Chronic myeloid leukemia (CML) is driven by the BCR-ABL fusion oncoprotein [102], which is involved, among other pathways, in the transcriptional regulation of STAT3 [103,104] and STAT5 [105,106]. In this context, the BCR-ABL/STAT3/STAT5 signaling pathway is mainly involved in tumor-initiating stem cell maintenance [107]. BCR-ABL is a HSP90 client protein that is destabilized by HSP90 inhibition, which leads to cell death [56]. In CML cells, BCR-ABL forms a high molecular weight network with JAK2, STAT3, and AKT. This network pushes disease progression, but could also be its Achilles' heel. Indeed, HSP90 directly binds to this signaling network, and its inhibition breaks the whole network apart [56]. As for other targeted therapies, resistance to BCR-ABL tyrosine kinase inhibitors can develop during the course of the treatment because of acquired *BCR-ABL* mutations. Hopefully, combination therapies involving HSP90 inhibitors and anti-JAK2 may overcome this resistance [57].

We are still unsure of how this mechanism of action can be extended to the interaction with other oncogenic mature kinases, but an important process for protein stabilization by HSP90 and CDC37 has been uncovered.

3.3.2. EML4-ALK

The echinoderm microtubule-associated protein-like 4-anaplastic lymphoma kinase (EML4-ALK) fusion gene is an oncogenic driver in about 5% of patients with non-small cell lung cancer (NSCLC). It is also an HSP90 client protein (one of the most sensitive), which is very rapidly degraded upon exposure to HSP90 inhibitors [40]. These results prompted the initiation of numerous clinical trials reviewed elsewhere [41,108]. Yet despite encouraging results, clinical response was weak, and so the development of HSP90 inhibitors was halted for NSCLC.

3.4. HSP90 and ErbB Family of Receptor Tyrosine Kinase (RTK)

STAT3 and STAT5 are also known to be phosphorylated by several receptor tyrosine kinases (RTK), such as the ErbB family, IGF-1R, or FGFR [109], most of which are HSP90 client proteins. Interestingly, they are also activated or captured for signal transduction by those RTK without phosphorylation [110]. Given their membrane localization, these RTK must go through a complex process of folding, maturation, and membrane insertion that requires significant chaperone cooperation. This is particularly true for mutated RTK, which is frequently observed in cancers. For instance, ErbB2 stability and maturation is regulated by its binding to HSP90 through its cytoplasmic tail [111] and is ATP dependent [112].

Accordingly, HSP90 inhibition leads to RTK destabilization and absence of STAT3 activation in different models of cancer [47,66,67]. Many drugs have been developed to inhibit mutated or rearranged RTK, but despite early success, most patients develop resistance and eventually relapse [113]. The strong HSP90/EGFR interaction has then been used to propose an alternative therapeutic strategy combining an HSP90 inhibitor with an EGFR inhibitor. Interestingly, this combination (with the EGFR inhibitor erlotinib) resulted in prolonged animal survival in nonmutated and erlotinib-resistant models [67,68,70,71,114,115].

3.5. STAT3/5 and HSP90

The STAT3/5 signaling pathway is also regulated downstream from the tyrosine kinases and RTK. Indeed, HSP90 is found within the cytosol, directly bound to dimers of STAT3 or STAT5 via its N-terminal regions [116]. However, in contrast to its role in TK or RTK folding and stabilization, HSP90 is not required for STAT3/5 maturation or total protein levels. They are therefore nonclassical HSP90 client proteins. The chaperone would rather change STAT conformation to ease the phosphorylation process and/or, once phosphorylated, maintain this active state for a prolonged period of time. Moulick et al. have suggested this pattern in chronic myeloid leukemia [69]. They showed that HSP90 directly binds to active pYSTAT5 (Tyr694), but not to inactive STAT5, and that pYSTAT5 acquires a conformation that is more susceptible to trypsine cleavage in the presence of HSP90. HSP90/STAT3 also protects pYSTAT3 from dephosphorylation by the phosphatase SHP-1 in gastric cancer cells. Luteolin (3,4,5,7-tetrahydroxyflavone), a natural flavonoid present in fruits and vegetables, inhibits STAT3 activation by disrupting the association of HSP90 to STAT3, which allows it to interact with SHP-1 [117].

In order to function as a transcription factor, STAT3/5 needs to translocate into the nucleus and form a stable interaction with DNA. In this context, as suggested by Longshaw et al., HSP90 appears to play a specific role in association with the cochaperone HOP [118]. They have shown that the depletion of HOP decreased the nuclear localization of STAT3. Although it is not yet clear how HSP90 promotes

STAT3/5 nuclear shuttling, it may involve the capacity of HSP90 to form molecular complexes with specific carriers, like importins alpha [119], that can transport STAT3 to the nucleus. This scheme would mimic what has been described for other molecular complexes implicating HSP90, such as the glucocorticoid receptor [120] or PKCZeta [121]. After entering the nucleus, HSP90 seems to promote STAT3/5 transcriptional activity as STAT3/5 interaction with promoters of target genes is enhanced by the presence of HSP90. Indeed, the STAT3/5 complexes and HSP90 have been shown to colocalize in MYC and in CCND2 promoters [69]. Furthermore, nuclear hormone receptors form multiprotein complexes with STAT3 and STAT5 [122,123], which together with HSPs could contribute to chromatin landscaping [124].

In conclusion, HSP90 appears to be a key chaperone for the STAT3/5 pathway. It operates at all levels of message transmission, from interaction with RTK in the cytoplasmic membrane, to the interaction with multiple kinases in the cytosol, and to favoring active STAT3/5 localization and binding of target genes the nucleus (Figure 1). This role is also central in the pathological overactivation of STAT signaling where HSP90 favors oncogenic proteins (Figure 1), promoting the development of several inhibitors for cancer treatment (Table 1, Figure 2). However so far, most clinical trials have yielded mixed results and frequent side effects that precluded the broad utilization of these treatments.

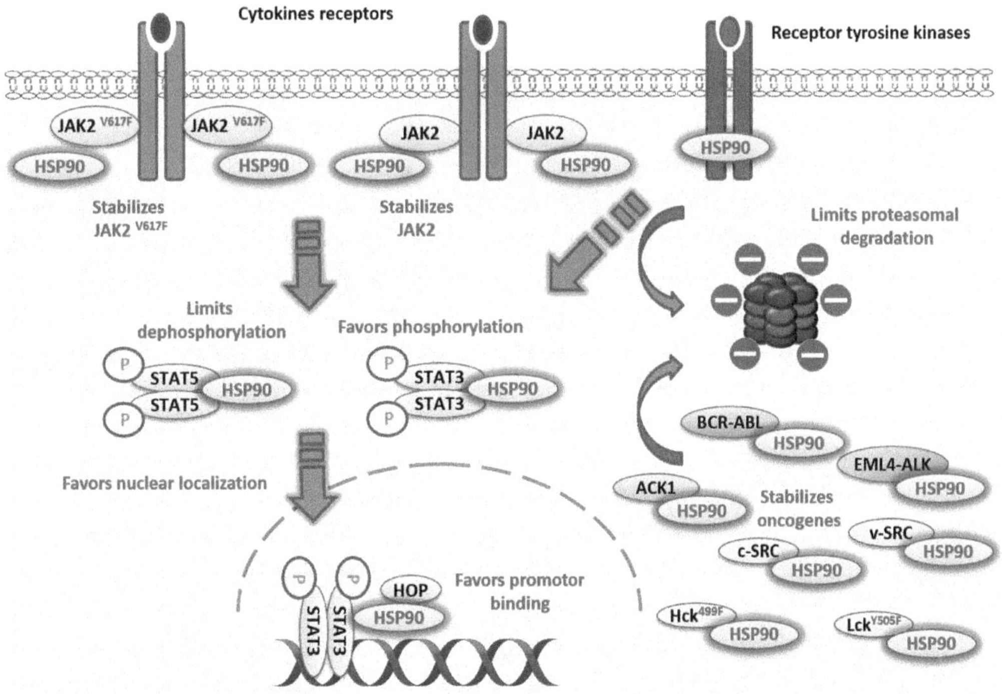

Figure 1. Localization and described functions of HSP90 within the STAT3 and STAT5 signaling pathways. HSP90 promotes those pathways through direct interaction with STAT3 or STAT5 dimers and favors their phosphorylation, nuclear localization, and promoter binding, but HSP90 also limits dephosphorylation and proteasomal degradation. Upstream of STAT3/STAT5 activation, HSP90 stabilizes several kinases, like JAK2, JAK2^{V617F}, c-Src, v-Src, ACK1, BCR-ABL, EML4-ALK, Lck^{Y505F}, and HCK^{499F}, and several receptor tyrosine kinases, such as the ErbB family.

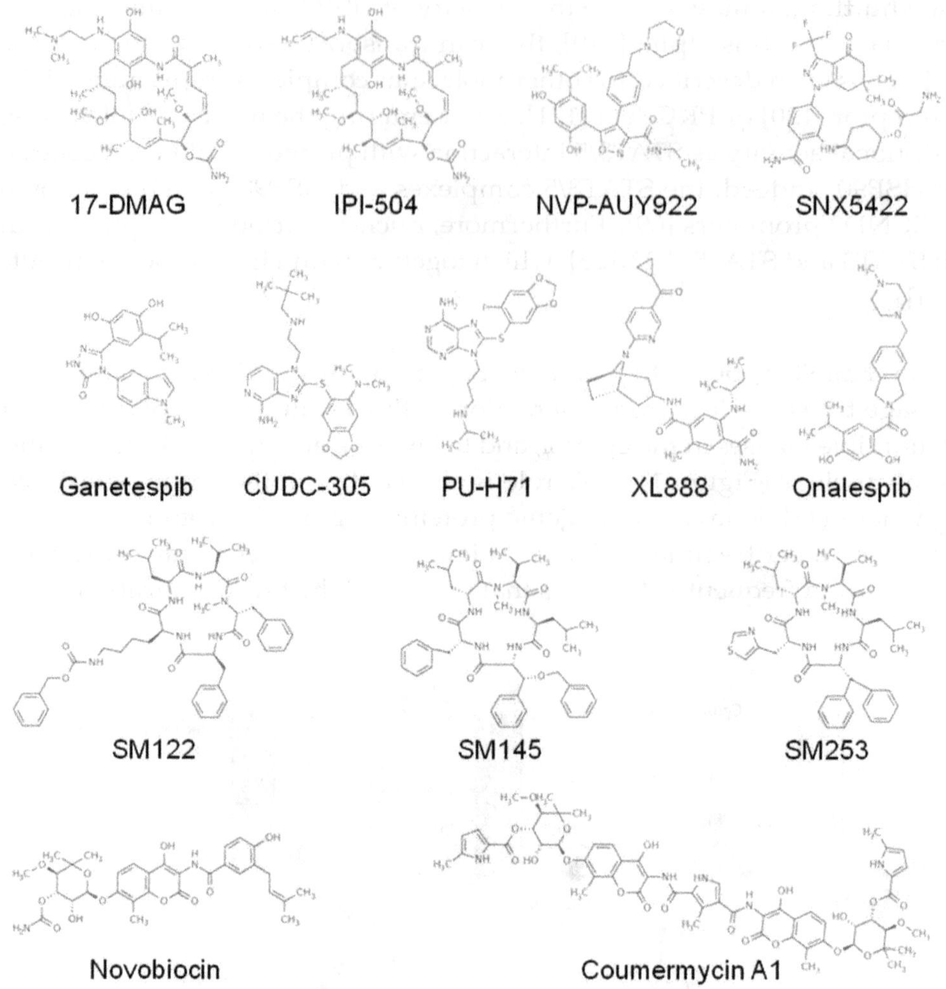

Figure 2. Structure of the main HSP90 inhibitors. The inhibitors targeting the ATP binding site at the N-terminus and the C-terminus of HSP90 are depicted in green and orange, respectively.

4. HSP27

4.1. HSP27 Structure

HSP27 (27 kDa), also known as HSPB1, belongs to the small HSP family. In contrast to other HSPs, it is an ATP-independent chaperone [125]. HSP27 shows a highly dynamic process of oligomerization that transforms proteins from dimers to large oligomers, which can culminate at 1000 kDa. This state of oligomerization dictates the affinity of HSP27 towards the proteins to be chaperoned, given that the multimer form is the most binding-competent state [126].

Four phosphorylation sites (S15, S78, S82 and T143) in the N-terminal domain regulate the assembly of oligomers [127]. Phosphorylation promotes the formation of small oligomers, while dephosphorylation promotes the formation of large oligomers [128]. Stressors such as anticancer agents, hydrogen peroxide, mitogens, inflammatory cytokines (TNF-α, IL-1b, etc.), and kinases (p38 MAPK, p70RSK, PKB, PKC, PKD and PKG) can promote HSP27 phosphorylation [129]. However, we have also shown that oligomerization can occur independently of phosphorylation through cell–cell contact, as observed in confluent cultures in vitro or solid tumors in vivo [130].

4.2. *HSP27 Functions*

HSP27 chaperone functions have been less described than higher molecular weight HSPs like HSP70 and HSP90. However, HSP27's main function as a chaperone is to stabilize denatured or aggregated proteins to bring them to a soluble and stable form [131,132].

HSP27 plays an important role and has been associated with poor prognosis in many cancers (for a recent review, see [127]). For instance, Rocchi et al. described the promotion of the STAT3 signaling pathway in prostate tumors [133]. They showed that HSP27 directly interacts with STAT3 and that total STAT3 levels correlated directly with HSP27 levels. Furthermore, the cytoprotective effect of HSP27 was attenuated by STAT3-reduced expression, underlying the importance of this pathway in prostate cancer [133]. HSP27 is also involved in the process of prostate cancer metastasis through the promotion of IL-6-mediated epithelial-to-mesenchymal transition (EMT), because reduced HSP27 expression decreases cell migration and invasion [134]. In the absence of HSP27, IL-6-induced phosphorylation of STAT3 is reduced, but not total STAT3 content. This inhibition leads naturally to reduced STAT3 nuclear localization and binding to the *TWIST* gene, which codes for a key EMT transcription factor.

Beside cancer development, HSP27 regulation of STAT3 has recently been implicated in placental implantation. HSP27 is indeed expressed during placenta formation and in the first two trimesters of pregnancy [135,136]. In particular, HSP27 is highly expressed during the differentiation of cytotrophoblast cells and extravillous trophoblast cells, and its silencing was found to significantly reduce total STAT3. Interestingly, the phosphorylation state of STAT3 was not altered in the absence of HSP27 in placental explants, suggesting a role in protein protection from proteasomal degradation [137]. This could be explained by the fact that STAT3 and STAT5 have a relatively low thermodynamic stability as isolated proteins and are thus more prone to aggregation, which would be limited by HSP27 [124]. Given the importance of STAT3 in embryonic development (STAT3 knock-out mice have a lethal embryonic phenotype) [138], this finding revealed the critical role of HSP27 in this process.

There has been little study of the state of HSP27 phosphorylation, and consequently the state of oligomerization, required for STAT3/5 binding, despite the fact that therapeutic targeting of specific kinases would logically impact HSP27 functions. We only know, from one study of prostate cancer, that IL-6 stimulation leads to HSP27 phosphorylation and correlates with the EMT, suggesting the phosphorylated form is required for STAT3 activation [134]. Other STAT family members are HSP27 client proteins, like STAT2 (a STAT family member involved in viral or interferon responses), which was also shown to be degraded upon HSP27 knockdown in Hela cells [139]. However, this process was reversed by proteasome inhibition. However, STAT3 and 5 were not or were only weakly reduced in this particular tumor cell line. Again, the discrepancy between tumoral contexts or cell lines could come from differences in HSP27 phosphorylation or oligomerization status.

As stated previously, HSP27/STAT3 interaction occurs also in nontumoral contexts. In normal liver cells under a high fat diet, the phosphorylated form of HSP27 stimulates autophagy and lipid droplet clearance through interaction with STAT3. In this particular situation, no STAT3 activation by HSP27 is described, but rather the disruption of STAT3/PKR complexes, facilitating PKR and eIF2α mediated autophagy [140]. These data suggest that dimers and multiprotein complexes can be displaced by the action of phospho-HSP27 on different binding partners, including STAT3, and this can also mediate critical cellular physiological processes.

Upstream from STAT3/5 activation, JAK2 plays a major role that can be modulated by HSP27 [141]. In the specific context of thrombopoietin- and JAK2^{V617F}-induced myelofibrosis (a chronic degenerative disorder of the hematopoietic system associated with the aberrant activation of the JAK/STAT pathway) [142], our team has recently shown that HSP27 interacts directly with JAK2/STAT5, stabilizing the complex. Neither total JAK2 nor STAT5 protein levels were affected, but we found that the state of phosphorylation of STAT5 (Tyr694) by JAK2 was HSP27 dependent. We demonstrated that HSP27, through interaction with STAT5, physically prevented its dephosphorylation by the phosphatase SHP2 in those cells.

In conclusion, HSP27 plays a very important role in STAT3/5 signaling, both in contexts of tumor development and others such as placenta development. In contrast with HSP90, whose functions depend on the client proteins, HSP27's different functions (protein stability, phosphorylation, disruption of complex of proteins, etc.) may rely on the phosphorylation and oligomerization status of the chaperone (Figure 3). The general vision is mainly dichotomous: on one side there are large phosphorylated oligomers, and on the other nonphosphorylated dimers. This simplistic description does not reflect the reality of cellular dynamics. Future studies will be needed to specify the proportion of each HSP27 oligomer within the cells and the associated function.

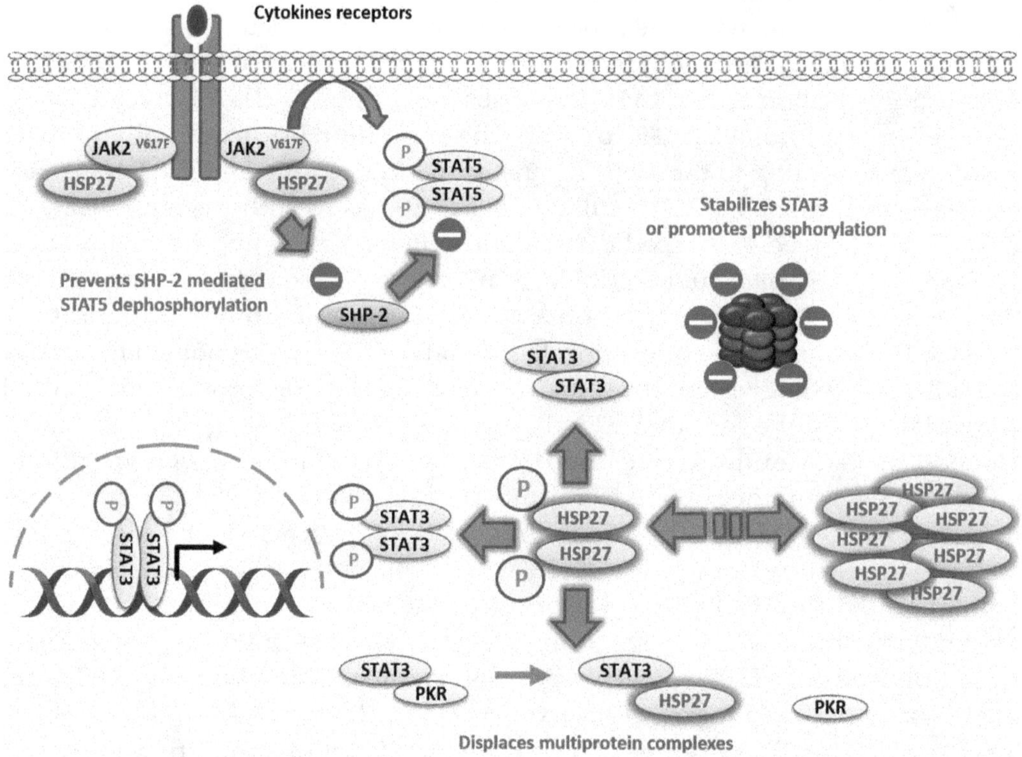

Figure 3. Localization and described functions of HSP27 within the STAT3 and STAT5 signaling pathways. HSP27 state of oligomerization varies dynamically to modulate its binding capacity to target proteins in a context-dependent way. HSP27 directly interacts with pYSTAT3 and total STAT3 to promote stabilization and phosphorylation. HSP27 directly binds to JAK2^{V617F}/STAT5 complexes to prevent STAT5 dephosphorylation by SHP-2 in MPN. HSP27 also displaces multiprotein complexes like the STAT3/PKR complex.

5. HSP110

5.1. HSP110 Structure

The 110 kDa heat shock protein (HSP110), also known as HSP105 or HSPH1, belongs to the members of the HSPH family. Although it appears to be distinct from other HSPs (HSP27, HSP40, HSP70, and HSP90) because of its molecular weight and the specificity of certain sequences, HSP110 is a member of the family of HSP70 proteins [143,144]. Until quite recently, HSP110 was considered as a mere nucleotide exchange factor of HSP70. However, it is now well established that HSP110 is able to act as an unfolding chaperone on its own using ATP hydrolysis to lead to conversion of stable misfolded polypeptide substrates into natively refolded products, even when HSP70 is not present [145]. In an ATP-independent manner, HSP110 also has the antiaggregating properties of unfolded or misfolded proteins.

5.2. HSP110 Functions

In contrast to other chaperones, like HSP90 or HSP70, little is known about the cellular and extracellular functions of HSP110. It is a ubiquitous and conserved chaperone with antiaggregation capabilities that act in synergy with the refolding activity of HSP70, which contributes to efficient protein homeostasis [145,146]. HSP110 expression is induced by a wide array of stress including hyperthermia, ethanol, oxidative stress, recovery from anoxia (i.e., reperfusion injury), some anticancer drugs, and inflammation. HSP110 is approximately four-fold more efficient at binding and stabilizing denatured protein substrates than HSP70 [147]. Due to its strong chaperone (or holder) function, HSP110 is a very good antigen carrier. It is therefore used in vaccine formulations [148] as a recombinant chaperone vaccine for antigen-targeted cancer immunotherapy. These vaccines have generated robust antigen-specific T lymphocyte responses in different preclinical cancer models [149]. Two forms of HSP110 exist: HSP110α and an alternatively spliced form called HSP110β, which contains 43 fewer amino acids [150]. While HSP110α is constitutively expressed in the cytoplasm of cells and can be induced in stressful conditions such as heat shock, HSP110β is strictly heat-inducible and specifically localized in the nucleus (Figure 4) [151]. Beyond the different localization of these two forms of HSP110, their differential roles remain unclear. Like other HSP proteins, the expression of HSP110 can be induced by a number of physical or chemical sources of stress and depends on the heat shock factor 1 (HSF1) transcription factor. Moreover, the presence STAT3 on the HSP110 promoter has been recently reported in humans, suggesting its regulation by STAT3 [152]. Conversely, HSP110β can induce HSP70 expression through STAT3 in mammalian cells (Figure 4) [153]. It is now clearly established that HSP110 favors several signaling pathways, including the Wnt/β-Catenin, MyD88/TLR, and STAT3 pathways [154,155].

Concerning the STAT signaling pathway, we have demonstrated that HSP110 directly binds to STAT3 and favors its phosphorylation (Tyr705) by JAK2 in the cytosol, thereby promoting cell proliferation (Figure 4) [156]. Colon cancer cells in which HSP110 has been shRNA-mediated and knocked down hardly proliferate, but proliferation is reactivated by the re-expression of HSP110 in these cells. Tumors from patients with high levels of HSP110 show high STAT3 phosphorylation levels and strong expression of proliferation markers [154,156]. Therefore, both the protein homeostasis function and the role of HSP110 on proliferative pathways may explain why this protein is linked to aggressive tumors. We suggest that HSP110 expression could be a surrogate prognostic marker and a potential therapeutic target, particularly for treatment of carcinomas, particularly colorectal cancer, for which there is strong evidence.

Given the emerging role of HSP110 in cancer and its role on STAT3 in particular, we selected two foldamers upon screening of a chemical library based on their ability to inhibit and block recombinant HSP110-mediated antiaggregation activity and to disrupt HSP110–STAT3 interaction [72]. These compounds, named 33 and 52 (Table 1, Figure 5), inhibit HSP110 chaperone function and colorectal cancer growth in vitro and in vivo [72]. Altogether these results confirm the interest of targeting HSP110, at least in colorectal cancers, and probably in other types of cancer, such as B-cell lymphoma.

Although HSPs are generally considered intracellular proteins, we now know that HSP110 can also be released to act extracellularly like HSP27, 70, and 90 [157–159]. The release of HSP110 from human intestinal epithelial cells has also been described, suggesting a role in the physiological process of epithelial renewal [160]. More recently, we have demonstrated that HSP110, like other HSPs, can be secreted by cancer cells and is abundantly observed in the cancer microenvironment [161]. Interestingly, extracellular HSP110 affects macrophage differentiation/polarization by favoring a protumor, anti-inflammatory profile and the formation of tumor-associated macrophages (TAMs), which are associated with immune suppression. Furthermore, we found a correlation between the level of extracellular HSP110 and the number of TAMs in patient biopsies [161], suggesting that the effect of extracellular HSP110 function on macrophages may also contribute to the poor outcomes that are associated with HSP110 expression.

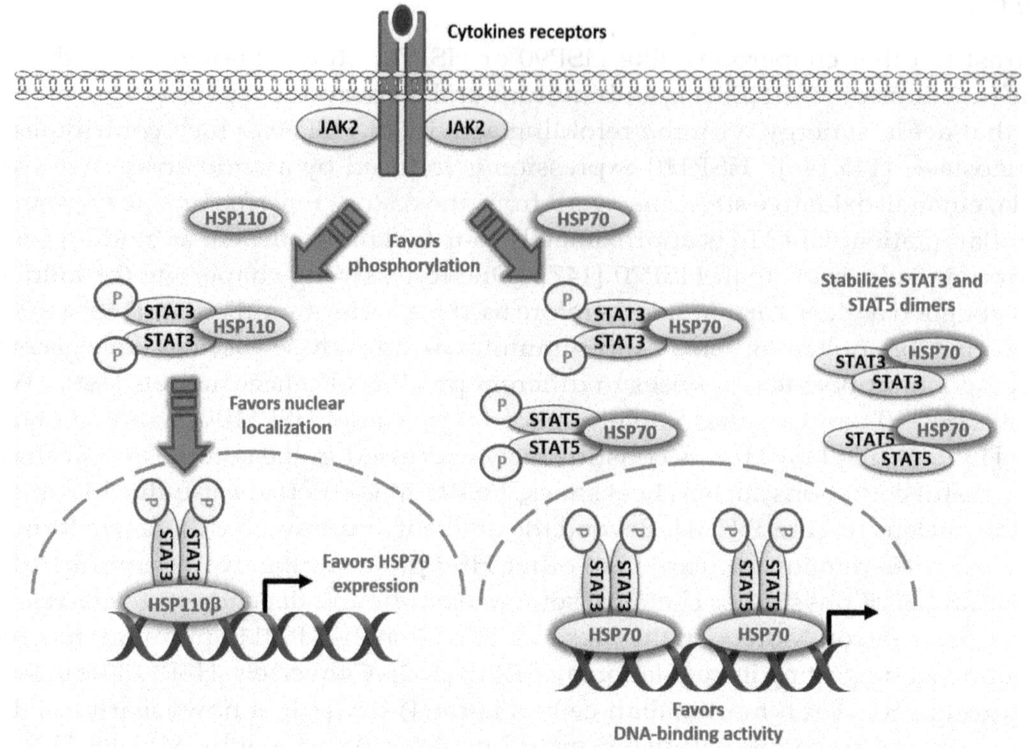

Figure 4. Localization and described functions of HSP110 and HSP70 within the STAT3 and STAT5 signaling pathways. HSP110 directly binds to STAT3 in the cytosol and favors its phosphorylation through JAK2, and, through this mechanism, participates in the promotion of cell proliferation. HSP110α and HSP110β localize to the cytoplasm and nucleus of cells, respectively. HSP110β induces the expression of HSP70 in mammalian cells. Overexpression of HSP110β stimulated the phosphorylation of STAT3 (Tyr705) and its translocation to the nucleus. STAT3 binds to the sequence of the HSP70 promoter at the level of a sequence (−206 to −187 base pair) whose mutation abrogated the activation of the HSP110β-mediated HSP70 promoter. HSP70 directly interacts with STAT3 and STAT5. It favors STAT3 phosphorylation and activity, and STAT5 levels, phosphorylation, and activity.

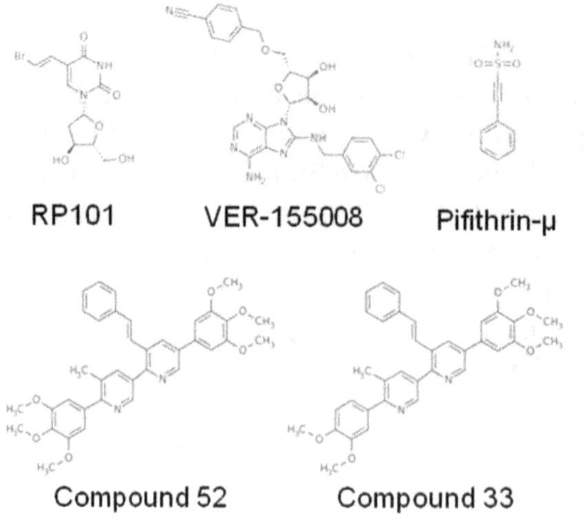

Figure 5. Structure of the main small-molecule inhibitors of HSP27 (blue), HSP70 (Green), and HSP110 (Orange). RP101 can inhibit HSP27 function via direct binding to Phe29 and Phe33. VER-155008 binds to the ATP-binding site at the N-terminus of HSP70. Pifithrin-μ inhibits specifically function of HSP70 via direct binding to its substrate binding domain. Compounds 52 and 33 bind to the ATP binding site at the N-terminus of HSP110.

6. HSP70

6.1. HSP70 Structure

Stress-inducible HSP70 (also called HSP72 or HSP1A, HSPA1B) is another member of the HSP superfamily that has emerged as a viable and very promising target for development of antitumor drugs to combat various forms of cancer [162–164]. As the most ubiquitous stress-inducible chaperone, HSP70 exhibits numerous chaperone functions that are critical for both the folding and proteasomal degradation of misfolded proteins. It thus participates in cellular protein quality control systems, leading to cell homeostasis and survival during stress conditions. The human HSP70 protein family consists of at least 13 members, including stress-induced HSP70 and the heat shock cognate protein 70 (HSC70) [165]. The chaperone mechanism of HSP70 has been extensively studied [166]. HSP70 associates with misfolded proteins in a manner controlled by its ATPase cycle and cochaperones such as HSP40, BAG family proteins, HIP, HOP, and HSPBP1 [165].

6.2. HSP70 Functions

Based on immunoprecipitation analysis from normal rat kidney interstitial fibroblast (NRK-49F) cells, HSP70 has been shown to directly interact with multiple STAT proteins, including STAT3/5 (Figure 4). This interaction increased when cells were stressed by exposure to advanced glycation end product [167]. Although the data are still limited, studies show that manipulating HSP70 expression or activity affects STAT protein activity within cells. HSP70 (HSP701A/B) knock down using siRNA further decreased constitutive STAT3 activity in an acute myeloid leukemia cell line (HEL) treated with arsenic trioxide and the HSP90 inhibitor 17-DMAG [168]. In addition, increasing HSP70 activity with administration of geranylgeranylacetone before and three days after intracerebral hemorrhage resulted in increased STAT3 and AKT phosphorylation and also cerebral levels of eNOS, which are collectively associated with preserved cerebral blood flow, decreased neuronal cell death, and improved functional recovery in rats [169]. Uchida et al. suggested that the mechanism for HSP70 induction through geranylgeranylacetone may be the result of geranylgeranylacetone-induced induction of protein kinase C [170].

It is worth noting that upregulation of HSP70 could stimulate cell proliferation through the control of tyrosine kinase functions. In cancer cells derived from chronic myeloid leukemia cells, BCR-ABL tyrosine kinase activity results in the phosphorylation and activation of AKT, and in the phosphorylation and DNA-binding activity of STAT5, which leads in turn to an increase in the expression of antiapoptotic protein Bcl-xL [171]. Interestingly, while increased expression of HSP70 results in the upregulation of STAT5 level and activity, inhibition of the BCR-ABL tyrosine kinase activity with imatinib or inhibition of PI-3K activity with wortmannin both result in a decrease in HSP70 expression and STAT5 activity. Thus, uncontrolled cell signaling may result in the transcription of HSP70, which in turn regulates the level and activity of STAT5 (Figure 4) [171]. Besides the intracellular role of HSP70 as a survival factor that promotes tumorigenesis [172], it is now known that HSP70 also regulates diverse immunoregulatory activities such as antigen cross presentation [173,174], dendritic cell maturation [175,176], and natural killer cell [177,178] and myeloid-derived suppressor cell [179] activities, by acting extracellularly as a cytokine [180,181]. We demonstrated that membrane-associated HSP70 is found extracellularly in tumor-derived exosomes and that it restrained tumor immune surveillance by promoting myeloid-derived suppressor cell functions in both mice and humans. Interestingly, tumor-derived exosomes harboring HSP70 were found to mediate the suppressive activity of the myeloid-derived suppressor cells via activation of STAT3 and ERK [179].

7. Regulation Mechanisms of HSF/HSPs by JAK/STAT Signaling: A Feedback Loop?

7.1. HSF and SOCS Regulation

The activation of the JAK family and the subsequent STAT signaling is regulated by the family of suppressors of cytokine signaling (SOCS) proteins. This system aims to protect organisms from permanent and/or overstimulation that could lead, for instance, to severe systemic inflammations mediated by IFN-γ signaling [182]. Conversely, a deficit in SOCS expression could play a pivotal role in the development and progression of cancers [183]. Expression of SOCS3, which has been shown to inhibit JAK1, JAK2, and TYK2 [184], is frequently reduced in cancer cells, thereby leading to a growth advantage. The effect of HSF1 and HSP on SOCS3 expression changes depending on the tissue and whether the context is normal or tumoral. In nontumoral microglia cells, the activation of HSF1 by paeoniflorin induced an indirect increase of SOCS3 expression mediated by HSP70 production and autocrine action [185]. Conversely, it was recently shown that HSF1 could directly bind to the SOCS3 promoter region and inhibit the transcriptional activity of its promoter [186]. This reinforces the role of HSF1 as a transcription factor for many genes not related to heat shock response [3]. SOCS3 repression of expression can also be mediated by HSPs, as seen in chronic lymphocytic leukemia. In this context, HSP90 inhibition by 17-DMAG induces the expression of SOCS3 through the activation of the p38 MAPK signaling [34]. This regulation is probably a specific to tumors, as normal B cells were not found to upregulate SOCS3 expression upon HSP90 inhibition, and their migration was not affected [34].

7.2. HSF/HSPs and JAK/STAT

The regulation of the STAT signaling pathway by HSPs has been largely described, but in a positive feedback loop STAT3/5 has been shown to regulate HSPs. Little is known about HSP regulation apart from the fundamental role of HSF. The STAT3/5 pathway modulates expression of HSP27, HSP70, HSP90, and HSP110 when faced with different stressful stimuli [187]. Among those, mild heat shock is a type of stress (that was used to first describe the function of HSPs) during which the induction of HSPs prevents cell death from intense heat shock that otherwise would have been lethal. This thermotolerance is also accompanied by STAT3 phosphorylation. Inhibition of STAT3 activation by STAT3 inhibitors AG490 or static partially suppresses thermotolerance and HSP110 expression, but not HSP70 and HSP27 [188]. This result underlines a particular role for HSP110 in the process, which is in line with its capacity to interact with α-tubulin. This mechanism could therefore protect microtubules from severe heat shock [189]. Besides STAT3, STAT1 is also involved in HSP70 and HSP90 regulation, particularly after IFN-γ activation [190,191] or under heat shock [192]. Together with HSF1, STAT1 is recruited to the first intron of the HSP90β gene to favor the recruitment of a chromatin-remodeling complex, leading to enhanced HSP90β expression.

Other types of stress, such as hypoxic stress (a common phenomenon in a majority of tumors), are strong inducers of HSP90α expression. In these conditions, STAT5 is one of the transcription factors that regulate HSP90α, and hypoxia increases the binding of STAT5 to the HSP90α promoter [193]. When the STAT signaling pathway is constitutively active in tumors, like in breast and colon cancer, STAT3 has also been shown to transcriptionally induce HSP27 expression [194]. Though less documented, this mechanism also exists in hematological malignancies like Burkitt's lymphoma, where pharmacological inhibition of STAT3 by AG490 downregulates HSP70 expression [195].

8. Conclusions

We have reviewed here the multiple roles of HSPs in the STAT3/5 activation network. From receptor tyrosine kinases to promoter binding of STAT3/5 target genes, HSPs appear to use several mechanisms to control this pathway. Among the various HSPs, HSP90 has a central role in the cellular machinery and is one of the most abundant cytosolic proteins. This role is clearly exposed when we describe both kinases and STAT proteins as HSP90 client proteins. The current literature points to the important role of HSP90/STAT3/STAT5 in cancer growth and the ability to thwart chemotherapy [3,196–198].

This discovery has prompted the development of drugs specifically targeting HSP90 and the subsequent initiation of clinical trials for cancer patients (Table 1, Figure 2). Almost all of the trials have reported clinical responses, confirming the relevance of this strategy. However, adverse side-effects were also frequent due to the inherent toxicity of HSP90 inhibitors. This could be explained by the number of client proteins that are simultaneously chaperoned by HSP90, which obviously endangers critical physiological processes in normal cells. A strategic question emerges from this observation: should we continue searching for more specific, less toxic drugs? Or should we limit studies to fundamental science to uncover new biological mechanisms? Other HSPs are gaining interest in the scientific community. In particular, HSP110 is a newly discovered player in the field of cancer aggressiveness that controls the STAT3 pathway in colon cancers and in B-cell lymphomas. Our team recently identified the first HSP110 inhibitors (Table 1, Figure 5) that block interaction with STAT3 and that effectively limit colon cancer progression in mouse models; we plan to bring these to the clinic within the next few years. No toxicity has been demonstrated so far, and we believe this new type of inhibitor is promising for cancers whose poor prognosis is associated with HSP110/pYSTAT3 expression. Both intra- and extracellular HSP70, which belong to a related HSP family, favor STAT3/5 phosphorylation by JAK2. This dual action is of particular therapeutic interest since targeting HSP70 would simultaneously blunt the macrophage-mediated immunosuppression and block the intratumoral growth signal. To reach the goal of bringing HSP70 inhibitors to the clinic, we have identified peptide aptamers that bind to the peptide-binding domain of HSP70. This has not been an easy task since HSP70, contrary to HSP90, is not a "druggable" protein. Despite this limitation, our HSP70 inhibitors are specific and have proven effective in xenograft models of colon cancer. Of course, further studies will be needed before clinical trials can be initiated. Finally, HSP27 has recently been found to play new roles in myeloproliferative neoplasms. In this disease, which is driven by the $JAK2^{V617F}$/STAT5 pathway, HSP27 inhibition destabilizes the protein complex and limits disease related myelofibrosis. Furthermore, in contrast to HSP90 inhibitors, the HSP27 inhibitor does not induce the compensatory expression of other HSPs that usually account for resistance to treatment. Targeting HSP27 would therefore be an alternative to the failed HSP90 therapy, which also targeted $JAK2^{V617F}$. An oligonucleotide antisense of second generation (OGX427) is currently under clinical evaluation (Table 1).

Furthermore, small molecules targeting STAT3/5 have been identified as enhancing protein degradation. The inhibition of STAT pathways is therefore likely highly amenable to HSP inhibition and presents a potential synergistic therapeutic strategy (Table 1).

In conclusion, we show in this review that the STAT3/5 pathways rely on multiple HSPs under physiological and pathological conditions. Targeting various members of the HSP family, alone or in combination, will probably improve the inhibition of this central pathway and should foster the development of new, more specific and less toxic HSP inhibitors to complete the existing therapeutic arsenal.

Author Contributions: G.J., F.H. co-wrote the paper, and designed and drew all the figures for this manuscript. C.G. proofread the manuscript, gave suggestions, and modified the language. F.G. contributed to the editing and critical reading of the manuscript. All authors have read and approved the final manuscript. All authors have read and agreed to the published version of the manuscript.

Acknowledgments: The authors thank S. Rankin (Dijon University Hospital) for proofreading the manuscript. F. Hermetet was supported by a fellowship from the University of Burgundy and Franche-Comté through the ISITE-BFC program. C Garrido's team is labeled by La Ligue Nationale contre le Cancer.

References

1. Sorger, P.K. Heat shock factor and the heat shock response. *Cell* **1991**, *65*, 363–366. [CrossRef]
2. Welch, W.J. Heat shock proteins functioning as molecular chaperones: Their roles in normal and stressed cells. *Philos. Trans. R. Soc. Lond. B Biol. Sci.* **1993** *339*, 327–333. [PubMed]

3. Jego, G.; Hazoumé, A.; Seigneuric, R.; Garrido, C. Targeting heat shock proteins in cancer. *Cancer Lett.* **2013**, *332*, 275–285. [CrossRef] [PubMed]
4. Schmitt, E.; Gehrmann, M.; Brunet, M.; Multhoff, G.; Garrido, C. Intracellular and extracellular functions of heat shock proteins: Repercussions in cancer therapy. *J. Leukoc. Biol.* **2007**, *81*, 15–27. [CrossRef] [PubMed]
5. Hanahan, D.; Weinberg, R.A. Hallmarks of cancer: The next generation. *Cell* **2011**, *144*, 646–674. [CrossRef] [PubMed]
6. Calderwood, S.K.; Gong, J. Heat Shock Proteins Promote Cancer: It's a Protection Racket. *Trends BioChem. Sci.* **2016**, *41*, 311–323. [CrossRef] [PubMed]
7. Lianos, G.D.; Alexiou, G.A.; Mangano, A.; Mangano, A.; Rausei, S.; Boni, L.; Dionigi, G.; Roukos, D.H. The role of heat shock proteins in cancer. *Cancer Lett.* **2015**, *360*, 114–118. [CrossRef]
8. Wu, J.; Liu, T.; Rios, Z.; Mei, Q.; Lin, X.; Cao, S. Heat Shock Proteins and Cancer. *Trends Pharmacol. Sci.* **2017**, *38*, 226–256. [CrossRef]
9. Guttmann, D.M.; Koumenis, C. The heat shock proteins as targets for radiosensitization and chemosensitization in cancer. *Cancer Biol. Ther.* **2011**, *12*, 1023–1031. [CrossRef]
10. Chi, K.N.; Yu, E.Y.; Jacobs, C.; Bazov, J.; Kollmannsberger, C.; Higano, C.S.; Mukherjee, S.D.; Gleave, M.E.; Stewart, P.S.; Hotte, S.J. A phase I dose-escalation study of apatorsen (OGX-427), an antisense inhibitor targeting heat shock protein 27 (Hsp27), in patients with castration-resistant prostate cancer and oTher. advanced cancers. *Ann. Oncol.* **2016**, *27*, 1116–1122. [CrossRef]
11. Trepel, J.; Mollapour, M.; Giaccone, G.; Neckers, L. Targeting the dynamic HSP90 complex in cancer. *Nat. Rev. Cancer* **2010**, *10*, 537–549. [CrossRef] [PubMed]
12. Yuno, A.; Lee, M.J.; Lee, S.; Tomita, Y.; Rekhtman, D.; Moore, B.; Trepel, J.B. Clinical Evaluation and Biomarker Profiling of Hsp90 Inhibitors. *Methods Mol. Biol.* **2018**, *1709*, 423–441. [PubMed]
13. Rocchi, P.; So, A.; Kojima, S.; Signaevsky, M.; Beraldi, E.; Fazli, L.; Hurtado-Coll, A.; Yamanaka, K.; Gleave, M. Heat shock protein 27 increases after androgen ablation and plays a cytoprotective role in hormone-refractory prostate cancer. *Cancer Res.* **2004**, *64*, 6595–6602. [CrossRef] [PubMed]
14. Song, T.F.; Zhang, Z.F.; Liu, L.; Yang, T.; Jiang, J.; Li, P. Small interfering RNA-mediated silencing of heat shock protein 27 (HSP27) Increases chemosensitivity to paclitaxel by increasing production of reactive oxygen species in human ovarian cancer cells (HO8910). *J. Int. Med. Res.* **2009**, *37*, 1375–1388. [CrossRef]
15. Baylot, V.; Andrieu, C.; Katsogiannou, M.; Taieb, D.; Garcia, S.; Giusianom, S.; Acunzo, J.; Iovanna, J.; Gleave, M.; Garrido, C.; et al. OGX-427 inhibits tumor progression and enhances gemcitabine chemotherapy in pancreatic cancer. *Cell Death Dis.* **2011**, *2*, e221. [CrossRef]
16. Lelj-Garolla, B.; Kumano, M.; Beraldi, E.; Nappi, L.; Rocchi, P.; Ionescu, D.N.; Fazlim, L.; Zoubeidi, A.; Gleave, M.E. Hsp27 Inhibition with OGX-427 Sensitizes Non-Small Cell Lung Cancer Cells to Erlotinib and Chemotherapy. *Mol. Cancer Ther.* **2015**, *14*, 1107–1116. [CrossRef]
17. Spigel, D.R.; Shipley, D.L.; Waterhouse, D.M.; Jones, S.F.; Ward, P.J.; Shih, K.C.; Hemphill, B.; McCleod, M.; Whorf, R.C.; Page, R.D.; et al. A Randomized Double-Blinded Phase II Trial of Carboplatin and Pemetrexed with or without Apatorsen (OGX-427) in Patients with Previously Untreated Stage IV Non-Squamous-Non-Small-Cell Lung Cancer: The SPRUCE Trial. *Oncologist* **2019**, *24*, e1409–e1416. [CrossRef]
18. Xia, Y.; Liu, Y.; Wan, J.; Wang, M.; Rocchi, P.; Qu, F.; Iovanna, J.L.; Peng, L. Novel triazole ribonucleoside down-regulates heat shock protein 27 and induces potent anticancer activity on drug-resistant pancreatic cancer. *J. Med. Chem.* **2009**, *52*, 6083–6096. [CrossRef]
19. Chauhan, D.; Li, G.; Shringarpure, R.; Podar, K.; Ohtake, Y.; Hideshima, T.; Anderson, K.C. Blockade of Hsp27 overcomes Bortezomib/proteasome inhibitor PS-341 resistance in lymphoma cells. *Cancer Res.* **2003**, *63*, 6174–6177.
20. Heinrich, J.C.; Tuukkanen, A.; Schroeder, M.; Fahrig, T.; Fahrig, R. RP101 (brivudine) binds to heat shock protein HSP27 (HSPB1) and enhances survival in animals and pancreatic cancer patients. *J. Cancer Res. Clin. Oncol.* **2011**, *137*, 1349–1361. [CrossRef]

21. Kaiser, M.; Kühnl, A.; Reins, J.; Fischer, S.; Ortiz-Tanchez, J.; Schlee, C.; Mochmann, L.H.; Heesch, S.; Benlasfer, O.; Hofmann, W.K.; et al. Antileukemic activity of the HSP70 inhibitor pifithrin-μ in acute leukemia. *Blood Cancer J.* **2011**, *1*, e28. [CrossRef] [PubMed]
22. Goloudina, A.R.; Demidov, O.N.; Garrido, C. Inhibition of HSP70: A challenging anti-cancer strategy. *Cancer Lett.* **2012**, *325*, 117–124. [CrossRef] [PubMed]
23. Massey, A.J.; Williamson, D.S.; Browne, H.; Murray, J.B.; Dokurno, P.; Shaw, T.; Macias, A.T.; Daniels, Z.; Geoffroy, S.; Dopson, M.; et al. A novel small molecule inhibitor of Hsc70/Hsp70 potentiates Hsp90 inhibitor induced apoptosis in HCT116 colon carcinoma cells. *Cancer ChemoTher. Pharmacol.* **2010**, *66*, 535–545. [CrossRef] [PubMed]
24. Brünnert, D.; Langer, C.; Zimmermann, L.; Bargou, R.C.; Burchardt, M.; Chatterjee, M.; Stope, M.B. The heat shock protein 70 inhibitor VER155008 suppresses the expression of HSP27, HOP and HSP90β and the androgen receptor, induces apoptosis, and attenuates prostate cancer cell growth. *J. Cell BioChem.* **2020**, *121*, 407–417. [CrossRef] [PubMed]
25. Chatterjee, M.; Andrulis, M.; Stühmer, T.; Müller, E.; Hofmann, C.; Steinbrunn, T.; Heimberger, T.; Schraud, H.; Kressmann, S.; Einsele, H.; et al. The PI3K/Akt signaling pathway regulates the expression of Hsp70, which critically contributes to Hsp90-chaperone function and tumor cell survival in multiple myeloma. *Haematologica* **2013**, *98*, 1132–1141. [CrossRef] [PubMed]
26. Rérole, A.L.; Gobbo, J.; De Thonel, A.; Schmitt, E.; Pais de Barros, J.P.; Hammann, A.; Lanneau, D.; Fourmaux, E.; Demidov, O.N.; Micheau, O.; et al. Peptides and aptamers targeting HSP70: A novel approach for anticancer chemotherapy. *Cancer Res.* **2011**, *71*, 484–495. [CrossRef]
27. Schmitt, E.; Maingret, L.; Puig, P.E.; Rerole, A.L.; Ghiringhelli, F.; Hammann, A.; Solary, E.; Kroemer, G.; Garrido, C. Heat shock protein 70 neutralization exerts potent antitumor effects in animal models of colon cancer and melanoma. *Cancer Res.* **2006**, *66*, 4191–4197. [CrossRef]
28. Stangl, S.; Gehrmann, M.; Riegger, J.; Kuhs, K.; Riederer, I.; Sievert, W.; Hube, K.; Mocikat, R.; Dressel, R.; Kremmer, E.; et al. Targeting membrane heat-shock protein 70 (Hsp70) on tumors by cmHsp70.1 antibody. *Proc. Natl. Acad. Sci. USA* **2011**, *108*, 733–738. [CrossRef]
29. Krause, S.W.; Gastpar, R.; Andreesen, R.; Gross, C.; Ullrich, H.; Thonigs, G.; Pfister, K.; Multhoff, G. Treatment of colon and lung cancer patients with ex vivo heat shock protein 70-peptide-activated, autologous natural killer cells: A clinical phase i trial. *Clin. Cancer Res.* **2004**, *10*, 3699–3707. [CrossRef]
30. Shiotsu, Y.; Neckers, L.M.; Wortman, I.; An, W.G.; Schulte, T.W.; Soga, S.; Murakata, C.; Tamaoki, T.; Akinaga, S. Novel oxime derivatives of radicicol induce erythroid differentiation associated with preferential G (1) phase accumulation against chronic myelogenous leukemia cells through destabilization of Bcr-Abl with Hsp90 complex. *Blood* **2000**, *96*, 2284–2291. [CrossRef]
31. Palacios, C.; López-Pérez, A.I.; López-Rivas, A. Down-regulation of RIP expression by 17-dimethylaminoethylamino-17-demethoxygeldanamycin promotes TRAIL-induced apoptosis in breast tumor cells. *Cancer Lett.* **2010**, *287*, 207–215. [CrossRef] [PubMed]
32. Schaefer, S.; Svenstrup, T.H.; Guerra, B. The small-molecule kinase inhibitor D11 counteracts 17-AAG-mediated up-regulation of HSP70 in brain cancer cells. *PLoS ONE* **2017**, *12*, e0177706. [CrossRef] [PubMed]
33. Ayrault, O.; Godeny, M.D.; Dillon, C.; Zindy, F.; Fitzgerald, P.; Roussel, M.F.; Beere, H.M. Inhibition of Hsp90 via 17-DMAG induces apoptosis in a p53-dependent manner to prevent medulloblastoma. *Proc. Natl. Acad. Sci. USA* **2009**, *106*, 17037–17042. [CrossRef] [PubMed]
34. Chen, T.L.; Gupta, N.; Lehman, A.; Ruppert, A.S.; Yu, L.; Oakes, C.C.; Claus, R.; Plass, C.; Maddocks, K.J.; Andritsos, L.; et al. Hsp90 inhibition increases SOCS3 transcript and regulates migration and cell death in chronic lymphocytic leukemia. *Oncotarget* **2016**, *7*, 28684–28696. [CrossRef]
35. Hanson, B.E.; Vesole, D.H. Retaspimycin hydrochloride (IPI-504): A novel heat shock protein inhibitor as an anticancer agent. *Expert Opin. Investig. Drugs* **2009**, *18*, 1375–1383. [CrossRef]
36. Song, D.; Chaerkady, R.; Tan, A.C.; García-García, E.; Nalli, A.; Suárez-Gauthier, A.; López-Ríos, F.; Zhang, X.F.; Solomon, A.; Tong, J.; et al. Antitumor activity and molecular effects of the novel heat shock protein 90 inhibitor, IPI-504, in pancreatic cancer. *Mol. Cancer Ther.* **2008**, *7*, 3275–3284. [CrossRef]

37. Wagner, A.J.; Chugh, R.; Rosen, L.S.; Morgan, J.A.; George, S.; Gordon, M.; Dunbar, J.; Normant, E.; Grayzel, D.; Demetri, G.D.; et al. A phase I study of the HSP90 inhibitor retaspimycin hydrochloride (IPI-504) in patients with gastrointestinal stromal tumors or soft-tissue sarcomas. *Clin. Cancer Res.* **2013**, *19*, 6020–6029. [CrossRef]
38. Scaltriti, M.; Serra, V.; Normant, E.; Guzman, M.; Rodriguez, O.; Lim, A.R.; Slocum, K.L.; West, K.A.; Rodriguez, V.; Prudkin, L.; et al. Antitumor activity of the Hsp90 inhibitor IPI-504 in HER2-positive trastuzumab-resistant breast cancer. *Mol. Cancer Ther.* **2011**, *10*, 817–824. [CrossRef]
39. Floris, G.; Debiec-Rychter, M.; Wozniak, A.; Stefan, C.; Normant, E.; Faa, G.; Machiels, K.; Vanleeuw, U.; Sciot, R.; Schöffski, P. The heat shock protein 90 inhibitor IPI-504 induces KIT degradation, tumor shrinkage, and cell proliferation arrest in xenograft models of gastrointestinal stromal tumors. *Mol. Cancer Ther.* **2011**, *10*, 1897–1908. [CrossRef]
40. Normant, E.; Paez, G.; West, K.A.; Lim, A.R.; Slocum, K.L.; Tunkey, C.; McDougall, J.; Wylie, A.A.; Robison, K.; Caliri, K.; et al. The Hsp90 inhibitor IPI-504 rapidly lowers EML4-ALK levels and induces tumor regression in ALK-driven NSCLC models. *Oncogene* **2011**, *30*, 2581–2586. [CrossRef]
41. Hendriks, L.E.L.; Dingemans, A.C. Heat shock protein antagonists in early stage clinical trials for NSCLC. *Expert Opin. Investig. Drugs* **2017**, *26*, 541–550. [CrossRef] [PubMed]
42. Wu, L.X.; Xu, J.H.; Zhang, K.Z.; Lin, Q.; Huang, X.W.; Wen, C.X.; Chen, Y.Z. Disruption of the Bcr-Abl/Hsp90 protein complex: A possible mechanism to inhibit Bcr-Abl-positive human leukemic blasts by novobiocin. *Leukemia* **2008**, *22*, 1402–1409. [CrossRef] [PubMed]
43. Shelton, S.N.; Shawgo, M.E.; Matthews, S.B.; Lu, Y.; Donnelly, A.C.; Szabla, K.; Tanol, M.; Vielhauer, G.A.; Rajewski, R.A.; Matts, R.L.; et al. KU135, a novel novobiocin-derived C-terminal inhibitor of the 90-kDa heat shock protein, exerts potent antiproliferative effects in human leukemic cells. *Mol. Pharmacol.* **2009**, *76*, 1314–1322. [CrossRef] [PubMed]
44. Matthews, S.B.; Vielhauer, G.A.; Manthe, C.A.; Chaguturu, V.K.; Szabla, K.; Matts, R.L.; Donnelly, A.C.; Blagg, B.S.; Holzbeierlein, J.M. Characterization of a novel novobiocin analogue as a putative C-terminal inhibitor of heat shock protein 90 in prostate cancer cells. *Prostate* **2010**, *70*, 27–36. [CrossRef]
45. Le, H.T.; Vielhauer, G.A.; Manthe, C.A.; Chaguturu, V.K.; Szabla, K.; Matts, R.L.; Donnelly, A.C.; Blagg, B.S.; Holzbeierlein, J.M. Panaxynol, a natural Hsp90 inhibitor, effectively targets both lung cancer stem and non-stem cells. *Cancer Lett.* **2018**, *412*, 297–307. [CrossRef]
46. Lin, S.F.; Lin, J.D.; Hsueh, C.; Chou, T.C.; Yeh, C.N.; Chen, M.H.; Wong, R.J. Efficacy of an HSP90 inhibitor, ganetespib, in preclinical thyroid cancer models. *Oncotarget* **2017**, *8*, 41294–41304. [CrossRef]
47. Lee, H.; Saini, N.; Howard, E.W.; Parris, A.B.; Ma, Z.; Zhao, Q.; Zhao, M.; Liu, B.; Edgerton, S.M.; Thor, A.D.; et al. Ganetespib targets multiple levels of the receptor tyrosine kinase signaling cascade and preferentially inhibits ErbB2-overexpressing breast cancer cells. *Sci. Rep.* **2018**, *8*, 6829. [CrossRef]
48. He, W.; Hu, H. BIIB021, an Hsp90 inhibitor: A promising therapeutic strategy for blood malignancies (Review). *Oncol. Rep.* **2018**, *40*, 3–15. [CrossRef]
49. Marubayashi, S.; Koppikar, P.; Taldone, T.; Abdel-Wahab, O.; West, N.; Bhagwat, N.; Caldas-Lopes, E.; Ross, K.N.; Gönen, M.; Gozman, A.; et al. HSP90 is a therapeutic target in JAK2-dependent myeloproliferative neoplasms in mice and humans. *J. Clin. Investig.* **2010**, *120*, 3578–3593. [CrossRef]
50. Fiskus, W.; Verstovsek, S.; Manshouri, T.; Rao, R.; Balusu, R.; Venkannagari, S.; Rao, N.N.; Ha, K.; Smith, J.E.; Hembruff, S.L.; et al. Heat shock protein 90 inhibitor is synergistic with JAK2 inhibitor and overcomes resistance to JAK2-TKI in human myeloproliferative neoplasm cells. *Clin. Cancer Res.* **2011**, *17*, 7347–7358. [CrossRef]
51. Park, K.S.; Hong, Y.S.; Choi, J.; Yoon, S.; Kang, J.; Kim, D.; Lee, K.P.; Im, H.S.; Lee, C.H.; Seo, S.; et al. HSP90 inhibitor, AUY922, debilitates intrinsic and acquired lapatinib-resistant HER2-positive gastric cancer cells. *BMB Rep.* **2018**, *51*, 660–665. [CrossRef] [PubMed]
52. Yang, H.; Lee, M.H.; Park, I.; Jeon, H.; Choi, J.; Seo, S.; Kim, S.W.; Koh, G.Y.; Park, K.S.; Lee, D.H. HSP90 inhibitor (NVP-AUY922) enhances the anti-cancer effect of BCL-2 inhibitor (ABT-737) in small cell lung cancer expressing BCL-2. *Cancer Lett.* **2017**, *411*, 19–26. [CrossRef] [PubMed]
53. Kim, S.H.; Kang, J.G.; Kim, C.S.; Ihm, S.H.; Choi, M.G.; Yoo, H.J.; Lee, S.J. The dipeptidyl peptidase-IV inhibitor gemigliptin alone or in combination with NVP-AUY922 has a cytotoxic activity in thyroid carcinoma cells. *Tumour. Biol.* **2017**, *39*, 1010428317722068. [CrossRef] [PubMed]

54. Brough, P.A.; Aherne, W.; Barril, X.; Borgognoni, J.; Boxall, K.; Cansfield, J.E.; Cheung, K.M.; Collins, I.; Davies, N.G.; Drysdale, M.J.; et al. 4,5-diarylisoxazole Hsp90 chaperone inhibitors: Potential therapeutic agents for the treatment of cancer. *J. Med. Chem.* **2008**, *51*, 196–218. [CrossRef] [PubMed]
55. Liu, J.; Sun, W.; Dong, W.; Wang, Z.; Qin, Y.; Zhang, T.; Zhang, H. HSP90 inhibitor NVP-AUY922 induces cell apoptosis by disruption of the survivin in papillary thyroid carcinoma cells. *Biochem. Biophys. Res. Commun.* **2017**, *487*, 313–319. [CrossRef]
56. Tao, W.; Chakraborty, S.N.; Leng, X.; Ma, H.; Arlinghaus, R.B. HSP90 inhibitor AUY922 induces cell death by disruption of the Bcr-Abl, Jak2 and HSP90 signaling network complex in leukemia cells. *Genes Cancer* **2015**, *6*, 19–29.
57. Chakraborty, S.N.; Leng, X.; Perazzona, B.; Sun, X.; Lin, Y.H.; Arlinghaus, R.B. Combination of JAK2 and HSP90 inhibitors: An effective therapeutic option in drug-resistant chronic myelogenous leukemia. *Genes Cancer* **2016**, *7*, 201–208.
58. Hobbs, G.S.; Hanasoge Somasundara, A.V.; Kleppe, M.; Litvin, R.; Arcila, M.; Ahn, J.; McKenney, A.S.; Knapp, K.; Ptashkin, R.; Weinstein, H.; et al. Hsp90 inhibition disrupts JAK-STAT signaling and leads to reductions in splenomegaly in patients with myeloproliferative neoplasms. *Haematologica* **2018**, *103*, e5–e9. [CrossRef]
59. Johnson, M.L.; Yu, H.A.; Hart, E.M.; Weitner, B.B.; Rademaker, A.W.; Patel, J.D.; Kris, M.G.; Riely, G.J. Phase I/II Study of HSP90 Inhibitor AUY922 and Erlotinib for EGFR-Mutant Lung Cancer with Acquired Resistance to Epidermal Growth Factor Receptor Tyrosine Kinase Inhibitors. *J. Clin. Oncol.* **2015**, *33*, 1666–1673. [CrossRef]
60. Weigert, O.; Lane, A.A.; Bird, L.; Kopp, N.; Chapuy, B.; van Bodegom, D.; Toms, A.V.; Marubayashi, S.; Christie, A.L.; McKeown, M.; et al. Genetic resistance to JAK2 enzymatic inhibitors is overcome by HSP90 inhibition. *J. Exp. Med.* **2012**, *209*, 259–273. [CrossRef]
61. Mahendrarajah, N.; Borisova, M.E.; Reichardt, S.; Godmann, M.; Sellmer, A.; Mahboobi, S.; Haitel, A.; Schmid, K.; Kenner, L.; Heinzel, T.; et al. HSP90 is necessary for the ACK1-dependent phosphorylation of STAT1 and STAT3. *Cell Signal.* **2017**, *39*, 9–17. [CrossRef] [PubMed]
62. Smyth, T.; Paraiso, K.H.T.; Hearn, K.; Rodriguez-Lopez, A.M.; Munck, J.M.; Haarberg, H.E.; Sondak, V.K.; Thompson, N.T.; Azab, M.; Lyons, J.F.; et al. Inhibition of HSP90 by AT13387 delays the emergence of resistance to BRAF inhibitors and overcomes resistance to dual BRAF and MEK inhibition in melanoma models. *Mol. Cancer Ther.* **2014**, *13*, 2793–2804. [CrossRef] [PubMed]
63. Graham, B.; Curry, J.; Smyth, T.; Fazal, L.; Feltell, R.; Harada, I.; Coyle, J.; Williams, B.; Reule, M.; Angove, H.; et al. The heat shock protein 90 inhibitor, AT13387, displays a long duration of action in vitro and in vivo in non-small cell lung cancer. *Cancer Sci.* **2012**, *103*, 522–527. [CrossRef] [PubMed]
64. Courtin, A.; Smyth, T.; Hearn, K.; Saini, H.K.; Thompson, N.T.; Lyons, J.F.; Wallis, N.G. Emergence of resistance to tyrosine kinase inhibitors in non-small-cell lung cancer can be delayed by an upFront combination with the HSP90 inhibitor onalespib. *Br. J. Cancer* **2016**, *115*, 1069–1077. [CrossRef] [PubMed]
65. Eroglu, Z.; Chen, Y.A.; Gibney, G.T.; Weber, J.S.; Kudchadkar, R.R.; Khushalani, N.I.; Markowitz, J.; Brohl, A.S.; Tetteh, L.F.; Ramadan, H.; et al. Combined BRAF and HSP90 Inhibition in Patients with Unresectable. *Clin. Cancer Res.* **2018**, *24*, 5516–5524. [CrossRef] [PubMed]
66. Friedman, J.A.; Wise, S.C.; Hu, M.; Gouveia, C.; Vander Broek, R.; Freudlsperger, C.; Kannabiran, V.R.; Arun, P.; Mitchell, J.B.; Chen, Z.; et al. HSP90 Inhibitor SNX5422/2112 Targets the Dysregulated Signal and Transcription Factor Network and Malignant Phenotype of Head and Neck Squamous Cell Carcinoma. *Transl. Oncol.* **2013**, *6*, 429–441. [CrossRef] [PubMed]
67. Rice, J.W.; Veal, J.M.; Barabasz, A.; Foley, B.; Fadden, P.; Scott, A.; Huang, K.; Steed, P.; Hall, S. Targeting of multiple signaling pathways by the Hsp90 inhibitor SNX-2112 in EGFR resistance models as a single agent or in combination with erlotinib. *Oncol. Res.* **2009**, *18*, 229–242. [CrossRef]
68. Smith, D.L.; Acquaviva, J.; Sequeira, M.; Jimenez, J.P.; Zhang, C.; Sang, J.; Bates, R.C.; Proia, D.A. The HSP90 inhibitor ganetespib potentiates the antitumor activity of EGFR tyrosine kinase inhibition in mutant and wild-type non-small cell lung cancer. *Target. Oncol.* **2015**, *10*, 235–245. [CrossRef]
69. Moulick, K.; Ahn, J.H.; Zong, H.; Rodina, A.; Cerchietti, L.; Gomes DaGama, E.M.; Caldas-Lopes, E.; Beebe, K.; Perna, F.; Hatzi, K.; et al. Affinity-based proteomics reveal cancer-specific networks coordinated by Hsp90. *Nat. Chem. Biol.* **2011**, *7*, 818–826. [CrossRef]

70. Hong, Y.S.; Jang, W.J.; Chun, K.S.; Jeong, C.H. Hsp90 inhibition by WK88-1 potently suppresses the growth of gefitinib-resistant H1975 cells harboring the T790M mutation in EGFR. *Oncol. Rep.* **2014**, *31*, 2619–2624. [CrossRef]
71. Kobayashi, N.; Toyooka, S.; Soh, J.; Yamamoto, H.; Dote, H.; Kawasaki, K.; Otani, H.; Kubo, T.; Jida, M.; Ueno, T.; et al. The anti-proliferative effect of heat shock protein 90 inhibitor, 17-DMAG, on non-small-cell lung cancers being resistant to EGFR tyrosine kinase inhibitor. *Lung Cancer* **2012**, *75*, 161–166. [CrossRef] [PubMed]
72. Gozzi, G.J.; Gonzalez, D.; Boudesco, C.; Dias, A.M.M.; Gotthard, G.; Uyanik, B.; Dondaine, L.; Marcion, G.; Hermetet, F.; Denis, C.; et al. Selecting the first chemical molecule inhibitor of HSP110 for colorectal cancer therapy. *Cell Death Differ.* **2019**. [CrossRef] [PubMed]
73. McClellan, A.J.; Xia, Y.; Deutschbauer, A.M.; Davis, R.W.; Gerstein, M.; Frydman, J. Diverse cellular functions of the Hsp90 molecular chaperone uncovered using systems approaches. *Cell* **2007**, *131*, 121–135. [CrossRef] [PubMed]
74. Taipale, M.; Krykbaeva, I.; Koeva, M.; Kayatekin, C.; Westover, K.D.; Karras, G.I.; Lindquist, S. Quantitative analysis of HSP90-client interactions reveals principles of substrate recognition. *Cell* **2012**, *150*, 987–1001. [CrossRef]
75. Makhnevych, T.; Houry, W.A. The role of Hsp90 in protein complex assembly. *Biochim. Biophys. Acta* **2012**, *1823*, 674–682. [CrossRef] [PubMed]
76. Boczek, E.E.; Reefschläger, L.G.; Dehling, M.; Struller, T.J.; Häusler, E.; Seidl, A.; Kaila, V.R.; Buchner, J. Conformational processing of oncogenic v-Src kinase by the molecular chaperone Hsp90. *Proc. Natl. Acad. Sci. USA* **2015**, *112*, E3189–E3198. [CrossRef] [PubMed]
77. Gray, P.J.; Prince, T.; Chengm, J.; Stevenson, M.A.; Calderwood, S.K. Targeting the oncogene and kinome chaperone CDC37. *Nat. Rev. Cancer* **2008**, *8*, 491–495. [CrossRef]
78. Shang, L.; Tomasi, T.B. The heat shock protein 90-CDC37 chaperone complex is required for signaling by types I and II interferons. *J. Biol. Chem.* **2006**, *281*, 1876–1884. [CrossRef]
79. Peeters, P.; Wlodarska, I.; Baens, M.; Criel, A.; Selleslag, D.; Hagemeijer, A.; Van den Berghe, H.; Marynen, P. Fusion of ETV6 to MDS1/EVI1 as a result of t(3;12)(q26;p13) in myeloproliferative disorders. *Cancer Res.* **1997**, *57*, 564–569.
80. Lacronique, V.; Boureux, A.; Valle, V.D.; Poirel, H.; Quang, C.T.; Mauchauffé, M.; Berthou, C.; Lessard, M.; Berger, R.; Ghysdael, J.; et al. A TEL-JAK2 fusion protein with constitutive kinase activity in human leukemia. *Science* **1997**, *278*, 1309–1312. [CrossRef]
81. Baxter, E.J.; Scott, L.M.; Campbell, P.J.; East, C.; Fourouclas, N.; Swanton, S.; Vassiliou, G.S.; Bench, A.J.; Boyd, E.M.; Curtin, N.; et al. Acquired mutation of the tyrosine kinase JAK2 in human myeloproliferative disorders. *Lancet* **2005**, *365*, 1054–1061. [CrossRef]
82. Levine, R.L.; Loriaux, M.; Huntly, B.J.; Loh, M.L.; Beran, M.; Stoffregen, E.; Berger, R.; Clark, J.J.; Willis, S.G.; Nguyen, K.T.; et al. The JAK2V617F activating mutation occurs in chronic myelomonocytic leukemia and acute myeloid Leukemia but not in acute lymphoblastic leukemia or chronic lymphocytic leukemia. *Blood* **2005**, *106*, 3377–3379. [CrossRef] [PubMed]
83. James, C.; Ugo, V.; Le Couédic, J.P.; Staerk, J.; Delhommeau, F.; Lacout, C.; Garçon, L.; Raslova, H.; Berger, R.; Bennaceur-Griscelli, A.; et al. A unique clonal JAK2 mutation leading to constitutive signalling causes polycythaemia vera. *Nature* **2005**, *434*, 1144–1148. [CrossRef] [PubMed]
84. Jones, A.V.; Kreil, S.; Zoi, K.; Waghorn, K.; Curtis, C.; Zhang, L.; Score, J.; Seear, R.; Chase, A.J.; Grand, F.H.; et al. Widespread occurrence of the JAK2 V617F mutation in chronic myeloproliferative disorders. *Blood* **2005**, *106*, 2162–2168. [CrossRef]
85. Bareng, J.; Jilani, I.; Gorre, M.; Kantarjian, H.; Giles, F.; Hannah, A.; Albitar, M. A potential role for HSP90 inhibitors in the treatment of JAK2 mutant-positive diseases as demonstrated using quantitative flow cytometry. *Leuk. Lymphoma* **2007**, *48*, 2189–2195. [CrossRef]
86. Sen, B.; Johnson, F.M. Regulation of SRC family kinases in human cancers. *J. Signal. Transduct* **2011**, *2011*, 865819. [CrossRef]
87. Garcia, R.; Bowman, T.L.; Niu, G.; Yu, H.; Minton, S.; Muro-Cacho, C.A.; Cox, C.E.; Falcone, R.; Fairclough, R.; Parsons, S.; et al. Constitutive activation of Stat3 by the Src and JAK tyrosine kinases participates in growth regulation of human breast carcinoma cells. *Oncogene* **2001**, *20*, 2499–2513. [CrossRef]

88. Frame, M.C. Src in cancer: Deregulation and consequences for cell behaviour. *Biochim. Biophys. Acta* **2002**, *1602*, 114–130. [CrossRef]
89. Takeya, T.; Hanafusa, H. Structure and sequence of the cellular gene homologous to the RSV src gene and the mechanism for generating the transforming virus. *Cell* **1983**, *32*, 881–890. [CrossRef]
90. Stehelin, D.; Varmus, H.E.; Bishop, J.M.; Vogt, P.K. DNA related to the transforming gene(s) of avian sarcoma viruses is present in normal avian DNA. *Nature* **1976**, *260*, 170–173. [CrossRef]
91. Hunter, T.; Sefton, B.M. Transforming gene product of Rous sarcoma virus phosphorylates tyrosine. *Proc. Natl. Acad. Sci. USA* **1980**, *77*, 1311–1315. [CrossRef]
92. Xu, Y.; Lindquist, S. Heat-shock protein hsp90 governs the activity of pp60v-src kinase. *Proc. Natl. Acad. Sci. USA* **1993**, *90*, 7074–7078. [CrossRef]
93. Whitesell, L.; Mimnaugh, E.G.; De Costa, B.; Myers, C.E.; Neckers, L.M. Inhibition of heat shock protein HSP90-pp60v-src heteroprotein complex formation by benzoquinone ansamycins: Essential role for stress proteins in oncogenic transformation. *Proc. Natl. Acad. Sci. USA* **1994**, *91*, 8324–8328. [CrossRef]
94. Nathan, D.F.; Lindquist, S. Mutational analysis of Hsp90 function: Interactions with a steroid receptor and a protein kinase. *Mol. Cell Biol.* **1995**, *15*, 3917–3925. [CrossRef]
95. Luo, Q.; Boczek, E.E.; Wang, Q.; Buchner, J.; Kaila, V.R. Hsp90 dependence of a kinase is determined by its conformational landscape. *Sci. Rep.* **2017**, *7*, 43996. [CrossRef]
96. Verba, K.A.; Wang, R.Y.; Arakawa, A.; Liu, Y.; Shirouzu, M.; Yokoyama, S.; Agard, D.A. Atomic structure of Hsp90-Cdc37-Cdk4 reveals that Hsp90 traps and stabilizes an unfolded kinase. *Science* **2016**, *352*, 1542–1547. [CrossRef]
97. Giannini, A.; Bijlmakers, M.J. Regulation of the Src family kinase Lck by Hsp90 and ubiquitination. *Mol. Cell Biol.* **2004**, *24*, 5667–5676. [CrossRef]
98. Scholz, G.M.; Hartson, S.D.; Cartledge, K.; Volk, L.; Matts, R.L.; Dunn, A.R. The molecular chaperone Hsp90 is required for signal transduction by wild-type Hck and maintenance of its constitutively active counterpart. *Cell Growth Differ.* **2001**, *12*, 409–417.
99. Mahajan, N.P.; Whang, Y.E.; Mohler, J.L.; Earp, H.S. Activated tyrosine kinase Ack1 promotes prostate tumorigenesis: Role of Ack1 in polyubiquitination of tumor suppressor Wwox. *Cancer Res.* **2005**, *65*, 10514–10523. [CrossRef]
100. Karachaliou, N.; Pilotto, S.; Teixidó, C.; Viteri, S.; González-Cao, M.; Riso, A.; Morales-Espinosa, D.; Molina, M.A.; Chaib, I.; Santarpia, M.; et al. Melanoma: Oncogenic drivers and the immune system. *Ann. Transl. Med.* **2015**, *3*, 265.
101. Da Rocha Dias, S.; Friedlos, F.; Light, Y.; Springer, C.; Workman, P.; Marais, R. Activated B-RAF is an Hsp90 client protein that is targeted by the anticancer drug 17-allylamino-17-demethoxygeldanamycin. *Cancer Res.* **2005**, *65*, 10686–10691. [CrossRef]
102. Jabbour, E.; Kantarjian, H. Chronic myeloid leukemia: 2018 update on diagnosis, therapy and monitoring. *Am. J. Hematol.* **2018**, *93*, 442–459. [CrossRef]
103. Coppo, P.; Friedlos, F.; Light, Y.; Springer, C.; Workman, P.; Marais, R. BCR-ABL activates STAT3 via JAK and MEK pathways in human cells. *Br. J. Haematol.* **2006**, *134*, 171–179. [CrossRef]
104. Nair, R.R.; Tolentino, J.H.; Hazlehurst, L.A. Role of STAT3 in Transformation and Drug Resistance in CML. *Front. Oncol.* **2012**, *2*, 30. [CrossRef]
105. De Groot, R.P.; Raaijmakers, J.A.; Lammers, J.W.; Jove, R.; Koenderman, L. STAT5 activation by BCR-Abl contributes to transformation of K562 leukemia cells. *Blood* **1999**, *94*, 1108–1112. [CrossRef]
106. Sillaber, C.; Gesbert, F.; Frank, D.A.; Sattler, M.; Griffin, J.D. STAT5 activation contributes to growth and viability in Bcr/Abl-transformed cells. *Blood* **2000**, *95*, 2118–2125. [CrossRef]
107. Hoelbl, A.; Schuster, C.; Kovacic, B.; Zhu, B.; Wickre, M.; Hoelzl, M.A.; Fajmann, S.; Grebien, F.; Warsch, W.; Stengl, G.; et al. Stat5 is indispensable for the maintenance of bcr/abl-positive leukaemia. *EMBO Mol. Med.* **2010**, *2*, 98–110. [CrossRef]
108. Esfahani, K.; Cohen, V. HSP90 as a novel molecular target in non-small-cell lung cancer. *Lung Cancer (Auckl)* **2016**, *7*, 11–17.
109. Lemmon, M.A.; Schlessinger, J. Cell Signaling by receptor tyrosine kinases. *Cell* **2010**, *141*, 1117–1134. [CrossRef]

110. Kosack, L.; Wingelhofer, B.; Popa, A.; Orlova, A.; Agerer, B.; Vilagos, B.; Majek, P.; Parapatics, K.; Lercher, A.; Ringler, A.; et al. The ERBB-STAT3 Axis Drives Tasmanian Devil Facial Tumor Disease. *Cancer Cell* **2019**, *35*, 125–139.e9. [CrossRef]
111. Xu, W.; Mimnaugh, E.G.; Kim, J.S.; Trepel, J.B.; Neckers, L.M. Hsp90, not Grp94, regulates the intracellular trafficking and stability of nascent ErbB2. *Cell Stress Chaperones* **2002**, *7*, 91–96. [CrossRef]
112. Peng, X.; Guo, X.; Borkan, S.C.; Bharti, A.; Kuramochi, Y.; Calderwood, S.; Sawyer, D.B. Heat shock protein 90 stabilization of ErbB2 expression is disrupted by ATP depletion in myocytes. *J. Biol. Chem.* **2005**, *280*, 13148–13152. [CrossRef] [PubMed]
113. Wee, P.; Wang, Z. Epidermal Growth Factor Receptor Cell Proliferation Signaling Pathways. *Cancers (Basel)* **2017**, *9*, 52. [CrossRef]
114. Bao, R.; Lai, C.J.; Wang, D.G.; Qu, H.; Yin, L.; Zifcak, B.; Tao, X.; Wang, J.; Atoyan, R.; Samson, M.; et al. Targeting heat shock protein 90 with CUDC-305 overcomes erlotinib resistance in non-small cell lung cancer. *Mol. Cancer Ther.* **2009**, *8*, 3296–3306. [CrossRef]
115. Ono, N.; Yamazaki, T.; Tsukaguchi, T.; Fujii, T.; Sakata, K.; Suda, A.; Tsukuda, T.; Mio, T.; Ishii, N.; Kondoh, O.; et al. Enhanced antitumor activity of erlotinib in combination with the Hsp90 inhibitor CH5164840 against non-small-cell lung cancer. *Cancer Sci.* **2013**, *104*, 1346–1352. [CrossRef]
116. Sato, N.; Yamamoto, T.; Sekine, Y.; Yumioka, T.; Junicho, A.; Fuse, H.; Matsuda, T. Involvement of heat-shock protein 90 in the interleukin-6-mediated signaling pathway through STAT3. *Biochem. Biophys. Res. Commun.* **2003**, *300*, 847–852. [CrossRef]
117. Song, S.; Su, Z.; Xu, H.; Niu, M.; Chen, X.; Min, H.; Zhang, B.; Sun, G.; Xie, S.; Wang, H.; et al. Luteolin selectively kills STAT3 highly activated gastric cancer cells through enhancing the binding of STAT3 to SHP-1. *Cell Death Dis.* **2017**, *8*, e2612. [CrossRef]
118. Longshaw, V.M.; Baxter, M.; Prewitz, M.; Blatch, G.L. Knockdown of the co-chaperone Hop promotes extranuclear accumulation of Stat3 in mouse embryonic stem cells. *Eur. J. Cell Biol.* **2009**, *88*, 153–166. [CrossRef]
119. Liu, L.; McBride, K.M.; Reich, N.C. STAT3 nuclear import is independent of tyrosine phosphorylation and mediated by importin-alpha3. *Proc. Natl. Acad. Sci. USA* **2005**, *102*, 8150–8155. [CrossRef]
120. Echeverría, P.C.; Mazaira, G.; Erlejman, A.; Gomez-Sanchez, C.; Piwien Pilipuk, G.; Galigniana, M.D. Nuclear import of the glucocorticoid receptor-hsp90 complex through the nuclear pore complex is mediated by its interaction with Nup62 and importin beta. *Mol. Cell Biol.* **2009**, *29*, 4788–4797. [CrossRef]
121. Adwan, T.S.; Ohm, A.M.; Jones, D.N.; Humphries, M.J.; Reyland, M.E. Regulated binding of importin-α to protein kinase Cδ in response to apoptotic signals facilitates nuclear import. *J. Biol. Chem.* **2011**, *286*, 35716–35724. [CrossRef] [PubMed]
122. Langlais, D.; Couture, C.; Balsalobre, A.; Drouin, J. The Stat3/GR interaction code: Predictive value of direct/indirect DNA recruitment for transcription outcome. *Mol. Cell* **2012**, *47*, 38–49. [CrossRef] [PubMed]
123. Engblom, D.; Kornfeld, J.W.; Schwake, L.; Tronche, F.; Reimann, A.; Beug, H.; Hennighausen, L.; Moriggl, R.; Schütz, G. Direct glucocorticoid receptor-Stat5 interaction in hepatocytes controls body size and maturation-related gene expression. *Genes Dev.* **2007**, *21*, 1157–1162. [CrossRef] [PubMed]
124. Wingelhofer, B.; Neubauer, H.A.; Valent, P.; Han, X.; Constantinescu, S.N.; Gunning, P.T.; Müller, M.; Moriggl, R. Implications of STAT3 and STAT5 signaling on gene regulation and chromatin remodeling in hematopoietic cancer. *Leukemia* **2018**, *32*, 1713–1726. [CrossRef]
125. Ehrnsperger, M.; Gräber, S.; Gaestel, M.; Buchner, J. Binding of non-native protein to Hsp25 during heat shock creates a reservoir of folding intermediates for reactivation. *EMBO J.* **1997**, *16*, 221–229. [CrossRef]
126. Shashidharamurthy, R.; Koteiche, H.A.; Dong, J.; McHaourab, H.S. Mechanism of chaperone function in small heat shock proteins: Dissociation of the HSP27 oligomer is required for recognition and binding of destabilized T4 lysozyme. *J. Biol. Chem.* **2005**, *280*, 5281–5289. [CrossRef]
127. Choi, S.K.; Kam, H.; Kim, K.Y.; Park, S.I.; Lee, Y.S. Targeting Heat Shock Protein 27 in Cancer: A Druggable Target for Cancer Treatment? *Cancers (Basel)* **2019**, *11*, 1195. [CrossRef]
128. Jehle, S.; van Rossum, B.; Stout, J.R.; Noguchi, S.M.; Falber, K.; Rehbein, K.; Oschkinat, H.; Klevit, R.E.; Rajagopal, P. alphaB-crystallin: A hybrid solid-state/solution-state NMR investigation reveals structural aspects of the heterogeneous oligomer. *J. Mol. Biol.* **2009**, *385*, 1481–1497. [CrossRef]
129. Kostenko, S.; Moens, U. Heat shock protein 27 phosphorylation: Kinases, phosphatases, functions and pathology. *Cell Mol. Life Sci.* **2009**, *66*, 3289–3307. [CrossRef]

130. Bruey, J.M.; Paul, C.; Fromentin, A.; Hilpert, S.; Arrigo, A.P.; Solary, E.; Garrido, C. Differential regulation of HSP27 oligomerization in tumor cells grown in vitro and in vivo. *Oncogene* **2000**, *19*, 4855–4863. [CrossRef]
131. Mymrikov, E.V.; Daake, M.; Richter, B.; Haslbeck, M.; Buchner, J. The Chaperone Activity and Substrate Spectrum of Human Small Heat Shock Proteins. *J. Biol. Chem.* **2017**, *292*, 672–684. [CrossRef] [PubMed]
132. Ungelenk, S.; Moayed, F.; Ho, C.T.; Grousl, T.; Scharf, A.; Mashaghi, A.; Tans, S.; Mayer, M.P.; Mogk, A.; Bukau, B. Small heat shock proteins sequester misfolding proteins in near-native conformation for cellular protection and efficient refolding. *Nat. Commun.* **2016**, *7*, 13673. [CrossRef] [PubMed]
133. Rocchi, P.; Beraldi, E.; Ettinger, S.; Fazli, L.; Vessella, R.L.; Nelson, C.; Gleave, M. Increased Hsp27 after androgen ablation facilitates androgen-independent progression in prostate cancer via signal transducers and activators of transcription 3-mediated suppression of apoptosis. *Cancer Res.* **2005**, *65*, 11083–11093. [CrossRef] [PubMed]
134. Shiota, M.; Bishop, J.L.; Nip, K.M.; Zardan, A.; Takeuchi, A.; Cordonnier, T.; Beraldi, E.; Bazov, J.; Fazli, L.; Chi, K.; et al. Hsp27 regulates epithelial mesenchymal transition, metastasis, and circulating tumor cells in prostate cancer. *Cancer Res.* **2013**, *73*, 3109–3119. [CrossRef]
135. Shah, M.; Stanek, J.; Handwerger, S. Differential localization of heat shock proteins 90, 70, 60 and 27 in human decidua and placenta during pregnancy. *HistoChem. J.* **1998**, *30*, 509–518. [CrossRef]
136. Matalon, S.T.; Drucker, L.; Fishman, A.; Ornoy, A.; Lishner, M. The Role of heat shock protein 27 in extravillous trophoblast differentiation. *J. Cell BioChem.* **2008**, *103*, 719–729. [CrossRef]
137. Shochet, G.E.; Komemi, O.; Sadeh-Mestechkin, D.; Pomeranz, M.; Fishman, A.; Drucker, L.; Lishner, M.; Matalon, S.T. Heat shock protein-27 (HSP27) regulates STAT3 and eIF4G levels in first trimester human placenta. *J. Mol. Histol.* **2016**, *47*, 555–563. [CrossRef]
138. Suman, P.; Malhotra, S.S.; Gupta, S.K. LIF-STAT signaling and trophoblast biology. *JAKSTAT* **2013**, *2*, e25155. [CrossRef]
139. Gibert, B.; Eckel, B.; Fasquelle, L.; Moulin, M.; Bouhallier, F.; Gonin, V.; Mellier, G.; Simon, S.; Kretz-Remy, C.; Arrigo, A.P.; et al. Knock down of heat shock protein 27 (HspB1) induces degradation of several putative client proteins. *PLoS ONE* **2012**, *7*, e29719. [CrossRef]
140. Shen, L.; Qi, Z.; Zhu, Y.; Song, X.; Xuan, C.; Ben, P.; Lan, L.; Luo, L.; Yin, Z. Phosphorylated heat shock protein 27 promotes lipid clearance in hepatic cells through interacting with STAT3 and activating autophagy. *Cell Signal.* **2016**, *28*, 1086–1098. [CrossRef]
141. Sevin, M.; Kubovcakova, L.; Pernet, N.; Causse, S.; Vitte, F.; Villeval, J.L.; Lacout, C.; Cordonnier, M.; Rodrigues-Lima, F.; Chanteloup, G.; et al. HSP27 is a partner of JAK2-STAT5 and a potential therapeutic target in myelofibrosis. *Nat. Commun.* **2018**, *9*, 1431. [CrossRef] [PubMed]
142. Kleppe, M.; Kwak, M.; Koppikar, P.; Riester, M.; Keller, M.; Bastian, L.; Hricik, T.; Bhagwat, N.; McKenney, A.S.; Papalexi, E.; et al. JAK-STAT pathway activation in malignant and nonmalignant cells contributes to MPN pathogenesis and therapeutic response. *Cancer Discov.* **2015**, *5*, 316–331. [CrossRef] [PubMed]
143. Lee-Yoon, D.; Easton, D.; Murawski, M.; Burd, R.; Subjeck, J.R. Identification of a major subfamily of large hsp70-like proteins through the cloning of the mammalian 110-kDa heat shock protein. *J. Biol. Chem.* **1995**, *270*, 15725–15733. [CrossRef] [PubMed]
144. Oh, H.J.; Easton, D.; Murawski, M.; Kaneko, Y.; Subjeck, J.R. The chaperoning activity of hsp110. Identification of functional domains by use of targeted deletions. *J. Biol. Chem.* **1999**, *274*, 15712–15718. [CrossRef] [PubMed]
145. Mattoo, R.U.; Sharma, S.K.; Priya, S.; Finka, A.; Goloubinoff, P. Hsp110 is a bona fide chaperone using ATP to unfold stable misfolded polypeptides and reciprocally collaborate with Hsp70 to solubilize protein aggregates. *J. Biol. Chem.* **2013**, *288*, 21399–21411. [CrossRef] [PubMed]
146. Rampelt, H.; Kirstein-Miles, J.; Nillegoda, N.B.; Chi, K.; Scholz, S.R.; Morimoto, R.I.; Bukau, B. Metazoan Hsp70 machines use Hsp110 to power protein disaggregation. *EMBO J.* **2012**, *31*, 4221–4235. [CrossRef]
147. Wang, L.; Duke, L.; Zhang, P.S.; Arlinghaus, R.B.; Symmans, W.F.; Sahin, A.; Mendez, R.; Dai, J.L. Alternative splicing disrupts a nuclear localization signal in spleen tyrosine kinase that is required for invasion suppression in breast cancer. *Cancer Res.* **2003**, *63*, 4724–4730.
148. Manjili, M.H.; Henderson, R.; Wang, X.Y.; Chen, X.; Li, Y.; Repasky, E.; Kazim, L.; Subjeck, J.R. Development of a recombinant HSP110-HER-2/neu vaccine using the chaperoning properties of HSP110. *Cancer Res.* **2002**, *62*, 1737–1742.

149. Guo, C.; Subjeck, J.R.; Wang, X.Y. Creation of Recombinant Chaperone Vaccine Using Large Heat Shock Protein for Antigen-Targeted Cancer Immunotherapy. *Methods Mol. Biol.* **2018**, *1709*, 345–357.
150. Yasuda, K.; Nakai, A.; Hatayama, T.; Nagata, K. Cloning and expression of murine high molecular mass heat shock proteins, HSP105. *J. Biol. Chem.* **1995**, *270*, 29718–29723.
151. Saito, Y.; Yamagishi, N.; Hatayama, T. Nuclear localization mechanism of Hsp105beta and its possible function in mammalian cells. *J. BioChem.* **2009**, *145*, 185–191. [CrossRef] [PubMed]
152. Olszak, T.; Neves, J.F.; Dowds, C.M.; Baker, K.; Glickman, J.; Davidson, N.O.; Lin, C.S.; Jobin, C.; Brand, S.; Sotlar, K.; et al. Protective mucosal immunity mediated by epithelial CD1d and IL-10. *Nature* **2014**, *509*, 497–502. [CrossRef] [PubMed]
153. Yamagishi, N.; Fujii, H.; Saito, Y.; Hatayama, T. Hsp105beta upregulates hsp70 gene expression through signal transducer and activator of transcription-3. *FEBS J.* **2009**, *276*, 5870–5880. [CrossRef] [PubMed]
154. Yu, N.; Kakunda, M.; Pham, V.; Lill, J.R.; Du, P.; Wongchenko, M.; Yan, Y.; Firestein, R.; Huang, X. HSP105 recruits protein phosphatase 2A to dephosphorylate beta-catenin. *Mol. Cell Biol.* **2015**, *35*, 1390–1400. [CrossRef]
155. Boudesco, C.; Verhoeyen, E.; Martin, L.; Chassagne-Clement, C.; Salmi, L.; Mhaidly, R.; Pangault, C.; Fest, T.; Ramla, S.; Jardin, F.; et al. HSP110 sustains chronic NF-kappaB signaling in activated B-cell diffuse large B-cell lymphoma through MyD88 stabilization. *Blood* **2018**, *132*, 510–520. [CrossRef]
156. Berthenet, K.; Bokhari, A.; Lagrange, A.; Marcion, G.; Boudesco, C.; Causse, S.; De Thonel, A.; Svrcek, M.; Goloudina, A.R.; Dumont, S.; et al. HSP110 promotes colorectal cancer growth through STAT3 activation. *Oncogene* **2017**, *36*, 2328–2336. [CrossRef]
157. Ono, K.; Eguchi, T.; Sogawa, C.; Calderwood, S.K.; Futagawa, J.; Kasai, T.; Seno, M.; Okamoto, K.; Sasaki, A.; Kozaki, K.I. HSP-enriched properties of extracellular vesicles involve survival of metastatic oral cancer cells. *J. Cell BioChem.* **2018**, *119*, 7350–7362. [CrossRef]
158. Eguchi, T.; Sogawa, C.; Okusha, Y.; Uchibe, K.; Iinuma, R.; Ono, K.; Nakano, K.; Murakami, J.; Itoh, M.; Arai, K.; et al. Organoids with cancer stem cell-like properties secrete exosomes and HSP90 in a 3D nanoenvironment. *PLoS ONE* **2018**, *13*, e0191109. [CrossRef]
159. Taha, E.A.; Ono, K.; Eguchi, T. Roles of Extracellular HSPs as Biomarkers in Immune Surveillance and Immune Evasion. *Int. J. Mol. Sci.* **2019**, *20*, 4588. [CrossRef]
160. Colgan, S.P.; Pitman, R.S.; Nagaishi, T.; Mizoguchi, A.; Mizoguchi, E.; Mayer, L.F.; Shao, L.; Sartor, R.B.; Subjeck, J.R.; Blumberg, R.S. Intestinal heat shock protein 110 regulates expression of CD1d on intestinal epithelial cells. *J. Clin. Investig.* **2003**, *112*, 745–754. [CrossRef]
161. Berthenet, K.; Boudesco, C.; Collura, A.; Svrcek, M.; Richaud, S.; Hammann, A.; Causse, S.; Yousfi, N.; Wanherdrick, K.; Duplomb, L.; et al. Extracellular HSP110 skews macrophage polarization in colorectal cancer. *Oncoimmunology* **2016**, *5*, e1170264. [CrossRef] [PubMed]
162. Goetz, M.P.; Toft, D.; Reid, J.; Ames, M.; Stensgard, B.; Safgren, S.; Adjei, A.A.; Sloan, J.; Atherton, P.; Vasile, V.; et al. Phase I trial of 17-allylamino-17-demethoxygeldanamycin in patients with advanced cancer. *J. Clin. Oncol.* **2005**, *23*, 1078–1087. [CrossRef] [PubMed]
163. Boudesco, C.; Cause, S.; Jego, G.; Garrido, C. Hsp70: A Cancer Target Inside and Outside the Cell. *Methods Mol. Biol.* **2018**, *1709*, 371–396. [PubMed]
164. Eguchi, T.; Lang, B.J.; Murshid, A.; Prince, T.; Gong, J.; Calderwood, S.K. Regulatory Roles for Hsp70 in Cancer Incidence and Tumor Progression. In *Role of Molecular Chaperones in Structural Folding, Biological Functions, and Drug Interactions of Client Proteins*; Galigniana, M.D., Ed.; Bentham Science Publishers: Sharjah, UAE, 2018; Volume 1, pp. 1–22.
165. Liu, T.; Daniels, C.K.; Cao, S. Comprehensive review on the HSC70 functions, interactions with related molecules and involvement in clinical diseases and therapeutic potential. *Pharmacol. Ther.* **2012**, *136*, 354–374. [CrossRef] [PubMed]
166. Zhuravleva, A.; Gierasch, L.M. Allosteric signal transmission in the nucleotide-binding domain of 70-kDa heat shock protein (Hsp70) molecular chaperones. *Proc. Natl. Acad. Sci. USA* **2011**, *108*, 6987–6992. [CrossRef] [PubMed]
167. Chen, S.C.; Guh, J.Y.; Chen, H.C.; Yang, Y.L.; Huang, J.S.; Chuang, L.Y. Advanced glycation end-product-induced mitogenesis is dependent on Janus kinase 2-induced heat shock protein 70 in normal rat kidney interstitial fibroblast cells. *Transl. Res.* **2007**, *149*, 274–281. [CrossRef] [PubMed]

168. Ghoshal, S.; Rao, I.; Earp, J.C.; Jusko, W.J.; Wetzler, M. Down-regulation of heat shock protein 70 improves arsenic trioxide and 17-DMAG effects on constitutive signal transducer and activator of transcription 3 activity. *Cancer ChemoTher. Pharmacol.* **2010**, *66*, 681–689. [CrossRef]
169. Sinn, D.I.; Kim, S.J.; Chu, K.; Jung, K.H.; Lee, S.T.; Song, E.C.; Kim, J.M.; Park, D.K.; Kun Lee, S.; Kim, M.; et al. Valproic acid-mediated neuroprotection in intracerebral hemorrhage via histone deacetylase inhibition and transcriptional activation. *NeuroBiol. Dis.* **2007**, *26*, 464–472. [CrossRef]
170. Uchida, S.; Fujiki, M.; Nagai, Y.; Abe, T.; Kobayashi, H. Geranylgeranylacetone, a noninvasive heat shock protein inducer, induces protein kinase C and leads to neuroprotection against cerebral infarction in rats. *NeuroSci. Lett.* **2006**, *396*, 220–224. [CrossRef]
171. Guo, F.; Sigua, C.; Bali, P.; George, P.; Fiskus, W.; Scuto, A.; Annavarapu, S.; Mouttaki, A.; Sondarva, G.; Wei, S.; et al. Mechanistic role of heat shock protein 70 in Bcr-Abl-mediated resistance to apoptosis in human acute leukemia cells. *Blood* **2005**, *105*, 1246–1255. [CrossRef]
172. Ciocca, D.R.; Arrigo, A.P.; Calderwood, S.K. Heat shock proteins and heat shock factor 1 in carcinogenesis and tumor development: An update. *Arch. Toxicol.* **2013**, *87*, 19–48. [CrossRef] [PubMed]
173. Binder, R.J.; Srivastava, P.K. Peptides chaperoned by heat-shock proteins are a necessary and sufficient source of antigen in the cross-priming of CD8+ T cells. *Nat. Immunol.* **2005**, *6*, 593–599. [CrossRef] [PubMed]
174. Li, Z.; Menoret, A.; Srivastava, P. Roles of heat-shock proteins in antigen presentation and cross-presentation. *Curr. Opin. Immunol.* **2002**, *14*, 45–51. [CrossRef]
175. Srivastava, P. Interaction of heat shock proteins with peptides and antigen presenting cells: Chaperoning of the innate and adaptive immune responses. *Annu. Rev. Immunol.* **2002**, *20*, 395–425. [CrossRef] [PubMed]
176. Srivastava, P. Roles of heat-shock proteins in innate and adaptive immunity. *Nat. Rev. Immunol.* **2002**, *2*, 185–194. [CrossRef]
177. Elsner, L.; Muppala, V.; Gehrmann, M.; Lozano, J.; Malzahn, D.; Bickeböller, H.; Brunner, E.; Zientkowska, M.; Herrmann, T.; Walter, L.; et al. The heat shock protein HSP70 promotes mouse NK cell activity against tumors that express inducible NKG2D ligands. *J. Immunol.* **2007**, *179*, 5523–5533. [CrossRef]
178. Gross, C.; Hansch, D.; Gastpar, R.; Multhoff, G. Interaction of heat shock protein 70 peptide with NK cells involves the NK receptor CD94. *Biol. Chem.* **2003**, *384*, 267–279. [CrossRef]
179. Chalmin, F.; Ladoire, S.; Mignot, G.; Vincent, J.; Bruchard, M.; Remy-Martin, J.P.; Boireau, W.; Rouleau, A.; Simon, B.; Lanneau, D.; et al. Membrane-associated Hsp72 from tumor-derived exosomes mediates STAT3-dependent immunosuppressive function of mouse and human myeloid-derived suppressor cells. *J. Clin. Investig.* **2010**, *120*, 457–471.
180. Asea, A.; Kraeft, S.K.; Kurt-Jones, E.A.; Stevenson, M.A.; Chen, L.B.; Finberg, R.W.; Koo, G.C.; Calderwood, S.K. HSP70 stimulates cytokine production through a CD14-dependant pathway, demonstrating its dual role as a chaperone and cytokine. *Nat. Med.* **2000**, *6*, 435–442. [CrossRef]
181. Chen, T.; Guo, J.; Han, C.; Yang, M.; Cao, X. Heat shock protein 70, released from heat-stressed tumor cells, initiates antitumor immunity by inducing tumor cell chemokine production and activating dendritic cells via TLR4 pathway. *J. Immunol.* **2009**, *182*, 1449–1459. [CrossRef]
182. Starr, R.; Metcalf, D.; Elefanty, A.G.; Brysha, M.; Willson, T.A.; Nicola, N.A.; Hilton, D.J.; Alexander, W.S. Liver degeneration and lymphoid deficiencies in mice lacking suppressor of cytokine signaling-1. *Proc. Natl. Acad. Sci. USA* **1998**, *95*, 14395–14399. [CrossRef] [PubMed]
183. Jiang, M.; Zhang, W.W.; Liu, P.; Yu, W.; Liu, T.; Yu, J. Dysregulation of SOCS-Mediated Negative Feedback of Cytokine Signaling in Carcinogenesis and Its Significance in Cancer Treatment. *Front. Immunol.* **2017**, *8*, 70. [CrossRef]
184. Babon, J.J.; Kershaw, N.J.; Murphy, J.M.; Varghese, L.N.; Laktyushin, A.; Young, S.N.; Lucet, I.S.; Norton, R.S.; Nicola, N.A. Suppression of cytokine signaling by SOCS3: Characterization of the mode of inhibition and the basis of its specificity. *Immunity* **2012**, *36*, 239–250. [CrossRef] [PubMed]
185. Fan, Y.X.; Qian, C.; Liu, B.; Wang, C.; Liu, H.; Pan, X.; Teng, P.; Hu, L.; Zhang, G.; Han, Y.; et al. Induction of suppressor of cytokine signaling 3 via HSF-1-HSP70-TLR4 axis attenuates neuroinflammation and ameliorates postoperative pain. *Brain Behav. Immun.* **2018**, *68*, 111–122. [CrossRef] [PubMed]
186. Wang, G.; Xiao, G.; Liu, H.; Chen, G.; Wang, X.; Wen, P.; Li, T.; Wen, J.; Xiao, X. Heat Shock Factor 1 Inhibits the Expression of Suppressor of Cytokine Signaling 3 in Cerulein-Induced Acute Pancreatitis. *Shock* **2018**, *50*, 465–471. [CrossRef] [PubMed]

187. Stephanou, A.; Isenberg, D.A.; Nakajima, K.; Latchman, D.S. Signal transducer and activator of transcription-1 and heat shock factor-1 interact and activate the transcription of the Hsp-70 and Hsp-90beta gene promoters. *J. Biol. Chem.* **1999**, *274*, 1723–1728. [CrossRef]
188. Matozaki, M.; Saito, Y.; Yasutake, R.; Munira, S.; Kaibori, Y.; Yukawa, A.; Tada, M.; Nakayama, Y. Involvement of Stat3 phosphorylation in mild heat shock-induced thermotolerance. *Exp. Cell Res.* **2019**, *377*, 67–74. [CrossRef]
189. Saito, Y.; Yamagishi, N.; Ishihara, K.; Hatayama, T. Identification of alpha-tubulin as an hsp105alpha-binding protein by the yeast two-hybrid system. *Exp. Cell Res.* **2003**, *286*, 233–240. [CrossRef]
190. Fritchley, S.J.; Kirby, J.A.; Ali, S. The antagonism of interferon-gamma (IFN-gamma) by heparin: Examination of the blockade of class II MHC antigen and heat shock protein-70 expression. *Clin. Exp. Immunol.* **2000**, *120*, 247–252. [CrossRef]
191. Chen, X.S.; Zhang, Y.; Wang, J.S.; Li, X.Y.; Cheng, X.K.; Wu, N.H.; Shen, Y.F. Diverse effects of Stat1 on the regulation of hsp90alpha gene under heat shock. *J. Cell BioChem.* **2007**, *102*, 1059–1066. [CrossRef]
192. Cheng, M.B.; Zhang, Y.; Zhong, X.; Sutter, B.; Cao, C.Y.; Chen, X.S.; Cheng, X.K.; Xiao, L.; Shen, Y.F. Stat1 mediates an auto-regulation of hsp90beta gene in heat shock response. *Cell Signal.* **2010**, *22*, 1206–1213. [CrossRef] [PubMed]
193. Pak, S.H.; Joung, Y.H.; Park, J.H.; Lim, E.J.; Darvin, P.; Na, Y.M.; Hong, D.Y.; Lee, B.; Hwang, T.S.; Park, T.; et al. Hypoxia upregulates Hsp90α expression via STAT5b in cancer cells. *Int. J. Oncol.* **2012**, *41*, 161–168. [PubMed]
194. Song, H.; Ethier, S.P.; Dziubinski, M.L.; Lin, J. Stat3 modulates heat shock 27kDa protein expression in breast epithelial cells. *Biochem. Biophys. Res. Commun.* **2004**, *314*, 143–150. [CrossRef] [PubMed]
195. Xu, N.W.; Chen, Y.; Liu, W.; Chen, Y.J.; Fan, Z.M.; Liu, M.; Li, L.J. Inhibition of JAK2/STAT3 Signaling Pathway Suppresses Proliferation of Burkitt's Lymphoma Raji Cells via Cell Cycle Progression, Apoptosis, and Oxidative Stress by Modulating HSP70. *Med. Sci. Monit.* **2018**, *24*, 6255–6263. [CrossRef]
196. Krawczyk, Z.; Gogler-Pigłowska, A.; Sojka, D.R.; Scieglinska, D. The Role of Heat Shock Proteins in Cisplatin Resistance. *Anticancer Agents Med. Chem.* **2018**, *18*, 2093–2109. [CrossRef]
197. Chatterjee, S.; Burns, T.F. Targeting Heat Shock Proteins in Cancer: A Promising Therapeutic Approach. *Int. J. Mol. Sci.* **2017**, *18*, 1978. [CrossRef]
198. Saini, J.; Sharma, P.K. Clinical, Prognostic and Therapeutic Significance of Heat Shock Proteins in Cancer. *Curr. Drug Targets* **2018**, *19*, 1478–1490. [CrossRef]

6

Nucleocytoplasmic Shuttling of STATs: A Target for Intervention?

Sabrina Ernst [1,2] and Gerhard Müller-Newen [1,*]

[1] Institute of Biochemistry and Molecular Biology, RWTH Aachen University, 52074 Aachen, Germany; sabrina.ernst@rwth-aachen.de
[2] Confocal Microscopy Facility, Interdisciplinary Center for Clinical Research IZKF, RWTH Aachen University, 52074 Aachen, Germany
[*] Correspondence: mueller-newen@rwth-aachen.de

Abstract: Signal transducer and activator of transcription (STAT) proteins are transcription factors that in the latent state are located predominantly in the cytoplasm. Activation of STATs through phosphorylation of a single tyrosine residue results in nuclear translocation. The requirement of tyrosine phosphorylation for nuclear accumulation is shared by all STAT family members but mechanisms of nuclear translocation vary between different STATs. These differences offer opportunities for specific intervention. To achieve this, the molecular mechanisms of nucleocytoplasmic shuttling of STATs need to be understood in more detail. In this review we will give an overview on the various aspects of nucleocytoplasmic shuttling of latent and activated STATs with a special focus on STAT3 and STAT5. Potential targets for cancer treatment will be identified and discussed.

Keywords: STAT3; STAT5; nuclear pore complex; nuclear transport receptors; nucleocytoplasmic shuttling; cancer; targeting

---

1. Aim and Scope

STAT (Signal transducer and activator of transcription) proteins can be seen as intracellular messengers that relay signals sensed at the plasma membrane to chromatin and genes in the nucleus. To achieve this, STATs must pass the nuclear envelope through nuclear pore complexes (NPCs). Thus, passage through the NPC is an essential step in the sequence of events from activation of STATs at cytokine receptors to DNA-binding and target gene induction. As detailed in the reviews and articles of this Special Issue of Cancers, deregulated activation of STAT3 and STAT5 contributes to various cancers in many ways. Thus, STAT3 and STAT5 proteins have emerged as promising therapeutic targets. Protein–protein interactions involved in nucleocytoplasmic transfer of STATs have not been exploited yet as molecular targets for intervention. Detailed knowledge of the involved molecules and mechanisms is an essential prerequisite for successful and specific targeting. In this review, we will first describe the general mechanisms involved in import and export of proteins in and out of the nucleus, concentrating on those which are most relevant for transcription factors. We will then focus on the mechanisms involved in nucleocytoplasmic shuttling of STAT3 and STAT5 and finally assess possible molecular targets for specific intervention.

2. General Mechanisms of Nucleocytoplasmic Transport of Proteins

2.1. The Nuclear Pore Complex

To enter or exit the nucleus, proteins must pass through the nuclear pore complex (NPC) [1]. NPCs are huge macromolecular assemblies (about 120 MDa in humans) made up of multiple copies of Nucleoporins (NUPs). More than 30 different NUPs have been identified that are built into the NPC as multiples of eight (8–64) resulting in the eight-fold rotational symmetry of the NPC [2]. The NPC can be seen as a channel that allows selective transfer of cargo and at the same time forms a soft barrier for free diffusion of macromolecules larger than about 30 kDa [3,4], preventing their access to the nucleus without permission. The barrier is formed by phenylalanine-glycine (FG)-repeats that protrude from certain NUPs into the lumen of the channel [5,6].

How exactly the FG-repeats form a selective permeability barrier is not completely understood and several models are currently being discussed [7,8]. One of the most prevalent is the selective phase model that relies on interactions between the FG-repeats creating a sieve-like meshwork with hydrogel-like properties, which would explain the observed mass exclusion limit [9]. Selectivity for cargo allowed to pass might result from phase separation that prevents passage of macromolecules that are unable to mix or interact with the selective phase made up by the FG-repeats [8].

The import/export pathways through the NPC involve soluble nuclear transport receptors (NTRs) that bind cargo in conjunction with the Ran-GTP/GDP cycle. NTRs can interact with FG-repeats [10] and facilitate passage of bound cargo through the NPC. According to the selective phase model, interaction of NTRs with the FG-repeats leads to local disturbance of the meshwork allowing the NTR/cargo complex to travel almost freely between cytoplasm and nucleoplasm [11]. The energy-consuming Ran-GTP/GDP cycle provides directionality of the transport through control of cargo to NTR binding which is differently regulated in nucleoplasm vs. cytoplasm [12,13].

NTRs, also known as Karyopherins, can be subdivided in Importins and Exportins, facilitating nuclear import and export, respectively. Biportins have also been described that support transport of cargo in both directions with imported and exported cargo being distinct [14].

2.2. *Importins*

The best characterized Karyopherin is Importin-β1 which either binds cargo directly or indirectly through interaction with adapters such as α-Importins or Snurportin-1 [15]. The transcription factors Snail1 [16,17] and SREBP2 [18] are among the cargoes directly bound by Importin-β1. Snurportin-1 is best known for its involvement in the nuclear import of spliceosomal snRNPs [19]. The interaction site of Importin-β1 with FG-repeats has been mapped [20,21] and is different from the well-defined interaction sites with cargoes and adapters [22]. This means that cargo-loaded Importin-β1 can interact with FG-Nucleoporins of the NPC and thereby facilitate passage of the Importin-β1/adapter/cargo complex.

The import pathway using α-Importins as adapters to connect cargo with Importin-β1 has been intensively studied and is now known as the classical import pathway [23]. Accordingly, the term classical nuclear localization signal (cNLS) refers to linear sequence motifs of cargoes that bind to α-Importins. The cNLS can be further subdivided into monopartite cNLS and bipartite cNLS. Monopartite cNLS consist of a short stretch of basic amino acid residues, the sequence PKKKRRV of the SV40 large T-antigen being the first identified [24]. The first bipartite cNLS has been found in the Nucleoplasmin protein of Xenopus laevis, consisting of two short stretches of basic amino acids, both essential for its function, separated by a few less relevant amino acids [25].

In humans, seven Importin-α isoforms have been identified: Importin-α1, -α3, -α4, -α5, -α6, -α7 [23] and the most recently discovered Importin-α8 [26]. All α-Importins are mainly built up of ten Armadillo (ARM)-repeats resulting in a structurally conserved solenoid protein domain [27]. The cNLS of cargo binds along a groove on the inner concave surface of α-Importins. Although the cNLS binding region is quite conserved, the α-Importins are specific for a set of cargo proteins with some considerable overlap. This specificity results in part from preferential binding of certain NLS but also from the three-dimensional context in which the NLS is presented by the cargo [28]. Apart from cNLS, so called non-classical or atypical NLS have been identified, which do not fit in the cNLS consensus motifs [29]. A short sequence termed Importin- β binding (IBB) domain precedes the solenoid domain of α-Importins. In the absence of cargo, the IBB folds back to the NLS binding site. Upon binding of cargo, the IBB is replaced by the cNLS of the cargo protein. The now exposed IBB binds to Importin-β1, facilitating the nuclear import of the ternary Importin-β1/Importin-α/cargo complex [30].

Transportin-1, also known as Importin-β2, also imports a broad spectrum of cargoes, including transcription factors and mRNA-binding proteins. Transportin-1 binds cargo directly through a broad range of loosely related NLS that are different from cNLS [31]. Symportin-1 has been identified as an adapter for nuclear import of some ribosomal proteins by Transportin-1 [32]. Transportin-3 recognizes arginine-serine (RS)-rich NLS found in proteins typically involved in mRNA metabolism such as the Splicing Factor 2 (SF2) [33]. Importin-13 is closely related to Transportin-3 but mediates both protein import and export. Therefore, by definition Importin-13 can be regarded as a Biportin. Some transcription factors are among the many cargoes identified so far. Cargo binding by Importin-13 relies on the recognition of folded domains rather than linear NLS [34,35].

Some import pathways have been identified that do not rely on β-Importins. Among those, Calmodulin-mediated nuclear import seems to be most relevant for transcription factors [36]. Finally, some proteins including transcription factors enter the nucleus independent of NTR or other carrier proteins, e.g., through direct interaction with NUPs [37].

2.3. *Exportins*

Crm1 (Chromosome region maintenance 1; in the systematic nomenclature of Karyopherins designated as Exportin-1 or Xpo1) is the most promiscuous Exportin, which mediates the export of about 1000 substrates, including various RNAs, ribonucleoproteins, and transcription factors [38]. Most protein cargo of Crm1 contains a leucine-rich nuclear export signal (NES) of 8–15 amino acids [39–41]. Like all β-Karyopherins, Crm1 is built up of HEAT-repeats (Huntingtin, Elongation factor 3, protein phosphatase 2A and TOR kinase). Each HEAT-repeat consists of two α-helices connected by loops of varying length. Similar to the ARM repeats of α-Importins, the HEAT-repeats of β-Karyopherins form a solenoid protein domain [42] with a slightly curved superhelical structure of high conformational flexibility. In all of the available structures of cargo bound to Crm1, the NES of the cargo fits into a hydrophobic groove between HEAT repeats 11 and 12 (out of 20) at the outer convex surface of Crm1 [43]. Binding at the outer convex surface of Crm1 allows accommodation of cargo of various sizes.

Other Exportins, such as Exportin-5, Exportin-t, and Exportin-2 (better known as CAS for Cellular Apoptosis Susceptibility), bind their cargo on the inner concave surface. Accordingly, these Exportins are more selective. Exportin-5 mediates export of several types of RNA but not protein; Exportin-t is even more restricted to the export of tRNAs [44]. CAS is specialized for the nuclear export of α-Importins, which is needed to maintain the classical import pathway [45]. Exportin-4, a distant member of the β-Karyopherin family, originally described to mediate nuclear export of the translation initiation factor 5A [46] and the transcription factor Smad3 [47], has more recently been described to be involved in the nuclear import of Sox transcription factors [48], thus acting as a Biportin. Another Biportin, Exportin-7 (also known as RanBP16) [49], has recently been described to act as a broad-spectrum Karyopherin with about 200 export and 30 import substrates [50].

2.4. *Ran-GTP/GDP Cycle*

In principle, through interaction with the FG-repeats, β-Karyopherins can pass the NPC in both directions [51]. Directionality of nucleocytoplasmic transport is achieved through coupling of cargo binding to β-Karyopherins with the energy-consuming Ran-GTP/GDP cycle [12,52]. Ran is a small G-protein of the Ras superfamily of GTPases that exists in a GTP- or GDP-bound state. The GTP-bound state results from replacement of bound GDP with GTP that requires the activity of a Ran-specific, chromatin-bound guanine-nucleotide exchange factor (RanGEF, also known as RCC1), which is predominantly located in the nucleus. Therefore, the Ran-GTP concentration in the nucleus is high. The GDP-bound state is generated through release of the β-phosphate of bound GTP by hydrolysis, which requires the activity of a ran-specific GTPase-activating protein (RanGAP). Through interaction with the Nucleoporin Ran-binding protein 2 (RanBP2, Nup358), RanGAP is located predominantly at the cytoplasmic surface of the NPC. Therefore, in the cytoplasm close to the nuclear envelope, the Ran-GTP concentration is low and the Ran-GDP concentration is high [53,54].

Cargo binding to Importins in the cytoplasm is Ran-independent. After passage of the Importin/cargo complex through the NPC, release of cargo in the nucleus is induced upon Ran-GTP binding. In contrast, Ran-GTP-binding to Exportins is required for cargo loading in the nucleus. After passage of the Exportin/cargo/Ran-GTP complex through the NPC, RanGAP-mediated GTP-hydrolysis results in release of RanGDP and cargo into the cytoplasm [55].

As mentioned above, NTRs facilitate nucleocytoplasmic transport through transient interactions with the hydrophobic FG-repeats within the channel of the NPC. In principle, other proteins with hydrophobic surface patches should to some extent also be able to disturb the FG-repeat meshwork, allowing them to shuttle between the cytoplasm and the nucleus in an NTR- and Ran-independent, passive manner. Indeed, in a recent study [56] it has been found that a continuum exists with gradual differences in the ability of proteins to pass the NPC in an active (i.e., NTR- and Ran-dependent) or passive manner. Depending on their surface properties, even large proteins can leak into the nucleus without involvement of NTRs.

3. STAT1—Using a Side Track for Nuclear Import

STAT1 is deeply involved in the antiviral response triggered by endogenously produced Interferons but also in the therapeutic responses to exogenously administered Interferons in cancer treatments and anti-viral therapies [57]. STAT1 is best known for its tumor suppressive activity in cancer but some tumor-promoting effects have also been documented [58]. IFNγ (type II Interferon) induces the activation of STAT1 through phosphorylation of Y701 by IFNγ receptor-associated Janus kinase 1 (JAK1) and JAK2, leading to dimerization and nuclear accumulation of the transcription factor [59]. Among all STAT family members, the mechanism of nuclear accumulation of the activated STAT1 dimer is best understood. Activated STAT1 interacts with Importin-α5 [60], but in an unconventional manner. Instead of employing the binding site for cNLS, which would involve ARM repeats 2–4 and 6–8 as the major and minor binding sites, respectively [22], STAT1 binds to the more C-terminally located ARM repeats of Importin-α5 [61] involving a critical tyrosine residue (Y476) located in ARM repeat 10 [62]. Accordingly, the activated STAT1 dimer does not expose a cNLS but a dimer-specific surface area to interact with Importin-α5. This surface has been termed dimer-specific NLS (dsNLS) and includes regions within the N-terminal domain (NTD) [63] and the DNA-binding domain (DBD; see Figure 1a for a scheme of the domain structure of STAT proteins; see Table 1 for an overview on putative NLS and NES sequences of STAT proteins; and see Figure 1b showing the corresponding putative NLS and NES motifs in the STAT1 dimer structure) [64]. Hence, Importin-α5 can be displaced from STAT1 by DNA-binding [65,66]. Most interestingly, this unique mode of interaction is exploited by the Ebola virus VP24 protein, that interacts with ARM repeats 8–10 of Importin-α5, preventing binding and nuclear translocation of activated STAT1 [67]. Interaction of Importin-α5 with cNLS cargo remains unaffected by VP24 binding. This example shows that, in principle, nuclear accumulation of a STAT transcription factor can be selectively blocked.

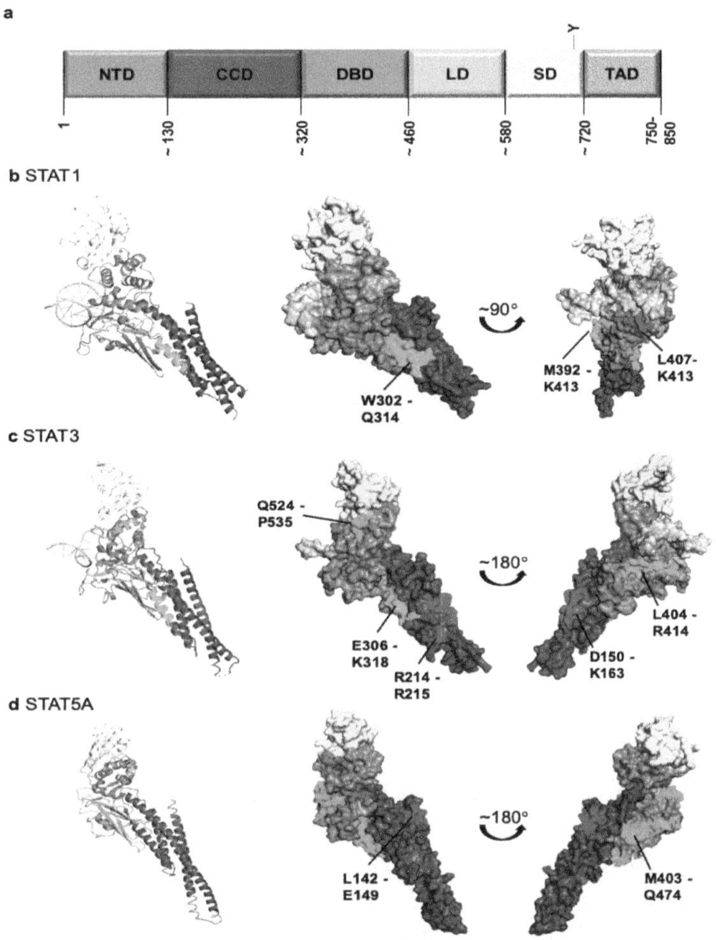

Figure 1. Structures of STAT proteins with putative nuclear localization signals (NLS) and nuclear export signals (NES) highlighted as listed in Table 1. In this context, "putative" means that the corresponding sequences do not fulfill classical NLS or NES functions but are required for nuclear transport or interaction with nuclear transport receptors. (**a**) General scheme of structural domains of STAT proteins (NTD, N-terminal domain; CCD coiled-coil domain; DBD, DNA-binding domain; LD, linker domain; SD, SH2 domain; TAD, transactivation domain). Numbers refer to amino acid positions. (**b–d**) Structures of individual STAT proteins as ribbon representations (left) or space-filling representations (middle and right). Domains are stained according to the coloring of the scheme in (**a**), DNA is shown in pink. Putative NLS and NES are marked in red and cyan, respectively. The corresponding references are listed in Table 1. PDB IDs: STAT1, 1BF5; STAT3, 1BG1; STAT5A, 1Y1U. Images of structures were generated with PyMOL.

Table 1. NTRs, NLS and NES involved in transport of STAT proteins.

	STAT1	STAT3	STAT5A	STAT5B
Importin	α5 [60,68], α7 [69]	α1 [70], α3 [70–72], α5 [69,70,72], α6 [71], α7 [69,72]	α3 [73]	n.d. [1]
Putative [2] NLS	L407-K413 [64,68] NTD involved [63,74]	D150-K163 [71] R214/R215 [69] NTD involved [75]	L142-E149 [76] intact CCD required [73]	n.d.
Exportin	Crm1 [66]	Crm1 [75]	Crm1 (not exclusively) [73]	Crm1 (not exclusively) [77]
Putative [2] NES	W302-Q314 [78] M392-K413 [66]	E306-K318 L404-R414 Q524-P535 [79]	L119/L133 M403-Q474 [73]	n.d.

[1] n.d., not determined; [2] in this context, "putative" means that the corresponding sequences do not fulfill classical NLS or NES functions but are required for nuclear transport or interaction with NTRs.

4. STAT3—Acting on Many Stages

4.1. Role of STAT3 in Cancer

STAT3 is involved in cancer in multiple ways, as detailed in several reviews and articles of this Special Issue of Cancers. Besides acting as an oncogene in a cell autonomous manner [80,81], activation of STAT3 in the tumor microenvironment contributes to metastasis, angiogenesis, cancer stem cell maintenance, and immune evasion [82]. Therefore, blocking STAT3 activity in cancer is of particular interest. However, it should also be noted that in some cancers STAT3 has tumor suppressive activities [83,84]. Several steps in the STAT3 activation pathway are potentially targetable and thus might be used to interfere with STAT3 activation. These include phosphorylation on Y705 by (oncogenic) tyrosine kinases, dimerization through phosphotyrosine/SH2 domain interactions, nuclear translocation, DNA-binding to GAS (γ-interferon activated sequence) elements, and interaction with cofactors. However, passage of the activated STAT3 dimer through the NPC has only rarely been considered as a potential molecular target for intervention [85,86]. This might be due to the fact that import mechanism for activated STAT3 and the interactions of STAT3 with NTRs have been analyzed in some detail but are not completely understood as outlined below.

4.2. Nuclear Import

Compared to STAT1, the mechanism of nuclear import of the activated STAT3 dimer is less well defined. Reports on interactions with α-Importins are contradictory in part. In one of the first studies, interactions of STAT3 with Importin-α1, -α3, and -α5 were detected [70]. Additionally, interactions of STAT3 with Importin-α5 and -α7, but not with Importin-α1, -α3, and -α4, were found [69]. In both reports, interactions were only seen upon activation of STAT3. Other groups found interactions with Importin-α3 and -α6 [71] and Importin-α3, -α5, and -α7 [72] being independent of activation of STAT3. In this context it is worth noting that expression of Importin-α6 is restricted to testis [87]. Therefore, Importin-α6 is less relevant for nuclear import of ubiquitously expressed transcription factors such as STAT3 and STAT5. Based on import assays with permeabilized cells, the MgcRacGAP protein bound to Rac has been suggested as an NLS-containing adapter for Importin-α-mediated nuclear import of STAT3 and STAT5 [88]. In another study using an MgcRacGAP inhibitor and siRNA mediated knockdown, the functional role of MgcRacGAP for nuclear import of STAT3 could not be substantiated [89]. These discrepancies might be the result of different experimental conditions and cellular systems used. For instance, subcellular distribution of STAT3 analyzed by immunofluorescence heavily depends on the applied fixation method [90]. Alternatively, the different findings indicate that STAT3 employs a different import system depending on the cytokines used for its activation, e.g., Interleukin-6 (IL-6) vs. Epidermal Growth Factor (EGF). Clearly, more work is required for better understanding.

Functional relevance of Importin-α5 in nuclear import of activated STAT3 has been demonstrated, as well as the involvement of Importin-β and Ran [70]. Moreover, it has been shown that the epitope on Importin-α5 employed in the interaction with STAT3 is different from the one used for the interaction with STAT1. A mutation in ARM repeat 10 of Importin-α5 blocks interaction with STAT1 [62] but does not affect the interaction with STAT3 [72]. Correspondingly, the epitopes used by STAT3 and STAT1 for interaction with Importin-α5 are also different. Amino acids in the DBD of STAT1 are involved in the recognition of Importin-α5 [65,66], connecting Importin-binding and nuclear presence with DNA-binding activity of STAT1 [91]. In STAT3, mutation of R214 and R215 located in the coiled-coil domain (CCD) abolishes binding of Importin-α5 and nuclear accumulation in response to stimulation [69]. Liu et al. [71] defined the sequence D150-K163 as being indispensable for nuclear import of STAT3, which is also located in the CCD. Nuclear accumulation of activated STAT3 does not require DNA-binding activity [72]. Common to STAT3 and STAT1 is the involvement of the N-terminal domain in stimulation-dependent nuclear import [63,74,75]. The essential role of the NTD for active nuclear import of STAT3 is further supported by the observation that the tumor suppressor ARHI

(A Ras Homologue member I/DIRAS3) blocks nuclear translocation of phosphorylated STAT3 via direct interaction with the NTD [92].

It seems that Importin-α5 is a major determinant for nuclear import of STAT3 and STAT1 but the molecular interfaces of the interactions are considerably different. This opens an avenue for specific interventions. However, to achieve a selective block of nuclear import of STAT3, the structural details of the interactions with NTRs must be characterized in more detail. It should be noted that no classical NLS or NES sequences have been identified in STAT3 in the sense of transferable functional motifs. The above-mentioned amino acids in the putative NLS and NES are most probably important for the integrity of domain structures or the part of epitopes that interact with NTRs in a non-conventional manner [75], as exemplified for STAT1 [62] (Figure 1c showing putative NLS and NES motifs in the activated STAT3 dimer structure).

4.3. Nuclear Export and Nucleocytoplasmic Shuttling

In the course of defining nuclear export signals of STAT3, it was found that pharmacological inhibition of Crm1-mediated nuclear export results in partial nuclear accumulation of STAT3 independent of cytokine stimulation [79]. From these and other studies it became evident that the subcellular distribution of latent STAT3 with high cytoplasmic and low nuclear concentration is the result of a steady-state of constitutive nuclear import and rapid export independent of phosphorylation at Y705 [93,94]. Determinants of stimulation-dependent nuclear import, such as the NTD and the R214/R215-motif, are not involved in basal nuclear import of latent STAT3 [75]. Unphosphorylated STAT1 has also been detected in the nucleus [95], and the mechanisms of nuclear import of latent and activated STAT1 were also found to be different [64]. Nuclear import of unphosphorylated STAT3 might be independent of NTRs and could involve direct interactions with FG-repeat NUPs of the NPC as observed for STAT1 [96]. From the three sequence motifs identified as putative NES in STAT3, at least the one containing L525/L528 is involved in constitutive nucleocytoplasmic shuttling [75,79].

It has been firmly established that STAT3 and STAT1 form homodimers in the absence of the activating tyrosine phosphorylation resulting in so called preformed dimers [97,98]. The phosphotyrosine-independent dimerization involves homotypic interactions of the NTDs [99]. However, basal nucleocytoplasmic shuttling of STAT3 does not require the NTD [75] and is not sensitive to a single point mutation (L78R) that prevents preformed dimer formation [100]. Not only is dimer formation not required for basal shuttling, but monomeric STAT3 shuttles even faster [75]. The increased shuttling rate of monomeric STAT3 could be attributed to the smaller size compared to the preformed dimer. However, a recent study suggests that hydrophobic surface patches that interact with FG-repeats of FG-Nups might be a stronger predictor for NTR-independent passage of the NPC than size [56]. Thus, accessible hydrophobic surfaces that are masked in the dimer may facilitate faster shuttling of monomeric STAT3 in an NTR-independent manner.

Since basal and stimulation-dependent nuclear import of STATs occur through different mechanisms, they could, in principle, be targeted independently of each other. However, the functional relevance of nuclear presence of unphosphorylated STAT3 and STAT1 (U-STATs) is not entirely clear. Nuclear U-STATs might be involved as cofactors in gene regulation independent of their DNA-binding activity [101].

5. STAT5—Leukemia and More

Important functional roles in hematopoiesis downstream of hematopoietic cytokines such as Erythropoietin and Thrombopoietin have been attributed to STAT5. Consequently, deregulated STAT5 signaling is prominently implicated in myeloproliferative diseases and leukemias [102]. The involvement of STAT5 in solid tumors is also well documented, both as an oncogene and a tumor suppressor [103], as detailed in several reviews and articles of this Special Issue of Cancers. In particular, in myeloproliferative neoplasms and leukemias, STAT5 has been identified as a promising therapeutic target.

Compared with the many reports on STAT3 and STAT1, only a handful of studies dedicated to the molecular mechanisms governing nucleocytoplasmic distribution of STAT5 exist. In one of the first studies [104], a sequence motif in the DNA-binding domain of STAT5B (V466-I469) was identified whose mutation prevented Growth Hormone-induced nuclear accumulation and DNA-binding. However, from this finding it cannot be concluded that DNA-binding per se is required for nuclear accumulation of activated STAT5B because the mutation could also interfere with binding of NTRs, which was not investigated in this study. We found that another mutation that affects DNA-binding of STAT5A does not impair nuclear accumulation upon Epo stimulation, in agreement with the DNA-binding independent nuclear accumulation of activated STAT3 [72]. In general, nuclear accumulation of STATs through retention by binding to nuclear structures seems to be of minor importance. If nuclear accumulation was dependent on retention of STATs on subnuclear structures, nuclear accumulation would be saturable, meaning that as soon as all binding sites are occupied, the remaining STAT molecules would not accumulate in the nucleus. This is not what is observed in the above-mentioned experiments with transfected cells: Even upon forced overexpression, STATs almost completely accumulate in the nucleus upon stimulation. This view is supported by the results of fluorescence recovery after photobleaching (FRAP) experiments that have been exemplarily performed on STAT1-GFP. Upon nuclear accumulation induced by Interferon treatment, STAT1-GFP freely diffuses through the nucleoplasm with the exclusion of nucleoli [105]. In this and another study [93], the involvement of the cytoskeleton in directed nuclear import of STAT1 and STAT3 was also excluded.

In one of the reports on binding of α-Importins to STAT3, binding to STAT5A and STAT5B was analyzed in parallel but no interaction could be detected [69]. This is in agreement with our own observations made with co-precipitation experiments from cellular lysates using Importins fused to GST (glutathione-S-transferase). Accordingly, no cNLS has been detected in STAT5. In another study, interaction of STAT5A with Importin-α3 has been shown [73], again supporting the notion that the outcomes of in vitro assays (mostly co-precipitations) for studying the interaction of STATs with NTRs are very sensitive to the experimental conditions used.

Similar to STAT3 and STAT1, constitutive nucleocytoplasmic shuttling of STAT5 independent of tyrosine phosphorylation has been detected [77]. A functional role of nuclear U-STAT5 has been established in megakaryocyte differentiation [106]. Basal and cytokine-induced import of STAT5A requires an intact CCD [73] and was abolished upon deletion of eight amino acids in the CCD (L142–E149, see Table 1 and Figure 1d showing putative NLS and NES motifs in the STAT5A monomer structure; a structure of the activated STAT5 dimer is not available) [76]. Interaction of STAT5A with Importin-α3 occurs in an unconventional manner involving the CCD [73]. As for STAT3 and STAT1, stimulation-dependent but not basal nuclear translocation requires the NTD of STAT5 [77]. In a study that tested all sequence motifs that might function as a cNES, a region between aa578 and aa675 in STAT5B comprising the SH2 domain was identified to be required for nuclear export [77]. Another region important for Crm1-mediated export was identified involving L119 and L133 in the NTD of STAT5A [73].

A discrepancy between nuclear localization of STAT5A and STAT5B was found in response to phosphorylation of the critical tyrosine residue (Y694 in STAT5A and Y699 in STAT5B) by Src family kinases (SFK). While phosphorylation of Y699 in STAT5B, as well as Y705 in STAT3, by Src leads to nuclear accumulation of the transcription factors, phosphorylation of Y694 in STAT5A by Src does not [107]. Later it was found that this observation is of some relevance in Bcr-Abl-positive chronic myeloid leukemia (CML), where STAT5 is predominantly localized in the cytoplasm despite being phosphorylated at the critical tyrosine residue [108]. Cytoplasmic retention of STAT5A in response to activation by Bcr-Abl was found to be mediated by SFK such as Src and Hck. Specific inhibition of SFKs resulted in nuclear accumulation of STAT5A, enhanced STAT5 target gene expression, and increased colony formation of CML cells [109]. SFK interfere specifically with dimerization of activated STAT5A and thereby prevent nuclear accumulation [110]. In another report, phosphorylation of S779 has been suggested as an additional requirement for the nuclear translocation of STAT5A in Bcr-Abl positive cells [111]. These intriguing examples show that even the highly homologous transcription factors STAT5A and STAT5B differ in some aspects of nucleocytoplasmic shuttling and these differences might be relevant for disease.

6. Perspectives for Therapeutic Interventions

Nucleocytoplasmic transport has been recognized as a target for cancer therapy [112,113] in part based on the observation that components of the nuclear transport machinery are differentially expressed in transformed cells [114]. These changes in expression can be relevant for disease. It has been shown that Crm1-mediated nuclear export contributes to drug resistance in multiple ways [115]. For example, Topoisomerase II inhibitors such as doxorubicin require nuclear Topoisomerase activity. As a drug resistance mechanism, Topoisomerase II can be exported from the nucleus in a Crm1-dependent manner [116]. Treatment with Crm1 inhibitors therefore sensitizes cancer cells to doxorubicin treatment [117]. Only recently, in two independent studies, was Crm1 convincingly identified as a synthetic lethality gene in different types of cancer [118,119]. Earlier clinical studies already showed that the highly efficient Crm1 inhibitor leptomycin B is not well tolerated by patients [120]. Less toxic small molecule inhibitors of Crm1 termed Selective Inhibitors of Nuclear Export (SINE), such as selinexor, are currently used in clinical trials in patients with hematological malignancies and solid tumors [112]. Most recently, selinexor in combination with dexamethasone has been approved for the treatment of refractory multiple myeloma. In this context, it has already been recognized that a more selective export inhibition could be more effective and reduce the serious side effects of treatment [115]. There are only a few inhibitors for some other NTRs available (Importin-β1, Importin-α/β1 heterodimers, and Transportin-1) which have not been explored in such detail yet [121].

For transcription factors that respond to extracellular cues, control of nucleocytoplasmic distribution contributes to efficient signal transduction. High cytoplasmic concentration of the latent transcription factor increases the sensitivity to stimulation, which usually occurs at the cytoplasmic face of membrane-bound receptors. After stimulation and activation of the transcription factor, high nuclear concentration increases the efficiency of DNA-binding and thus gene induction. However, the molecular mechanisms involved in the control of subcellular distribution vary considerably between different transcription factors. For instance, the NLS of NF-κB is masked by IκB [122]. Upon sensing of inflammatory mediators by membrane-bound receptors, IκB is phosphorylated and ubiquitinated resulting in its proteasomal degradation. The now exposed NLS facilitates α-Importin-mediated nuclear accumulation of NF-κB [123]. Likewise, the NLS of some nuclear receptors, such as the Glucocorticoid Receptor and Estrogen Receptor, are masked. Ligand binding to the receptors results in release of Hsp90 and exposition of the NLS [124]. Similar cytoplasmic interaction partners that would prevent nuclear translocation of latent STAT proteins have not been identified yet. Various other mechanisms involving SUMOylation, ubiquitination, mono-ADP-ribosylation, and acetylation have been described to control the nucleocytoplasmic distribution of transcription factors [22,125]. Beside the well-established Karyopherin-mediated transport, other non-conventional transport mechanisms exist [126]. Nucleocytoplasmic transport of β-Catenin is regulated independently of Ran and classical transport factors [127,128]. At the same time, β-Catenin acts as a transport receptor for its transcription factor partner LEF-1 [129]. The regulatory 14-3-3 proteins have also been suggested to be involved in such a piggy-back mechanism of nuclear transport [130,131], which has so far not been described for STAT proteins. Because of these various transport modes, specific interference should be feasible in principle.

In the case of STAT transcription factors, a non-classical NLS is generated through phosphorylation-induced dimerization (dsNLS) that is recognized by α-Importins or other yet uncharacterized factors. Because this mechanism is quite unique, a specific intervention through targeting the dsNLS/α-Importin interface should in general be feasible, as exemplified for STAT1 [67]. Extended surface areas are involved in the STAT/NTR interaction, including the NTD, DBD and CCD. The NTD seems to be a common denominator of nuclear import of activated STATs. Because the NTDs of different STATs vary in their sequence, a specific targeting should be possible, as exemplified by the specific interaction of ARHI with the NTD of STAT3, but not with the NTDs of STAT5 and STAT1 [92]. The impact of the ARHI/STAT3-NTD interaction on Importin-binding was further analyzed in a study applying homology modelling and molecular dynamics simulation [132]. Indeed, targeting of the NTD of STAT3 has been reported, however, with a focus on the role of the NTD as a tetramerization domain at enhancers and its function in gene induction [133]. The more recently solved structure of the NTD of STAT3 [134] should pave the way for a closer inspection of the surface areas involved in nuclear import by mutation of surface-exposed residues. For nuclear import of STAT5 and STAT3, data concerning use of the classical α-Importin/β1-Importin pathway are conflicting. Here, a fresh approach is needed to identify the NTRs involved, including α-Importins/Importin-β1-independent pathways mentioned in Section 2 of this review.

Reduced export of the imported STAT dimers also contributes to nuclear accumulation because export and redistribution requires dephosphorylation [135]. Blocking nuclear export of STATs as a therapeutic approach seems at first sight counterintuitive because it might trap the activated transcription factor in the nucleus where it could drive persistent oncogenic gene induction. However, STAT1, STAT3, and STAT5 are dephosphorylated in the nucleus by TC45, the 45 kDa isoform of T Cell Protein Tyrosine Phosphatase (TC-PTP) [136–138]. For STAT1 it has been shown that only the dephosphorylated form leaves the nucleus for reactivation at the receptor [91]. Exit out of the nucleus is required for reactivation of STATs [139]. Thus, STAT-mediated gene induction could be inhibited by trapping the dephosphorylated, transcriptionally inactive STATs (in the sense of canonical STAT signaling) in the nucleus through blockade of nuclear export. This has already been shown for STAT3 using the Crm1 inhibitor ratjadone A [140]. The more clinically advanced SINE have also been tested. These compounds inhibited STAT3-mediated Survivin expression in breast cancer models [141]. To interfere more specifically with export of STATs, the Exportins or Biportins involved must be characterized and the molecular interfaces of the STAT/NTR interaction mapped in detail.

Mechanistically, the export of dephosphorylated STATs most probably corresponds to the exit mechanisms involved in constitutive shuttling of latent STATs. Treatment of cells with specific inhibitors for the Exportin Crm1 often leads to only a partial nuclear accumulation of latent STATs. This means that either passive NTR-independent export occurs through direct hydrophobic interactions with FG-Nups or that other Exportins are involved. Specifically for STAT5A, a Crm1-independent export signal in the DBD has been mapped [73]. Redundancy in cargo recognition by NTRs in the sense that one cargo is transported by several NTRs is not unusual. As outlined in Section 2, broad-specificity Exportins or Biportins have been identified in recent years, worthy of analysis with respect to their involvement in the nuclear export of STAT proteins.

7. Conclusions

Taken together, effective inhibition of oncogenic STAT activity can in principle be achieved through specific blockade of both nuclear import of phosphorylated STAT dimers and export of dephosphorylated STATs out of the nucleus. However, this requires a deeper understanding of the protein–protein interactions involved in these processes. First, all NTRs facilitating nucleocytoplasmic shuttling need to be unambiguously identified for the individual STAT proteins. These studies should also include those NTRs that have been characterized in more detail in recent years. Then, the interactions between STATs and NTRs must be mapped in more detail using all structural data available. The ultimate goal would be to solve the structure of STAT/NTR complexes by X-ray crystallography or cryo-electron microscopy. Furthermore, the contribution of NTR-independent nuclear transport of STATs should be evaluated through assessment and mutation of surface patches as delineated by Frey et al. [56]. Based on solid structural and functional data, specific blockade of nucleocytoplasmic transport of individual STATs by tailor-made inhibitory molecules might be feasible.

Author Contributions: S.E. edited the manuscript, prepared figure and table and searched the literature, G.M.-N. drew the conceptual framework and wrote the manuscript.

Acknowledgments: This work was supported by the Confocal Microscopy Facility, a core facility of the Interdisciplinary Center for Clinical Research (IZKF) Aachen within the Faculty of Medicine at RWTH Aachen University. The authors would like to thank Wolfram Antonin and Bernhard Lüscher for critical reading of the manuscript. The authors apologize to all colleagues whose contributions have not been cited.

References

1. Feldherr, C.M. The uptake and accumulation of proteins by the cell nucleus. *Bioessays* **1985**, *3*, 52–55. [CrossRef] [PubMed]
2. Beck, M.; Hurt, E. The nuclear pore complex: Understanding its function through structural insight. *Nat. Rev. Mol. Cell Biol.* **2017**, *18*, 73–89. [CrossRef] [PubMed]
3. Mohr, D.; Frey, S.; Fischer, T.; Guttler, T.; Gorlich, D. Characterisation of the passive permeability barrier of nuclear pore complexes. *EMBO J.* **2009**, *28*, 2541–2553. [CrossRef] [PubMed]
4. Timney, B.L.; Raveh, B.; Mironska, R.; Trivedi, J.M.; Kim, S.J.; Russel, D.; Wente, S.R.; Sali, A.; Rout, M.P. Simple rules for passive diffusion through the nuclear pore complex. *J. Cell Biol.* **2016**, *215*, 57–76. [CrossRef]
5. Frey, S.; Gorlich, D. A saturated FG-repeat hydrogel can reproduce the permeability properties of nuclear pore complexes. *Cell* **2007**, *130*, 512–523. [CrossRef]
6. Lim, R.Y.; Huang, N.P.; Koser, J.; Deng, J.; Lau, K.H.; Schwarz-Herion, K.; Fahrenkrog, B.; Aebi, U. Flexible phenylalanine-glycine nucleoporins as entropic barriers to nucleocytoplasmic transport. *Proc. Natl. Acad. Sci. USA* **2006**, *103*, 9512–9517. [CrossRef]
7. Rout, M.P.; Aitchison, J.D.; Magnasco, M.O.; Chait, B.T. Virtual gating and nuclear transport: The hole picture. *Trends Cell Biol.* **2003**, *13*, 622–628. [CrossRef]
8. Schmidt, H.B.; Gorlich, D. Transport selectivity of nuclear pores, phase separation, and membraneless organelles. *Trends Biochem. Sci.* **2016**, *41*, 46–61. [CrossRef]
9. Hulsmann, B.B.; Labokha, A.A.; Gorlich, D. The permeability of reconstituted nuclear pores provides direct evidence for the selective phase model. *Cell* **2012**, *150*, 738–751. [CrossRef]
10. Radu, A.; Moore, M.S.; Blobel, G. The peptide repeat domain of nucleoporin Nup98 functions as a docking site in transport across the nuclear pore complex. *Cell* **1995**, *81*, 215–222. [CrossRef]
11. Ribbeck, K.; Gorlich, D. Kinetic analysis of translocation through nuclear pore complexes. *EMBO J.* **2001**, *20*, 1320–1330. [CrossRef] [PubMed]
12. Rexach, M.; Blobel, G. Protein import into nuclei: Association and dissociation reactions involving transport substrate, transport factors, and nucleoporins. *Cell* **1995**, *83*, 683–692. [CrossRef]
13. Wente, S.R.; Rout, M.P. The nuclear pore complex and nuclear transport. *Cold Spring Harb Perspect. Biol.* **2010**, *2*, a000562. [CrossRef] [PubMed]
14. Aksu, M.; Trakhanov, S.; Vera Rodriguez, A.; Gorlich, D. Structural basis for the nuclear import and export functions of the biportin Pdr6/Kap122. *J. Cell Biol.* **2019**, *218*, 1839–1852. [CrossRef] [PubMed]
15. Harel, A.; Forbes, D.J. Importin beta: Conducting a much larger cellular symphony. *Mol. Cell* **2004**, *16*, 319–330. [CrossRef] [PubMed]
16. Choi, S.; Yamashita, E.; Yasuhara, N.; Song, J.; Son, S.Y.; Won, Y.H.; Hong, H.R.; Shin, Y.S.; Sekimoto, T.; Park, I.Y.; et al. Structural basis for the selective nuclear import of the C2H2 zinc-finger protein Snail by importin beta. *Acta Crystallogr. D Biol. Crystallogr.* **2014**, *70*, 1050–1060. [CrossRef]
17. Yamasaki, H.; Sekimoto, T.; Ohkubo, T.; Douchi, T.; Nagata, Y.; Ozawa, M.; Yoneda, Y. Zinc finger domain of Snail functions as a nuclear localization signal for importin beta-mediated nuclear import pathway. *Genes Cells* **2005**, *10*, 455–464. [CrossRef]
18. Lee, S.J.; Sekimoto, T.; Yamashita, E.; Nagoshi, E.; Nakagawa, A.; Imamoto, N.; Yoshimura, M.; Sakai, H.; Chong, K.T.; Tsukihara, T.; et al. The structure of importin-beta bound to SREBP-2: Nuclear import of a transcription factor. *Science* **2003**, *302*, 1571–1575. [CrossRef]
19. Huber, J.; Cronshagen, U.; Kadokura, M.; Marshallsay, C.; Wada, T.; Sekine, M.; Luhrmann, R. Snurportin1, an m3G-cap-specific nuclear import receptor with a novel domain structure. *EMBO J.* **1998**, *17*, 4114–4126. [CrossRef]

20. Bayliss, R.; Littlewood, T.; Strawn, L.A.; Wente, S.R.; Stewart, M. GLFG and FxFG nucleoporins bind to overlapping sites on importin-beta. *J. Biol. Chem.* **2002**, *277*, 50597–50606. [CrossRef]
21. Bednenko, J.; Cingolani, G.; Gerace, L. Importin beta contains a COOH-terminal nucleoporin binding region important for nuclear transport. *J. Cell Biol.* **2003**, *162*, 391–401. [CrossRef] [PubMed]
22. Christie, M.; Chang, C.W.; Rona, G.; Smith, K.M.; Stewart, A.G.; Takeda, A.A.; Fontes, M.R.; Stewart, M.; Vertessy, B.G.; Forwood, J.K.; et al. Structural biology and regulation of protein import into the nucleus. *J. Mol. Biol.* **2016**, *428*, 2060–2090. [CrossRef] [PubMed]
23. Pumroy, R.A.; Cingolani, G. Diversification of importin-alpha isoforms in cellular trafficking and disease states. *Biochem. J.* **2015**, *466*, 13–28. [CrossRef] [PubMed]
24. Kalderon, D.; Richardson, W.D.; Markham, A.F.; Smith, A.E. Sequence requirements for nuclear location of simian virus 40 large-T antigen. *Nature* **1984**, *311*, 33–38. [CrossRef]
25. Dingwall, C.; Robbins, J.; Dilworth, S.M.; Roberts, B.; Richardson, W.D. The nucleoplasmin nuclear location sequence is larger and more complex than that of SV-40 large T antigen. *J. Cell Biol.* **1988**, *107*, 841–849. [CrossRef] [PubMed]
26. Kelley, J.B.; Talley, A.M.; Spencer, A.; Gioeli, D.; Paschal, B.M. Karyopherin alpha7 (KPNA7), a divergent member of the importin alpha family of nuclear import receptors. *BMC Cell Biol.* **2010**, *11*, 63. [CrossRef]
27. Andrade, M.A.; Perez-Iratxeta, C.; Ponting, C.P. Protein repeats: Structures, functions, and evolution. *J. Struct. Biol.* **2001**, *134*, 117–131. [CrossRef]
28. Sankhala, R.S.; Lokareddy, R.K.; Begum, S.; Pumroy, R.A.; Gillilan, R.E.; Cingolani, G. Three-dimensional context rather than NLS amino acid sequence determines importin alpha subtype specificity for RCC1. *Nat. Commun.* **2017**, *8*, 979. [CrossRef]
29. Kosugi, S.; Hasebe, M.; Matsumura, N.; Takashima, H.; Miyamoto-Sato, E.; Tomita, M.; Yanagawa, H. Six classes of nuclear localization signals specific to different binding grooves of importin alpha. *J. Biol. Chem.* **2009**, *284*, 478–485. [CrossRef]
30. Cingolani, G.; Petosa, C.; Weis, K.; Muller, C.W. Structure of importin-beta bound to the IBB domain of importin-alpha. *Nature* **1999**, *399*, 221–229. [CrossRef]
31. Lee, B.J.; Cansizoglu, A.E.; Suel, K.E.; Louis, T.H.; Zhang, Z.; Chook, Y.M. Rules for nuclear localization sequence recognition by karyopherin beta 2. *Cell* **2006**, *126*, 543–558. [CrossRef] [PubMed]
32. Kressler, D.; Bange, G.; Ogawa, Y.; Stjepanovic, G.; Bradatsch, B.; Pratte, D.; Amlacher, S.; Strauss, D.; Yoneda, Y.; Katahira, J.; et al. Synchronizing nuclear import of ribosomal proteins with ribosome assembly. *Science* **2012**, *338*, 666–671. [CrossRef] [PubMed]
33. Lai, M.C.; Lin, R.I.; Tarn, W.Y. Transportin-SR2 mediates nuclear import of phosphorylated SR proteins. *Proc. Natl. Acad. Sci. USA* **2001**, *98*, 10154–10159. [CrossRef] [PubMed]
34. Baade, I.; Spillner, C.; Schmitt, K.; Valerius, O.; Kehlenbach, R.H. Extensive identification and in-depth validation of importin 13 cargoes. *Mol. Cell. Proteomics* **2018**, *17*, 1337–1353. [CrossRef] [PubMed]
35. Grunwald, M.; Lazzaretti, D.; Bono, F. Structural basis for the nuclear export activity of Importin13. *EMBO J.* **2013**, *32*, 899–913. [CrossRef]
36. Hanover, J.A.; Love, D.C.; Prinz, W.A. Calmodulin-driven nuclear entry: Trigger for sex determination and terminal differentiation. *J. Biol. Chem.* **2009**, *284*, 12593–12597. [CrossRef]
37. Sharma, M.; Jamieson, C.; Lui, C.; Henderson, B.R. Distinct hydrophobic "patches" in the N- and C-tails of beta-catenin contribute to nuclear transport. *Exp. Cell Res.* **2016**, *348*, 132–145. [CrossRef]
38. Kirli, K.; Karaca, S.; Dehne, H.J.; Samwer, M.; Pan, K.T.; Lenz, C.; Urlaub, H.; Gorlich, D. A deep proteomics perspective on CRM1-mediated nuclear export and nucleocytoplasmic partitioning. *eLife* **2015**, *4*, e11466. [CrossRef]
39. Fornerod, M.; Ohno, M.; Yoshida, M.; Mattaj, I.W. CRM1 is an export receptor for leucine-rich nuclear export signals. *Cell* **1997**, *90*, 1051–1060. [CrossRef]
40. Kosugi, S.; Hasebe, M.; Tomita, M.; Yanagawa, H. Nuclear export signal consensus sequences defined using a localization-based yeast selection system. *Traffic* **2008**, *9*, 2053–2062. [CrossRef]
41. Stade, K.; Ford, C.S.; Guthrie, C.; Weis, K. Exportin 1 (Crm1p) is an essential nuclear export factor. *Cell* **1997**, *90*, 1041–1050. [CrossRef]
42. Andrade, M.A.; Petosa, C.; O'Donoghue, S.I.; Muller, C.W.; Bork, P. Comparison of ARM and HEAT protein repeats. *J. Mol. Biol.* **2001**, *309*, 1–18. [CrossRef] [PubMed]

43. Matsuura, Y. Mechanistic insights from structural analyses of Ran-GTPase-Driven nuclear export of proteins and RNAs. *J. Mol. Biol.* **2016**, *428*, 2025–2039. [CrossRef] [PubMed]
44. Kohler, A.; Hurt, E. Exporting RNA from the nucleus to the cytoplasm. *Nat. Rev. Mol. Cell Biol.* **2007**, *8*, 761–773. [CrossRef]
45. Kutay, U.; Bischoff, F.R.; Kostka, S.; Kraft, R.; Gorlich, D. Export of importin alpha from the nucleus is mediated by a specific nuclear transport factor. *Cell* **1997**, *90*, 1061–1071. [CrossRef]
46. Lipowsky, G.; Bischoff, F.R.; Schwarzmaier, P.; Kraft, R.; Kostka, S.; Hartmann, E.; Kutay, U.; Gorlich, D. Exportin 4: A mediator of a novel nuclear export pathway in higher eukaryotes. *EMBO J.* **2000**, *19*, 4362–4371. [CrossRef]
47. Kurisaki, A.; Kurisaki, K.; Kowanetz, M.; Sugino, H.; Yoneda, Y.; Heldin, C.H.; Moustakas, A. The mechanism of nuclear export of Smad3 involves exportin 4 and Ran. *Mol. Cell. Biol.* **2006**, *26*, 1318–1332. [CrossRef]
48. Gontan, C.; Guttler, T.; Engelen, E.; Demmers, J.; Fornerod, M.; Grosveld, F.G.; Tibboel, D.; Gorlich, D.; Poot, R.A.; Rottier, R.J. Exportin 4 mediates a novel nuclear import pathway for Sox family transcription factors. *J. Cell Biol.* **2009**, *185*, 27–34. [CrossRef]
49. Mingot, J.M.; Bohnsack, M.T.; Jakle, U.; Gorlich, D. Exportin 7 defines a novel general nuclear export pathway. *EMBO J.* **2004**, *23*, 3227–3236. [CrossRef]
50. Aksu, M.; Pleiner, T.; Karaca, S.; Kappert, C.; Dehne, H.J.; Seibel, K.; Urlaub, H.; Bohnsack, M.T.; Gorlich, D. Xpo7 is a broad-spectrum exportin and a nuclear import receptor. *J. Cell Biol.* **2018**, *217*, 2329–2340. [CrossRef]
51. Nachury, M.V.; Weis, K. The direction of transport through the nuclear pore can be inverted. *Proc. Natl. Acad. Sci. USA* **1999**, *96*, 9622–9627. [CrossRef] [PubMed]
52. Kuersten, S.; Ohno, M.; Mattaj, I.W. Nucleocytoplasmic transport: Ran, beta and beyond. *Trends Cell Biol.* **2001**, *11*, 497–503. [CrossRef]
53. Macara, I.G. Transport into and out of the nucleus. *Microbiol. Mol. Biol. Rev.* **2001**, *65*, 570–594. [CrossRef] [PubMed]
54. Melchior, F.; Gerace, L. Two-way trafficking with Ran. *Trends Cell Biol.* **1998**, *8*, 175–179. [CrossRef]
55. Guttler, T.; Gorlich, D. Ran-dependent nuclear export mediators: A structural perspective. *EMBO J.* **2011**, *30*, 3457–3474. [CrossRef] [PubMed]
56. Frey, S.; Rees, R.; Schunemann, J.; Ng, S.C.; Funfgeld, K.; Huyton, T.; Gorlich, D. Surface properties determining passage rates of proteins through nuclear pores. *Cell* **2018**, *174*, 202–217. [CrossRef]
57. Parker, B.S.; Rautela, J.; Hertzog, P.J. Antitumour actions of interferons: Implications for cancer therapy. *Nat. Rev. Cancer* **2016**, *16*, 131–144. [CrossRef]
58. Meissl, K.; Macho-Maschler, S.; Muller, M.; Strobl, B. The good and the bad faces of STAT1 in solid tumours. *Cytokine* **2017**, *89*, 12–20. [CrossRef]
59. Green, D.S.; Young, H.A.; Valencia, J.C. Current prospects of type II interferon gamma signaling and autoimmunity. *J. Biol. Chem.* **2017**, *292*, 13925–13933. [CrossRef]
60. Sekimoto, T.; Imamoto, N.; Nakajima, K.; Hirano, T.; Yoneda, Y. Extracellular signal-dependent nuclear import of Stat1 is mediated by nuclear pore-targeting complex formation with NPI-1, but not Rch1. *EMBO J.* **1997**, *16*, 7067–7077. [CrossRef]
61. Melen, K.; Fagerlund, R.; Franke, J.; Kohler, M.; Kinnunen, L.; Julkunen, I. Importin alpha nuclear localization signal binding sites for STAT1, STAT2, and influenza A virus nucleoprotein. *J. Biol. Chem.* **2003**, *278*, 28193–28200. [CrossRef] [PubMed]
62. Nardozzi, J.; Wenta, N.; Yasuhara, N.; Vinkemeier, U.; Cingolani, G. Molecular basis for the recognition of phosphorylated STAT1 by importin alpha5. *J. Mol. Biol.* **2010**, *402*, 83–100. [CrossRef] [PubMed]
63. Meissner, T.; Krause, E.; Lodige, I.; Vinkemeier, U. Arginine methylation of STAT1: A reassessment. *Cell* **2004**, *119*, 587–589. [CrossRef] [PubMed]
64. Meyer, T.; Begitt, A.; Lodige, I.; van Rossum, M.; Vinkemeier, U. Constitutive and IFN-gamma-induced nuclear import of STAT1 proceed through independent pathways. *EMBO J.* **2002**, *21*, 344–354. [CrossRef]
65. Fagerlund, R.; Melen, K.; Kinnunen, L.; Julkunen, I. Arginine/lysine-rich nuclear localization signals mediate interactions between dimeric STATs and importin alpha 5. *J. Biol. Chem.* **2002**, *277*, 30072–30078. [CrossRef]
66. McBride, K.M.; McDonald, C.; Reich, N.C. Nuclear export signal located within theDNA-binding domain of the STAT1transcription factor. *EMBO J.* **2000**, *19*, 6196–6206. [CrossRef]

67. Xu, W.; Edwards, M.R.; Borek, D.M.; Feagins, A.R.; Mittal, A.; Alinger, J.B.; Berry, K.N.; Yen, B.; Hamilton, J.; Brett, T.J.; et al. Ebola virus VP24 targets a unique NLS binding site on karyopherin alpha 5 to selectively compete with nuclear import of phosphorylated STAT1. *Cell Host Microbe* **2014**, *16*, 187–200. [CrossRef]
68. McBride, K.M.; Banninger, G.; McDonald, C.; Reich, N.C. Regulated nuclear import of the STAT1 transcription factor by direct binding of importin-alpha. *EMBO J.* **2002**, *21*, 1754–1763. [CrossRef]
69. Ma, J.; Cao, X. Regulation of Stat3 nuclear import by importin alpha5 and importin alpha7 via two different functional sequence elements. *Cell. Signal.* **2006**, *18*, 1117–1126. [CrossRef]
70. Ushijima, R.; Sakaguchi, N.; Kano, A.; Maruyama, A.; Miyamoto, Y.; Sekimoto, T.; Yoneda, Y.; Ogino, K.; Tachibana, T. Extracellular signal-dependent nuclear import of STAT3 is mediated by various importin alphas. *Biochem. Biophys. Res. Commun.* **2005**, *330*, 880–886. [CrossRef]
71. Liu, L.; McBride, K.M.; Reich, N.C. STAT3 nuclear import is independent of tyrosine phosphorylation and mediated by importin-alpha3. *Proc. Natl. Acad. Sci. USA* **2005**, *102*, 8150–8155. [CrossRef] [PubMed]
72. Martincuks, A.; Fahrenkamp, D.; Haan, S.; Herrmann, A.; Kuster, A.; Muller-Newen, G. Dissecting functions of the N-terminal domain and GAS-site recognition in STAT3 nuclear trafficking. *Cell. Signal.* **2016**, *28*, 810–825. [CrossRef] [PubMed]
73. Shin, H.Y.; Reich, N.C. Dynamic trafficking of STAT5 depends on an unconventional nuclear localization signal. *J. Cell Sci.* **2013**, *126*, 3333–3343. [CrossRef] [PubMed]
74. Strehlow, I.; Schindler, C. Amino-terminal signal transducer and activator of transcription (STAT) domains regulate nuclear translocation and STAT deactivation. *J. Biol. Chem.* **1998**, *273*, 28049–28056. [CrossRef]
75. Vogt, M.; Domoszlai, T.; Kleshchanok, D.; Lehmann, S.; Schmitt, A.; Poli, V.; Richtering, W.; Muller-Newen, G. The role of the N-terminal domain in dimerization and nucleocytoplasmic shuttling of latent STAT3. *J. Cell Sci.* **2011**, *124*, 900–909. [CrossRef]
76. Iyer, J.; Reich, N.C. Constitutive nuclear import of latent and activated STAT5a by its coiled coil domain. *FASEB J.* **2008**, *22*, 391–400. [CrossRef]
77. Zeng, R.; Aoki, Y.; Yoshida, M.; Arai, K.; Watanabe, S. Stat5B shuttles between cytoplasm and nucleus in a cytokine-dependent and -independent manner. *J. Immunol.* **2002**, *168*, 4567–4575. [CrossRef]
78. Begitt, A.; Meyer, T.; van Rossum, M.; Vinkemeier, U. Nucleocytoplasmic translocation of Stat1 is regulated by a leucine-rich export signal in the coiled-coil domain. *Proc. Natl. Acad. Sci. USA* **2000**, *97*, 10418–10423. [CrossRef]
79. Bhattacharya, S.; Schindler, C. Regulation of Stat3 nuclear export. *J. Clin. Investig.* **2003**, *111*, 553–559. [CrossRef]
80. Bromberg, J.F.; Wrzeszczynska, M.H.; Devgan, G.; Zhao, Y.; Pestell, R.G.; Albanese, C.; Darnell, J.E., Jr. Stat3 as an oncogene. *Cell* **1999**, *98*, 295–303. [CrossRef]
81. Pilati, C.; Zucman-Rossi, J. Mutations leading to constitutive active gp130/JAK1/STAT3 pathway. *Cytokine Growth Factor Rev.* **2015**, *26*, 499–506. [CrossRef] [PubMed]
82. Yu, H.; Lee, H.; Herrmann, A.; Buettner, R.; Jove, R. Revisiting STAT3 signalling in cancer: New and unexpected biological functions. *Nat. Rev. Cancer* **2014**, *14*, 736–746. [CrossRef] [PubMed]
83. Aigner, P.; Mizutani, T.; Horvath, J.; Eder, T.; Heber, S.; Lind, K.; Just, V.; Moll, H.P.; Yeroslaviz, A.; Fischer, M.J.M.; et al. STAT3beta is a tumor suppressor in acute myeloid leukemia. *Blood Adv.* **2019**, *3*, 1989–2002. [CrossRef]
84. Pencik, J.; Schlederer, M.; Gruber, W.; Unger, C.; Walker, S.M.; Chalaris, A.; Marie, I.J.; Hassler, M.R.; Javaheri, T.; Aksoy, O.; et al. STAT3 regulated ARF expression suppresses prostate cancer metastasis. *Nat. Commun.* **2015**, *6*, 7736. [CrossRef] [PubMed]
85. Fagard, R.; Metelev, V.; Souissi, I.; Baran-Marszak, F. STAT3 inhibitors for cancer therapy: Have all roads been explored? *JAKSTAT* **2013**, *2*, e22882. [CrossRef]
86. Meyer, T.; Vinkemeier, U. STAT nuclear translocation: Potential for pharmacological intervention. *Expert Opin. Ther. Targets* **2007**, *11*, 1355–1365. [CrossRef]
87. Kohler, M.; Ansieau, S.; Prehn, S.; Leutz, A.; Haller, H.; Hartmann, E. Cloning of two novel human importin-alpha subunits and analysis of the expression pattern of the importin-alpha protein family. *FEBS Lett.* **1997**, *417*, 104–108. [CrossRef]
88. Kawashima, T.; Bao, Y.C.; Minoshima, Y.; Nomura, Y.; Hatori, T.; Hori, T.; Fukagawa, T.; Fukada, T.; Takahashi, N.;

Nosaka, T.; et al. A Rac GTPase-activating protein, MgcRacGAP, is a nuclear localizing signal-containing nuclear chaperone in the activation of STAT transcription factors. *Mol. Cell. Biol.* **2009**, *29*, 1796–1813. [CrossRef]

89. van Adrichem, A.J.; Wennerberg, K. MgcRacGAP inhibition stimulates JAK-dependent STAT3 activity. *FEBS Lett.* **2015**, *589*, 3859–3865. [CrossRef]
90. Martincuks, A.; Andryka, K.; Kuster, A.; Schmitz-Van de Leur, H.; Komorowski, M.; Muller-Newen, G. Nuclear translocation of STAT3 and NF-kappaB are independent of each other but NF-kappaB supports expression and activation of STAT3. *Cell. Signal.* **2017**, *32*, 36–47. [CrossRef]
91. Meyer, T.; Marg, A.; Lemke, P.; Wiesner, B.; Vinkemeier, U. DNA binding controls inactivation and nuclear accumulation of the transcription factor Stat1. *Genes Dev.* **2003**, *17*, 1992–2005. [CrossRef] [PubMed]
92. Nishimoto, A.; Yu, Y.; Lu, Z.; Mao, X.; Ren, Z.; Watowich, S.S.; Mills, G.B.; Liao, W.S.; Chen, X.; Bast, R.C., Jr.; et al. A Ras homologue member I directly inhibits signal transducers and activators of transcription 3 translocation and activity in human breast and ovarian cancer cells. *Cancer Res.* **2005**, *65*, 6701–6710. [CrossRef] [PubMed]
93. Mohr, A.; Chatain, N.; Domoszlai, T.; Rinis, N.; Sommerauer, M.; Vogt, M.; Muller-Newen, G. Dynamics and non-canonical aspects of JAK/STAT signalling. *Eur. J. Cell Biol.* **2012**, *91*, 524–532. [CrossRef] [PubMed]
94. Pranada, A.L.; Metz, S.; Herrmann, A.; Heinrich, P.C.; Muller-Newen, G. Real time analysis of STAT3 nucleocytoplasmic shuttling. *J. Biol. Chem.* **2004**, *279*, 15114–15123. [CrossRef] [PubMed]
95. Meyer, T.; Gavenis, K.; Vinkemeier, U. Cell type-specific and tyrosine phosphorylation-independent nuclear presence of STAT1 and STAT3. *Exp. Cell Res.* **2002**, *272*, 45–55. [CrossRef] [PubMed]
96. Marg, A.; Shan, Y.; Meyer, T.; Meissner, T.; Brandenburg, M.; Vinkemeier, U. Nucleocytoplasmic shuttling by nucleoporins Nup153 and Nup214 and CRM1-dependent nuclear export control the subcellular distribution of latent Stat1. *J. Cell Biol.* **2004**, *165*, 823–833. [CrossRef] [PubMed]
97. Braunstein, J.; Brutsaert, S.; Olson, R.; Schindler, C. STATs dimerize in the absence of phosphorylation. *J. Biol. Chem.* **2003**, *278*, 34133–34140. [CrossRef]
98. Haan, S.; Kortylewski, M.; Behrmann, I.; Muller-Esterl, W.; Heinrich, P.C.; Schaper, F. Cytoplasmic STAT proteins associate prior to activation. *Biochem. J.* **2000**, *345 Pt 3*, 417–421. [CrossRef]
99. Wenta, N.; Strauss, H.; Meyer, S.; Vinkemeier, U. Tyrosine phosphorylation regulates the partitioning of STAT1 between different dimer conformations. *Proc. Natl. Acad. Sci. USA* **2008**, *105*, 9238–9243. [CrossRef]
100. Domoszlai, T.; Martincuks, A.; Fahrenkamp, D.; Schmitz-Van de Leur, H.; Kuster, A.; Muller-Newen, G. Consequences of the disease-related L78R mutation for dimerization and activity of STAT3. *J. Cell Sci.* **2014**, *127*, 1899–1910. [CrossRef]
101. Majoros, A.; Platanitis, E.; Kernbauer-Holzl, E.; Rosebrock, F.; Muller, M.; Decker, T. Canonical and non-canonical aspects of JAK-STAT signaling: Lessons from interferons for cytokine responses. *Front. Immunol.* **2017**, *8*, 29. [CrossRef] [PubMed]
102. Wingelhofer, B.; Neubauer, H.A.; Valent, P.; Han, X.; Constantinescu, S.N.; Gunning, P.T.; Muller, M.; Moriggl, R. Implications of STAT3 and STAT5 signaling on gene regulation and chromatin remodeling in hematopoietic cancer. *Leukemia* **2018**, *32*, 1713–1726. [CrossRef] [PubMed]
103. Ferbeyre, G.; Moriggl, R. The role of Stat5 transcription factors as tumor suppressors or oncogenes. *Biochim. Biophys. Acta* **2011**, *1815*, 104–114. [CrossRef] [PubMed]
104. Herrington, J.; Rui, L.; Luo, G.; Yu-Lee, L.Y.; Carter-Su, C. A functional DNA binding domain is required for growth hormone-induced nuclear accumulation of Stat5B. *J. Biol. Chem.* **1999**, *274*, 5138–5145. [CrossRef] [PubMed]
105. Lillemeier, B.F.; Koster, M.; Kerr, I.M. STAT1 from the cell membrane to the DNA. *EMBO J.* **2001**, *20*, 2508–2517. [CrossRef]
106. Park, H.J.; Li, J.; Hannah, R.; Biddie, S.; Leal-Cervantes, A.I.; Kirschner, K.; Flores Santa Cruz, D.; Sexl, V.; Gottgens, B.; Green, A.R. Cytokine-induced megakaryocytic differentiation is regulated by genome-wide loss of a uSTAT transcriptional program. *EMBO J.* **2016**, *35*, 580–594. [CrossRef]
107. Kazansky, A.V.; Kabotyanski, E.B.; Wyszomierski, S.L.; Mancini, M.A.; Rosen, J.M. Differential effects of prolactin and src/abl kinases on the nuclear translocation of STAT5B and STAT5A. *J. Biol. Chem.* **1999**, *274*, 22484–22492. [CrossRef]
108. Harir, N.; Pecquet, C.; Kerenyi, M.; Sonneck, K.; Kovacic, B.; Nyga, R.; Brevet, M.; Dhennin, I.; Gouilleux-Gruart, V.; Beug, H.; et al. Constitutive activation of Stat5 promotes its cytoplasmic localization and association with PI3-kinase in myeloid leukemias. *Blood* **2007**, *109*, 1678–1686. [CrossRef]

109. Chatain, N.; Ziegler, P.; Fahrenkamp, D.; Jost, E.; Moriggl, R.; Schmitz-Van de Leur, H.; Muller-Newen, G. Src family kinases mediate cytoplasmic retention of activated STAT5 in BCR-ABL-positive cells. *Oncogene* **2013**, *32*, 3587–3597. [CrossRef]
110. Fahrenkamp, D.; de Leur, H.S.; Kuster, A.; Chatain, N.; Muller-Newen, G. Src family kinases interfere with dimerization of STAT5A through a phosphotyrosine-SH2 domain interaction. *Cell Commun. Signal.* **2015**, *13*, 10. [CrossRef]
111. Berger, A.; Hoelbl-Kovacic, A.; Bourgeais, J.; Hoefling, L.; Warsch, W.; Grundschober, E.; Uras, I.Z.; Menzl, I.; Putz, E.M.; Hoermann, G.; et al. PAK-dependent STAT5 serine phosphorylation is required for BCR-ABL-induced leukemogenesis. *Leukemia* **2014**, *28*, 629–641. [CrossRef] [PubMed]
112. Cagatay, T.; Chook, Y.M. Karyopherins in cancer. *Curr. Opin. Cell Biol.* **2018**, *52*, 30–42. [CrossRef] [PubMed]
113. Mahipal, A.; Malafa, M. Importins and exportins as therapeutic targets in cancer. *Pharmacol. Ther.* **2016**, *164*, 135–143. [CrossRef] [PubMed]
114. Poon, I.K.; Jans, D.A. Regulation of nuclear transport: Central role in development and transformation? *Traffic* **2005**, *6*, 173–186. [CrossRef]
115. Turner, J.G.; Dawson, J.; Cubitt, C.L.; Baz, R.; Sullivan, D.M. Inhibition of CRM1-dependent nuclear export sensitizes malignant cells to cytotoxic and targeted agents. *Semin. Cancer Biol.* **2014**, *27*, 62–73. [CrossRef]
116. Turner, J.G.; Engel, R.; Derderian, J.A.; Jove, R.; Sullivan, D.M. Human topoisomerase IIalpha nuclear export is mediated by two CRM-1-dependent nuclear export signals. *J. Cell Sci.* **2004**, *117*, 3061–3071. [CrossRef]
117. Turner, J.G.; Dawson, J.; Emmons, M.F.; Cubitt, C.L.; Kauffman, M.; Shacham, S.; Hazlehurst, L.A.; Sullivan, D.M. CRM1 inhibition sensitizes drug resistant human myeloma cells to topoisomerase II and proteasome inhibitors both In Vitro and Ex Vivo. *J. Cancer* **2013**, *4*, 614–625. [CrossRef]
118. Hong, A.L.; Tseng, Y.Y.; Cowley, G.S.; Jonas, O.; Cheah, J.H.; Kynnap, B.D.; Doshi, M.B.; Oh, C.; Meyer, S.C.; Church, A.J.; et al. Integrated genetic and pharmacologic interrogation of rare cancers. *Nat. Commun.* **2016**, *7*, 11987. [CrossRef]
119. Kim, J.; McMillan, E.; Kim, H.S.; Venkateswaran, N.; Makkar, G.; Rodriguez-Canales, J.; Villalobos, P.; Neggers, J.E.; Mendiratta, S.; Wei, S.; et al. XPO1-dependent nuclear export is a druggable vulnerability in KRAS-mutant lung cancer. *Nature* **2016**, *538*, 114–117. [CrossRef]
120. Newlands, E.S.; Rustin, G.J.; Brampton, M.H. Phase I trial of elactocin. *Br. J. Cancer* **1996**, *74*, 648–649. [CrossRef]
121. Stelma, T.; Chi, A.; van der Watt, P.J.; Verrico, A.; Lavia, P.; Leaner, V.D. Targeting nuclear transporters in cancer: Diagnostic, prognostic and therapeutic potential. *IUBMB Life* **2016**, *68*, 268–280. [CrossRef] [PubMed]
122. Beg, A.A.; Ruben, S.M.; Scheinman, R.I.; Haskill, S.; Rosen, C.A.; Baldwin, A.S., Jr. I kappa B interacts with the nuclear localization sequences of the subunits of NF-kappa B: A mechanism for cytoplasmic retention. *Genes Dev.* **1992**, *6*, 1899–1913. [CrossRef] [PubMed]
123. Fagerlund, R.; Kinnunen, L.; Kohler, M.; Julkunen, I.; Melen, K. NF-κB is transported into the nucleus by importin α3 and importin α4. *J. Biol. Chem.* **2005**, *280*, 15942–15951. [CrossRef] [PubMed]
124. Echeverria, P.C.; Picard, D. Molecular chaperones, essential partners of steroid hormone receptors for activity and mobility. *Biochim. Biophys. Acta* **2010**, *1803*, 641–649. [CrossRef] [PubMed]
125. Verheugd, P.; Forst, A.H.; Milke, L.; Herzog, N.; Feijs, K.L.; Kremmer, E.; Kleine, H.; Luscher, B. Regulation of NF-kappaB signalling by the mono-ADP-ribosyltransferase ARTD10. *Nat. Commun.* **2013**, *4*, 1683. [CrossRef] [PubMed]
126. Wagstaff, K.M.; Jans, D.A. Importins and beyond: Non-conventional nuclear transport mechanisms. *Traffic* **2009**, *10*, 1188–1198. [CrossRef]
127. Wiechens, N.; Fagotto, F. CRM1- and Ran-independent nuclear export of beta-catenin. *Curr. Biol.* **2001**, *11*, 18–27. [CrossRef]
128. Yokoya, F.; Imamoto, N.; Tachibana, T.; Yoneda, Y. Beta-catenin can be transported into the nucleus in a Ran-unassisted manner. *Mol. Biol. Cell* **1999**, *10*, 1119–1131. [CrossRef]
129. Asally, M.; Yoneda, Y. Beta-catenin can act as a nuclear import receptor for its partner transcription factor, lymphocyte enhancer factor-1 (lef-1). *Exp. Cell Res.* **2005**, *308*, 357–363. [CrossRef]
130. Brunet, A.; Kanai, F.; Stehn, J.; Xu, J.; Sarbassova, D.; Frangioni, J.V.; Dalal, S.N.; DeCaprio, J.A.; Greenberg, M.E.; Yaffe, M.B. 14-3-3 transits to the nucleus and participates in dynamic nucleocytoplasmic transport. *J. Cell Biol.* **2002**, *156*, 817–828. [CrossRef]

131. van Hemert, M.J.; Niemantsverdriet, M.; Schmidt, T.; Backendorf, C.; Spaink, H.P. Isoform-specific differences in rapid nucleocytoplasmic shuttling cause distinct subcellular distributions of 14-3-3 sigma and 14-3-3 zeta. *J. Cell Sci.* **2004**, *117*, 1411–1420. [CrossRef] [PubMed]
132. Muthu, K.; Panneerselvam, M.; Topno, N.S.; Jayaraman, M.; Ramadas, K. Structural perspective of ARHI mediated inhibition of STAT3 signaling: An insight into the inactive to active transition of ARHI and its interaction with STAT3 and importinbeta. *Cell. Signal.* **2015**, *27*, 739–755. [CrossRef] [PubMed]
133. Timofeeva, O.A.; Tarasova, N.I.; Zhang, X.; Chasovskikh, S.; Cheema, A.K.; Wang, H.; Brown, M.L.; Dritschilo, A. STAT3 suppresses transcription of proapoptotic genes in cancer cells with the involvement of its N-terminal domain. *Proc. Natl. Acad. Sci. USA* **2013**, *110*, 1267–1272. [CrossRef] [PubMed]
134. Hu, T.; Yeh, J.E.; Pinello, L.; Jacob, J.; Chakravarthy, S.; Yuan, G.C.; Chopra, R.; Frank, D.A. Impact of the N-Terminal Domain of STAT3 in STAT3-Dependent Transcriptional Activity. *Mol. Cell. Biol.* **2015**, *35*, 3284–3300. [CrossRef]
135. Haspel, R.L.; Darnell, J.E., Jr. A nuclear protein tyrosine phosphatase is required for the inactivation of Stat1. *Proc. Natl. Acad. Sci. USA* **1999**, *96*, 10188–10193. [CrossRef]
136. Aoki, N.; Matsuda, T. A nuclear protein tyrosine phosphatase TC-PTP is a potential negative regulator of the PRL-mediated signaling pathway: Dephosphorylation and deactivation of signal transducer and activator of transcription 5a and 5b by TC-PTP in nucleus. *Mol. Endocrinol.* **2002**, *16*, 58–69. [CrossRef]
137. ten Hoeve, J.; de Jesus Ibarra-Sanchez, M.; Fu, Y.; Zhu, W.; Tremblay, M.; David, M.; Shuai, K. Identification of a nuclear Stat1 protein tyrosine phosphatase. *Mol. Cell. Biol.* **2002**, *22*, 5662–5668. [CrossRef]
138. Yamamoto, T.; Sekine, Y.; Kashima, K.; Kubota, A.; Sato, N.; Aoki, N.; Matsuda, T. The nuclear isoform of protein-tyrosine phosphatase TC-PTP regulates interleukin-6-mediated signaling pathway through STAT3 dephosphorylation. *Biochem. Biophys. Res. Commun.* **2002**, *297*, 811–817. [CrossRef]
139. Lodige, I.; Marg, A.; Wiesner, B.; Malecova, B.; Oelgeschlager, T.; Vinkemeier, U. Nuclear export determines the cytokine sensitivity of STAT transcription factors. *J. Biol. Chem.* **2005**, *280*, 43087–43099. [CrossRef]
140. Herrmann, A.; Vogt, M.; Monnigmann, M.; Clahsen, T.; Sommer, U.; Haan, S.; Poli, V.; Heinrich, P.C.; Muller-Newen, G. Nucleocytoplasmic shuttling of persistently activated STAT3. *J. Cell Sci.* **2007**, *120*, 3249–3261. [CrossRef]
141. Cheng, Y.; Holloway, M.P.; Nguyen, K.; McCauley, D.; Landesman, Y.; Kauffman, M.G.; Shacham, S.; Altura, R.A. XPO1 (CRM1) inhibition represses STAT3 activation to drive a survivin-dependent oncogenic switch in triple-negative breast cancer. *Mol. Cancer Ther.* **2014**, *13*, 675–686. [CrossRef] [PubMed]

7

Targeting STAT3 in Cancer with Nucleotide Therapeutics

Yue-Ting K. Lau, Malini Ramaiyer, Daniel E. Johnson and Jennifer R. Grandis *

Department Otolaryngology—Head and Neck Surgery, University of California at San Francisco, 1450 3rd Street, Room HD268, Box 3111, San Francisco, CA 94143, USA; Kara.Lau@ucsf.edu (Y.-T.K.L.); malinisramaiyer@gmail.com (M.R.); daniel.johnson@ucsf.edu (D.E.J.)

* Correspondence: jennifer.grandis@ucsf.edu

Abstract: Signal transducer and activator of transcription 3 (STAT3) plays a critical role in promoting the proliferation and survival of tumor cells. As a ubiquitously-expressed transcription factor, STAT3 has commonly been considered an "undruggable" target for therapy; thus, much research has focused on targeting upstream pathways to reduce the expression or phosphorylation/activation of STAT3 in tumor cells. Recently, however, novel approaches have been developed to directly inhibit STAT3 in human cancers, in the hope of reducing the survival and proliferation of tumor cells. Several of these agents are nucleic acid-based, including the antisense molecule AZD9150, CpG-coupled STAT3 siRNA, G-quartet oligodeoxynucleotides (GQ-ODNs), and STAT3 decoys. While the AZD9150 and CpG-STAT3 siRNA interfere with STAT3 expression, STAT3 decoys and GQ-ODNs target constitutively activated STAT3 and modulate its ability to bind to target genes. Both STAT3 decoy and AZD9150 have advanced to clinical testing in humans. Here we will review the current understanding of the structures, mechanisms, and potential clinical utilities of the nucleic acid-based STAT3 inhibitors.

Keywords: hedging; transaction costs; dynamic programming; risk management; post-decision state variable

---

1. Introduction

Signal transducer and activator of transcription 3 (STAT3) is a transcription factor that is overexpressed and/or hyperactivated in multiple human cancers, where it enhances tumor cell survival and invasion through transcription of anti-apoptotic and pro-proliferative genes [1]. STAT3 has also been shown to directly interact with mitochondrial DNA to contribute to Ras-dependent malignant transformation and cancer progression by amplifying electron transport chain function [2–6]. STAT3 overexpression has been significantly associated with poor overall survival rates in patients with solid tumors [7]. Aberrant phosphorylation of STAT3 on Tyrosine 705 or Serine 727 by upstream kinases results in hyperactivation of the STAT3 protein [8,9]. In addition, genome silencing of phosphatases that play a role in dephosphorylation/inactivation of STAT3, such as those encoded by *PTPR* genes [10], can also result in constitutive activation of STAT3 in cancer [11].

In addition to contributing to the proliferation and survival of tumor cells, STAT3 hyperactivation plays an important role in the resistance of tumors to conventional chemotherapy drugs, as well as molecular targeting agents [12]. Moreover, STAT3 promotes immunosuppression in the tumor microenvironment [13]. STAT3 activation in tumor cells leads to increased production of immunosuppressive cytokines, including interleukin-6 (IL-6), IL-10, vascular endothelial growth factor (VEGF), and transforming growth factor-b1 (TGF-β1) [14–17]. Cytokines and growth factors produced by tumor cells also commonly lead to STAT3 activation in tumor-infiltrating immune cells. The activation of STAT3 in infiltrating immune cells, in general, inhibits anti-tumor immunity. Specifically, STAT3 exhibits cell-autonomous inhibitory activities against cytotoxic T cells (CTLs), natural killer (NK) cells, and dendritic cells (DCs), while increasing the levels of immunosuppressive T regulatory (Treg) and myeloid-derived suppressor cells (MDSCs) [18].

Activation of STAT3 is known to occur via several general pathways (Figure 1). Ligand binding initiates engagement of receptor tyrosine kinases, such as the receptors for epidermal growth factor (EGF) or vascular endothelial growth factor (VEGF), or receptors that lack intrinsic tyrosine kinase activity, such as the IL-6 receptor/gp130 complex, and/or nonreceptor tyrosine kinases such as c-Src [19].

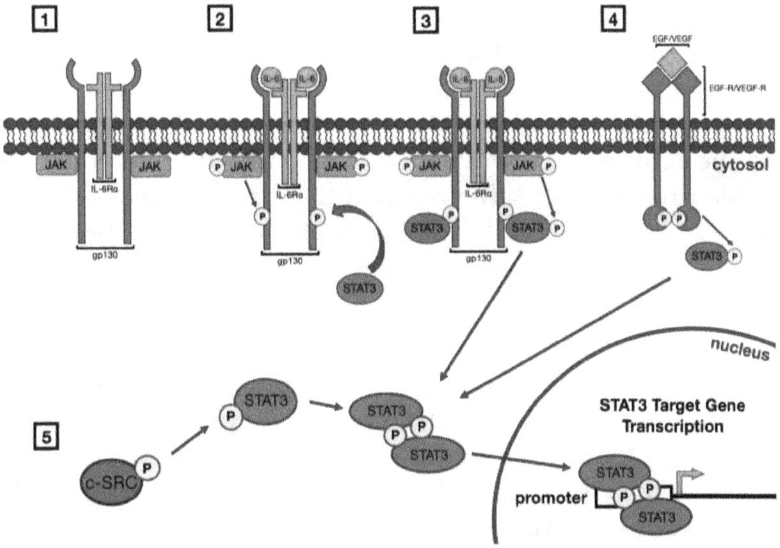

Figure 1. Pathways of signal transducer and activator of transcription 3 (STAT3) activation. Activation of STAT3 occurs via the initial phosphorylation of tyrosine 705 on the STAT3 molecule. This can occur in several ways: (1) Ligand binding activates Janus kinases (JAKs) that are associated with a receptor that lacks intrinsic tyrosine kinase activity, such as the interleukin-6 (IL-6) receptor/gp130 complex; (2) Activated JAKs phosphorylate the cytoplasmic region of the receptor molecule which then serves as a recruitment site for STAT3; (3) STAT3 is then phosphorylated by JAK. (4) Receptor tyrosine kinases (RTK) such as epidermal growth factor receptor (EGFR) or vascular endothelial growth factor receptor (VEGFR) have intrinsic kinase capabilities which directly phosphorylate STAT3 following ligand binding. (5) Receptor independent tyrosine kinases such as c–SRC can phosphorylate JAK without receptor activation. Once Y705 of STAT3 is phosphorylated the SH2 domain of each STAT3 molecule binds the phospho-tyrosine of another, resulting in dimerization of the two proteins. These homodimers are then able to translocate to the nucleus and bind the promoter regions of target genes and induce their transcription. Many of these target genes encode proteins that drive cellular proliferation and survival.

In the case of receptors lacking kinase activity, the binding of ligand leads to activation of receptor-associated Janus kinases (JAK). Activated JAKs phosphorylate the cytoplasmic region of the receptor molecule, which then serves as a docking site for STAT3. Recruitment of STAT3 then results in the direct phosphorylation of Tyrosine 705 by JAK. Phosphorylation of Serine 727 has also been shown to occur secondarily as a mechanism for maximal activation [20]. Receptor tyrosine kinases, including EGFR and VEGFR, harbor intrinsic kinase activity. Following ligand binding, the cytoplasmic regions of receptor tyrosine kinases are subjected to autophosphorylation, and these sites of phosphorylation then serve to recruit STAT3 which is subsequently phosphorylated/activated by the activated receptor.

Phosphorylation of Tyrosine 705 leads to homodimerization of STAT3 proteins, but STAT3 can also participate in alternative heterodimerization with STAT1α [21,22]. Canonical homodimerization occurs via SH2 recognition of the phospho-tyrosine residue in another STAT3 molecule and subsequent binding. STAT3 homodimers translocate to the nucleus where they bind to the promoter regions of STAT3 target genes, inducing the transcription of a broad number of genes whose products drive cellular proliferation and survival. STAT3 has also been demonstrated in an important role as a transcription modulator for mitochondrial respiration and oxidative metabolism [23–25].

1.1. Peptide and Small Molecule Inhibitors of STAT3

The critical role that overexpression and/or hyperactivation of STAT3 plays in the development of multiple cancers has spawned considerable effort to develop inhibitory molecules with potential for clinical application. Early efforts were focused on the development of peptide-based inhibitors. Additional pursuits have led to the identification of several small molecule inhibitors of STAT3. Furthermore, natural derivatives and natural models as lead scaffolds for molecular design have recently gained ground and demonstrated efficacy in STAT3 inhibition. In general, both peptidic inhibitors and small molecule inhibitors of STAT3 suffer from issues of low potency, poor cell penetrance, or undesirable nonspecific activities. However, a few recently discovered small molecule inhibitors are showing considerable promise and have reached the stage of advancing to clinical testing [26]. Hence, before describing nucleic acid-based inhibitors of STAT3 we will briefly review progress made in developing peptide and small molecule inhibitors.

1.2. Peptide Inhibitors

Critical to the function of STAT3 is the recognition of phosphotyrosine residues on activated cell surface receptors (for the purpose of binding to the receptor), or phosphotyrosine 705 on another STAT3 molecule (for the purpose of homodimerization). Recognition of these phosphotyrosine residues occurs via the STAT3 SH2 domain, which recognizes the consensus sequence PY*LKTK (where Y* represents phosphotyrosine). Early studies by Turkson et al. demonstrated that phosphorylated PY*LKTK peptide inhibited the DNA binding activity of STAT3 in nuclear extracts, but had no activity against STAT5 [27]. However, exceptionally high concentrations of the phosphorylated peptide were required to inhibit STAT3 activity in cells. In an effort to generate a more stable version of the PY*LKTK peptide, a peptidomimetic version named ISS-610 was developed and shown to have improved capacity for inhibiting STAT3 DNA binding activity in NIH3T3 cells (IC_{50} = 42 μM) [28]. ISS-610 also demonstrated growth inhibitory activity against several different cancer cell lines characterized by hyperactivation of STAT3. Mandal et al. made additional modifications in an effort to prevent dephophorylation of the peptide, generating the agent PM-73G [29]. This novel drug disrupted STAT3 DNA binding activity in the nanomolar range and also slowed the growth of MDA-MB-468 tumor xenografts.

Despite advances made in generating peptide and peptidomimetic inhibitors of STAT3, further development is hindered by multiple factors. The potencies, stabilities, and cellular uptake of peptides and peptidomimetic compounds will need to be improved and specificities will need to be closely evaluated. Moreover, there is concern that peptides and peptidomimetics may stimulate an immune response, which could limit their effectiveness.

1.3. Small Molecule Inhibitors

Given the obstacles associated with developing clinically relevant peptide inhibitors, greater emphasis has been placed on identifying small molecule inhibitors of STAT3 [30,31]. A number of small-molecule STAT3 inhibitors have been discovered using either experimental screening strategies or virtual screening approaches. Among the best characterized small molecule inhibitors are STATTIC, OPB-31121, OPB-51602, OPB-111077, and C188-9. In addition, an alternative pathway in mitochondrial STAT3 inhibition has been described in the small molecule MDC-1112.

STATTIC was identified in a screen of 17,000 compounds that used a fluorescence polarization assay to detect compounds capable of dissociating phosphopeptide from the SH2 domain of STAT3 [29,32]. STATTIC inhibits STAT3 DNA binding activity and slows the growth of xenograft tumors representing breast cancer and head and neck squamous cell carcinoma [31]. Treatment with STATTIC has been shown to enhance the activities of chemotherapy and radiation against cancer cells in vitro [33].

OPB-31121, OPB-51602, and OPB-111077 are orally bioavailable inhibitors developed by Otsuka Pharmaceutical Company. Molecular and computational modeling studies indicate high affinity binding (Kd = 10 nM) of these compounds to the SH2 domain of STAT3 [34]. Preclinical studies with OPB-31121 have shown that the compound inhibits STAT3 DNA binding and enhances the activities of chemotherapy drugs in leukemia and gastric cancer models [34–36]. OPB-31121 also slows the growth of primary leukemia and SNU484 gastric cancer xenograft tumors in mice [35,36]. A Phase I trial of OPB-31121 revealed considerable dose-limiting toxicities at doses below those needed for STAT3 inhibition. Furthermore, limited anti-tumor activity was observed [37]. Similar to OPB-31121, OPB-51602 demonstrates growth inhibition against xenograft tumors in preclinical models [38]. Phase I testing of OPB-51602 demonstrated reduction of phosphorylated STAT3 in monocytic cells and regression of tumors in two patients with non-small cell lung cancer [39]. One of these patients exhibited complete regression of target lesions and a progression-free survival interval of 6.9 months, while the other responder showed a 41 percent decline in tumor burden. However, treatment with OPB-51602 for multiple cycles resulted in substantial toxicities in multiple patients, including diarrhea, nausea, lactic acidosis, and peripheral neuropathy, leading to necessary discontinuation of treatment [38,39]. Phase I evaluation of OPB-111077 has demonstrated greater tolerability, with only mild or moderate side effects. To date, only modest anti-tumor responses have been observed with the OPB-111077, and further clinical investigation is necessary.

The small molecule C188 was discovered via virtual screening of a large compound library (920,000 compounds) to identify small molecules that may bind to the STAT3 SH2 domain [40]. Evaluation of C188 in cell-free assays demonstrated its ability to abrogate binding of phosphopeptide to the SH2 domain. Chemical optimization of C188 yielded the compound C188-9, also called TTI-101, with the capacity to inhibit phosphopeptide/SH2 interactions in the low nanomolar range [41]. C188-9 is an orally bioavailable STAT3 inhibitor that exhibits in vitro and in vivo activity against preclinical models of non-small cell lung cancer [42], head and neck squamous cell carcinoma [41], liver cancer [43], and breast cancer [44]. Phase I evaluation of C188-9 in patients with solid tumors is currently underway.

MDC-1112, or V-P, is a valporic acid derivative synthesized by Medicon Pharmaceuticals Inc. which blocks the translocation of STAT3 into the mitochondria and promotes the production of reactive oxygen species, inducing cellular apoptosis [45]. In combination with cimetidine, MDC-1112 inhibited the growth of pancreatic cancer xenografts in mice by 60–70% [45] and glioblastoma multiforme xenografts by 78.2% [46].

Napabucasin (2-acetylfuro-1,4-naphthoquinone or BBI-608) is a small molecule in clinical development that has been reported to abrogate STAT3 signaling. A small Phase I trial in combination with paclitaxel for advanced/recurrent gastric cancer reported that the combination was well tolerated [47]. A Phase III trial investigating napabucassin in combination with nab-paclitaxel and gemcitabine for metastatic pancreatic cancer is ongoing [48].

1.4. Future Directions for Small Molecule STAT3 Inhibitors

Although current small molecule inhibitors of STAT3, particularly C188-9, hold potential promise, the success rate for generating clinically viable small molecule inhibitors is low. A number of factors contribute to this low rate of success, including poor pharmacokinetic properties, insufficient potency, and unacceptable levels of nonspecific activity leading to toxic side effects. In addition, acquired resistance to small molecule inhibitors frequently occurs via mutation of the binding site on the target protein. A new approach towards drug development using molecules called proteolysis-targeting chimeras (PROTACs) may help to overcome some of the issues related to nonspecific activity and acquired resistance [49,50]. In the PROTACs approach, a fusion molecule is generated in which a linker sequence is used to covalently connect a small molecule inhibitor to a molecule (eg., thalidomide) that can attract an intracellular E3 ubiquitin ligase. By bringing the E3 ligase in close proximity to the target protein (eg., STAT3), the inhibitor fusion molecule serves to facilitate ubiquitination and subsequent proteasomal degradation of the target protein [49,50]. This process is both rapid and catalytic, meaning the fusion inhibitor molecule is released after destruction of the target protein and can subsequently interact and facilitate destruction of additional target molecules. Because of this catalytic nature, low doses of the inhibitor fusion molecule can be used to achieve efficient removal of the target protein, minimizing the impact of nonspecific activities that would be seen when higher doses are required for target inhibition by a conventional small molecule inhibitor. Moreover, because elimination of the target protein occurs rapidly, the risk of developing acquired resistance via target protein mutation is minimized. The application of PROTACs to inhibition of STAT3 signaling is an exciting opportunity, although, to date, no STAT3 PROTACs inhibitors have been reported.

1.5. Natural Inhibitors

Of the currently available anti-cancer agents used in current practice, it was found that over 40% were based in or derived from natural products [51]. Because of the increasing development of resistance to chemotherapeutic drugs, novel strategies to generate treatments can be sought in bioprospecting and combinatorial biosynthesis from nature-based derivatives. Recent advances in plant-based derivatives that have shown efficacy in STAT3 inhibition include Erasin, bruceanitol, and Curcumin.

Erasin is a chromone-based STAT5 inhibitor that has hydrophobic substituents at the 6-position, resulting in STAT3 specific inhibition. Derived from the natural product classes of flavones and isoflavones, it was determined to decrease STAT3 Y705 phosphorylation and increase apoptosis in MDA-MB-231 breast cancer cells, as well as Erlotinib resistant non-small cell lung cancer (NSCLC) cells [52].

From Brucinea javanica, a chinese plant used to treat cancer, bruceanitol (BOL) was isolated and discovered to have potent anti-leukemic activity. BOL demonstrated the ability to suppress cell growth in a wide range of genetically varied colorectal cancer cell lines, all of which were equally sensitive to the compound [53]. Reductions in phosphorylated STAT3 and downstream target expression of Mcl-1, c-Myc, and Survivin was observed at BOL concentrations of 30 nM. In vivo experiments on mice with CRC xenografts reflected similar results with doses of 2 mg/kg and 4 mg/kg reducing p-STAT3 32% and 80%, and tumor volume 35% and 58% respectively [53].

Diferuloylmethane is a polyphenol derived from the plant Curcuma longa, known as Curcumin, and has been shown to downregulate the expression of downstream STAT3 expression targets such as BCL-2 and Bcl-xL. Curcumin was shown to downregulate the production of Survivin mRNA concordant with the reduction of p-STAT3 in pancreatic cancer cell lines [54].

Collectively these naturally-derived small molecules offer considerable potential in the inhibition of STAT3 hyperactivation in pre-clinical studies. Natural products serve as promising starting points for development of inhibitors, but represent a platform that is currently difficult to exploit because of the lack of knowledge surrounding lead structures for rational design.

1.6. Nucleic Acid-Based Agents to Inhibit Expression of STAT3

1.6.1. AZD9150

Inhibition of STAT3-mediated gene expression can be achieved by directly inhibiting STAT3 expression using antisense oligonucleotides that promote the destruction or inhibit the translation of STAT3 mRNA. These oligonucleotides are short, 12–25 nucleotide strands with sequences such as 5′-GCTCCAGCATCTGCTGCTTC-3′, designed to pair with complementary STAT3 mRNA sequence(s) [55,56]. Binding results in cleavage of the target via RNAse H, alteration of post-transcriptional RNA splicing, or arrest of translation, leading to downmodulation of STAT3 expression [55]. Several generations of antisense oligonucleotides have been developed and chemically modified to optimize stability and allow systemic administration [57].

Early versions of the STAT3 antisense molecule contained 2′-O-methyl or 2′-O-methoxyethyl moieties to prevent free-end degradation and were synthesized with phosphorothioate chemistry to provide further stability [58]. Expression of STAT3 and STAT3 response genes was reduced when prostate (DU145) [59,60], breast (SCK), and melanoma (B16) [61] cell lines were treated with STAT3 antisense oligonucleotides. Furthermore, in vivo experiments showed inhibition of STAT3-mediated tumorigenesis, angiogenesis, and tumor growth in xenograft mouse models of prostate and hepatocellular carcinoma [59,62]. Sensitivity to STAT3 antisense treatment was also demonstrated in androgen-resistant models of prostate cancer and in lung metastases arising from hepatocellular primary tumors [59,62].

The second-generation antisense oligonucleotide AZD9150 (ISIS 481464) is the only STAT3 antisense molecule to enter clinical trials. The ASO structure of AZD9150 was optimized by replacing the 2′-O-methoxyethyl groups with constrained ethyl modifications to increase stability and potency [63]. Preclinical studies with AZD9150 in lymphoma (KARPAS299 and SUP-M2) [63] and neuroblastoma (IMR 32) [64] cell lines with aberrantly activated STAT3 showed a decrease in the expression of STAT3 protein and downstream signaling targets. AZD9150 also demonstrated selective uptake and promoted impaired growth in hematopoietic myelodysplastic and leukemic stem cells [65]. Systemic administration of AZD9150 to immunodeficient mice harboring lymphoma, neuroblastoma, or non-small-cell lung cancer xenografts resulted in decreased STAT3 expression in tumor cells and reduction of tumor initiating potential following serial implantation of the AZD9150-treated tumors [63,64]. While established tumor growth in neuroblastoma xenografts was not inhibited, systemic administration of AZD9150 led to reversal of STAT3-mediated resistance to cisplatin, as indicated by a twofold decrease in the IC_{50} for cisplatin [64].

As a class, antisense oligonucleotides have demonstrated nonspecific immune system activation due to the presence of unmethylated CpG motifs that are recognized as pathogenic [66]. Toxicological effects of AZD9150 have been studied in cynomolgus monkeys and in mice, with key findings being transient prolongation of intrinsic pathway clotting times, elevation of serum transaminase levels, and accumulation in renal and liver tissue [67]. However, these effects occur in doses far greater than those used in clinical trials, with no signs of end organ damage in kidneys or liver until doses of 40 mg/kg or 70 mg/kg, respectively, are reached.

A sufficient therapeutic margin provided the foundation to carry out a phase I clinical trial (NCT01563302) in 30 patients with hematologic and solid malignancies refractory to at least one prior systemic therapy [68]. All responses following administration of AZD9150 at dose levels of 2 mg/kg and 3 mg/kg occurred in diffuse large B cell lymphoma (DLBCL) patients. Two DLBCL patients achieved complete response, two achieved a partial response, and one maintained stable disease for a response rate of 13% [69]. There was no significant difference in progression-free and overall survival between the dose levels. Notable adverse events included transaminitis, fatigue, thrombocytopenia, nausea, and anemia with thrombocytopenia being more closely related to high-grade events. Based on the observed toxicities, the recommended phase 2 trial dose was 3 mg/kg. Peripheral blood analyses in this trial suggested an impact of AZD9150 in selected immune cell populations [69].

The heterogeneity in the response of lymphoma patients, five of whom had DLBCL, to AZD9150 highlights the importance of exploring what factors predict the efficacy of treatment. STAT3 inhibition may increase the immunogenicity of malignant cells through cell-autonomous processes and prevent tumor microenvironment immunosuppression [70]. Immunogenic biomarkers could thus offer clues to treatment progress and success. This could also lay the foundation for examining how combination with immunotherapeutic agents may further increase therapeutic benefits.

1.6.2. CpG-coupled STAT3 siRNA

While STAT3 is known to be overexpressed in tumor cells, it is also dysregulated in tumor-associated myeloid cells, such as dendritic cells and macrophages [71]. This results in a loss of MHCII expression and the accumulation of inactive antigen presenting cells. In effect, there is reduced Th-1-mediated CD8+ cytotoxicity towards tumor cells [13], and an increased presence of myeloid-derived suppressor cells (MDSCs) and regulatory T-cells (Tregs) generating an immunosuppressive tumor microenvironment [72].

Selective targeting of STAT3 in tumor-associated myeloid-derived cells is possible with siRNA conjugation to a CpG TLR-9 ligand [73]. TLR9 is a cell-surface transmembrane receptor that is upregulated under conditions of cellular or environmental stress and is known to be expressed on myeloid-derived cells in the tumor microenvironment [74]. It is also upregulated in acute myeloid leukemia, multiple myeloma, and B-cell lymphoma, and activation has been shown to increase antigenicity of primary malignant B-cells and induce apoptosis [75]. When activated by CpG binding, TLR9 facilitates immunostimulatory signaling and the release of proinflammatory cytokines as well as the presentation of tumor-specific antigens [74]. However, STAT3 signaling abrogates CpG-activated immunostimulation.

Therefore, to generate a sufficient immune response, it is necessary to both stimulate TLR9 and deactivate STAT3 [76,77]. RNA interference or RNAi, is another mechanism by which STAT3 mRNA can be degraded, thereby silencing its expression. A segment of double-stranded RNA targeted to the STAT3 sequence is introduced to cells where it is metabolized to 20-21 base-pair fragments and incorporated into an RNA-induced silencing complex, RISC. The two strands of RNA are unwound, and this allows the antisense strand to bind STAT3 mRNA. RISC endonuclease activity can then cleave the target [78]. CpG-conjugated STAT3 siRNA is internalized into cells by endocytosis, where TLR9 binding facilitates the release of Dicer-uncoupled siRNA from the early endosome before acidification [79].

In vitro studies of CpG-conjugated STAT3 siRNA showed that more than 80% of mouse dendritic cells, macrophages, and B-cells were positive for uptake without transfection reagent within 60 minutes [73]. STAT3 gene silencing was achieved with maximal reduction in expression at high siRNA concentrations of 1 μM [73]. In mice bearing B16 melanoma, C4 melanoma, or CT26 colon xenograft tumors, peritumoral injection with CpG-Stat3 siRNA led to tumor regression and systemic administration reduced the number of B16 lung cancer metastases [73]. This was associated with enhanced CD8+ T-Cell recruitment and cytolytic activity in association with increased immunostimulatory cytokine and chemokine production [73,80].

Studies with murine models of AML have shown that intravenous delivery of CpG-STAT3 siRNA results in a 70-80%, CD8+ T-cell-mediated, reduction of leukemic cells in bone marrow, spleen, lymph nodes, and peripheral blood while simultaneously reducing Treg levels [81]. Combination intratumoral administration of CpG-STAT3 siRNA with radiotherapy has demonstrated complete rejection of A20 lymphoma tumors in mice and generates long-term protective immunity against the primary tumor [75]. Similarly, CpG-STAT3 siRNA inhibits tumor growth of androgen-independent prostate cancer, with a concomitant reduction of immunosuppressive MDSC levels in peritumoral lymph nodes [82].

1.7. Nucleic Acid-Based Agents that Act as Competitive Inhibitors of STAT3

1.7.1. G-Quartet Oligodeoxynucleotides (GQ-ODNs)

G-quartet oligodeoxynucleotides are macrocycles composed of four guanosine bases that, upon hydrogen-bonding, form a polyguanylate, tetrad-helical structure in the presence of a monovalent cation, usually potassium (Figure 2) [83,84]. In vivo, G-quartets are found in telomeric regions of chromosomes and in the transcriptional regulatory regions of some oncogenes [85]. Their therapeutic potential has been demonstrated as direct competitive inhibitors of HIV-1 integrase, blocking HIV-1 DNA integration into the host genome [86–88].

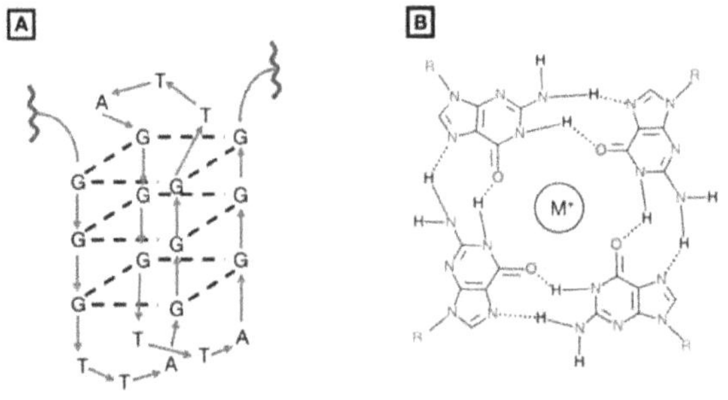

Figure 2. Structure of G-quartet inhibitor. (**A**) The G-quartet oligodeoxynucleotide is comprised of four guanosine macrocycles stacked on top of one another. To date, they have been used as competitive inhibitors of HIV-1 integrase but have demonstrated potential STAT3 dimerization inhibitors. With i.p. and i.v. injection in vivo mouse xenografts, G-quartets have been shown to reduce tumor growth in breast, prostate, and non-small cell lung cancers [85–87]. (**B**) An overhead view of a G-macrocycle demonstrates how hydrogen bonding generates a tetrad-helical structure with a monovalent cation at its core.

GQ-ODNs have been proposed as a class of unique, anti-cancer STAT3 inhibitors which directly destabilize the homodimerization of STAT3, and thus interfere with its DNA binding activity [89,90]. Computational analyses revealed a GQ-ODN interacts with residues Q643, Q644, N646, and N647 of the SH2 domain [91]. The folds that comprise the G-quartet intramolecular structure significantly occlude single-stranded endonuclease access to its phosphodiester linkages, resulting in an inability to cleave the oligonucleotide and in a long serum half-life [92]. However, because of their large size and charge, G-quartets cannot penetrate cell membranes and must be delivered via a polyethyleneimine (PEI) complex [93], or another delivery vehicle.

Preliminary in vitro assays demonstrated inhibition of IL-6-induced STAT3 activation in Hep2G cells incubated with the PEI-GQ-ODN T40214 (90% inhibition at 50 ng/mL). Preincubation at the same concentration inhibited the expression of anti-apoptotic mediators, with complete blockage of Bcl-XL mRNA upregulation and 50% blockage of Mcl-1 mRNA upregulation [89].

Treatment with GQ-ODN has demonstrated inhibition of STAT3 and tumor growth in in vivo models of head and neck squamous cell carcinoma (HNSCC) [94], breast cancer, prostate cancer [93], and non-small cell lung cancer (NSCLC) [95]. Inhibition of tumor growth was accompanied by a reduction in anti-apoptotic Bcl-2, Bcl-XL, Mcl-1, survivin, and VEGF in tumor tissue, and decreased expression of proliferation mediators cyclin D and c-Myc.

GQ-ODN selectivity for STAT3 is a potential consideration, given that a negative regulator of cell growth, STAT1 shares 50% sequence similarity with STAT3. Computational docking models indicate a two- to four-fold greater IC_{50} for STAT1 over STAT3 [89], and 3D analysis showed different residues involved in STAT1 and STAT3 surface interaction [96]. However, optimization of specificity for the therapeutic target is necessary for further development of clinical potential.

1.7.2. STAT3 Decoys

Once activated, STAT3 acts as a transcription factor, binding to a response element in the promoter regions of target genes to induce gene expression. Early investigation determined that STAT3 bound to a 15-base pair (bp) response element termed human serum-inducible element (hSIE) in the promoter region of the c-*fos* gene. We derived the decoy by systematically shortening the double stranded oligonucleotide to determine the smallest formulation that retained binding activity to STAT3 on gel shift assays. Figure 3A depicts the sequence of the SIE from the murine, feline, canine, and human c-*fos* genes, demonstrating nearly perfect conservation across these species. Subsequent studies identified optimal binding of STAT3 to a variant sequence of hSIE, specifically the sequence 5′-CATTTCCCGTAAATC-3′ (Figure 3A; [97]).

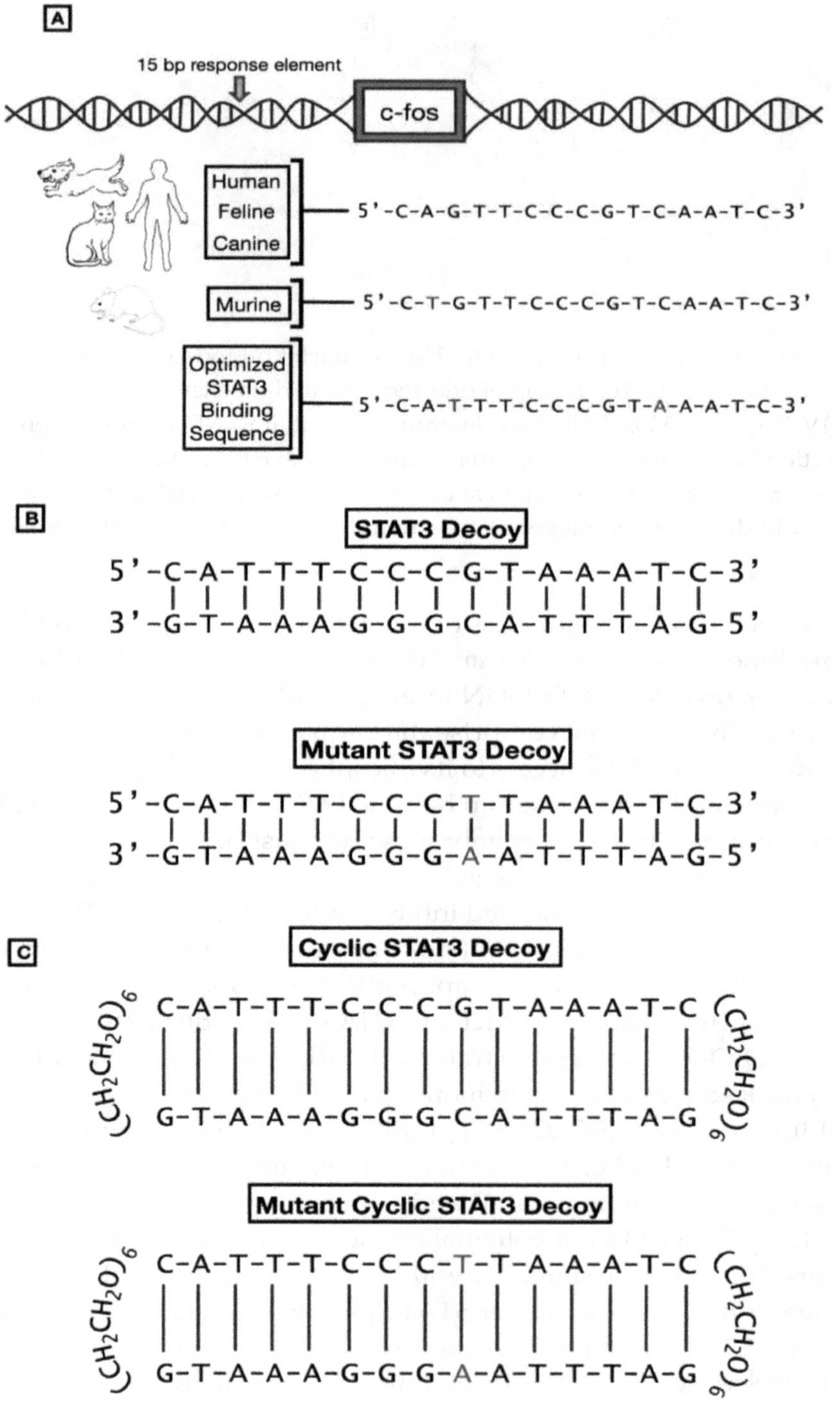

Figure 3. STAT3 response elements and STAT3 decoys. (**A**) The human serum-inducible element (hSIE) sequence is located upstream in the promoter region of the human *c-fos* gene and is almost perfectly conserved across human, feline, canine, and murine species. Sequence differences are shown in red. Below these is an optimized STAT3 binding sequence, with sequence modifications, shown in red, at positions 3 and 11 [52]. This allows the decoy to act as a direct competitive inhibitor of the DNA binding domain in the STAT3 molecule and thus inhibit expression of downstream anti-apoptotic and pro-proliferative signals. (**B**) Structure of the first-generation linear STAT3 decoy and mutant STAT3 decoy which is administered intratumorally in in vivo models. The location of the mutation in the mutant decoy is shown in blue. (**C**) Structure of the cyclic STAT3 decoy and mutant cyclic STAT3 decoy, with hexaethylene glycol linkages which is administered intravenously in in vivo models. The location of the mutation in the mutant decoy is shown in blue.

We synthesized the double-stranded STAT3 binding sequence (Figure 3B), the STAT3 decoy, and determined its effects when added to cells in culture [98]. As a negative control for binding specificity, a point mutant version of the decoy (termed mutant STAT3 decoy, or MT STAT3 decoy) was also synthesized and evaluated. In cell-free assays the STAT3 decoy, but not the MT STAT3 decoy, was confirmed to competitively inhibit binding of STAT3 protein to a radiolabeled decoy sequence. Fluorescently-tagged versions of both the STAT3 decoy and the MT STAT3 decoy were readily incorporated into the cytosol and nucleus of HNSCC cells and normal oral keratinocytes (NOKs) within 6 hours after treatment. Remarkably, the STAT3 decoy potently inhibited the growth of STAT3-dependent cancer cell lines, but not NOKs. The control molecule, MT STAT3, was largely ineffective against either the cancer cell lines or the NOKs.

The inhibition of cell growth resulting from treatment with STAT3 decoy was associated with induction of apoptosis and downregulation of the STAT3 target gene encoding Bcl-XL. Further investigation [99] revealed that the STAT3 decoy also inhibits the STAT1 protein, which is known to form heterodimers with STAT3. This finding was initially troubling, as STAT1 is known to have tumor suppressor activity. However, the expression or activation of STAT1 did not alter the apoptosis-inducing activity of the STAT3 decoy, indicating that the therapeutic activity of the decoy is independent of STAT1 signaling. Subsequent studies have determined that the STAT3 decoy can inhibit the growth of a broad variety of cancer cell lines, including cells representing melanoma and cancers of the bladder, brain, breast, colon, liver, lung, and ovary [98,100–103]. In addition, intratumoral injection of STAT3 decoy has been shown to inhibit the in vivo growth of breast, glioma, head and neck, lung, and ovarian xenograft tumors [98,100–103]. The anti-tumor effects of STAT3 decoy are associated with reduced tumor cell proliferation, induction of apoptosis, and reduced expression of STAT3 target genes, including the genes encoding Bcl-XL and cyclin D1 [101,104]. Based on these promising preclinical results, we conducted a Phase 0 clinical trial to assess the pharmacodynamic impact of the STAT3 decoy [105]. Intratumoral delivery of the decoy was found to downmodulate expression of Bcl-XL and cyclin D1 in the tumors of patients undergoing surgical resection of their head and neck cancer.

Fusion of the STAT3 decoy with CpG has enabled targeting of the decoy to TLR9-expressing leukemia and lymphoma cells [106,107]. Treatment of preclinical mouse models harboring leukemia or lymphoma with the CpG-STAT3 decoy resulted in potent in vivo growth inhibition of the malignant cells. Hence, the CpG-STAT3 decoy fusion molecule may provide an effective means for treating TLR9-expressing hematologic malignancies.

A major limitation of the first generation STAT3 decoy was the necessity for intratumoral injection. In preclinical studies, systemic delivery of STAT3 decoy failed to inhibit xenograft tumor growth [105], presumably due to rapid degradation of the molecule by nucleases in the blood. In an effort to produce a more stable molecule, we [105] generated a cyclic version of the decoy, using hexaethylene glycol linkages to cyclize the free ends (Figure 3C). The cyclic STAT3 decoy exhibited markedly enhanced thermal stability and a longer half-life in human serum (8 hours; [105,108]). Intravenous delivery of the cyclic STAT3 decoy has been shown to potently inhibit the growth of both non-small cell lung cancer and head and neck cancer xenograft tumors [105,109,110]. The cyclic STAT3 decoy was well tolerated in wild-type mice, with no apparent toxicity, even when delivered at a dose 20-fold higher than the maximal effective dose for growth inhibition of HNSCC xenograft tumors [110].

The therapeutic value of treatment with the STAT3 decoy, as with other STAT3 inhibitors, may be best realized in combination with other anticancer agents. In this regard, STAT3 decoy treatment has been shown to enhance the sensitivity of head and neck xenograft tumors to cisplatin [101,104]. Similarly, the STAT3 decoy heightens the response of head and neck cancer cells to bortezomib [111], ovarian cancer cells to paclitaxel [112], and leukemia cells to Adriamycin [113]. FDA IND-directed pharmacologic and toxicity studies are planned to enable a Phase I trial of the cyclic STAT3 decoy in cancer patients.

2. Conclusion

Hyperactivation of STAT3 in malignant cells and tumor-associated immune cells promotes cancer survival and growth through expression of anti-apoptotic and proliferative genes, as well as cytokines that generate an immunosuppressive tumor microenvironment. Understanding the role of STAT3 in cancer has led to promising strides in the development of specific nucleotide inhibitors (Figure 4). These anti-STAT3 agents can silence STAT3 expression or directly inhibit STAT3 DNA-binding ability and have shown promising in vivo results beyond proof-of-principle studies. However, optimization of specificity, potency, stability, and delivery of these nucleotide therapeutics will be important for enhancing their therapeutic benefits in the clinic.

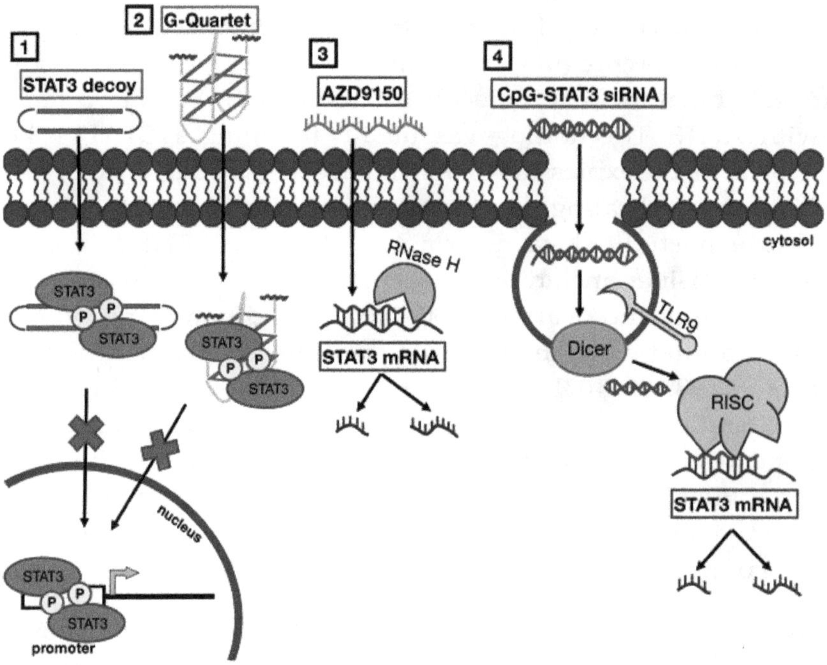

Figure 4. Mechanisms of STAT3 inhibition with nucleic acid-based agents. STAT3 inhibition with nucleic acid-based agents occurs via two main mechanisms: (1 and 2) STAT3 decoys or G-quartet deoxyoligonucleotides prevent STAT3 binding to promoter regions of target genes, and (3 and 4) antisense or siRNAs promote degradation of STAT3 mRNA.

3. Discussion

Currently, there are four classes of drugs that directly inhibit STAT3: SH2 domain inhibitors; DNA-binding domain inhibitors (eg., STAT3 decoys); N-terminal domain inhibitors; and STAT3 antisense and siRNA [114]. As a whole, translation to the clinical setting has been challenged by intracellular drug delivery, selectivity for the target, and minimization of systemic toxicity. More clinical success has been achieved with inhibition of upstream Janus kinases (JAKs) using Tofacitinib, developed for the treatment of rheumatoid arthritis [115], and Ruxolitinib, developed for the treatment of myelofibrosis [116]. Still, small molecule drugs often exhibit off-target effects and a lack of targeted potency; thus, the limitations of monotherapy should be considered.

The future of STAT3 inhibition is likely to be in combination therapy with currently existing or newly developed drugs. Blocking STAT3 could result in synergistic anti-tumor effects in combination with inhibition of EGFR or other tumorigenic dysregulated transcription factors [117,118]. STAT3 inhibition resulting in resensitization to immunotherapies could pave the way for more efficacious responses to checkpoint inhibitors or other immunotherapies [119–123]. Looking to the future, while the role of STAT3 has been established in cancer biology, identifying tumor biomarkers that can indicate patient sensitivity to STAT3 inhibition and expansion of target inhibition to other aberrant transcription factors will be important. Research efforts should continue to search for novel applications of STAT3 inhibition and continue to pursue clinical validation of efficacy.

Author Contributions: D.E.J. and J.R.G. are co-inventors of cyclic STAT3 decoy and have financial interests in STAT3 Therapeutics. STAT3 Therapeutics holds an interest in cyclic STAT3 decoy.

References

1. Huynh, J.; Chand, A.; Gough, D.; Ernst, M. Therapeutically Exploiting STAT3 Activity in Cancer—Using Tissue Repair as a Road Map. *Nat. Rev. Cancer* **2019**, *19*, 82–96. [CrossRef]
2. Gough, D.J.; Corlett, A.; Schlessinger, K.; Wegrzyn, J.; Larner, A.C.; Levy, D.E. Mitochondrial STAT3 Supports Ras-Dependent Oncogenic Transformation. *Science* **2009**, *324*, 1713–1716. [CrossRef]
3. Gough, D.J.; Koetz, L.; Levy, D.E. The MEK-ERK Pathway is Necessary for Serine Phosphorylation of Mitochondrial STAT3 and Ras-Mediated Transformation. *PLoS ONE* **2013**, *8*, e83395. [CrossRef]
4. Du, W.; Hong, J.; Wang, Y.-C.; Zhang, Y.-J.; Wang, P.; Su, W.-Y.; Lin, Y.-W.; Lu, R.; Zou, W.-P.; Xiong, H.; et al. Inhibition of JAK2/STAT3 Signalling Induces Colorectal Cancer Cell Apoptosis via Mitochondrial Pathway. *J. Cell. Mol. Med.* **2012**, *16*, 1878–1888. [CrossRef]
5. Yu, H.; Lee, H.; Herrmann, A.; Buettner, R.; Jove, R. Revisiting STAT3 Signalling in Cancer: New and Unexpected Biological Functions. *Nat. Rev. Cancer* **2014**, *14*, 736–746. [CrossRef]
6. Yang, R.; Rincon, M. Mitochondrial Stat3, the Need for Design Thinking. *Int. J. Biol. Sci.* **2016**, *12*, 532–544. [CrossRef]
7. Wu, P.; Wu, D.; Zhao, L.; Huang, L.; Shen, G.; Huang, J.; Chai, Y. Prognostic role of STAT3 in Solid Tumors: A Systematic Review and Meta-Analysis. *Oncotarget* **2016**, *7*, 19863–19883. [CrossRef]
8. Sakaguchi, M.; Oka, M.; Iwasaki, T.; Fukami, Y.; Nishigori, C. Role and Regulation of STAT3 Phosphorylation at Ser727 in Melanocytes and Melanoma Cells. *J. Investig. Dermatol.* **2012**, *132*, 1877–1885. [CrossRef]
9. Schuringa, J.-J.; Wierenga, A.T.J.; Kruijer, W.; Vellenga, E. Constitutive Stat3, Tyr705, and Ser727 Phosphorylation in Acute Myeloid Leukemia Cells Caused by the Autocrine Secretion of Interleukin-6. *Blood* **2000**, *95*, 3765–3770. [CrossRef]
10. Lui, V.W.Y.; Peyser, N.D.; Ng, P.K.-S.; Hritz, J.; Zeng, Y.; Lu, Y.; Li, H.; Wang, L.; Gilbert, B.R.; General, I.J.; et al. Frequent Mutation of Receptor Protein Tyrosine Phosphatases Provides a Mechanism for STAT3 Hyperactivation in Head and Neck Cancer. *Proc. Natl. Acad. Sci. USA* **2014**, *111*, 1114–1119. [CrossRef]
11. Geiger, J.L.; Grandis, J.R.; Bauman, J.E. The STAT3 Pathway as a Therapeutic Target in Head and Neck Cancer: Barriers and Innovations. *Oral Oncol.* **2016**, *56*, 84–92. [CrossRef]
12. Spitzner, M.; Ebner, R.; Wolff, H.A.; Michael Ghadimi, B.; Wienands, J.; Grade, M. STAT3: A Novel Molecular Mediator of Resistance to Chemoradiotherapy. *Cancers* **2014**, *6*, 1986–2011. [CrossRef]
13. Lee, H.; Pal, S.K.; Reckamp, K.; Figlin, R.A.; Yu, H. STAT3: A Target to Enhance Antitumor Immune Response. *Curr. Top. Microbiol. Immunol.* **2011**, *344*, 41–59.
14. Chen, M.-F.; Chen, P.-T.; Lu, M.S.; Lin, P.Y.; Chen, W.-C.; Lee, K.-D. IL-6 Expression Predicts Treatment Response and Outcome in Squamous Cell Carcinoma of the Esophagus. *Mol. Cancer* **2013**, *12*, 26. [CrossRef]
15. Wu, C.-T.; Chen, M.-F.; Chen, W.-C.; Hsieh, C.-C. The Role of IL-6 in the Radiation Response of Prostate Cancer. *Radiat. Oncol.* **2013**, *8*, 159. [CrossRef]
16. Cho, B.C.; Kim, S.M.; Solca, F.; Kim, J.-H. Abstract 1886: Activation of IL-6R/JAK1/STAT3 Signaling Induces de Novo Resistance to Irreversible EGFR Inhibitors in Non-Small Cell Lung Cancer with T790M Resistance Mutation. *Exp. Mol. Ther.* **2012**, *11*, 2254–2264.
17. Yu, H.; Kortylewski, M.; Pardoll, D. Crosstalk between Cancer and Immune Cells: Role of STAT3 in the Tumour Microenvironment. *Nat. Rev. Immunol.* **2007**, *7*, 41–51. [CrossRef]
18. Ferguson, S.D.; Srinivasan, V.M.; Heimberger, A.B. The Role of STAT3 in Tumor-Mediated Immune Suppression. *J. Neurooncol.* **2015**, *123*, 385–394. [CrossRef]
19. Rébé, C.; Végran, F.; Berger, H.; Ghiringhelli, F. STAT3 Activation: A Key Factor in Tumor Immunoescape. *JAKSTAT* **2013**, *2*, e23010. [CrossRef]

20. Wen, Z.; Zhong, Z.; Darnell, J.E., Jr. Maximal Activation of Transcription by Stat1 and Stat3 Requires both Tyrosine and Serine Phosphorylation. *Cell* **1995**, *82*, 241–250. [CrossRef]
21. Zhong, Z.; Wen, Z.; Darnell, J.E., Jr. Stat3: A STAT Family Member Activated by Tyrosine Phosphorylation in Response to Epidermal Growth Factor and Interleukin-6. *Science* **1994**, *264*, 95–98. [CrossRef]
22. Delgoffe, G.M.; Vignali, D.A.A. STAT Heterodimers in Immunity: A Mixed Message or a Unique Signal? *JAKSTAT* **2013**, *2*, e23060. [CrossRef]
23. Wegrzyn, J.; Potla, R.-J.; Chwae, Y.; Sepuri, N.B.V.; Zhang, Q.; Koeck, T.; Derecka, M.; Szczepanek, K.; Szelag, M.; Gornicka, A.; et al. Function of Mitochondrial Stat3 in Cellular Respiration. *Science* **2009**, *323*, 793–797. [CrossRef]
24. Carbognin, E.; Betto, R.M.; Soriano, M.E.; Smith, A.G.; Martello, G. Stat3 Promotes Mitochondrial Transcription and Oxidative Respiration during Maintenance and Induction of Naive Pluripotency. *EMBO J.* **2016**, *35*, 618–634. [CrossRef]
25. Liu, F.; Jia, L.; Farren, T.; Gribben, J.; Agrawal, S. 2.21 Autocrine Interleukin-6 Production Correlated with Survival of Chronic Lymphocytic Leukaemia Cells. *Clin. Lymphoma Myeloma Leuk.* **2011**, *11*, S172. [CrossRef]
26. Wake, M.S.; Watson, C.J. STAT3 the oncogene - still eluding therapy? *FEBS J.* **2015**, *282*, 2600–2611. [CrossRef]
27. Turkson, J.; Ryan, D.; Kim, J.S.; Zhang, Y.; Chen, Z.; Haura, E.; Laudano, A.; Sebti, S.; Hamilton, A.D.; Jove, R. Phosphotyrosyl Peptides Block Stat3-mediated DNA Binding Activity, Gene Regulation, and Cell Transformation. *J. Biol. Chem.* **2001**, *276*, 45443–45455. [CrossRef]
28. Turkson, J.; Kim, J.S.; Zhang, S.; Yuan, J.; Huang, M.; Glenn, M.; Haura, E.; Sebti, S.; Hamilton, A.D.; Jove, R. Novel Peptidomimetic Inhibitors of Signal Transducer and Activator of Transcription 3 Dimerization and Biological Activity. *Mol. Cancer Ther.* **2004**, *3*, 261–269.
29. Mandal, P.K.; Gao, F.; Lu, Z.; Ren, Z.; Ramesh, R.; Birtwistle, J.S.; Kaluarachchi, K.K.; Chen, X.; Bast, R.C., Jr.; Liao, W.S.; et al. Potent and Selective Phosphopeptide Mimetic Prodrugs Targeted to the Src Homology 2 (SH2) Domain of Signal Transducer and Activator of Transcription 3. *J. Med. Chem.* **2011**, *54*, 3549–3563. [CrossRef]
30. Schust, J.; Berg, T. A High-Throughput Fluorescence Polarization Assay for Signal Transducer and Activator of Transcription 3. *Anal. Biochem.* **2004**, *330*, 114–118. [CrossRef]
31. Schust, J.; Sperl, B.; Hollis, A.; Mayer, T.U.; Berg, T. Stattic: A Small-Molecule Inhibitor of STAT3 Activation and Dimerization. *Chem. Biol.* **2006**, *13*, 1235–1242. [CrossRef]
32. Auzenne, E.J.; Klostergaard, J.; Mandal, P.K.; Liao, W.S.; Lu, Z.; Gao, F.; Bast, R.C., Jr.; Robertson, F.M.; McMurray, J.S. A Phosphopeptide Mimetic Prodrug Targeting the SH2 Domain of Stat3 Inhibits Tumor Growth and Angiogenesis. *J. Exp. Ther. Oncol.* **2012**, *10*, 155–162.
33. Pan, Y.; Zhou, F.; Zhang, R.; Claret, F.X. Stat3 inhibitor Stattic Exhibits Potent Antitumor Activity and Induces Chemo- and Radio-Sensitivity in Nasopharyngeal Carcinoma. *PLoS ONE* **2013**, *8*, e54565. [CrossRef]
34. Brambilla, L.; Genini, D.; Laurini, E.; Merulla, J.; Perez, L.; Fermeglia, M.; Carbone, G.M.; Pricl, S.; Catapano, C.V. Hitting the right spot: Mechanism of action of OPB-31121, a novel and potent inhibitor of the Signal Transducer and Activator of Transcription 3 (STAT3). *Mol. Oncol.* **2015**, *9*, 1194–1206. [CrossRef]
35. Hayakawa, F.; Sugimoto, K.; Harada, Y.; Hashimoto, N.; Ohi, N.; Kurahashi, S.; Naoe, T. A novel STAT inhibitor, OPB-31121, has a Significant Antitumor Effect on Leukemia with STAT-Addictive Oncokinases. *Blood Cancer J.* **2013**, *3*, e166. [CrossRef]
36. Kim, M.-J.; Nam, H.-J.; Kim, H.-P.; Han, S.-W.; Im, S.-A.; Kim, T.-Y.; Oh, D.-Y.; Bang, Y.-J. OPB-31121, a Novel Small Molecular Inhibitor, Disrupts the JAK2/STAT3 Pathway and Exhibits an Antitumor Activity in Gastric Cancer Cells. *Cancer Lett.* **2013**, *335*, 145–152. [CrossRef]
37. Bendell, J.C.; Hong, D.S.; Burris, H.A., 3rd; Naing, A.; Jones, S.F.; Falchook, G.; Bricmont, P.; Elekes, A.; Rock, E.P.; Kurzrock, R. Phase 1, Open-Label, Dose-Escalation, and Pharmacokinetic Study of STAT3 Inhibitor OPB-31121 in Subjects with Advanced Solid Tumors. *Cancer Chemother. Pharmacol.* **2014**, *74*, 125–130. [CrossRef]
38. Ogura, M.; Uchida, T.; Terui, Y.; Hayakawa, F.; Kobayashi, Y.; Taniwaki, M.; Takamatsu, Y.; Naoe, T.; Tobinai, K.; Munakata, W.; et al. Phase I Study of OPB-51602, an Oral Inhibitor of Signal Transducer and Activator of Transcription 3, in Patients with Relapsed/Refractory Hematological Malignancies. *Cancer Sci.* **2015**, *106*, 896–901. [CrossRef]
39. Wong, A.L.; Soo, R.A.; Tan, D.S.; Lee, S.C.; Lim, J.S.; Marban, P.C.; Kong, L.R.; Lee, Y.J.; Wang, L.Z.; Thuya, W.L.; et al. Phase I and Biomarker Study of OPB-51602, a Novel Signal Transducer and Activator

of Transcription (STAT) 3 Inhibitor, in Patients with Refractory Solid Malignancies. *Ann. Oncol.* **2015**, *26*, 998–1005. [CrossRef]

40. Xu, X.; Kasembeli, M.M.; Jiang, X.; Tweardy, B.J.; Tweardy, D.J. Chemical Probes that Competitively and Selectively Inhibit Stat3 Activation. *PLoS ONE* **2009**, *4*, e4783. [CrossRef]
41. Bharadwaj, U.; Eckols, T.K.; Xu, X.; Kasembeli, M.M.; Chen, Y.; Adachi, M.; Song, Y.; Mo, Q.; Lai, S.Y.; Tweardy, D.J. Small-Molecule Inhibition of STAT3 in Radioresistant Head and Neck Squamous Cell Carcinoma. *Oncotarget* **2016**, *7*, 26307–26330. [CrossRef]
42. Lewis, K.M.; Bharadwaj, U.; Eckols, T.K.; Kolosov, M.; Kasembeli, M.M.; Fridley, C.; Siller, R.; Tweardy, D.J. Small-Molecule Targeting of Signal Transducer and Activator of Transcription (STAT) 3 to Treat Non-Small Cell Lung Cancer. *Lung Cancer* **2015**, *90*, 182–190. [CrossRef]
43. Jung, K.H.; Yoo, W.; Stevenson, H.L.; Deshpande, D.; Shen, H.; Gagea, M.; Yoo, S.-Y.; Wang, J.; Kris Eckols, T.; Bharadwaj, U.; et al. Multifunctional Effects of a Small-Molecule STAT3 Inhibitor on NASH and Hepatocellular Carcinoma in Mice. *Clin. Cancer Res.* **2017**, *23*, 5537–5546. [CrossRef]
44. Kettner, N.M.; Vijayaraghavan, S.; Durak, M.G.; Bui, T.; Kohansal, M.; Ha, M.J.; Liu, B.; Rao, X.; Wang, J.; Yi, M.; et al. Combined Inhibition of STAT3 and DNA Repair in Palbociclib-Resistant ER-Positive Breast Cancer. *Clin. Cancer Res.* **2019**, *25*, 3996–4013. [CrossRef]
45. Mackenzie, G.G.; Huang, L.; Alston, N.; Ouyang, N.; Vrankova, K.; Mattheolabakis, G.; Constantinides, P.P.; Rigas, B. Targeting Mitochondrial STAT3 with the Novel Phospho-Valproic Acid (MDC-1112) Inhibits Pancreatic Cancer Growth in Mice. *PLoS ONE* **2013**, *8*, e61532. [CrossRef]
46. Luo, D.; Fraga-Lauhirat, M.; Millings, J.; Ho, C.; Villarreal, E.M.; Fletchinger, T.C.; Bonfiglio, J.V.; Mata, L.; Nemesure, M.D.; Bartels, L.E.; et al. Phospho-Valproic Acid (MDC-1112) Suppresses Glioblastoma Growth in Preclinical Models Through the Inhibition of STAT3 Phosphorylation. *Carcinogenesis* **2019**. [CrossRef]
47. Shitara, K.; Yodo, Y.; Iino, S. A Phase I Study of Napabucasin Plus Paclitaxel for Japanese Patients with Advanced/Recurrent Gastric Cancer. *In Vivo* **2019**, *33*, 933–937. [CrossRef]
48. Sonbol, M.B.; Ahn, D.H.; Goldstein, D.; Okusaka, T.; Tabernero, J.; Macarulla, T.; Reni, M.; Li, C.-P.; O'Neil, B.; Van Cutsem, E.; et al. CanStem111P trial: A Phase III Study of Napabucasin Plus Nab-Paclitaxel with Gemcitabine. *Future Oncol.* **2019**, *15*, 1295–1302. [CrossRef]
49. Pettersson, M.; Crews, C.M. PROteolysis TArgeting Chimeras (PROTACs)—Past, Present and Future. *Drug Discov. Today Technol.* **2019**, *31*, 15–27. [CrossRef]
50. Paiva, S.-L.; Crews, C.M. Targeted Protein Degradation: Elements of PROTAC Design. *Curr. Opin. Chem. Biol.* **2019**, *50*, 111–119. [CrossRef]
51. Demain, A.L.; Vaishnav, P. Natural Products for Cancer Chemotherapy. *Microb. Biotechnol.* **2011**, *4*, 687–699. [CrossRef]
52. Lis, C.; Rubner, S.; Roatsch, M.; Berg, A.; Gilcrest, T.; Fu, D.; Nguyen, E.; Schmidt, A.-M.; Krautscheid, H.; Meiler, J.; et al. Development of Erasin: A Chromone-Based STAT3 Inhibitor which Induces Apoptosis in Erlotinib-Resistant Lung Cancer Cells. *Sci. Rep.* **2017**, *7*, 17390. [CrossRef]
53. Wei, N.; Li, J.; Fang, C.; Chang, J.; Xirou, V.; Syrigos, N.K.; Marks, B.J.; Chu, E.; Schmitz, J.C. Targeting Colon Cancer with the Novel STAT3 Inhibitor Bruceantinol. *Oncogene* **2019**, *38*, 1676–1687. [CrossRef]
54. Glienke, W.; Maute, L.; Wicht, J.; Bergmann, L. Curcumin Inhibits Constitutive STAT3 Phosphorylation in Human Pancreatic Cancer Cell Lines and Downregulation of Survivin/BIRC5 Gene Expression. *Cancer Investig.* **2010**, *28*, 166–171. [CrossRef]
55. Dean, N.M.; Bennett, C.F. Antisense Oligonucleotide-Based Therapeutics for Cancer. *Oncogene* **2003**, *22*, 9087–9096. [CrossRef]
56. Engelhard, H.H. Antisense Oligodeoxynucleotide Technology: Potential Use for the Treatment of Malignant Brain Tumors. *Cancer Control* **1998**, *5*, 163–170. [CrossRef]
57. Khvorova, A.; Watts, J.K. The Chemical Evolution of Oligonucleotide Therapies of Clinical Utility. *Nat. Biotechnol.* **2017**, *35*, 238–248. [CrossRef]
58. Shen, X.; Corey, D.R. Chemistry, Mechanism and Clinical Status of Antisense Oligonucleotides and Duplex RNAs. *Nucleic Acids Res.* **2018**, *46*, 1584–1600. [CrossRef]
59. Barton, B.E.; Karras, J.G.; Murphy, T.F.; Barton, A.; Huang, H.F.-S. Signal Transducer and Activator of Transcription 3 (STAT3) Activation in Prostate Cancer: Direct STAT3 Inhibition Induces Apoptosis in Prostate Cancer Lines. *Mol. Cancer Ther.* **2004**, *3*, 11–20.

60. Mora, L.B.; Buettner, R.; Seigne, J.; Diaz, J.; Ahmad, N.; Garcia, R.; Bowman, T.; Falcone, R.; Fairclough, R.; Cantor, A.; et al. Constitutive Activation of Stat3 in Human Prostate Tumors and Cell Lines: Direct Inhibition of Stat3 Signaling Induces Apoptosis of Prostate Cancer Cells. *Cancer Res.* **2002**, *62*, 6659–6666.
61. Niu, G.; Wright, K.L.; Huang, M.; Song, L.; Haura, E.; Turkson, J.; Zhang, S.; Wang, T.; Sinibaldi, D.; Coppola, D.; et al. Constitutive Stat3 Activity Up-Regulates VEGF Expression and Tumor Angiogenesis. *Oncogene* **2002**, *21*, 2000–2008. [CrossRef]
62. Li, W.-C.; Ye, S.-L.; Sun, R.-X.; Liu, Y.-K.; Tang, Z.-Y.; Kim, Y.; Karras, J.G.; Zhang, H. Inhibition of Growth and Metastasis of Human Hepatocellular Carcinoma by Antisense Oligonucleotide Targeting Signal Transducer and Activator of Transcription 3. *Clin. Cancer Res.* **2006**, *12*, 7140–7148. [CrossRef]
63. Hong, D.; Kurzrock, R.; Kim, Y.; Woessner, R.; Younes, A.; Nemunaitis, J.; Fowler, N.; Zhou, T.; Schmidt, J.; Jo, M.; et al. AZD9150, a Next-Generation Antisense Oligonucleotide Inhibitor of STAT3 with Early Evidence of Clinical Activity in Lymphoma and Lung Cancer. *Sci. Transl. Med.* **2015**, *7*, 314ra185. [CrossRef]
64. Odate, S.; Veschi, V.; Yan, S.; Lam, N.; Woessner, R.; Thiele, C.J. Inhibition of STAT3 with the Generation 2.5 Antisense Oligonucleotide, AZD9150, Decreases Neuroblastoma Tumorigenicity and Increases Chemosensitivity. *Clin. Cancer Res.* **2017**, *23*, 1771–1784. [CrossRef]
65. Shastri, A.; Choudhary, G.; Teixeira, M.; Gordon-Mitchell, S.; Ramachandra, N.; Bernard, L.; Bhattacharyya, S.; Lopez, R.; Pradhan, K.; Giricz, O.; et al. Antisense STAT3 Inhibitor Decreases Viability of Myelodysplastic and Leukemic Stem Cells. *J. Clin. Investig.* **2018**, *128*, 5479–5488. [CrossRef]
66. Barton, B.E.; Murphy, T.F.; Shu, P.; Huang, H.F.; Meyenhofer, M.; Barton, A. Novel Single-Stranded Oligonucleotides that Inhibit Signal Transducer and Activator of Transcription 3 Induce Apoptosis in Vitro and in Vivo in Prostate Cancer Cell Lines. *Mol. Cancer Ther.* **2004**, *3*, 1183–1191.
67. Burel, S.A.; Han, S.-R.; Lee, H.-S.; Norris, D.A.; Lee, B.-S.; Machemer, T.; Park, S.-Y.; Zhou, T.; He, G.; Kim, Y.; et al. Preclinical Evaluation of the Toxicological Effects of a Novel Constrained Ethyl Modified Antisense Compound Targeting Signal Transducer and Activator of Transcription 3 in Mice and Cynomolgus Monkeys. *Nucleic Acid Ther.* **2013**, *23*, 213–227. [CrossRef]
68. Phase 1/2, Open-Label, Dose-Escalation Study of IONIS-STAT3Rx, Administered to Patients with Advanced Cancers—Full Text View—ClinicalTrials.gov. Available online: https://clinicaltrials.gov/ct2/show/NCT01563302 (accessed on 10 July 2019).
69. Reilley, M.J.; McCoon, P.; Cook, C.; Lyne, P.; Kurzrock, R.; Kim, Y.; Woessner, R.; Younes, A.; Nemunaitis, J.; Fowler, N.; et al. STAT3 Antisense Oligonucleotide AZD9150 in a Subset of Patients with Heavily Pretreated Lymphoma: Results of a Phase 1b Trial. *J. Immunother. Cancer* **2018**, *6*, 119. [CrossRef]
70. Kroemer, G.; Galluzzi, L.; Zitvogel, L. STAT3 Inhibition for Cancer Therapy: Cell-Autonomous Effects Only? *Oncoimmunology* **2016**, *5*, e1126063. [CrossRef]
71. Kortylewski, M.; Moreira, D. Myeloid Cells as a Target for Oligonucleotide Therapeutics: Turning Obstacles into Opportunities. *Cancer Immunol. Immunother.* **2017**, *66*, 979–988. [CrossRef]
72. Hossain, D.M.S.; Pal, S.K.; Moreira, D.; Duttagupta, P.; Zhang, Q.; Won, H.; Jones, J.; D'Apuzzo, M.; Forman, S.; Kortylewski, M. TLR9-Targeted STAT3 Silencing Abrogates Immunosuppressive Activity of Myeloid-Derived Suppressor Cells from Prostate Cancer Patients. *Clin. Cancer Res.* **2015**, *21*, 3771–3782. [CrossRef]
73. Kortylewski, M.; Swiderski, P.; Herrmann, A.; Wang, L.; Kowolik, C.; Kujawski, M.; Lee, H.; Scuto, A.; Liu, Y.; Yang, C.; et al. In Vivo Delivery of siRNA to Immune Cells by Conjugation to a TLR9 Agonist Enhances Antitumor Immune Responses. *Nat. Biotechnol.* **2009**, *27*, 925–932. [CrossRef] [PubMed]
74. Kortylewski, M.; Kuo, Y.-H. Push and Release: TLR9 activation plus STAT3 Blockade for Systemic Antitumor Immunity. *Oncoimmunology* **2014**, *3*, e27441. [CrossRef] [PubMed]
75. Zhang, Q.; Hossain, D.M.S.; Nechaev, S.; Kozlowska, A.; Zhang, W.; Liu, Y.; Kowolik, C.M.; Swiderski, P.; Rossi, J.J.; Forman, S.; et al. TLR9-Mediated siRNA Delivery for Targeting of Normal and Malignant Human Hematopoietic Cells in Vivo. *Blood* **2013**, *121*, 1304–1315. [CrossRef] [PubMed]
76. Zhu, G.; Mei, L.; Vishwasrao, H.D.; Jacobson, O.; Wang, Z.; Liu, Y.; Yung, B.C.; Fu, X.; Jin, A.; Niu, G.; et al. Intertwining DNA-RNA Nanocapsules Loaded with Tumor Neoantigens as Synergistic Nanovaccines for Cancer Immunotherapy. *Nat. Commun.* **2017**, *8*, 1482. [CrossRef]
77. Kortylewski, M.; Kujawski, M.; Herrmann, A.; Yang, C.; Wang, L.; Liu, Y.; Salcedo, R.; Yu, H. Toll-like Receptor 9 Activation of Signal Transducer and Activator of Transcription 3 Constrains Its Agonist-Based Immunotherapy. *Cancer Res.* **2009**, *69*, 2497–2505. [CrossRef] [PubMed]

78. Setten, R.L.; Rossi, J.J.; Han, S.-P. The Current State and Future Directions of RNAi-Based Therapeutics. *Nat. Rev. Drug Discov.* **2019**, *18*, 421–446. [CrossRef]
79. Nechaev, S.; Gao, C.; Moreira, D.; Swiderski, P.; Jozwiak, A.; Kowolik, C.M.; Zhou, J.; Armstrong, B.; Raubitschek, A.; Rossi, J.J.; et al. Intracellular Processing of Immunostimulatory CpG-siRNA: Toll-like Receptor 9 Facilitates siRNA Dicing and Endosomal Escape. *J. Control. Release* **2013**, *170*, 307–315. [CrossRef]
80. Herrmann, A.; Kortylewski, M.; Kujawski, M.; Zhang, C.; Reckamp, K.; Armstrong, B.; Wang, L.; Kowolik, C.; Deng, J.; Figlin, R.; et al. Targeting Stat3 in the Myeloid Compartment Drastically Improves the in vivo Antitumor Functions of Adoptively Transferred T cells. *Cancer Res.* **2010**, *70*, 7455–7464. [CrossRef]
81. Hossain, D.M.S.; Dos Santos, C.; Zhang, Q.; Kozlowska, A.; Liu, H.; Gao, C.; Moreira, D.; Swiderski, P.; Jozwiak, A.; Kline, J.; et al. Leukemia Cell-Targeted STAT3 Silencing and TLR9 Triggering Generate Systemic Antitumor Immunity. *Blood* **2014**, *123*, 15–25. [CrossRef]
82. Moreira, D.; Zhang, Q.; Hossain, D.M.S.; Nechaev, S.; Li, H.; Kowolik, C.M.; D'Apuzzo, M.; Forman, S.; Jones, J.; Pal, S.K.; et al. TLR9 Signaling Through NF-κB/RELA and STAT3 Promotes Tumor-Propagating Potential of Prostate Cancer Cells. *Oncotarget* **2015**, *6*, 17302–17313. [CrossRef] [PubMed]
83. Davis, J.T. G-quartets 40 years later: From 5′-GMP to Molecular Biology and Supramolecular Chemistry. *Angew. Chem. Int. Ed. Engl.* **2004**, *43*, 668–698. [CrossRef] [PubMed]
84. Jing, N.; Gao, X.; Rando, R.F.; Hogan, M.E. Potassium-Induced Loop Conformational Transition of a Potent anti-HIV Oligonucleotide. *J. Biomol. Struct. Dyn.* **1997**, *15*, 573–585. [CrossRef] [PubMed]
85. McMicken, H.W.; Bates, P.J.; Chen, Y. Antiproliferative activity of G-Quartet-Containing Oligonucleotides Generated by a Novel Single-Stranded DNA Expression System. *Cancer Gene Ther.* **2003**, *10*, 867–869. [CrossRef] [PubMed]
86. Jing, N.; Marchand, C.; Guan, Y.; Liu, J.; Pallansch, L.; Lackman-Smith, C.; De Clercq, E.; Pommier, Y. Structure–Activity of Inhibition of HIV-1 Integrase and Virus Replication by G-quartet Oligonucleotides. *DNA Cell Biol.* **2001**, *20*, 499–508. [CrossRef]
87. Mazumder, A.; Neamati, N.; Ojwang, J.O.; Sunder, S.; Rando, R.F.; Pommier, Y. Inhibition of the human immunodeficiency virus type 1 integrase by guanosine quartet structures. *Biochemistry* **1996**, *35*, 13762–13771. [CrossRef]
88. Rando, R.F.; Ojwang, J.; Elbaggari, A.; Reyes, G.R.; Tinder, R.; McGrath, M.S.; Hogan, M.E. Suppression of Human Immunodeficiency Virus Type 1 Activity in Vitro by Oligonucleotides which Form Intramolecular Tetrads. *J. Biol. Chem.* **1995**, *270*, 1754–1760. [CrossRef]
89. Jing, N.; Tweardy, D.J. Targeting Stat3 in Cancer Therapy. *Anticancer Drugs* **2005**, *16*, 601–607. [CrossRef]
90. Sen, M.; Grandis, J.R. Nucleic Acid-Based Approaches to STAT Inhibition. *JAKSTAT* **2012**, *1*, 285–291. [CrossRef]
91. Zhu, Q.; Jing, N. Computational Study on Mechanism of G-Quartet Oligonucleotide T40214 Selectively Targeting Stat3. *J. Comput.-Aided Mol. Des.* **2007**, *21*, 641–648. [CrossRef]
92. Bishop, J.S.; Guy-Caffey, J.K.; Ojwang, J.O.; Smith, S.R.; Hogan, M.E.; Cossum, P.A.; Rando, R.F.; Chaudhary, N. Intramolecular G-quartet Motifs Confer Nuclease Resistance to a Potent anti-HIV Oligonucleotide. *J. Biol. Chem.* **1996**, *271*, 5698–5703. [CrossRef] [PubMed]
93. Weerasinghe, P.; Li, Y.; Guan, Y.; Zhang, R.; Tweardy, D.J.; Jing, N. T40214/PEI complex: A Potent Therapeutics for Prostate Cancer that Targets STAT3 Signaling. *Prostate* **2008**, *68*, 1430–1442. [CrossRef]
94. Jing, N.; Zhu, Q.; Yuan, P.; Li, Y.; Mao, L.; Tweardy, D.J. Targeting Signal Transducer and Activator of Transcription 3 with G-quartet Oligonucleotides: A Potential Novel Therapy for Head and Neck Cancer. *Mol. Cancer Ther.* **2006**, *5*, 279–286. [CrossRef] [PubMed]
95. Weerasinghe, P.; Garcia, G.E.; Zhu, Q.; Yuan, P.; Feng, L.; Mao, L.; Jing, N. Inhibition of Stat3 Activation and Tumor Growth Suppression of Non-Small Cell Lung Cancer by G-Quartet Oligonucleotides. *Int. J. Oncol.* **2007**, *31*, 129–136. [CrossRef] [PubMed]
96. Fagard, R.; Metelev, V.; Souissi, I.; Baran-Marszak, F. STAT3 Inhibitors for Cancer Therapy: Have All Roads Been Explored? *JAKSTAT* **2013**, *2*, e22882. [CrossRef]
97. Wagner, B.J.; Hayes, T.E.; Hoban, C.J.; Cochran, B.H. The SIF Binding Element Confers sis/PDGF Inducibility onto the c-fos Promoter. *EMBO J.* **1990**, *9*, 4477–4484. [CrossRef]
98. Leong, P.L.; Andrews, G.A.; Johnson, D.E.; Dyer, K.F.; Xi, S.; Mai, J.C.; Robbins, P.D.; Gadiparthi, S.; Burke, N.A.; Watkins, S.F.; et al. Targeted Inhibition of Stat3 with a Decoy Oligonucleotide Abrogates Head and Neck Cancer Cell Growth. *Proc. Natl. Acad. Sci. USA* **2003**, *100*, 4138–4143. [CrossRef]

99. Lui, V.W.Y.; Boehm, A.L.; Koppikar, P.; Leeman, R.J.; Johnson, D.; Ogagan, M.; Childs, E.; Freilino, M.; Grandis, J.R. Antiproliferative Mechanisms of a Transcription Factor Decoy Targeting Signal Transducer and Activator of Transcription (STAT) 3: The Role of STAT1. *Mol. Pharmacol.* **2007**, *71*, 1435–1443. [CrossRef]
100. Shen, J.; Li, R.; Li, G. Inhibitory Effects of Decoy-ODN Targeting Activated STAT3 on Human Glioma Growth in Vivo. *In Vivo* **2009**, *23*, 237–243.
101. Xi, S.; Gooding, W.E.; Grandis, J.R. In Vivo Antitumor Efficacy of STAT3 Blockade Using a Transcription Factor Decoy Approach: Implications for Cancer Therapy. *Oncogene* **2005**, *24*, 970–979. [CrossRef]
102. Zhang, X.; Liu, P.; Zhang, B.; Mao, H.; Shen, L.; Ma, Y. Inhibitory Effects of STAT3 Decoy Oligodeoxynucleotides on Human Epithelial Ovarian Cancer Cell Growth in Vivo. *Int. J. Mol. Med.* **2013**, *32*, 623–628. [CrossRef] [PubMed]
103. Zhang, X.; Zhang, J.; Wang, L.; Wei, H.; Tian, Z. Therapeutic Effects of STAT3 Decoy Oligodeoxynucleotide on Human Lung Cancer in Xenograft Mice. *BMC Cancer* **2007**, *7*, 149. [CrossRef] [PubMed]
104. Sen, M.; Joyce, S.; Panahandeh, M.; Li, C.; Thomas, S.M.; Maxwell, J.; Wang, L.; Gooding, W.E.; Johnson, D.E.; Grandis, J.R. Targeting Stat3 Abrogates EGFR Inhibitor Resistance in Cancer. *Clin. Cancer Res.* **2012**, *18*, 4986–4996. [CrossRef] [PubMed]
105. Sen, M.; Thomas, S.M.; Kim, S.; Yeh, J.I.; Ferris, R.L.; Johnson, J.T.; Duvvuri, U.; Lee, J.; Sahu, N.; Joyce, S.; et al. First-in-Human Trial of a STAT3 Decoy Oligonucleotide in Head and Neck Tumors: Implications for Cancer Therapy. *Cancer Discov.* **2012**, *2*, 694–705. [CrossRef]
106. Zhang, Q.; Hossain, D.M.S.; Duttagupta, P.; Moreira, D.; Zhao, X.; Won, H.; Buettner, R.; Nechaev, S.; Majka, M.; Zhang, B.; et al. Serum-Resistant CpG-STAT3 Decoy for Targeting Survival and Immune Checkpoint Signaling in Acute Myeloid Leukemia. *Blood* **2016**, *127*, 1687–1700. [CrossRef]
107. Zhao, X.; Zhang, Z.; Moreira, D.; Su, Y.-L.; Won, H.; Adamus, T.; Dong, Z.; Liang, Y.; Yin, H.H.; Swiderski, P.; et al. B Cell Lymphoma Immunotherapy Using TLR9-Targeted Oligonucleotide STAT3 Inhibitors. *Mol. Ther.* **2018**, *26*, 695–707. [CrossRef]
108. Lee, D.S.; O'Keefe, R.A.; Ha, P.K.; Grandis, J.R.; Johnson, D.E. Biochemical Properties of a Decoy Oligodeoxynucleotide Inhibitor of STAT3 Transcription Factor. *Int. J. Mol. Sci.* **2018**, *19*, 1608. [CrossRef]
109. Njatcha, C.; Farooqui, M.; Kornberg, A.; Johnson, D.E.; Grandis, J.R.; Siegfried, J.M. STAT3 Cyclic Decoy Demonstrates Robust Antitumor Effects in Non-Small Cell Lung Cancer. *Mol. Cancer Ther.* **2018**, *17*, 1917–1926. [CrossRef]
110. Sen, M.; Paul, K.; Freilino, M.L.; Li, H.; Li, C.; Johnson, D.E.; Wang, L.; Eiseman, J.; Grandis, J.R. Systemic Administration of a Cyclic Signal Transducer and Activator of Transcription 3 (STAT3) Decoy Oligonucleotide Inhibits Tumor Growth without Inducing Toxicological Effects. *Mol. Med.* **2014**, *20*, 46–56. [CrossRef]
111. Li, C.; Zang, Y.; Sen, M.; Leeman-Neill, R.J.; Man, D.S.K.; Grandis, J.R.; Johnson, D.E. Bortezomib up-Regulates Activated Signal Transducer and Activator of Transcription-3 and Synergizes with Inhibitors of Signal Transducer and Activator of Transcription-3 to Promote Head and Neck Squamous Cell Carcinoma Cell Death. *Mol. Cancer Ther.* **2009**, *8*, 2211–2220. [CrossRef]
112. Zhang, X.; Liu, P.; Zhang, B.; Wang, A.; Yang, M. Role of STAT3 Decoy Oligodeoxynucleotides on Cell Invasion and chemosensitivity in Human Epithelial Ovarian Cancer Cells. *Cancer Genet. Cytogenet.* **2010**, *197*, 46–53. [CrossRef] [PubMed]
113. Zhang, X.; Xiao, W.; Wang, L.; Tian, Z.; Zhang, J. Deactivation of Signal Transducer and Activator of Transcription 3 Reverses Chemotherapeutics Resistance of Leukemia Cells via Down-Regulating P-gp. *PLoS ONE* **2011**, *6*, e20965. [CrossRef] [PubMed]
114. Bharadwaj, U.; Kasembeli, M.M.; Tweardy, D.J. STAT3 Inhibitors in Cancer: A Comprehensive Update. In *Cancer Drug Discovery and Development*; Springer: Berlin/Heidelberg, Germany, 2016; pp. 95–161.
115. Hodge, J.A.; Kawabata, T.T.; Krishnaswami, S.; Clark, J.D.; Telliez, J.-B.; Dowty, M.E.; Menon, S.; Lamba, M.; Zwillich, S. The Mechanism of Action of Tofacitinib—An Oral Janus Kinase Inhibitor for the Treatment of Rheumatoid Arthritis. *Clin. Exp. Rheumatol.* **2016**, *34*, 318–328. [PubMed]
116. Zhou, T.; Georgeon, S.; Moser, R.; Moore, D.J.; Caflisch, A.; Hantschel, O. Specificity and Mechanism-of-Action of the JAK2 Tyrosine Kinase Inhibitors Ruxolitinib and SAR302503 (TG101348). *Leukemia* **2014**, *28*, 404–407. [CrossRef]
117. Eiring, A.M.; Page, B.D.G.; Kraft, I.L.; Mason, C.C.; Vellore, N.A.; Resetca, D.; Zabriskie, M.S.; Zhang, T.Y.; Khorashad, J.S.; Engar, A.J.; et al. Combined STAT3 and BCR-ABL1 Inhibition Induces Synthetic Lethality in Therapy-Resistant Chronic Myeloid Leukemia. *Leukemia* **2017**, *31*, 1253–1254. [CrossRef]

118. Wen, W.; Wu, J.; Liu, L.; Tian, Y.; Buettner, R.; Hsieh, M.-Y.; Horne, D.; Dellinger, T.H.; Han, E.S.; Jove, R.; et al. Synergistic Anti-Tumor Effect of Combined Inhibition of EGFR and JAK/STAT3 Pathways in Human Ovarian Cancer. *Mol. Cancer* **2015**, *14*, 100. [CrossRef]
119. Woessner, R.; McCoon, P.; Bell, K.; DuPont, R.; Collins, M.; Lawson, D.; Nadella, P.; Pablo, L.; Reimer, C.; Sah, V.; et al. Abstract A93: STAT3 Inhibition Enhances the Activity of Immune Checkpoint Inhibitors in Murine Syngeneic Tumor Models by Creating a More Immunogenic Tumor Microenvironment. *Tumor Microenviron.* **2015**, *3*. [CrossRef]
120. Gao, Y.; Li, Y.; Hsu, E.; Wang, Y.; Huang, J.; Brooks, E.; Li, C.J. Abstract LB-140: Inhibition of Cancer Stemness Sensitizes Colorectal Cancer to Immune Checkpoint Inhibitors. *Tumor Biol.* **2017**. [CrossRef]
121. Weber, R.; Fleming, V.; Hu, X.; Nagibin, V.; Groth, C.; Altevogt, P.; Utikal, J.; Umansky, V. Myeloid-Derived Suppressor Cells Hinder the Anti-Cancer Activity of Immune Checkpoint Inhibitors. *Front. Immunol.* **2018**, *9*, 1310. [CrossRef]
122. Johnson, D.E.; O'Keefe, R.A.; Grandis, J.R. Targeting the IL-6/JAK/STAT3 signalling axis in cancer. *Nat. Rev. Clin. Oncol.* **2018**, *15*, 234. [CrossRef]
123. Yang, H.; Yamazaki, T.; Pietrocola, F.; Zhou, H.; Zitvogel, L.; Ma, Y.; Kroemer, G. STAT3 Inhibition Enhances the Therapeutic Efficacy of Immunogenic Chemotherapy by Stimulating Type 1 Interferon Production by Cancer Cells. *Cancer Res.* **2015**, *75*, 3812–3822. [CrossRef] [PubMed]

8

Hepatic Stress Response in HCV Infection Promotes STAT3-Mediated Inhibition of HNF4A-*miR-122* Feedback Loop in Liver Fibrosis and Cancer Progression

Yucel Aydin [1], Ramazan Kurt [2], Kyoungsub Song [1], Dong Lin [1], Hanadi Osman [1], Brady Youngquist [1], John W. Scott [1], Nathan J. Shores [2], Paul Thevenot [3], Ari Cohen [3] and Srikanta Dash [1,2,*]

[1] Department of Pathology and Laboratory Medicine, Tulane University Health Sciences Center, New Orleans, LA 70112, USA; yaydin@tulane.edu (Y.A.); ksong@tulane.edu (K.S.); dlin6@tulane.edu (D.L.); hosman1@tulane.edu (H.O.); byoungquist@tulane.edu (B.Y.); jscottmd@tulane.edu (J.W.S.)

[2] Section of Gastroenterology and Hepatology, Tulane University Health Sciences Center, New Orleans, LA 70112, USA; rkurt@tulane.edu (R.K.); shoresnj@gmail.com (N.J.S.)

[3] Liver Transplant Surgery Section, Ochsner Medical Center, New Orleans, LA 70121, USA; paul.thevenot@oschner.org (P.T.); acohen@oschner.org (A.C.)

* Correspondence: sdash@tulane.edu

Abstract: Hepatitis C virus (HCV) infection compromises the natural defense mechanisms of the liver leading to a progressive end stage disease such as cirrhosis and hepatocellular carcinoma (HCC). The hepatic stress response generated due to viral replication in the endoplasmic reticulum (ER) undergoes a stepwise transition from adaptive to pro-survival signaling to improve host cell survival and liver disease progression. The minute details of hepatic pro-survival unfolded protein response (UPR) signaling that contribute to HCC development in cirrhosis are unknown. This study shows that the UPR sensor, the protein kinase RNA-like ER kinase (PERK), mediates the pro-survival signaling through nuclear factor erythroid 2-related factor 2 (NRF2)-mediated signal transducer and activator of transcription 3 (STAT3) activation in a persistent HCV infection model of Huh-7.5 liver cells. The NRF2-mediated STAT3 activation in persistently infected HCV cell culture model resulted in the decreased expression of hepatocyte nuclear factor 4 alpha (HNF4A), a major liver-specific transcription factor. The stress-induced inhibition of HNF4A expression resulted in a significant reduction of liver-specific *microRNA-122* (*miR-122*) transcription. It was found that the reversal of hepatic adaptive pro-survival signaling and restoration of *miR-122* level was more efficient by interferon (IFN)-based antiviral treatment than direct-acting antivirals (DAAs). To test whether *miR-122* levels could be utilized as a biomarker of hepatic adaptive stress response in HCV infection, serum *miR-122* level was measured among healthy controls, and chronic HCV patients with or without cirrhosis. Our data show that serum *miR-122* expression level remained undetectable in most of the patients with cirrhosis (stage IV fibrosis), suggesting that the pro-survival UPR signaling increases the risk of HCC through STAT3-mediated suppression of *miR-122*. In conclusion, our data indicate that hepatic pro-survival UPR signaling suppresses the liver-specific HNF4A and its downstream target *miR-122* in cirrhosis. These results provide an explanation as to why cirrhosis is a risk factor for the development of HCC in chronic HCV infection.

Keywords: hepatitis C virus (HCV); cirrhosis; hepatocellular carcinoma (HCC); endoplasmic reticulum (ER) stress; oxidative stress (OS); unfolded protein response (UPR); *microRNA-122* (*miR-122*); nuclear factor erythroid 2-related factor 2 (NRF2); signal transducer and activator of transcription 3 (STAT3); hepatocyte nuclear factor 4 alpha (HNF4A)

1. Introduction

Hepatocellular Carcinoma (HCC) is the most common form of liver cancer. The incidence of HCC has nearly doubled over the past decade, making it the third-leading cause of cancer-related death worldwide [1]. In most cases, HCC develops on the background of cirrhosis that is mainly associated with viral hepatitis, overconsumption of alcohol, and non-alcoholic fatty liver disease. In the USA, chronic hepatitis C virus (HCV) infection is the leading cause of HCC and liver transplantation [2]. Globally, 30% of all HCC cases are related with persistent HCV infection representing a significant public health problem [3]. Recently approved direct-acting antivirals (DAAs) enable to cure the majority of patients with chronic HCV infection, including patients with cirrhosis [4]. HCV treatment reduces liver inflammation, fibrosis, and HCC thus drops liver-disease related mortality in the future [5,6]. Although the causal relationship between HCV infection and chronic liver disease has been well documented, the minute mechanisms of the virus-cell interactions involved in disease progression are not well understood. Since there is no effective treatment option of cirrhosis and late-stage HCC, it is hoped that understanding the mechanisms of HCV-induced liver cancer will allow developing better treatment. This knowledge will also result in a better understanding of the pathogenesis of other agents such as hepatitis B virus (HBV) and non-viral agents leading to cirrhosis and HCC.

The mechanisms of how chronic HCV infections are associated with the development of cirrhosis and HCC have been a challenge to study due to lack of an appropriate small animal model [7]. In a previous publication, it was shown that hepatocytes develop an integrated stress response (ISR) due to the combination of endoplasmic reticulum (ER) and oxidative stress occurring with persistent HCV replication in the cell culture model [8]. The increased cellular stress response activates nuclear factor erythroid 2-related factor 2 (NRF2) signaling as a primary cell survival pathway by inducing the transcription of genes with the antioxidant response [9]. Previously published data have shown that the stress-induced NRF2 signaling induces expression of LAMP2A and HSC70 leading to induction of chaperone-mediated autophagy (CMA). The CMA induction in HCV-infected culture improves cell survival through degradation of major tumor suppressors: p53 and retinoblastoma protein (pRB) [10–12]. A recent publication from this laboratory has shown that HCV-mediated NRF2 signaling impairs autophagy process at the level of autophagosome-lysosome fusion leading to activation of the oncogenic signaling through the tyrosine kinase receptors (EGFR) [13]. These data support that the activation of NRF2 signaling contributes to cell survival and progression of preneoplastic lesion to HCC malignancies [14]. Signal transducer and activator of transcription 3 (STAT3) pathway is also activated as a major cell survival mechanism in HCC but not in the surrounding non-tumor tissue or normal liver [15,16]. The mechanisms of STAT3 activation could be due to an elevated expression of cytokines such as interleukin 6 (IL6), interleukin 22 (IL22), oxidative stress or epigenetic regulation [17,18]. However, studies have shown that STAT3 activation is transient in non-transformed cells even in the presence of cytokines and STAT3 activating mutations are rarely present in HCC [17,18]. The HCC-specific mechanism of STAT3 activation has not been addressed thoroughly.

Due to these reasons, we investigated the contribution of NRF2-related pro-survival mechanism through STAT3 activation in HCV culture in response to oxidative stress. We investigated STAT3-mediated pro-survival mechanism via a circuit that involves downregulation of hepatocyte nuclear factor 4 alpha (HNF4A) through *miR-24* and *miR-619* [19,20]. STAT3-inducible *miR-24* and *miR-629* destabilize the *HNF4A* mRNA leading to the permanent suppression of protein expression. HNF4A is a principal transcription factor required for liver development, hepatocyte differentiation, and hepatic function [21]. Many recent publications claim that the expression level of HNF4A and its target genes are impaired in cirrhosis and diminished in HCC [22–27]. HNF4A has been shown to play a role in HCC development related to chronic inflammation processes of the liver through regulating the transcription of *miR-122* [28–30]. The importance of this circuit in HCC development is further supported by previous reports showing that *miR-122* knockout mice develop HCC [31–33]. Based on these evidences, we hypothesized that the NRF2-mediated activation of the STAT3-HNF4A inflammatory loop could lead to a long-term suppression of *miR-122* that increases the HCC risk among patients with cirrhosis.

Using a persistently HCV-infected Huh-7.5 liver cell culture model, it was found that the hepatic adaptive response through the protein kinase RNA-like ER kinase (PERK)-NRF2 axis activates the STAT3-HNF4A inflammatory loop as a cell survival mechanism. Data presented in this manuscript suggest that the NRF2-mediated STAT3 activation silence the expression of HNF4A through *miR-24* upregulation. The activation of the STAT3-HNF4A loop leads to suppression of *miR-122* in persistently infected HCV cell culture and chronic HCV patients with cirrhosis. Finally, the data in the study demonstrate that the serum *miR-122* levels are depleted during advanced liver disease such as cirrhosis, which may explain why HCC develops most of the time on the background of chronic liver disease with cirrhosis.

2. Results

2.1. Persistent HCV Replication Leads to ER Stress and Oxidative Stress

To understand the hepatic adaptive response to ER stress, highly permissive Huh-7.5 cells were infected with JFH-AM120 at a multiplicity of infection (MOI) of 0.1, and HCV replication was studied over 21 days with a regular passage of infected cells at three-day intervals. Infected cells collected at different time points were examined for HCV protein expression by Western blotting. The efficiency of replication and spread of HCV-green fluorescence protein (GFP) chimera virus were examined by fluorescence microscopy with nuclear DAPI staining. Quantifications of GFP positive cells by ImageJ software show that the level of HCV replication increased with time. The percentage of Huh-7.5 cells expressing HCV-GFP fusion was quantified by flow cytometry analysis at 0, 3, 6, 9, and 12 days. These results indicated that about 5.9% of cells were GFP positive on day 3, and the number increased to 85.4% and 93.5% on day 9 and day 12, respectively. All these results support our previously published data showing that the high-level replication and spread of JFH-AM120 chimera HCV clone in Huh-7.5 liver cells (Figure S1). When the accumulation of viral protein exceeds the folding capacity of the ER resident chaperones, it creates a stress response called UPR. The activation of UPR secondary to ER stress response is the major driver of liver disease progression in HCV infection. Sustained HCV infection in Huh-7.5 liver cells is expected to accumulate unfolded protein load and expansion of the ER membranes [34]. We measured misfolded protein stress in the ER membranes by live cell imaging using a transient transfection-based approach with a commercially available ER-red fluorescence protein (RFP) construct (ER-RFP, BacMam 2.0, ThermoFisher, Waltham, MA, USA). The staining was intense in the rough ER around the nucleus of infected cells and the RFP signal overlapped with HCV-GFP. The ER staining of the uninfected cells was diffuse, whereas it was more intense in HCV-infected cells, suggesting the evidence of ER membrane expansion and accumulation of ER-RFP protein in the cells replicating HCV. Interestingly, HCV infection induced an extensive colocalization of ER-Tracker RFP with NS5A-GFP. The yellow fluorescence signal due to colocalization was markedly increased after overlaying the images of NS5A-GFP and ER-tracker RFP fluorescence. Quantitative assessment of the colocalization of HCV-GFP with ER-tracker RFP was significantly higher in HCV-infected cells than uninfected Huh-7.5 cells. These data confirm that HCV-GFP fusion protein is colocalized with the RFP protein in the perinuclear region of the infected cells that is consistent with increased GRP78/94 protein localization in the ER [35–37]. These results are also consistent with previous reports confirming the activation of the stress sensor of the UPR by Western blot analysis [13]. During HCV replication, several reactive oxidants such as reactive oxygen species (ROS) or reactive nitrogen species (RNS) are generated as a by-product of biochemical reactions in mitochondria, peroxisomes, and ER membranes leading to oxidative stress. A fluorescent-based assay was used to quantify ROS levels between uninfected and persistently infected cultures with HCV-Renilla luciferase virus (JFH1-V3-Rluc) on day 9. In this assay, H2DCFDA is converted to the highly fluorescent 2′7′-dichlorofluorescein due to the presence of ROS. The fluorescence intensity is proportional to the amount of oxidant present in HCV-infected cells. A flow cytometric analysis based on the quantitative approach displays that the majority of cells infected with HCV show punctate fluorescence staining due to the presence of high oxidative stress response. The oxidative stress was significantly higher in HCV-infected culture than uninfected Huh-7.5 cells by three repeated analyses. Results shown in Figure S2 confirm previously published data indicating that HCV infection also induces ER stress and oxidative stress [38–41].

2.2. Oxidative Stress and ER Stress Activate NRF2 Signaling in HCV Infection

A transcription factor, NRF2, plays a critical role in the control of genes involved in cell survival pathway. It induces varieties of cytoprotective genes harboring a short cis-acting sequence called the antioxidant response element (ARE) in their promoters [42,43]. First, we demonstrated that robust expression of the HCV core protein in the infected culture could be detected on day 3 and the number of positive cells was increased from day 6 and over 90% cells became core positive on day 12. The intensity of cytoplasmic core staining was quantified by ImageJ software (NIH, Bethesda, MD, USA) and found to increase over time, indicating active viral replication and spread after HCV infection. The activation of NRF2 signaling was examined by measuring the nuclear translocation in persistently infected culture in a kinetic study by immunocytochemical staining. NRF2 nuclear accumulation was found in 100% of HCV-infected culture starting from day 9 that correlated well with HCV core expression (Figure S3). These results indicate that HCV infection results in NRF2 activation and nuclear translocation.

2.3. Persistent HCV Infection Activates the NRF2-STAT3-HNF4A Regulatory Axis as a Cytoprotective Response

Increased oxidative stress has been reported in liver disease caused by viral hepatitis and non-viral etiologies [43]. The NRF2 has a crucial role in enabling adaptation to oxidative stress and ER stress by transcribing more than 2000 cytoprotective genes involved in cell survival [44,45]. The induction of NRF2 genes requires a common NRF2-binding motif on the promoter called ARE [46,47]. Five ARE (TGnnnnGC) and six ARE-like (TGAnnnnGC or TGAnnnnnGC) binding sites were identified in STAT3 promoter region (Figure 1A). Western blot analysis using phosphorylated and unphosphorylated specific antibodies revealed that HCV infection induced NRF2 and STAT3 activation (Figure 1B). The expression of STAT3 was found to be increased more in HCV-infected culture as compared to uninfected Huh-7.5 cells. The expression level of β-Actin did not change, suggesting that the differences were not at the level of protein loading or variation in the protein content of the lysates used. The inflammatory circuit consisting of HNF4A, *miR-122*, IL6, STAT3, and *miR-24* is implicated in hepatocellular transformation and liver oncogenesis [19]. The first component of the circuit is the STAT3-mediated suppression of HNF4A through *miR-24*. The second component is HNF4A-mediated suppression of *miR-122* transcription. Western blot analysis was performed to verify whether HCV-induced STAT3 also modulates expression of HNF4A pathway. Results shown in Figure 1B demonstrate that HCV infection increased NRF2, and STAT3 expression but suppressed HNF4A expression. Western blot data were quantified using ImageJ software to compare relative expression of NRF2, STAT3 and HNF4A (Figure S4). To determine the statistical correlation coefficient between the expression levels of STAT3 and HNF4A, the fraction of variance denoted by R^2 values was calculated (Figure 1C,D). Analyses of the R^2 values are very close to 1.0, suggesting that the induction of STAT3 correctly predicts the decrease of HNF4A in HCV-infected cells. Immunolocalization of STAT3 and HNF4A was performed in uninfected and infected cell culture on day 9 by confocal microscopy. This analysis showed that STAT3 activation in HCV-infected culture was associated with negative HNF4A expression in the nucleus (Figure 1E,F). These results indicate an inverse relationship between STAT3 activation and the expression of liver-specific transcription factor, HNF4A. Real-time RT-PCR data showed that HCV infection induced *NRF2* and *STAT3* mRNA levels, whereas *HNF4A* mRNA levels were decreased (Figure 1G–I), suggesting that activation of NRF2 signaling decreases the HNF4A protein level by altering the stability or reduced transcription of HNF4A. These results are consistent with the previous report suggesting that HCV infection inhibits *HNF4A* expression by reducing its mRNA levels [48]. To test the consistency of this observation, results were verified using an HCV infection model of primary human hepatocytes (PHHs). It was found that the expression of total NRF2 and phosphorylated NRF2 (pNRF2) were induced in HCV-infected PHHs by Western blot analysis measured for 12 days (Figure 2A). It was observed that the expression of total STAT3 and phosphorylated STAT3 (pSTAT3) were increased in a time-dependent manner in the infected PHHs culture as compared to uninfected PHHs, indicating that HCV replication induces STAT3 pathway. The expression level of HNF4A was decreased after HCV infection and inversely correlated with expression of HCV NS3 by Western blot analysis. All the Western blot bands were quantified using ImageJ software to compare the relative expression of NRF2, STAT3 and HNF4A (Figure S5). Total RNA isolated from infected PHHs was used to quantify the mRNA levels of STAT3, NRF2, HNF4A and internal control β-Actin by real-time RT-PCR. First, Ct values of individual gene were normalized with *β-Actin*mRNA and then, the fold change was determined as compared to uninfected PHHs. The mRNA levels of NRF2 and STAT3 were found to be increased in a time-dependent manner in HCV-infected PHHs (Figure 2B,C), whereas *HNF4A* mRNA levels were decreased over time post infection (Figure 2D).

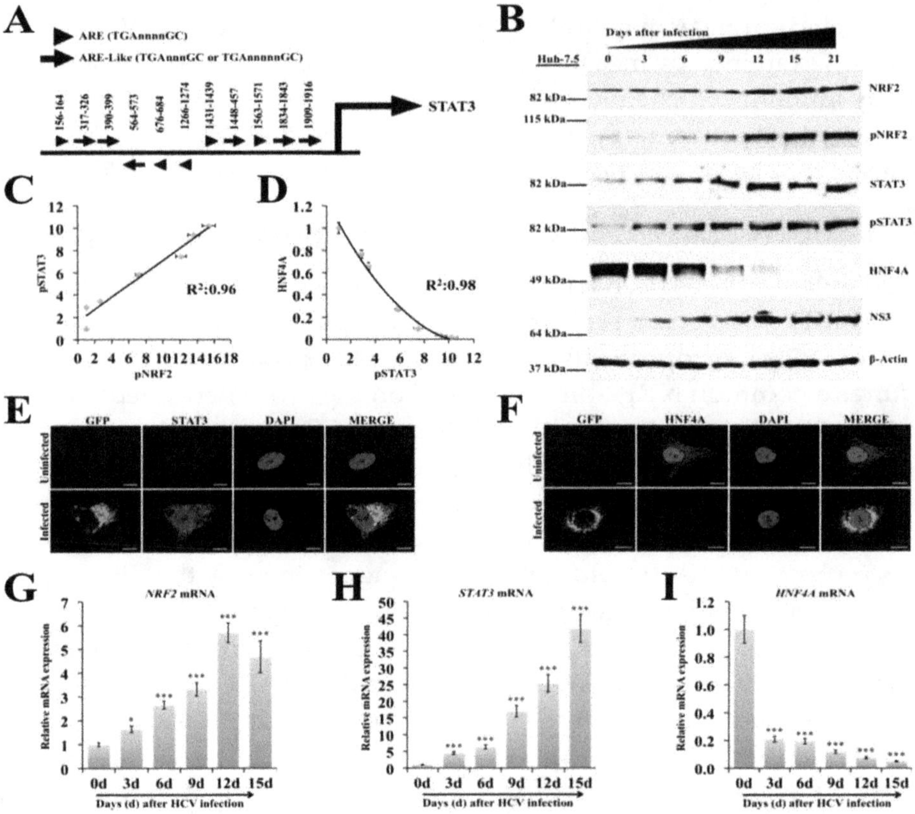

Figure 1. Persistent hepatitis C virus (HCV) replication induces nuclear factor erythroid 2-related factor 2 (NRF2)-mediated regulation of the signal transducer and activator of transcription 3 (STAT3)-hepatocyte nuclear factor 4 alpha (HNF4A) inflammatory loop. (**A**) STAT3-promoter region was examined for the presence of antioxidant response element (ARE) consensus sequences indicated by arrowheads, and ARE-like sequences shown by arrows. (**B**) Western blot analysis shows expression levels of total and phosphorylated NRF2, and STAT3 in infected Huh-7.5 liver cells over 21 days. The expression of HNF4A was inversely correlated with HCV NS3 protein expression. (**C**) Band intensities of phosphorylated STAT3 (pSTAT3) and phosphorylated NRF2 (pNRF2) were quantified using ImageJ software and R^2 values were determined by excel software. (**D**) Western blot results of HNF4A and pSTAT3 were quantified using ImageJ software and R^2 values were determined by excel software. (**E**) Colocalization of HCV-green fluorescence protein (GFP) with STAT3 by confocal microscopy. (**F**) Colocalization studies by confocal microscopy between HCV-GFP and nuclear expression of HNF4A in infected Huh-7.5 cells on day 9. Messenger RNA (mRNA) levels of individual genes in HCV-infected Huh-7.5 culture were measured by real-time RT-PCR analysis. (**G**) *NRF2* mRNA levels (**H**) *STAT3* mRNA levels. (**I**) *HNF4A* mRNA levels. The results are expressed as the mean ± standard deviation (SD) of three experiments. Error bars represent SD. *p*-values were calculated by ANOVA as compared to uninfected control. * *p*-value < 0.05, *** *p*-value < 0.001. Original magnification ×60. Scale bars = 10 μm.

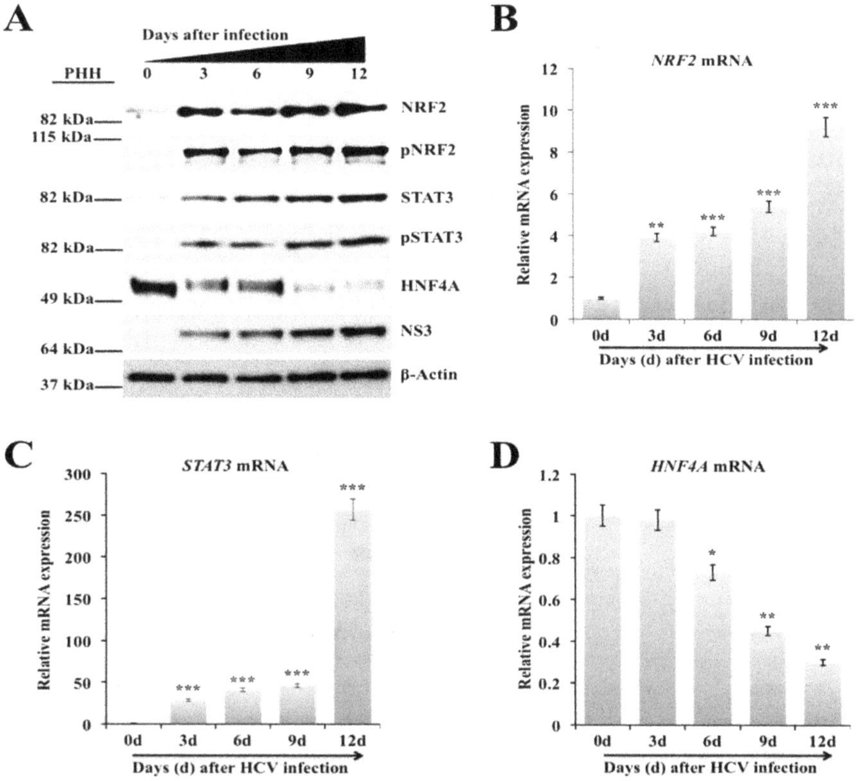

Figure 2. Hepatic stress response due to hepatitis C virus (HCV) replication in non-proliferative primary human hepatocytes (PHHs) activates nuclear factor erythroid 2-related factor 2 (NRF2)-mediated signal transducer and activator of transcription 3 (STAT3)-hepatocyte nuclear factor 4 alpha (HNF4A) expression. (**A**) Cell lysate was examined for the presence of HCV NS3 protein and for the activation of NRF2, STAT3 and HNF4A protein expression in infected PHH by Western blot. Messenger RNA (mRNA) levels of individual genes in HCV-infected PHHs were measured by real-time RT-PCR analysis. (**B**) Shown are *NRF2* mRNA levels (**C**) Shown are *STAT3* mRNA levels. (**D**) Shown are *HNF4A* mRNA levels. The results are expressed as the mean ± standard deviation (SD) of three experiments. Error bars represent SD. p-values were calculated by ANOVA as compared to day 0. * p-value < 0.05, ** p-value < 0.01, *** p-value < 0.001.

2.4. Silencing PERK and NRF2 Pathway Restore Expression of STAT3-HNF4A Inflammatory Loop

Persistent HCV replication induces sustained nuclear translocation of NRF2 as a cytoprotective mechanism in response to ER stress and oxidative stress. The impact of silencing NRF2 by a small interfering RNA (siRNA) treatment on the expression of STAT3 and HNF4A was examined in a persistently infected HCV cell culture model. The siRNA treatment was performed using day 9 infected HCV-GFP culture. As shown in Figure 3A, knockdown of NRF2 by siRNA decreased STAT3 expression and restored HNF4A expression indicating that the increased NRF2-mediated STAT3 expression is inversely associated with HNF4A downregulation. Uninfected Huh-7.5 cells treated with the NRF2 activator (sulforaphane) shows increased expression of phosphorylated NRF2 and STAT3 but suppressed HNF4A expression (Figure 3B). These data suggest that NRF2 mediated the regulation of the STAT3-HNF4A inflammatory loop in HCV infection. The relationship of NRF2 activation and ER stress was determined in infected Huh-7.5 cells after treatment with the ER stress inhibitor. As shown in Figure 3C, tauroursodeoxycholic acid (TUDCA) treatment decreased NRF2 expression, and STAT3 activation but restored HNF4A expression. To test the role of the PERK pathway on regulating the STAT3-HNF4A axis, the expression levels of STAT3 and HNF4A were measured after treatment with the PERK inhibitor by Western blotting. As shown in Figure 3D, inhibiting PERK blocked STAT3 induction and restored HNF4A expression. It is well known that STAT3 is phosphorylated by Janus kinases (JAK) leading to homo- or heterodimers, and nuclear translocation for transcriptional regulation of microRNAs and numerous genes involved in cell survival pathway. We examined whether treatment with the JAK inhibitor could also prevent STAT3 phosphorylation and restore HNF4A expression in HCV-infected culture. Day 9 infected Huh-7.5 cell culture was treated with increasing concentrations of the JAK inhibitor (pyridone 6, Calbiochem, San Diego, CA, USA) and after 72 hours, the expression of STAT3, pSTAT3 and HNF4A levels was examined by Western blotting (Figure 3E). Treatment with the JAK inhibitor did not block HCV replication but prevented STAT3 phosphorylation and restored expression of HNF4A. Furthermore, we found that infected Huh-7.5 cells treated with a highly specific STAT3 inhibitor (S3I-201, Selleck Chemicals, Houston, TX, USA) restored the expression level of HNF4A without altering level of HCV core expression (Figure 3F). Over expression of STAT3 by transient transfection downregulated the expression of HNF4A (Figure 3G). Huh-7.5 cells transfected with a control plasmid did not alter expression of HNF4A, suggesting specific regulation by STAT3 expression. The relative expression of NRF2, STAT3 and HNF4A under different siRNA or chemical treatments was compared by quantifying the Western blot results by ImageJ software (Figure S6).

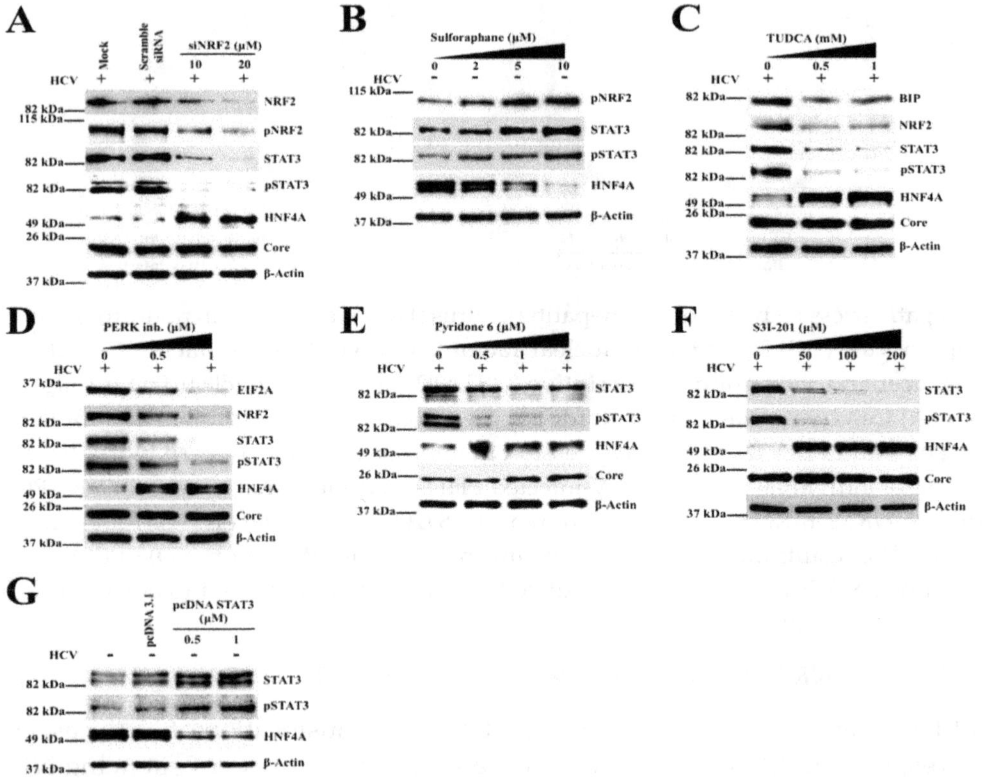

Figure 3. Shown is the nuclear factor erythroid 2-related factor 2 (NRF2)-dependent regulation of signal transducer and activator of transcription 3 (STAT3)-hepatocyte nuclear factor 4 alpha (HNF4A) expression in hepatitis C infection (HCV) infection. (**A**) The effect of NRF2 silencing on the expression of STAT3-HNF4A in HCV culture. Infected Huh-7.5 cells on day 9 were treated with combination of two small interfering RNAs (siRNAs) targeted to NRF2 and scrambled siRNA and the expression of STAT3 and HNF4A was measured by Western blot analysis after 72 hours. The effect of siRNA transfection did not alter HCV core expression or expression of β-Actin levels. (**B**) Shown is the effect of treatment with the NRF2 activator, sulforaphane, on the expression of STAT3 and HNF4A in uninfected Huh-7.5 liver cells. (**C**) Western blot analysis shows the effect of the endoplasmic reticulum (ER) stress inhibitor (TUDCA) on the expression levels of STAT3 and HNF4A in HCV-infected culture. (**D**) Shown are the expression levels of STAT3-HNF4A in HCV culture treated with the protein kinase RNA-like ER kinase (PERK) inhibitor by Western blot analysis. (**E**) Shown is the effect of the Janus kinase (JAK) inhibitor on the phosphorylation of STAT3 and HNF4A regulation. (**F**) Shown is the effect of the STAT3 inhibitor on the expression levels of HNF4A in infected cells by Western blot analysis. (**G**) The effect of STAT3 over expression by plasmid transfection on the levels of HNF4A in uninfected Huh-7.5 cells.

2.5. Persistent HCV Replication in Huh-7.5 Cells Leads to STAT3-HNF4A-Mediated Silencing of miR-122

STAT3-inducible *miR-24* controls the stability of *HNF4A* mRNA through binding to the 3′untranslated region (3′UTR). This regulation is linked to HNF4A-mediated induction of liver-specific *miR-122* transcription [49]. We tested the hypothesis whether NRF2-mediated cellular adaptive response to HCV infection regulates the STAT3-HNF4A inflammatory loop through modulating the expression of *miR-122* and *miR-24* (Figure 4A). We used a RT-PCR assay to measure their expression in persistently infected culture over 21 days. As expected, the *miR-24* expression progressively increased whereas the copy number of *miR-122* decreased in HCV culture over time (Figure 4B). We found that HCV-induced stress response decreased the expression of *miR-122* in a time-dependent manner. The appearance of control unrelated *miR-16* did not change under similar assay conditions. The impact of HCV replication on the regulation of microRNA loop was also verified in infected PHHs model. It was observed that *miR-122* level decreased in HCV-infected PHHs whereas *miR-24* level increased (Figure 4C). The level of *miR-16* did not change due to HCV infection in PHHs. We conducted as a next step a series of antiviral treatment using HCV-infected Huh-7.5 cells on day 9 to see whether HCV clearance could reverse the expression of *miR-122*. As shown in Figure 4D, IFN alpha (IFNα), sofosbuvir, ledipasvir and a combination of sofosbuvir and ledipasvir effectively cleared HCV-GFP expression without affecting viability of infected Huh-7.5 cells. Furthermore, it was found that inducing HCV clearance by treatment with either IFNα or DAAs (sofosbuvir, and ledipasvir) inhibited STAT3 protein levels and restored HNF4A expression (Figure 4E). The relative expression of STAT3, HNF4A and HCV core was determined quantifying the Western blot results by ImageJ software (Figure S7). The cross-regulatory effect of STAT3 and HNF4A expression on *miR-24* and *miR-122* levels was also verified after HCV clearance with different antivirals. It was found that restoration of *miR-122* level was more efficient by IFN-based antiviral treatment than by DAAs (Figure 4F,G). These results suggest that cellular adaptive response to HCV-induced stress activated the STAT3-HNF4A inflammatory loop that leads to decreased expression of *miR-122*.

It was also of interest to determine if the treatment of HCV-infected cells with small molecule inhibitors of PERK and ER stress could show a differential effect on the expression of microRNA that controls the expression of STAT3 and HNF4A. Infected Huh-7.5 cells on day 9 was treated with a PERK inhibitor or TUDCA for 72 hours and then the expression levels of *miR-24* and *miR-122* were examined by real-time RT-PCR. Data shown in Figure 5A indicate that this treatment did not inhibit HCV replication since no difference of HCV-GFP expression between untreated and treated groups was seen. The expression level of *miR-24* was higher in HCV-infected untreated culture due to STAT3 activation (Figure 5B). As expected, persistently infected HCV culture treated with the PERK inhibitor and TUDCA inhibited *miR-24* expression but restored expression of *miR-122* in HCV culture (Figure 5B,C). Taken together, these results suggested that HCV-induced PERK activation promoted the activation of the STAT3-HNF4A inflammatory loop to overcome the stress response associated with persistent HCV replication.

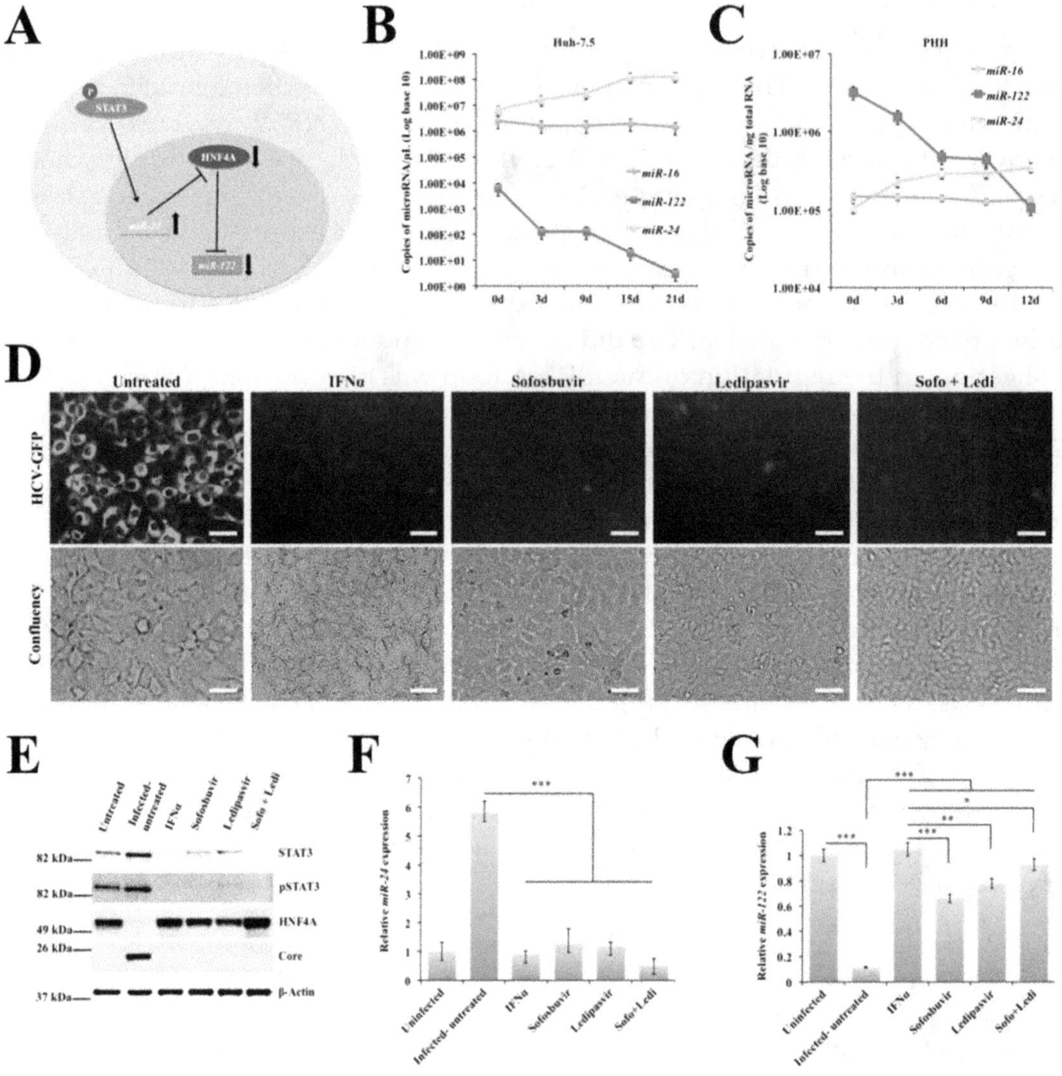

Figure 4. Relationship between signal transducer and activator of transcription 3 (STAT3)-hepatocyte nuclear factor 4 alpha (HNF4A) and the expression *miR-122* and *miR-24* in hepatitis C virus (HCV)-infected Huh-7.5 cells and primary human hepatocytes (PHHs). (**A**) Shown is the proposed hypothesis relating stress and STAT3-*miR-24*-HNF4A-*miR-122* feedback circuit in HCV infection. (**B**) Quantification of *miR-16*, *miR-24* and *miR-122* levels in persistently HCV-infected Huh-7.5 cells by real-time RT-PCR analysis over 21 days. (**C**) Quantification of *miR-122*, *miR-24* and *miR-122* levels in HCV-infected PHHs over 12 days. (**D**) Shown is the HCV-green fluorescence protein (GFP) expression in the infected culture with or without antiviral treatment. (**E**) Western blot shows the expression levels of STAT3 and HNF4A in the infected culture on day 9 with or without antiviral treatment. (**F**) The expression levels of *miR-24* in HCV-infected culture before and after treatment with interferon alpha (IFNα) or direct-acting antivirals (DAAs). (**G**) The expression levels of *miR-122* in HCV-infected culture treated with IFNα or DAA. The results are expressed as the mean ± standard deviation (SD) of three experiments. Error bars represent SD. *p*-values were calculated by ANOVA between different groups as compared to untreated HCV-infected group. * *p*-value < 0.05, ** *p*-value < 0.01, *** *p*-value < 0.001. Original magnification ×40, scale bars = 25 μm.

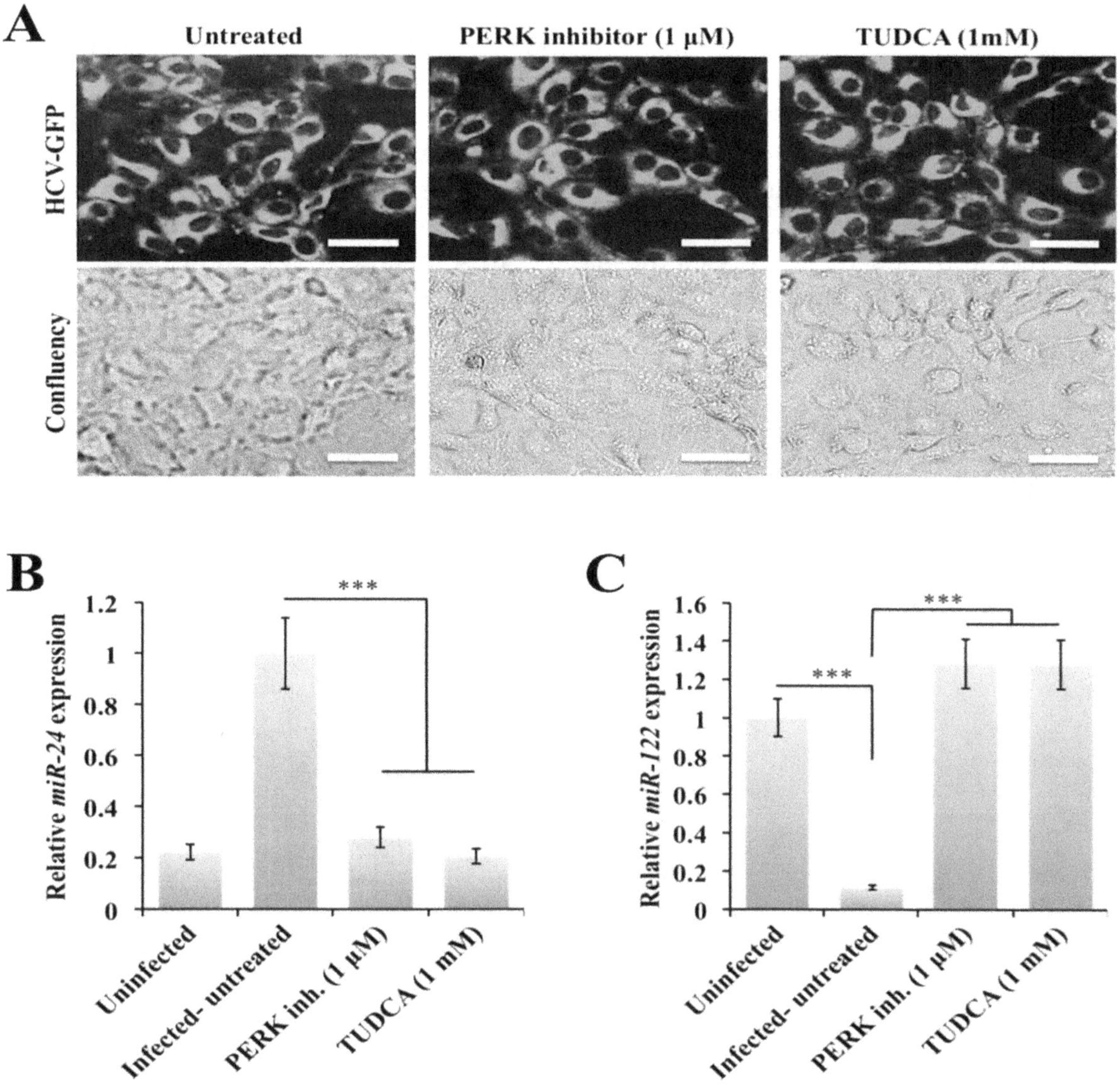

Figure 5. The expression of *miR-24* and *miR-122* in hepatitis C virus (HCV)-infected Huh-7.5 cells in the presence of the protein kinase RNA-like ER kinase (PERK) or endoplasmic reticulum (ER) stress inhibitors. Infected Huh-7.5 cells were treated with the PERK inhibitor or ER stress inhibitors for 72 hours. Cells were collected and the expression levels of *miR-24* and *miR-122* were measured by real-time RT-PCR. (**A**) Shown are expression levels of HCV-green fluorescence protein (GFP) chimera in day 9 HCV culture by fluorescence microscopy before and after treatment. (**B**) Shown are amounts of *miR-24* before and after treatment with the PERK inhibitor or tauroursodeoxycholic (TUDCA) in HCV-infected culture. (**C**) The levels of *miR-122* in uninfected and infected Huh-7.5 cells treated with either with the PERK inhibitor or TUDCA. The results are expressed as the mean ± standard deviation (SD) of three experiments. Error bars represent SD. *p*-values were calculated by ANOVA between different groups as compared to untreated HCV-infected group. *** *p*-value < 0.001. Original magnification ×40. Scale bars = 25 μm.

2.6. *Persistent HCV Replication in Huh-7.5 Cells Leads to Decreased miR-122 Promoter Activity*

An in vitro assessment was done to study the impact of the HCV-induced stress response on *miR-122* promoter activity in the presence and absence of a stress inducer and inhibitors. A firefly luciferase reporter construct p-(5.7/3.8k) that contained the 3726 to 5645 bp region (chr18: 54263641-54265560) in the pGL3-basic vector (Figure 6A) was used to measure the impact of HCV replication on *miR-122* promoter activity [50]. First, we measured the impact of HCV replication on the promoter activity in a transient transfection assay. The infected culture on day 9 was transfected with microRNA-promoter plasmid for 24 hours, and then firefly luciferase activity was measured. The firefly activity was normalized with protein concentration. It was found that HCV infection significantly suppressed *miR-122* promoter activity in Huh-7.5 cells. The promoter activity in the HCV culture was enhanced in the presence of the STAT3 inhibitor and JAK inhibitor. The PERK inhibitor as well as the ER stress inhibitor, TUDCA, also increased *miR-122* promoter activity (Figure 6B). Second, we tested whether uninfected Huh-7.5 cells treated with the ER stress or the NRF2 activator could modulate *miR-122* promoter activity. As shown in Figure 6C, inducing ER stress in uninfected Huh-7.5 cells by thapsigargin (TG) treatment suppressed the *miR-122* promoter activity significantly. The treatment with sulforaphane also suppressed the *miR-122* promoter activity in a concentration-dependent manner (Figure 6D). All these results support the conclusion that the ER stress and the NRF2 activators actively suppressed the *miR-122* promoter activity, but ER stress inhibitors restored the HCV-induced suppression of the promoter activity.

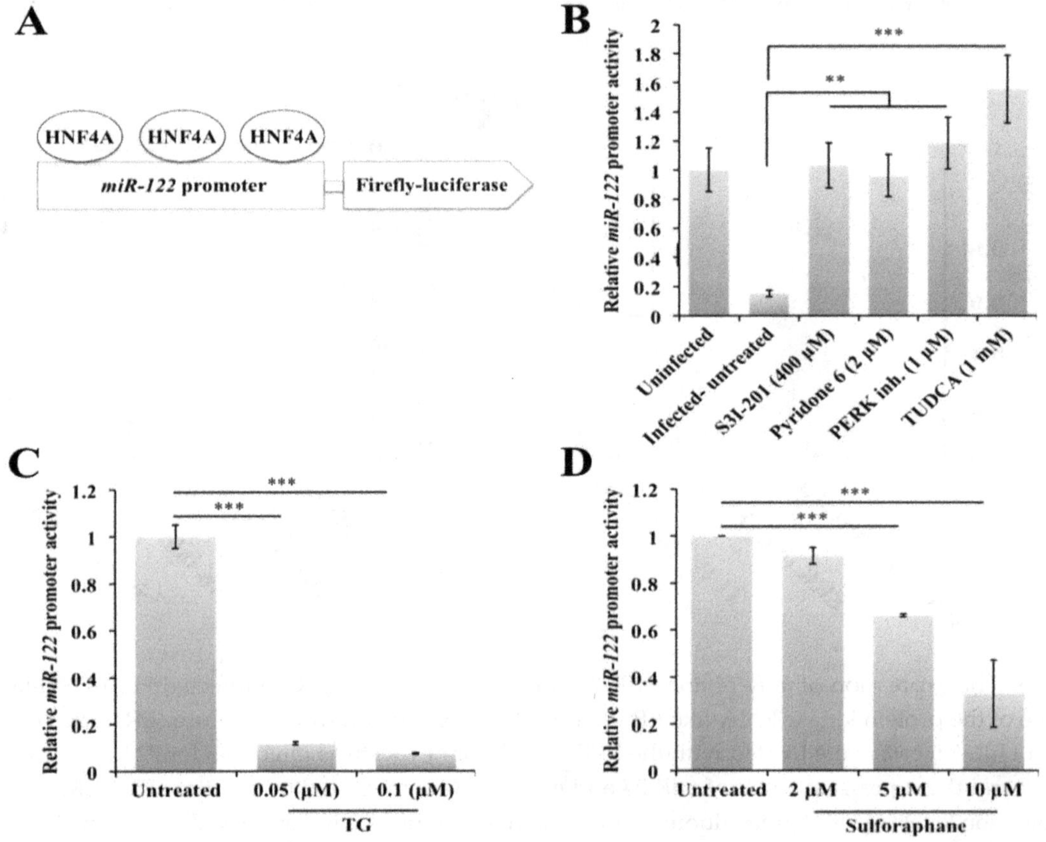

Figure 6. The impact of endoplasmic reticulum (ER) stress on the *miR-122* promoter activity. (**A**) There are three hepatocyte nuclear factor 4 alpha (HNF4A) binding sites present in *miR-122* promoter, which explain why HNF4A downregulation decreases *miR-122* under stress. (**B**) The effect of small molecule inhibitors targeted to signal transducer and activator of transcription 3 (STAT3), Janus kinase 1 (JAK1), protein kinase RNA-like ER kinase (PERK) and ER stress on the *miR-122* promoter regulation. Uninfected and hepatitis C virus (HCV)-infected Huh-7.5 cells were transfected with one microgram of *miR-122* promoter-construct with firefly luciferase. Following the transfection step, cells were treated with or without thapsigargin (TG), sulforaphane, and STAT3 inhibitors. After 24 hours, the luciferase activity was measured. Luciferase assays were performed three times. (**C**) ER stress activator regulation of *miR-122* promoter. (**D**) The impact of a nuclear factor erythroid 2-related factor 2 (NRF2) activator, sulforaphane, on *miR-122* promoter activity. The results are expressed as the mean ± standard deviation (SD) of three experiments. Error bars represent SD. *p*-values were calculated by ANOVA between different groups as compared to untreated HCV-infected group. ** *p*-value < 0.01, *** *p*-value < 0.001.

2.7. Decreased Expression of *miR-122* Correlates with Patients with Cirrhosis

MiR-122 is the most abundant liver-specific microRNA that accounts for about 70% of the total microRNA population in the adult liver [49]. Based on the fact that the cellular stress adaptation to HCV infection leads to decreased expression of *miR-122* through STAT3-HNF4A, suggesting this liver-specific microRNA may be used as a biomarker for assessing the stress response among patients with chronic HCV infection. We next sought to test this hypothesis to see whether the expression levels of *miR-122* would be different among chronically infected patient with or without cirrhosis. A retrospective analysis of serum *miR-122* levels of patients with chronic hepatitis C patients with or without cirrhosis was performed. The analysis measured the levels of *miR-122* in sera from 18 healthy, 18 chronic HCV without cirrhosis and 18 chronic HCV patients with cirrhosis samples. Total RNA was isolated from 200 µL of serum samples, and *miR-122* levels were quantified by real-time reverse transcription. Tables 1 and 2 summarize the characteristics of the subjects included in this study. The viral titer was available in all patients with chronic hepatitis C. All cirrhotic patients were positive for HCV but viral titer was available for only 6 individuals (Table 2). It was found that the copy number of *miR-122* was higher in chronic HCV infection as compared to healthy control ($p < 0.01$). Interestingly, the *miR-122* levels were almost undetectable among most of the patients with stage IV fibrosis (Figure 7A). To rule out the possibility that decreased *miR-122* expression relates to the poor hepatic biosynthetic capability of cirrhotic liver, we measured expression of two additional microRNAs. A liver-specific *miR-221* level was measured in the same serum samples by real-time RT-PCR. These data showed that *miR-221* levels were detectable adequately (Figure 7B). The levels of STAT3-induced *miR-24* level were found to be increased in serum samples of patients with cirrhosis as compared to healthy individuals (Figure 7C). There was no difference in the serum *miR-16* level between the healthy control and chronic HCV infection without cirrhosis. The copy number of *miR-16* was comparable between normal, chronic HCV infection with or without cirrhosis (Figure 7D). There was no correlation between the age and serum *miR-122*, and *miR-16* levels, suggesting that the differences in the expression level are not related to the mean differences in the age of these patients. Serum *miR-122* levels were consistently found to be decreased since we found a significant difference between chronic HCV infection with or without cirrhosis ($p < 0.001$). Since the microRNA is required for HCV replication, we found that serum *miR-122* level was increased during chronic HCV infection. Difference in the *miR-122* level between normal healthy control and chronic HCV infection was also significant ($p < 0.05$) (Figure 8A). These data from serum testing suggest that *miR-122* expression is decreased in the advanced stage of liver disease. It was of interest to know if serum *miR-122* could be a potential diagnostic marker for cirrhosis through Receiver Operating Characteristic (ROC) plot analysis. The data showed that levels of serum *miR-122* are a potential marker for discriminating cirrhosis patients from chronic HCV and healthy control (Figure 8B). The ROC curve analysis revealed a strong separation between chronic HCV patients with cirrhosis versus without cirrhosis with an area under the curve (AUC) value 1.0, suggesting that *miR-122* is a handy marker for discriminating patients with chronic HCV from the cirrhosis group. The analysis revealed that serum levels of other *miR-16*, *miR-221* and *miR-24* were not reliable to accurately differentiate chronic HCV patients with or without cirrhosis. In summary, these results suggested that hepatic adaptive response to cellular stress associated with chronic HCV infection was stronger among the cirrhotic group that leads to depletion of serum *miR-122* level.

Table 1. Characteristics of the patients with Chronic Hepatitis C.

Patient Number	Age	Sex	HCV RNA IU/mL	HCV Genotype	Metavir Score
1	48	F	2,000,000	1a	3
2	46	F	29,000	1b	0
3	27	F	1,000,000	1a	1.2
4	59	F	553,000	1a	3
5	64	M	2,000,000	1a	1.2
6	34	F	7,000,000	1a	1,2
7	59	F	1,000,000	2	0
8	44	M	170,000	1a	1
9	50	M	3,000,000	1b	0
10	33	F	3,000,000	1b	1.2
11	53	M	904,000	1a	1
12	62	M	7062	1a	0
13	69	F	10,000,000	1b	2
14	47	F	2,000,000	2	0
15	50	F	10,000,000	1a	0
16	48	M	2,000,000	1b	0
17	40	F	200,000	1a	1
18	53	M	815,000	1a	1

HCV: hepatitis C virus; M: male; F: female.

Table 2. Characteristics of the patients with cirrhosis.

Patient Number	Age	Sex	HCV RNA IU/mL	HCV Genotype	Metavir Score
1	57	M	+	1a	4
2	52	M	+	1a	4
3	60	F	+	1a	4
4	64	F	+	N/A	4
5	62	M	2,700,000	1b	4
6	66	M	+	1a	4
7	54	M	+	1a	4
8	54	M	+	1b	4
9	62	F	+	N/A	4
10	58	F	+	N/A	4
11	64	M	+	2	4
12	66	F	+	N/A	4
13	57	F	+	N/A	4
14	56	M	5,940,000	N/A	4
15	60	M	7,200,000	1a	4
16	60	M	2,860,000	1b	4
17	56	M	4,310,000	1a	4
18	60	M	6,480,000	1a	4

HCV: hepatitis C virus; M: male; F: female; N/A: Not Available.

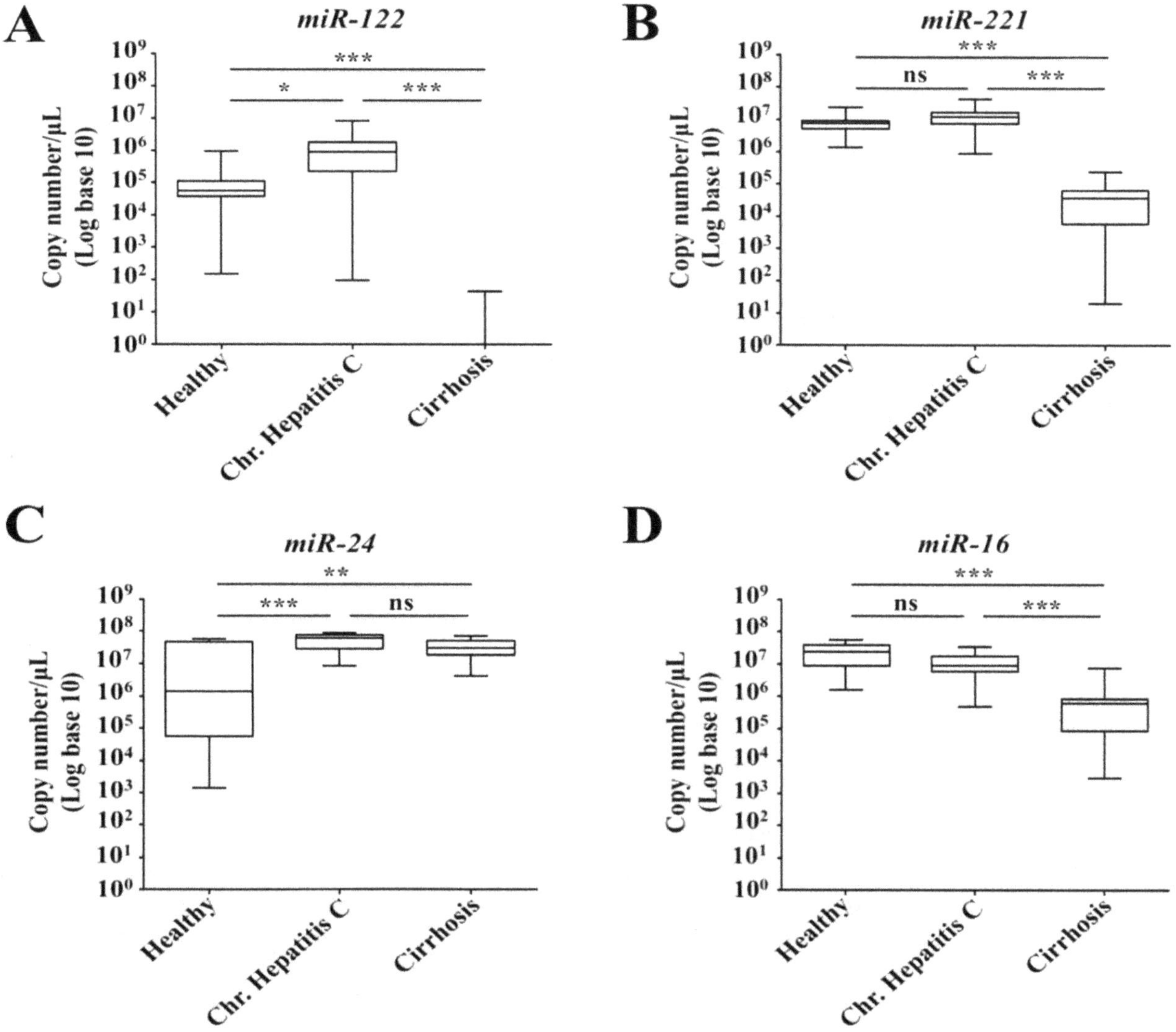

Figure 7. Serum levels of *miR-122*, *miR-221*, *miR-24* and *miR-16* were measured by real-time RT-PCR. (**A**) Shown are the serum *miR-122* levels among healthy individuals, chronic hepatitis C virus (HCV) infection and chronic HCV infection with cirrhosis. (**B**) Serum levels of another control liver-specific *miR-221* using the same set of serum samples. (**C**) Serum levels of signal transducer and activator of transcription 3 (STAT3)-regulated *miR-24* among healthy, chronic HCV infection and chronic HCV with cirrhosis. (**D**) Shown are the serum levels of *miR-16* in the same sets of serum samples. The results are expressed as the box plot with the median bar in triplicates. Whiskers represent minimum and maximum observed values. *p*-values were calculated by ANOVA between different groups. ns: not significant, * p-value < 0.05, ** p-value < 0.01, *** p-value < 0.001.

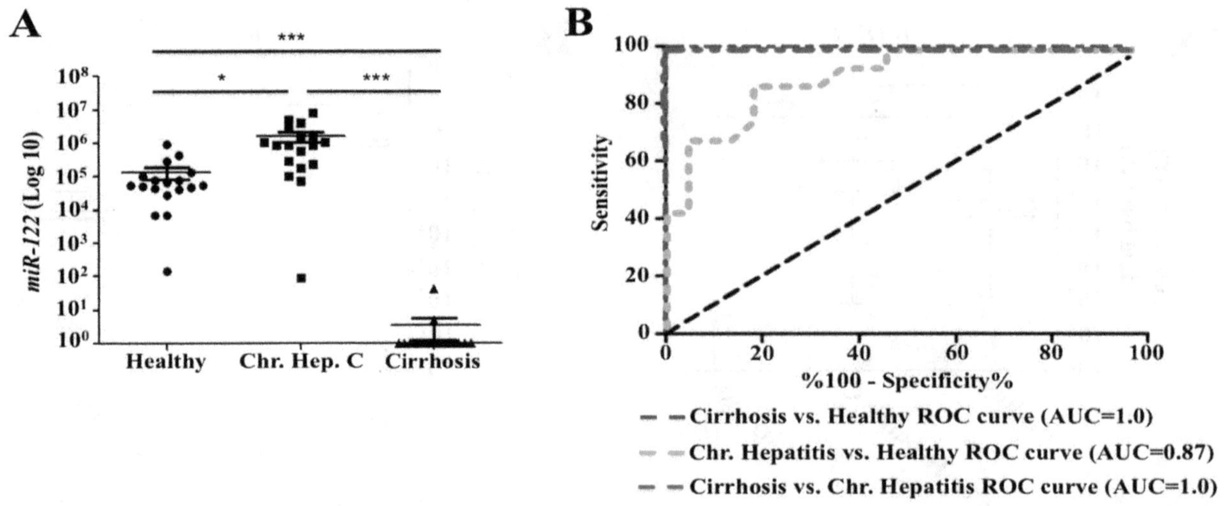

Figure 8. Serum *miR-122* levels discriminating patients with hepatitis C virus (HCV) infection and cirrhosis from healthy controls. (**A**) Copy number variation of serum *miR-122* levels between healthy, chronic HCV and chronic HCV with cirrhosis. (**B**) Receiver Operating Characteristic (ROC) plot. Data shown in panel A was used to generate ROC plot for determining diagnostic value of serum *miR-122* of cirrhosis and chronic HCV patients from healthy controls. The results are expressed as the mean ± standard error of mean (SEM). Error bars represent SEM. *p*-values were calculated by ANOVA between different groups. * *p*- value < 0.05, *** *p*-value < 0.001.

3. Discussion

Chronic liver disease associated with hepatitis virus infection (HBV and HCV) contributes to more than 70% of the HCC worldwide, whereas the remaining cases relate to non-viral etiologies [51]. Since most of HCC cases related to viral and non-viral etiology develop on the background of cirrhosis, suggesting that there may be a common pathway involved in the development of cirrhosis and HCC. However, approximately 20–35% of HCC develop without cirrhosis in the liver, suggesting that inflammation is not an absolute requirement for HCC development [52,53]. The molecular mechanism that drives the progression of cirrhosis and HCC during the chronic stage of liver disease is not understood. Whole genome sequencing found that many driver genes show altered expression in HCC related to chronic HCV infection [51]. Our hypothesis was that the ISR triggered in the infected cells as a compensatory mechanism to deal with varieties of stress responses such as oxidative stress, and ER stress during HCV replication in hepatocytes. The presence of long-lasting unresolved stress response also compromises liver function that leads to cirrhosis and HCC development. In support of this hypothesis, we showed that HCV infection induced ISR in response to ER stress and oxidative stress in cirrhosis that lead to decreased expression of type I IFN receptors [54]. Due to this reason, patients with cirrhosis frequently remain as non-responders to IFN/ribavirin-based antiviral therapy. Some of the pro-survival outcomes of the stress response are the inhibition of host protein synthesis due to increased eukaryotic translation initiation factor 2 subunit alpha (EIF2A) phosphorylation, degradation of misfolded proteins by ER-assisted protein degradation (ERAD) and autophagy. We showed that prolonged stress response activated PERK-induced NRF2 signaling as a cell survival mechanism [13]. The NRF2 signaling activates tumor-promoting autophagy (CMA) that degrades tumor suppressors and induces oncogenic signaling implicated in cell survival.

In this study, we found that NRF2 signaling activated transcription of STAT3, a member of STAT protein family, that is known to be induced by IL6 and participates in inflammation, tumorigenesis and autophagy [55]. STAT3 protein gets phosphorylated at tyrosine 705 by JAK2/tyrosine kinase 2 (TYK2), resulting from dissociation from the cytoplasmic tail of the cytokine receptors, homodimerizes and enters to the nucleus to activate gene transcriptions. Increased STAT3 activity is associated with HCC development and poor prognosis [56]. Activated STAT3 can inhibit cellular autophagy for promoting cell survival, supporting our hypothesis that prolonged stress response inhibits autophagy [57]. Other investigators and we have demonstrated that chronic inflammatory responses can activate STAT3 through IFN and IL6 production [22,58]. Although STAT3 activation occurs due to a variety of mechanisms, HCC-specific STAT3 activation mechanisms have not been well established. Our data show that cellular adaptive response to HCV-induced stress activates STAT3-mediated programing through HNF4A that leads to silencing of liver-specific *miR-122*. To support this hypothesis, we demonstrated that the excessive cellular stress associated with HCV infection activated STAT3 transcription through NRF2 signaling. Our data showed that persistent HCV infection promotes STAT3-inducible expression of *miR-24*. The increased *miR-24* transcription correlated with the decreased expression of *HNF4A* mRNA through binding to the 3′UTR in HCV infection. Our data support the data published earlier showing that an oncogenic circuit consisting of HNF4A, *miR-122*, IL6, STAT3, and *miR-24* is involved in the cell survival mechanisms under stress during HCV infection [19]. Cellular adaptive stress response associated with chronic HCV infection activated STAT3 and utilized this oncogenic circuit, which resulted in a decrease in the expression of HNF4A through *miR-24*. Nuclear HNF4A is a crucial transcription factor during embryonic liver development and hepatocyte differentiation [59]. Available literature shows that HNF4A is the major liver-specific transcription factor that modulates the expression of nearly 42% of the genes expressed in hepatocytes, involved in glucose, fatty acid metabolism, synthesis of blood coagulation factor VII, enzymes involved in hepatic detoxification, and hepatic differentiation [60–65]. Our data is supported by studies showing that mRNA and protein expression levels of HNF4A decreased severely patients with cirrhosis. Hemorrhage is a common cause of death in cirrhosis, especially variceal bleeding. Hepatocytes synthesize both clotting factors and endogenous anticoagulants. The levels of these proteins are reduced in cirrhosis. HNF4A induces the production of blood coagulation factor VII. Our study also provides an explanation why HNF4A targeted protein expression is decreased in cirrhosis due to severe hepatic stress response. Studies have shown that decreased expression of HNF4A is also correlate with bleeding disorder due to decreased expression of the blood coagulation factor VII [64,65].

The activation of this STAT3- HNF4A inflammatory loop led to decreased expression of *miR-124* and *miR-122*. These two microRNAs were regulated by HNF4A. This study was extended to see whether this hypothesis could be validated using prospectively collected serum samples of patients chronically infected with HCV. The measurement of serum *miR-122* levels was performed using serum samples from patients with chronic HCV infection with or without cirrhosis. It was observed that *miR-122* levels increased during the chronic stage of HCV infection, the levels significantly decreased in advanced liver disease, particularly with stage IV fibrosis called cirrhosis. Interestingly, we report here that the levels of *miR-122* have remained below the detection limit among all cirrhotic patients tested in our hand. Our study results are in agreement with a prior report showing that the serum *miR-122* level was found to decreased in liver injury in humans with advanced liver disease, including cirrhosis [66–69]. Decreased expression of *miR-122* has also been found in cirrhosis related to non-alcoholic steatohepatitis (NASH) [33]. The decreased production of *miR-122* in the cirrhotic patients was not related to the impaired biosynthetic capability of hepatocytes present in the cirrhotic liver because the expressions of a liver-specific *miR-221* and STAT3-specific *miR-24* were detectable. Our data show that unresolved stress response depletes the expression of *miR-122*, suggesting that the activation of the STAT3-HNF4A inflammatory loop also occurs in the chronic HCV infection in humans. Our findings provide two pieces of new information that may increase our understanding why STAT3 is transcriptionally activated in liver cancer. First, we show that cellular adaptive response to ER stress in HCV infection causes STAT3 activation at the level of mRNA transcription as well as phosphorylation. Second, we found that STAT3 activation is involved in silencing of *miR-122* in the cirrhosis. Recent studies have observed that epigenetic programing is involved in HCC development in cirrhosis after the viral cure [70–72]. We claimed that our results show a novel adaptive cell survival mechanism related to excessive ER stress that involved STAT3-mediated silencing of *miR-122* expression, the major liver-specific microRNA. The early silencing of *miR-122* and HNF4A increases the risk for HCC development in cirrhosis.

The *miR-122* plays a central role in liver development, differentiation, hepatic metabolic function and innate immunity in the liver [66]. However, the studies investigating the regulation of *miR-122* expression during chronic liver diseases related to viral and non-viral etiologies have produced conflicting results. A handful of publications showed that the levels of *miR-122* decreased during advanced liver disease related to hepatitis C, hepatitis B, and non-viral etiologies [73–75]. *MiR-122* expression in the liver is critical for maintaining innate immunity and IFN production. A report by Xu et al. [73] shows that *miR-122* supports innate immunity by removing the negative regulation of STAT3 signaling on IFN expression. They showed that *miR-122* targets three receptor tyrosine kinases that directly modulate STAT3 phosphorylation. Another recent study by Van Renne et al. [74] showed that the tumor suppressor protein, tyrosine phosphatase receptor delta (PTPRD) controls the STAT3 activation in HCV-induced HCC. They found that high expression of PTPRD suppressed STAT3 phosphorylation in healthy liver tissue but low expression of this protein in HCC resulted in the tumor-specific STAT3 activation. These reports are consistent with our results that HCV infection activates STAT3. On the other hand, the liver-specific *miR-122* expression is also decreased during chronic HBV infection. Chen et al. [75] demonstrated that miR-122 binds to the highly conserved region of HBV pre genomic RNA, causing RNA degradation and reduce viral replication. Data presented in this study show that HCV replication decreased the expression of miR-122 in a time-dependent manner. Data from recent publications confirm the downregulated expression of miR-122 during HCV-induced advanced liver disease [69,76,77]. A report by Luna et al. [76] showed that HCV RNA replication occurs independent of the miR-122 interaction. The team showed that the Argonaute (AGO) protein directly binds to the miR-122 binding sites in HCV RNA, specifically sequesters *miR-122*, to repress the liver target genes during chronic HCV infection. Their investigation showed that the *miR-122* expression decreased in the liver tissues of humans and chimpanzees infected with HCV. Another report by Hyrina et al. [77] showed that the plasma level of *miR-122* decreased during chronic HCV infection and were not restored after HCV cure. Their study also showed that serum *miR-122* level was correlated with liver transaminases. These results suggest that *miR-122* could be a liver-specific biomarker after HCV cure. A study by Sarasin-Filipowicz et al. [69] reported decreased expression of *miR-122* in the liver of HCV-infected patients who were resistant to IFN-based therapy. Our data showed that severe ER stress response during high-level HCV replication in cell culture models of Huh-7.5 liver cells and PHHs decreased the expression of *miR-122*. The expression levels of *miR-122* were depleted among patients with cirrhosis, but not with chronic HCV infections. The reason why the *miR-122* level decreased only in cirrhosis, but not during chronic liver disease is unclear. The host factor requirement for HCV replication has not been fully established. It is possible that HCV replication in the cirrhotic liver may depend on factors other than *miR-122*. The potential liver regeneration and increased inflammation may contribute to increased *miR-122* level in the blood during chronic HCV infection. However, data presented in this report do not support the earlier finding showing that *miR-122* is an essential host factor for HCV replication. The *miR-122* binding does not cause HCV RNA degradation instead it stabilizes the viral genome and promotes translation [78]. In summary, the results presented in this article are consistent with many earlier publications suggesting that decreased expression of *miR-122* in cirrhosis may be associated with loss of hepatic function and innate immunity, impaired liver regeneration, as well as increased risk of HCC development.

4. Materials and Methods

4.1. Cell Culture, Antibodies and Chemicals

The highly permissive transformed liver Huh-7.5 cell line was obtained from the laboratory of Charles M. Rice (Rockefeller University, New York). The Huh-7.5 cells were cultured in Dulbecco's Modified Eagle's Medium (DMEM) supplemented with two mM L-glutamine, sodium pyruvate, nonessential amino acids, 100 U/mL penicillin, 100 μg/mL streptomycin, and 10% fetal bovine serum. Huh-7.5 cell culture was infected with either JFH1-GFP chimera HCV or JFH-dV3-Rluc HCV using a protocol developed in our laboratory as previously described [79]. PHHs were obtained from XenoTech LLC (Kansas City, MO, USA) and cultured with hepatocyte culture media supplemented with 10% human serum (Invitrogen, Brown Deer, WI, USA). After 24 hours they were infected with cell culture grown HCV-GFP chimera virus with a MOI of 0.1. After 18 hours of infection, hepatocytes were replaced with fresh hepatocyte culture media (XenoTech, LLC, Kansas City, MO, USA) supplemented with 10% human serum (Invitrogen, Brown Deer, WI, USA). Uninfected or infected PHH were harvested every 3 days and cell pellets were used for RNA isolation and Western blot analysis. The success of HCV replication in the infected PHHs was assessed by Western blot analysis of HCV NS3 protein. Sofosbuvir was purchased from Acme Biosciences, Inc. (Palo Alto, CA, USA), and IFNα was purchased from EMD Merck (Billerica, MA, USA). Ledipasvir and S3I-201 were obtained from Selleck Chemicals (Houston, TX, USA). TG, TUDCA, the PERK inhibitor (GSK 2606414), and sulforaphane were obtained from Sigma-Aldrich (St. Louis, CA, USA). Pyridone 6 was obtained from Calbiochem (San Diego, CA, USA). The antibody to phospho-NRF2 was purchased from Abcam (Cambridge, MA, USA). Antibodies to GRP78 (BIP), EIF2A, STAT3, phospho-STAT3, HNF4A, and β-Actin were obtained from Cell Signaling (Beverly, MA, USA). Antibody to NS3 was purchased from Virogen Inc. (Boston, MA, USA). The Antibody to HCV core was purchased from Thermo Scientific (Waltham, MA, USA). The antibody to total NRF2 was purchased from Santa Cruz Biotechnology (Santa Cruz, CA, USA).

4.2. Quantitative Assessment of Misfolded Protein Burden in the ER

The accumulation of misfolded proteins in the infected culture was quantified using an ER-tracker reagent called CellLight ER-RFP BacMam 2.0 (Thermo Fisher Scientific). This construct expresses Red Fluorescence Protein (RFP) fused to the ER signal sequence of calreticulin and KDEL (ER-retention signal) to quantitate the extent of red fluorescence by fluorescence microscopy and flow cytometry. This reagent allows live multiplexing imaging of ER in the cells expressing HCV-GFP fusion protein. Briefly, HCV-infected Huh-7.5 cells, and uninfected Huh-7.5 cells were incubated with ER-tracker dye overnight at 37 °C, and the next day, cells were observed using a fluorescence microscope. For nuclear staining and imaging, the cells were directly incubated with Hoechst 33342 (Thermo Fisher Scientific) at 1 μg/mL for 5 minutes in 6-well plate. Then, the cells were washed with PBS and fixed for 10 minutes with 2% paraformaldehyde. The colocalization of GFP and RFP between infected and uninfected cells quantified by flow analysis.

4.3. Detection of Reactive Oxygen Species (ROS)

ROS in HCV culture measured using the cell-permeant H2DCFDA reagent (Thermo Fisher Scientific) according to the manufacturer's instructions. The 2′,7′-dichlorodihydrofluorescein diacetate (H2DCFDA) fluorescent probe reacts with ROS including hydrogen peroxide and hydroxyl radicals. The cell-permeant H2DCFDA diffuses into cells and is retained in the cytoplasm after cleavage by intracellular esterase. The ROS converts the non-fluorescent H2DCFDA to the highly fluorescent 2′,7′-dichlorofluorescein (DCF), which could be detected by fluorescent microscopy or quantified by flow cytometry.

4.4. Immunohistochemical Staining

Tissue culture cells were immobilized onto a glass slide by cytospin. Immunostaining of the cytospin slides of infected cells was performed using a standard protocol established in our laboratory [11]. We used a monoclonal antibody to HCV core protein (Thermo Scientific) and monoclonal antibody that detects phosphorylated NRF2 (Abcam).

4.5. SDS-PAGE and Western Blotting

Infected Huh-7.5 cells were harvested by the treatment of trypsin-ethylenediaminetetraacetic acid (EDTA). Cells were lysed in ice-cold lysis buffer (50 mM Tris HCl pH 8.0, 140 mM NaCl, 1.5 mM MgCl2, 0.5% NP-40 with complete protease inhibitor from Invitrogen) for 10 minutes in ice (about 1×10^6 cells/200 µL). Infected Huh-7.5 cells pelleted by low-speed centrifugation. The detergent compatible (DC) protein assay determined protein concentration. Samples were boiled for 10 minutes at 80 °C in the presence of 1× sodium dodecyl sulfate-polyacrylamide gel electrophoresis (SDS-PAGE)-loading buffer (250 mM Tris-HCl pH 6.8, 40% glycerol, 8% SDS, 0.57M β-mercaptoethanol, 0.12% bromophenol blue). Approximately 20 µg of protein was loaded onto 12% SDS-PAGE and transferred into a nitrocellulose membrane (Thermo Scientific). The membrane was blocked using a solution containing 5% of blotting-grade milk power (Bio-Rad, Hercules, CA, USA) for 2 hours then incubated with a primary antibody. After overnight incubation, the antigen-antibody complex was visualized with horseradish peroxidase (HRP)-conjugated goat anti-rabbit or anti-mouse Immunoglobulin G (IgG) and the ECL detection system (Amersham ECL, GE Healthcare Bio-Sciences, Pittsburgh, PA, USA).

4.6. siRNA Transfection

Persistently HCV-infected Huh-7.5 cells were cultured in 6-well plates (up to 60% confluence in DMEM supplemented with 10% FBS media) without antibiotics. The next day, culture media was replaced with fresh DMEM with 2% FBS and cells were then transfected with siRNA to either NRF2 (siRNA1: 5′-GAAUGGUCCUAAAACACCA-3′, siRNA2: 5′-UGACAGAAGUUGACAAUUA-3′) synthesized by Invitrogen [80,81] or scramble siRNA (Invitrogen) using Lipofectamine (Life Technology, Grand Island, NY, USA). The knockdown efficiency was analyzed by Western blot.

4.7. Quantification of mRNA Levels by RT-qPCR

Infected Huh-7.5 cells and PHHs were harvested at different time points post HCV infection, and total RNA was isolated using the RNeasy Mini kit (Qiagen, Germantown, MD, USA). The first strand of complementary DNA (cDNA) was synthesized from 1 µg of total RNA using an iScript Reverse Transcription Supermix (Bio-Rad). 100 ng cDNA was amplified using iTaq Universal SYBR Green Supermix (Bio-Rad) with gene-specific forward and reverse primers following instructions in the kit. The mRNA levels of NRF2, STAT3, HNF4A, and β-Actin (as internal control) were quantified using quantitative RT-PCR (RT-qPCR). Amplification, data acquisition, and analysis were carried out on the CFX96 real-time instrument using CFX manager software version 3.0 (Bio-Rad). The expression levels of mRNAs were normalized with the *β-Actin*mRNA level using the comparative threshold cycle method. The nucleotide sequences of oligonucleotide primers for *NRF2* mRNA (sense primer 5′-TACTCCCAGGTTGCCCACA-3′ and antisense primer 5′-CATCTACAAACGGGAATGTCTGC-3′) [82], for *STAT3* mRNA (sense primer 5′-TGGAAGAGGCGGCAGCAGATAGC-3′ and antisense primer 5′-CACGGCCCCCATTCCCACAT-3′) [83], for *HNF4A* mRNA (sense primer 5′-TGTCCCGACAGATCACCTC-3′ and antisense primer 5′-CACTCAACGAGAACCAGCAG-3′) [20], and for *β-Actin*mRNA (sense primer 5′-GAGCTACGAGCTGCCTGAC-3′ and antisense primer 5′-AGCACTGTGTTGGCGTACAG-3′) [84] were derived from previous published reports.

4.8. Quantification of Serum microRNA Levels by RT-qPCR

Total microRNA was isolated from 20 µL serum samples using miRNeasy kit (Qiagen) used for reverse transcription using miScript II RT kit (Qiagen) as described previously [85]. Individual microRNA was amplified using cDNA templates, a universal primer and a PCR kit using a recommended protocol. A standard curve was generated for determining microRNA copy number and Ct values using serially diluted synthetic microRNA. The copy number of each microRNA in the serum was calculated from the Ct values.

4.9. Statistical Analysis

Western blot, immunostaining, and immunofluorescence images were quantified using a computer image analysis software package (ImageJ version 1.52c, NIH, Bethesda, MD, USA). All measurements were made at least in triplicate ($n = 3$). To compare means within groups, we performed a one-factor analysis of variance (ANOVA) using the GraphPad Prism software (GraphPad Company, San Diego, CA, USA). Data were tested and found to be normally distributed. To determine the statistical correlation coefficient between protein expressions, fraction of variance denoted by R^2 values was determined using excel software. Sensitivity of the assays was plotted against the false positivity (1-specificity) using ROC curves using the GraphPad Prism software. Comparison of AUC was performed, which compares the AUC to the diagonal line of no information (AUC 0.5). We applied Dunnet's post hoc test to compare control samples with experimental samples when the overall p-value for the ANOVA analysis was significant ($p < 0.05$). When performing comparisons between multiple groups, each analyzed with ANOVA; we used the Bonferroni correction to determine a revised cutoff for statistical significance that gives a combined 5% type I error probability. * p-value < 0.05, ** p-value < 0.01, *** p-value < 0.001.

5. Conclusions

Hepatic adaptive response to HCV-induced stress reduces liver-specific miR-122 through activation of STAT3-HNF4A inflammatory feedback loop. Serum *miR-122* could be used as a biomarker to monitor the activation of this inflammatory loop in liver cirrhosis.

Author Contributions: Y.A., R.K., K.S., D.L., H.O. and B.Y. performed various experiments included in this manuscript. Y.A. prepared figures for publication. J.W.S., N.J.S., P.T. and A.C. supplied serum samples and helped in data analysis. Y.A. and S.D. involved in analysis and wrote the paper.

Acknowledgments: We acknowledge Professor Shi-Mei Zhuang, Key Laboratory of Gene Engineering of the Ministry of Education, State Key Laboratory of Biocontrol, School of Life Sciences Guangzhou, People's Republic of China who have supplied microRNA-122 promoter construct.

References

1. Sayiner, M.; Golabi, P.; Younossi, Z.M. Disease Burden of Hepatocellular Carcinoma: A Global Perspective. *Dig. Dis. Sci.* **2019**, *64*, 910–917. [CrossRef] [PubMed]
2. Allison, R.D.; Tong, X.; Moorman, A.C.; Ly, K.N.; Rupp, L.; Xu, F.; Gordon, S.C.; Holmberg, S.D.; Chronic Hepatitis Cohort Study (CHeCS) Investigators. Increased incidence of cancer and cancer-related mortality among persons with chronic hepatitis C infection, 2006–2010. *J. Hepatol.* **2015**, *63*, 822–828. [CrossRef] [PubMed]
3. Thrift, A.P.; El-Serag, H.B.; Kanwal, F. Global epidemiology and burden of HCV infection and HCV-related disease. *Nat. Rev. Gastroenterol. Hepatol.* **2016**, *14*, 122. [CrossRef] [PubMed]
4. Pawlotsky, J.M.; Feld, J.J.; Zeuzem, S.; Hoofnagle, J.H. From non-A, non-B hepatitis to hepatitis C virus cure. *J. Hepatol.* **2015**, *62*, S87–S99. [CrossRef] [PubMed]
5. Lombardi, A.; Mondelli, M.U.; ESCMID Study Group for Viral Hepatitis (ESGVH). Hepatitis C: Is eradication possible? *Liver Int.* **2019**, *39*, 416–426. [CrossRef] [PubMed]
6. Thomas, D.L. Global control of hepatitis C: Where challenge meets opportunity. *Nat. Med.* **2013**, *19*, 850–858. [CrossRef] [PubMed]
7. Bartosch, B.; Thimme, R.; Blum, H.E.; Zoulim, F. Hepatitis C virus-induced hepatocarcinogenesis. *J. Hepatol.* **2009**, *51*, 810–820. [CrossRef]

8. Dash, S.; Aydin, Y.; Wu, T. Integrated stress response in hepatitis C promotes Nrf2-related chaperone-mediated autophagy: A novel mechanism for host-microbe survival and HCC development in liver cirrhosis. *Semin. Cell Dev. Biol.* **2019**. [CrossRef]
9. Raghunath, A.; Sundarraj, K.; Nagarajan, R.; Arfuso, F.; Bian, J.; Kumar, A.P.; Sethi, G.; Perumal, E. Antioxidant response elements: Discovery, classes, regulation and potential applications. *Redox Biol.* **2018**, *17*, 297–314. [CrossRef]
10. Chava, S.; Lee, C.; Aydin, Y.; Chandra, P.K.; Dash, A.; Chedid, M.; Thung, S.N.; Moroz, K.; Wu, T.; Nayak, N.C.; et al. Chaperone-mediated autophagy compensates for impaired macroautophagy in the cirrhotic liver to promote hepatocellular carcinoma. *Oncotarget* **2017**, *8*, 40019–40036. [CrossRef]
11. Aydin, Y.; Chatterjee, A.; Chandra, P.K.; Chava, S.; Chen, W.; Tandon, A.; Dash, A.; Chedid, M.; Moehlen, M.W.; Regenstein, F.; et al. Interferon-alpha-induced hepatitis C virus clearance restores p53 tumor suppressor more than direct-acting antivirals. *Hepatol. Commun.* **2017**, *1*, 256–269. [CrossRef] [PubMed]
12. Aydin, Y.; Chedid, M.; Chava, S.; Danielle Williams, D.; Liu, S.; Hagedorn, C.H.; Sumitran-Holgersson, S.; Reiss, K.; Moroz, K.; Lu, H.; et al. Activation of PERK-Nrf2 oncogenic signaling promotes Mdm2-mediated Rb degradation in persistently infected HCV culture. *Sci. Rep.* **2017**, *7*, 9223. [CrossRef] [PubMed]
13. Aydin, Y.; Stephens, C.M.; Chava, S.; Heidari, Z.; Panigrahi, R.; Williams, D.D.; Wiltz, K.; Bell, A.; Wilson, W.; Reiss, K.; et al. Chaperone-Mediated Autophagy Promotes Beclin1 Degradation in Persistently Infected Hepatitis C Virus Cell Culture. *Am. J. Pathol.* **2018**, *188*, 2339–2355. [CrossRef] [PubMed]
14. Petrelli, A.; Perra, A.; Cora, D.; Sulas, P.; Menegon, S.; Manca, C.; Migliore, C.; Kowalik, M.A.; Ledda-Columbano, G.M.; Giordano, S.; et al. MicroRNA/gene profiling unveils early molecular changes and nuclear factor erythroid related factor 2 (NRF2) activation in a rat model recapitulating human hepatocellular carcinoma (HCC). *Hepatology* **2014**, *59*, 228–241. [CrossRef] [PubMed]
15. Yang, S.F.; Wang, S.N.; Wu, C.F.; Yeh, Y.T.; Chai, C.Y.; Chunag, S.C.; Sheen, M.C.; Lee, K.T. Altered p-STAT3 (tyr705) expression is associated with histological grading and intratumour microvessel density in hepatocellular carcinoma. *J. Clin. Pathol.* **2007**, *60*, 642–648. [CrossRef] [PubMed]
16. He, G.; Yu, G.Y.; Temkin, V.; Ogata, H.; Kuntzen, C.; Sakurai, T.; Sieghart, W.; Peck-Radosavljevic, M.; Leffert, H.L.; Karin, M. Hepatocyte IKKbeta/NF-kappaB inhibits tumor promotion and progression by preventing oxidative stress-driven STAT3 activation. *Cancer Cell* **2010**, *17*, 286–297. [CrossRef] [PubMed]
17. He, G.; Karin, M. NF-kappaB and STAT3—Key players in liver inflammation and cancer. *Cell Res.* **2011**, *21*, 159–168. [CrossRef]
18. Svinka, J.; Mikulits, W.; Eferl, R. STAT3 in hepatocellular carcinoma: New perspectives. *Hepat. Oncol.* **2014**, *1*, 107–120. [CrossRef] [PubMed]
19. Hatziapostolou, M.; Polytarchou, C.; Aggelidou, E.; Drakaki, A.; Poultsides, G.A.; Jaeger, S.A.; Ogata, H.; Karin, M.; Struhl, K.; Hadzopoulou-Cladaras, M.; et al. An HNF4alpha-miRNA inflammatory feedback circuit regulates hepatocellular oncogenesis. *Cell* **2011**, *147*, 1233–1247. [CrossRef]
20. Aboulnasr, F.; Hazari, S.; Nayak, S.; Chandra, P.K.; Panigrahi, R.; Ferraris, P.; Chava, S.; Kurt, R.; Song, K.; Dash, A.; et al. IFN-lambda Inhibits MiR-122 Transcription through a Stat3-HNF4alpha Inflammatory Feedback Loop in an IFN-alpha Resistant HCV Cell Culture System. *PLoS ONE* **2015**, *10*, e0141655. [CrossRef]
21. Lau, H.H.; Ng, N.H.J.; Loo, L.S.W.; Jasmen, J.B.; Teo, A.K.K. The molecular functions of hepatocyte nuclear factors—In and beyond the liver. *J. Hepatol.* **2018**, *68*, 1033–1048. [CrossRef] [PubMed]
22. Yue, H.Y.; Yin, C.; Hou, J.L.; Zeng, X.; Chen, Y.X.; Zhong, W.; Hu, P.F.; Deng, X.; Tan, Y.X.; Zhang, J.P.; et al. Hepatocyte nuclear factor 4alpha attenuates hepatic fibrosis in rats. *Gut* **2010**, *59*, 236–246. [CrossRef] [PubMed]
23. Lazarevich, N.L.; Cheremnova, O.A.; Varga, E.V.; Ovchinnikov, D.A.; Kudrjavtseva, E.I.; Morozova, O.V.; Fleishman, D.I.; Engelhardt, N.V.; Duncan, S.A. Progression of HCC in mice is associated with a downregulation in the expression of hepatocyte nuclear factors. *Hepatology* **2004**, *39*, 1038–1047. [CrossRef] [PubMed]
24. Lazarevich, N.L.; Shavochkina, D.A.; Fleishman, D.I.; Kustova, I.F.; Morozova, O.V.; Chuchuev, E.S.; Patyutko, Y.I. Deregulation of hepatocyte nuclear factor 4 (HNF4) as a marker of epithelial tumors progression. *Exp. Oncol.* **2010**, *32*, 167–171. [PubMed]

25. Ning, B.F.; Ding, J.; Yin, C.; Zhong, W.; Wu, K.; Zeng, X.; Yang, W.; Chen, Y.X.; Zhang, J.P.; Zhang, X.; et al. Hepatocyte nuclear factor 4 alpha suppresses the development of hepatocellular carcinoma. *Cancer Res.* **2010**, *70*, 7640–7651. [CrossRef] [PubMed]
26. Tanaka, T.; Jiang, S.; Hotta, H.; Takano, K.; Iwanari, H.; Sumi, K.; Daigo, K.; Ohashi, R.; Sugai, M.; Ikegame, C.; et al. Dysregulated expression of P1 and P2 promoter-driven hepatocyte nuclear factor-4alpha in the pathogenesis of human cancer. *J. Pathol.* **2006**, *208*, 662–672. [CrossRef] [PubMed]
27. Yin, C.; Lin, Y.; Zhang, X.; Chen, Y.X.; Zeng, X.; Yue, H.Y.; Hou, J.L.; Deng, X.; Zhang, J.P.; Han, Z.G.; et al. Differentiation therapy of hepatocellular carcinoma in mice with recombinant adenovirus carrying hepatocyte nuclear factor-4alpha gene. *Hepatology* **2008**, *48*, 1528–1539. [CrossRef]
28. Li, C.; Deng, M.; Hu, J.; Li, X.; Chen, L.; Ju, Y.; Hao, J.; Meng, S. Chronic inflammation contributes to the development of hepatocellular carcinoma by decreasing miR-122 levels. *Oncotarget* **2016**, *7*, 17021–17034. [CrossRef]
29. Yang, Y.M.; Lee, C.G.; Koo, J.H.; Kim, T.H.; Lee, J.M.; An, J.; Kim, K.M.; Kim, S.G. Galpha12 overexpressed in hepatocellular carcinoma reduces microRNA-122 expression via HNF4alpha inactivation, which causes c-Met induction. *Oncotarget* **2015**, *6*, 19055–19069. [CrossRef]
30. Li, M.; Tang, Y.; Wu, L.; Mo, F.; Wang, X.; Li, H.; Qi, R.; Zhang, H.; Srivastava, A.; Ling, C. The hepatocyte-specific HNF4alpha/miR-122 pathway contributes to iron overload-mediated hepatic inflammation. *Blood* **2017**, *130*, 1041–1051. [CrossRef]
31. Hsu, S.H.; Wang, B.; Kota, J.; Yu, J.; Costinean, S.; Kutay, H.; Yu, L.; Bai, S.; La Perle, K.; Chivukula, R.R.; et al. Essential metabolic, anti-inflammatory, and anti-tumorigenic functions of miR-122 in liver. *J. Clin. Investig.* **2012**, *122*, 2871–2883. [CrossRef] [PubMed]
32. Tsai, W.C.; Hsu, S.D.; Hsu, C.S.; Lai, T.C.; Chen, S.J.; Shen, R.; Huang, Y.; Chen, H.C.; Lee, C.H.; Tsai, T.F.; et al. MicroRNA-122 plays a critical role in liver homeostasis and hepatocarcinogenesis. *J. Clin. Investig.* **2012**, *122*, 2884–2897. [CrossRef] [PubMed]
33. Takaki, Y.; Saito, Y.; Takasugi, A.; Toshimitsu, K.; Yamada, S.; Muramatsu, T.; Kimura, M.; Sugiyama, K.; Suzuki, H.; Arai, E.; et al. Silencing of microRNA-122 is an early event during hepatocarcinogenesis from non-alcoholic steatohepatitis. *Cancer Sci.* **2014**, *105*, 1254–1260. [CrossRef] [PubMed]
34. Shim, S.H.; Xia, C.; Zhong, G.; Babcock, H.P.; Vaughan, J.C.; Huang, B.; Wang, X.; Xu, C.; Bi, G.Q.; Zhuang, X. Super-resolution fluorescence imaging of organelles in live cells with photoswitchable membrane probes. *Proc. Natl. Acad. Sci. USA* **2012**, *109*, 13978–13983. [CrossRef] [PubMed]
35. Burban, A.; Sharanek, A.; Guguen-Guillouzo, C.; Guillouzo, A. Endoplasmic reticulum stress precedes oxidative stress in antibiotic-induced cholestasis and cytotoxicity in human hepatocytes. *Free Radic. Biol. Med.* **2018**, *115*, 166–178. [CrossRef]
36. Beriault, D.R.; Werstuck, G.H. Detection and quantification of endoplasmic reticulum stress in living cells using the fluorescent compound, Thioflavin T. *Biochim. Biophys. Acta* **2013**, *1833*, 2293–2301. [CrossRef] [PubMed]
37. Xing, Y.; Higuchi, K. Amyloid fibril proteins. *Mech. Ageing Dev.* **2002**, *123*, 1625–1636. [CrossRef]
38. Tardif, K.D.; Waris, G.; Siddiqui, A. Hepatitis C virus, ER stress, and oxidative stress. *Trends Microbiol.* **2005**, *13*, 159–163. [CrossRef]
39. Waris, G.; Turkson, J.; Hassanein, T.; Siddiqui, A. Hepatitis C virus (HCV) constitutively activates STAT-3 via oxidative stress: Role of STAT-3 in HCV replication. *J. Virol.* **2005**, *79*, 1569–1580. [CrossRef]
40. Gong, G.; Waris, G.; Tanveer, R.; Siddiqui, A. Human hepatitis C virus NS5A protein alters intracellular calcium levels, induces oxidative stress, and activates STAT-3 and NF-kappa B. *Proc. Natl. Acad. Sci. USA* **2001**, *98*, 9599–9604. [CrossRef]
41. Joyce, M.A.; Walters, K.A.; Lamb, S.E.; Yeh, M.M.; Zhu, L.F.; Kneteman, N.; Doyle, J.S.; Katze, M.G.; Tyrrell, D.L. HCV induces oxidative and ER stress, and sensitizes infected cells to apoptosis in SCID/Alb-uPA mice. *PLoS Pathog.* **2009**, *5*, e1000291. [CrossRef] [PubMed]
42. Burdette, D.; Olivarez, M.; Waris, G. Activation of transcription factor Nrf2 by hepatitis C virus induces the cell-survival pathway. *J. Gen. Virol.* **2010**, *91*, 681–690. [CrossRef]

43. Xu, D.; Xu, M.; Jeong, S.; Qian, Y.; Wu, H.; Xia, Q.; Kong, X. The Role of Nrf2 in Liver Disease: Novel Molecular Mechanisms and Therapeutic Approaches. *Front. Pharmacol.* **2018**, *9*, 1428. [CrossRef] [PubMed]
44. Ma, Q. Role of nrf2 in oxidative stress and toxicity. *Annu. Rev. Pharmacol. Toxicol.* **2013**, *53*, 401–426. [CrossRef] [PubMed]
45. Nguyen, T.; Nioi, P.; Pickett, C.B. The Nrf2-antioxidant response element signaling pathway and its activation by oxidative stress. *J. Biol. Chem.* **2009**, *284*, 13291–13295. [CrossRef]
46. Mitsuishi, Y.; Motohashi, H.; Yamamoto, M. The Keap1-Nrf2 system in cancers: Stress response and anabolic metabolism. *Front. Oncol.* **2012**, *2*, 200. [CrossRef] [PubMed]
47. Kansanen, E.; Kuosmanen, S.M.; Leinonen, H.; Levonen, A.L. The Keap1-Nrf2 pathway: Mechanisms of activation and dysregulation in cancer. *Redox Biol.* **2013**, *1*, 45–49. [CrossRef]
48. Vallianou, I.; Dafou, D.; Vassilaki, N.; Mavromara, P.; Hadzopoulou-Cladaras, M. Hepatitis C virus suppresses Hepatocyte Nuclear Factor 4 alpha, a key regulator of hepatocellular carcinoma. *Int. J. Biochem. Cell Biol.* **2016**, *78*, 315–326. [CrossRef]
49. Li, Z.Y.; Xi, Y.; Zhu, W.N.; Zeng, C.; Zhang, Z.Q.; Guo, Z.C.; Hao, D.L.; Liu, G.; Feng, L.; Chen, H.Z.; et al. Positive regulation of hepatic miR-122 expression by HNF4alpha. *J. Hepatol.* **2011**, *55*, 602–611. [CrossRef]
50. Zeng, C.; Wang, R.; Li, D.; Lin, X.J.; Wei, Q.K.; Yuan, Y.; Wang, Q.; Chen, W.; Zhuang, S.M. A novel GSK-3 beta-C/EBP alpha-miR-122-insulin-like growth factor 1 receptor regulatory circuitry in human hepatocellular carcinoma. *Hepatology* **2010**, *52*, 1702–1712. [CrossRef]
51. Kanda, T.; Goto, T.; Hirotsu, Y.; Moriyama, M.; Omata, M. Molecular Mechanisms Driving Progression of Liver Cirrhosis towards Hepatocellular Carcinoma in Chronic Hepatitis B and C Infections: A Review. *Int. J. Mol. Sci.* **2019**, *20*, 1358. [CrossRef] [PubMed]
52. Desai, A.; Sandhu, S.; Lai, J.P.; Sandhu, D.S. Hepatocellular carcinoma in non-cirrhotic liver: A comprehensive review. *World J. Hepatol.* **2019**, *11*, 1. [CrossRef] [PubMed]
53. Stine, J.G.; Wentworth, B.J.; Zimmet, A.; Rinella, M.E.; Loomba, R.; Caldwell, S.H.; Argo, C.K. Systematic review with meta-analysis: Risk of hepatocellular carcinoma in non-alcoholic steatohepatitis without cirrhosis compared to other liver diseases. *Aliment. Pharmacol. Ther.* **2018**, *48*, 696–703. [CrossRef] [PubMed]
54. Chandra, P.K.; Gunduz, F.; Hazari, S.; Kurt, R.; Panigrahi, R.; Poat, B.; Bruce, D.; Cohen, A.J.; Bohorquez, H.E.; Carmody, I.; et al. Impaired expression of type I and type II interferon receptors in HCV-associated chronic liver disease and liver cirrhosis. *PLoS ONE* **2014**, *9*, e108616. [CrossRef] [PubMed]
55. Levy, D.E.; Darnell, J.E., Jr. Stats: Transcriptional control and biological impact. *Nat. Rev. Mol. Cell Biol.* **2002**, *3*, 651–662. [CrossRef] [PubMed]
56. Calvisi, D.F.; Ladu, S.; Gorden, A.; Farina, M.; Conner, E.A.; Lee, J.S.; Factor, V.M.; Thorgeirsson, S.S. Ubiquitous activation of Ras and Jak/Stat pathways in human HCC. *Gastroenterology* **2006**, *130*, 1117–1128. [CrossRef]
57. Shen, S.; Niso-Santano, M.; Adjemian, S.; Takehara, T.; Malik, S.A.; Minoux, H.; Souquere, S.; Marino, G.; Lachkar, S.; Senovilla, L.; et al. Cytoplasmic STAT3 represses autophagy by inhibiting PKR activity. *Mol. Cell* **2012**, *48*, 667–680. [CrossRef]
58. Tacke, R.S.; Tosello-Trampont, A.; Nguyen, V.; Mullins, D.W.; Hahn, Y.S. Extracellular hepatitis C virus core protein activates STAT3 in human monocytes/macrophages/dendritic cells via an IL-6 autocrine pathway. *J. Biol. Chem.* **2011**, *286*, 10847–10855. [CrossRef]
59. Kyrmizi, I.; Hatzis, P.; Katrakili, N.; Tronche, F.; Gonzalez, F.J.; Talianidis, I. Plasticity and expanding complexity of the hepatic transcription factor network during liver development. *Genes Dev.* **2006**, *20*, 2293–2305. [CrossRef]
60. Odom, D.T.; Zizlsperger, N.; Gordon, D.B.; Bell, G.W.; Rinaldi, N.J.; Murray, H.L.; Volkert, T.L.; Schreiber, J.; Rolfe, P.A.; Gifford, D.K.; et al. Control of pancreas and liver gene expression by HNF transcription factors. *Science* **2004**, *303*, 1378–1381. [CrossRef]
61. Guzman-Lepe, J.; Cervantes-Alvarez, E.; Collin de l'Hortet, A.; Wang, Y.; Mars, W.M.; Oda, Y.; Bekki, Y.; Shimokawa, M.; Wang, H.; Yoshizumi, T.; et al. Liver-enriched transcription factor expression relates to chronic hepatic failure in humans. *Hepatol. Commun.* **2018**, *2*, 582–594. [CrossRef] [PubMed]
62. Liu, L.; Yannam, G.R.; Nishikawa, T.; Yamamoto, T.; Basma, H.; Ito, R.; Nagaya, M.; Dutta-Moscato, J.; Stolz, D.B.; Duan, F.; et al. The microenvironment in hepatocyte regeneration and function in rats with advanced cirrhosis. *Hepatology* **2012**, *55*, 1529–1539. [CrossRef] [PubMed]

63. Safdar, H.; Cheung, K.L.; Vos, H.L.; Gonzalez, F.J.; Reitsma, P.H.; Inoue, Y.; van Vlijmen, B.J. Modulation of mouse coagulation gene transcription following acute in vivo delivery of synthetic small interfering RNAs targeting HNF4alpha and C/EBPalpha. *PLoS ONE* **2012**, *7*, e38104. [CrossRef] [PubMed]
64. Zheng, X.W.; Kudaravalli, R.; Russell, T.T.; DiMichele, D.M.; Gibb, C.; Russell, J.E.; Margaritis, P.; Pollak, E.S. Mutation in the factor VII hepatocyte nuclear factor 4alpha-binding site contributes to factor VII deficiency. *Blood Coagul. Fibrinolysis* **2011**, *22*, 624–627. [CrossRef] [PubMed]
65. Inoue, Y.; Peters, L.L.; Yim, S.H.; Inoue, J.; Gonzalez, F.J. Role of hepatocyte nuclear factor 4alpha in control of blood coagulation factor gene expression. *J. Mol. Med.* **2006**, *84*, 334–344. [CrossRef] [PubMed]
66. Bandiera, S.; Pfeffer, S.; Baumert, T.F.; Zeisel, M.B. miR-122-a key factor and therapeutic target in liver disease. *J. Hepatol.* **2015**, *62*, 448–457. [CrossRef] [PubMed]
67. Trebicka, J.; Anadol, E.; Elfimova, N.; Strack, I.; Roggendorf, M.; Viazov, S.; Wedemeyer, I.; Drebber, U.; Rockstroh, J.; Sauerbruch, T.; et al. Hepatic and serum levels of miR-122 after chronic HCV-induced fibrosis. *J. Hepatol.* **2013**, *58*, 234–239. [CrossRef]
68. Cermelli, S.; Ruggieri, A.; Marrero, J.A.; Ioannou, G.N.; Beretta, L. Circulating microRNAs in patients with chronic hepatitis C and non-alcoholic fatty liver disease. *PLoS ONE* **2011**, *6*, e23937. [CrossRef]
69. Sarasin-Filipowicz, M.; Krol, J.; Markiewicz, I.; Heim, M.H.; Filipowicz, W. Decreased levels of microRNA miR-122 in individuals with hepatitis C responding poorly to interferon therapy. *Nat. Med.* **2009**, *15*, 31–33. [CrossRef]
70. Hamdane, N.; Juhling, F.; Crouchet, E.; El Saghire, H.; Thumann, C.; Oudot, M.A.; Bandiera, S.; Saviano, A.; Ponsolles, C.; Roca Suarez, A.A.; et al. HCV-Induced Epigenetic Changes Associated with Liver Cancer Risk Persist After Sustained Virologic Response. *Gastroenterology* **2019**, *156*, 2313–2329. [CrossRef]
71. Perez, S.; Kaspi, A.; Domovitz, T.; Davidovich, A.; Lavi-Itzkovitz, A.; Meirson, T.; Alison Holmes, J.; Dai, C.Y.; Huang, C.F.; Chung, R.T.; et al. Hepatitis C virus leaves an epigenetic signature post cure of infection by direct-acting antivirals. *PLoS Genet.* **2019**, *15*, e1008181. [CrossRef] [PubMed]
72. El-Araby, R.E.; Khalifa, M.A.; Zoheiry, M.M.; Zahran, M.Y.; Rady, M.I.; Ibrahim, R.A.; El-Talkawy, M.D.; Essawy, F.M. The interaction between microRNA-152 and DNA methyltransferase-1 as an epigenetic prognostic biomarker in HCV-induced liver cirrhosis and HCC patients. *Cancer Gene Ther.* **2019**. [CrossRef] [PubMed]
73. Chandra, P.K.; Bao, L.; Song, K.; Aboulnasr, F.M.; Baker, D.P.; Shores, N.; Wimley, W.C.; Liu, S.; Hagedorn, C.H.; Fuchs, S.Y.; et al. HCV infection selectively impairs type I but not type III IFN signaling. *Am. J. Pathol.* **2014**, *184*, 214–229. [CrossRef] [PubMed]
74. Florczyk, U.; Czauderna, S.; Stachurska, A.; Tertil, M.; Nowak, W.; Kozakowska, M.; Poellinger, L.; Jozkowicz, A.; Loboda, A.; Dulak, J. Opposite effects of HIF-1alpha and HIF-2alpha on the regulation of IL-8 expression in endothelial cells. *Free Radic. Biol. Med.* **2011**, *51*, 1882–1892. [CrossRef]
75. Son, Y.O.; Pratheeshkumar, P.; Roy, R.V.; Hitron, J.A.; Wang, L.; Zhang, Z.; Shi, X. Nrf2/p62 signaling in apoptosis resistance and its role in cadmium-induced carcinogenesis. *J. Biol. Chem.* **2014**, *289*, 28660–28675. [CrossRef] [PubMed]
76. Xu, H.; Xu, S.J.; Xie, S.J.; Zhang, Y.; Yang, J.H.; Zhang, W.Q.; Zheng, M.N.; Zhou, H.; Qu, L.H. MicroRNA-122 supports robust innate immunity in hepatocytes by targeting the RTKs/STAT3 signaling pathway. *Elife* **2019**, *8*. [CrossRef] [PubMed]
77. Van Renne, N.; Suarez, A.A.R.; Duong, F.H.T.; Gondeau, C.; Calabrese, D.; Fontaine, N.; Ababsa, A.; Bandiera, S.; Croonenborghs, T.; Pochet, N.; et al. miR-135a-5p-mediated downregulation of protein tyrosine phosphatase receptor delta is a candidate driver of HCV-associated hepatocarcinogenesis. *Gut* **2018**, *67*, 953–962. [CrossRef] [PubMed]
78. Chen, Y.N.; Shen, A.; Rider, P.J.; Yu, Y.; Wu, K.L.; Mu, Y.X.; Hao, Q.; Liu, Y.L.; Gong, H.; Zhu, Y.; et al. A liver-specific microRNA binds to a highly conserved RNA sequence of hepatitis B virus and negatively regulates viral gene expression and replication. *FASEB J.* **2011**, *25*, 4511–4521. [CrossRef] [PubMed]
79. Luna, J.M.; Scheel, T.K.H.; Danino, T.; Shaw, K.S.; Mele, A.; Fak, J.J.; Nishiuchi, E.; Takacs, C.N.; Catanese, M.T.; de Jong, Y.P.; et al. Hepatitis C Virus RNA Functionally Sequesters miR-122. *Cell* **2015**, *160*, 1099–1110. [CrossRef]
80. Hyrina, A.; Olmstead, A.D.; Steven, P.; Krajden, M.; Tam, E.; Jean, F. Treatment-Induced Viral Cure of Hepatitis C Virus-Infected Patients Involves a Dynamic Interplay among three Important Molecular Players

in Lipid Homeostasis: Circulating microRNA (miR)-24, miR-223, and Proprotein Convertase Subtilisin/Kexin Type 9. *Ebiomedicine* **2017**, *23*, 68–78. [CrossRef]
81. Jopling, C.L.; Yi, M.K.; Lancaster, A.M.; Lemon, S.M.; Sarnow, P. Modulation of hepatitis C virus RNA abundance by a liver-specific microRNA. *Science* **2005**, *309*, 1577–1581. [CrossRef] [PubMed]
82. Mylroie, H.; Dumont, O.; Bauer, A.; Thornton, C.C.; Mackey, J.; Calay, D.; Hamdulay, S.S.; Choo, J.R.; Boyle, J.J.; Samarel, A.M.; et al. PKC epsilon-CREB-Nrf2 signalling induces HO-1 in the vascular endothelium and enhances resistance to inflammation and apoptosis. *Cardiovasc. Res.* **2015**, *106*, 509–519. [CrossRef] [PubMed]
83. Gao, H.W.; Guo, R.F.; Speyer, C.L.; Reuben, J.; Neff, T.A.; Hoesel, L.M.; Riedemann, N.C.; McClintock, S.A.; Sarma, J.V.; Van Rooijen, N.; et al. Stat3 activation in acute lung injury. *J. Immunol.* **2004**, *172*, 7703–7712. [CrossRef] [PubMed]
84. Tawani, A.; Amanullah, A.; Mishra, A.; Kumar, A. Evidences for Piperine inhibiting cancer by targeting human G-quadruplex DNA sequences. *Sci. Rep.* **2016**, *6*, 39239. [CrossRef] [PubMed]
85. Mitchell, P.S.; Parkin, R.K.; Kroh, E.M.; Fritz, B.R.; Wyman, S.K.; Pogosova-Agadjanyan, E.L.; Peterson, A.; Noteboom, J.; O'Briant, K.C.; Allen, A.; et al. Circulating microRNAs as stable blood-based markers for cancer detection. *Proc. Natl. Acad. Sci. USA* **2008**, *105*, 10513–10518. [CrossRef] [PubMed]

9

mTOR and STAT3 Pathway Hyper-Activation is Associated with Elevated Interleukin-6 Levels in Patients with Shwachman-Diamond Syndrome: Further Evidence of Lymphoid Lineage Impairment

Antonio Vella [1,†], Elisabetta D'Aversa [2,†], Martina Api [3], Giulia Breveglieri [2], Marisole Allegri [3], Alice Giacomazzi [1], Elena Marinelli Busilacchi [4], Benedetta Fabrizzi [3], Tiziana Cestari [1], Claudio Sorio [5], Gloria Bedini [6], Giovanna D'Amico [6], Vincenzo Bronte [1], Antonella Poloni [4], Antonio Benedetti [7], Chiara Bovo [8], Seth J. Corey [9], Monica Borgatti [2,10,‡], Marco Cipolli [11,‡] and Valentino Bezzerri [3,*,‡]

1 Unit of Immunology, Azienda Ospedaliera Universitaria Integrata, 37134 Verona, Italy; antonio.vella@univr.it (A.V.); alice.giacomazzi@univr.it (A.G.); tiziana.cestari@univr.it (T.C.); vincenzo.bronte@univr.it (V.B.)
2 Department of Life Sciences and Biotechnology, University of Ferrara, 44100 Ferrara, Italy; elisabetta.daversa@unife.it (E.D.); giulia.breveglieri@unife.it (G.B.); monica.borgatti@unife.it (M.B.)
3 Cystic Fibrosis Center, Azienda Ospedaliero Universitaria Ospedali Riuniti, 60126 Ancona, Italy; martina.api@ospedaliriuniti.marche.it (M.A.); marisole.allegri@ospedaliriuniti.marche.it (M.A.); benedetta.fabrizzi@ospedaliriuniti.marche.it (B.F.)
4 Hematology Clinic, Università Politecnica delle Marche -AOU Ospedali Riuniti, 60126 Ancona, Italy; e.busilacchi@staff.univpm.it (E.M.B.); Antonella.poloni@ospedaliriuniti.marche.it (A.P.)
5 Department of Medicine, University of Verona, 37134 Verona, Italy; claudio.sorio@univr.it
6 Immunology and Cell Therapy Unit, Tettamanti Research Center, University of Milano-Bicocca, 20900 Monza, Italy; gloriabedini85@gmail.com (G.B.); giovanna.damico@asst-monza.it (G.D.)
7 Department of Gastroenterology and Hepatology, Università Politecnica delle Marche, 60126 Ancona, Italy; antonio.benedetti@ospedaliriuniti.marche.it
8 Hospital Health Direction, Azienda Ospedaliera Universitaria Integrata, 37126 Verona, Italy; Direzione.sanitaria@aovr.veneto.it
9 Department of Pediatric Hematology/Oncology and Stem Cell Transplantation, Cleveland Clinic, Cleveland, OH 44195, USA; coreys2@ccf.org
10 Biotechnology Center, University of Ferrara, 44100 Ferrara, Italy
11 Cystic Fibrosis Center, Azienda Ospedaliera Universitaria Integrata, 37126 Verona, Italy; marco.cipolli@aovr.veneto.it
* Correspondence: valentino.bezzerri@ospedaliriuniti.marche.it
† These authors contributed equally to this work.
‡ Shared seniorship.

Abstract: Shwachman–Diamond syndrome (SDS) is a rare inherited bone marrow failure syndrome, resulting in neutropenia and a risk of myeloid neoplasia. A mutation in a ribosome maturation factor accounts for almost all of the cases. Lymphoid involvement in SDS has not been well characterized. We recently reported that lymphocyte subpopulations are reduced in SDS patients. We have also shown that the mTOR-STAT3 pathway is hyper-activated in SDS myeloid cell populations. Here we show that mTOR-STAT3 signaling is markedly upregulated in the lymphoid compartment of SDS patients. Furthermore, our data reveal elevated IL-6 levels in cellular supernatants obtained from lymphoblasts, bone marrow mononuclear and mesenchymal stromal cells, and plasma samples obtained from a cohort of 10 patients. Of note, everolimus-mediated inhibition of mTOR signaling is associated with basal state of phosphorylated STAT3. Finally, inhibition of mTOR-STAT3 pathway activation leads to normalization of IL-6 expression in SDS cells. Altogether, our data strengthen the hypothesis that SDS affects both lymphoid and myeloid blood compartment and suggest everolimus as a potential therapeutic agent to reduce excessive mTOR-STAT3 activation in SDS.

Keywords: STAT3; mTOR; Bone Marrow Failure Syndromes; lymphocytes

1. Introduction

Shwachman–Diamond syndrome (SDS) is one of the most common inherited bone marrow failure syndromes (IBMFS), occurring in almost 1 out of 75,000 live births [1]. SDS results from biallelic mutations in the Shwachman–Bodian–Diamond syndrome gene (*SBDS*), which encode the SBDS protein. SBDS protein cooperates with its partner elongation factor-like GTPase 1 (EFL1) to catalyze the release of the ribosomal anti-association factor eIF6, facilitating the assembly of the functional 80S ribosome [2–4]. The IBMFS are also cancer predisposition syndromes, in particular myelodysplastic syndrome (MDS) and acute myeloid leukemia (AML). In the general population, MDS has an incidence ranging from 2–12 cases per 100,000 people, which increases as individuals age [5]. Patients with SDS demonstrate a risk of evolution to MDS of 8.1% and 36% at 10 and 30 years, respectively [6]. A recent genomic analysis of 1514 patients with MDS who underwent a stem cell transplant showed that 4% of the young adult patients had undiagnosed compound heterozygous mutations in *SBDS*, suggesting that SDS prevalence among MDS/AML patients may be underestimated [7]. AML derives from dysregulated proliferation and accumulation of immature myeloid progenitor cells into the bone marrow and peripheral blood, which finally leads to a severe impairment of the hematopoietic system. Acute leukemias rapidly disseminate after initial inception, escaping the anti-leukemic immunity process. Regulatory T cells play a key role in the maintenance of immune tolerance, which acts as a regulator of the tumor immunity [8]. CD4/CD8 double negative (DN) T cells have gained prominence among T regulatory cell subsets engaged in immunosurveillance. DN T cells are mature T cells representing almost 3–5% of the total peripheral T cell population [9]. Most human and murine T cells express and rearrange the α and β chains of the T cell receptor (TCR) and are recognized as TCR$\alpha\beta$ T cells, whereas a small part of T cells do express the γ and δ chains, which are mostly DN T cells [9]. Interestingly, DN T cells showed anti-leukemic activity and synergy with conventional chemotherapies both in vitro and in patient-derived xenograft models of AML [10,11]. In a mouse model of AML, the leukemic cells promoted T cell tolerance with suppression of anti-tumor $CD8^+$ T cells [12]. Failure of T cell-mediated anti-cancer immune response is associated with disease progression and poor outcome in MDS and AML.

IL-6/JAK/STAT3 signaling axis plays a key role in leukemogenesis [13]. The genes encoding the kinase protein JAK2 have indeed often been mutated in myeloproliferative disorders, leading to constitutive hyper-activation of its downstream effector, the transcription factor STAT3 [14]. Of note, STAT3 hyper-activation has been found in tumor-infiltrating leukocytes, in which STAT3 orchestrates the crosstalk between cancer and immune cells [15]. Furthermore, it has been reported that STAT3 inhibition in the myeloid compartment remarkably induces the anti-tumor capabilities of T cells and promotes their expansion in vivo [16,17]. STAT3 activation in immune cells is indeed associated with suppression of anti-tumor immunity. STAT3 excessive activation may be triggered by elevated levels of IL-6 present in the serum or released within the tumor microenvironment [14]. IL-6 may act in an autocrine or paracrine manner. IL-6 binds to its IL-6 receptor (IL-6R), localized onto the plasma membrane (membrane bound (mb)IL-6R), which is physically associated with the gp130 protein. This process is recognized as classical IL-6 signaling and leads to gp130 homodimerization, resulting in the activation of the IL-6 receptor complex [18]. In addition, IL-6 binds to the small extracellular secretory soluble IL-6 receptor (sIL-6R), which is generated by alternative splicing of IL-6R gene or by metalloproteinase-dependent cleavage of mbIL-6R. The sIL-6R mediates JAK-STAT3 activation in gp130 positive cells, which do not express mbIL-6R through a process termed IL-6 trans-signaling [19]. The

IL-6 trans-signaling pathway has been reported in murine hematopoietic stem cells [20,21]. Elevated levels of IL-6 have been found in adult bone marrow niche of patients with AML [22]. Serum IL-6 levels were found to be significantly increased in pediatric patients with AML [23]. Moreover, increased IL-6 serum levels are associated with poor prognosis in several types of cancer, including AML [23]. Interestingly, *IL-6* gene expression can be regulated by STAT3 itself, resulting in a feedforward autocrine feedback loop [14].

STAT3 has been reported as a direct substrate for the mammalian target of rapamycin (mTOR), which induces STAT3 S727 phosphorylation [24,25]. In addition, the mTOR-inhibitor rapamycin inhibits STAT3 S727 phosphorylation [24]. Moreover, we have previously shown that mTOR can promote STAT3 phosphorylation both at residue tyrosine 705 and serine 727 in SDS leukocytes [26]. Previous studies have reported that relapse of AML is associated with the gain of additional mutations in the mTOR gene, often due to the cytotoxic chemotherapy received by patients [27,28]. Inhibition of the mTOR pathway using rapamycin or other analogue molecules, including everolimus (RAD001) as anti-leukemic agent, has shown potent anti-cancer capabilities both in vitro and in vivo [29,30].

To date, no pharmacological therapy has been developed for IBMFS-related MDS or AML, and allogeneic hematopoietic stem cell transplantation remains the unique option in these cases. Unfortunately, its efficacy is limited by the morbidity and mortality associated with graft-versus-host disease.

Here we show further analysis of the mTOR-STAT3 axis in an extended panel of lymphocytic populations including $CD4^+$, $CD8^+$, T cells, DN T cells, $\gamma\delta$T cells, and Natural Killer cells (NK). Moreover, we have checked the expression of IL-6 in different cellular and clinical SDS models, including lymphoblastoid cell lines (LCL), bone marrow mononuclear hematopoietic progenitors (BM-MNC) and mesenchymal stromal cells (BM-MSC), and plasma obtained from an enlarged cohort of 31 patients with SDS (Table 1). Our data indicate that everolimus can restore a normal level of mTOR and STAT3 activation in primary SDS lymphocytes. Importantly, mTOR-STAT3 inhibition was paralleled by a downregulation of IL-6 expression in hematopoietic SDS cells. Our results suggest the existence of a mTOR-STAT3-IL-6 loop of activation in hematopoietic SDS cells, which may affect both myeloid and lymphoid compartment, thus contributing to malignant transformation over the time. Taken together, these data strengthen the hypothesis of the involvement of lymphoid lineage in SDS and suggest everolimus or new rapalogs as potential therapeutic agents in SDS patients.

Table 1. Clinical data and genetics of patients with SDS recruited in this study.

UPN	Gender	Age	Genotype	PMN (Cell/mm^3)	Phenotype	Cytogenetics
1	M	27	258+2T>C/183–184TA>CT	1460	PI, FTT, recurrent infections, HbF > 2%, bone malformation, thrombocytopenia	46, XY, i(7)(q10)
6	M	27	258+2T>C/101A>T	3515	PI, FTT, recurrent infections, bone malformation, thrombocytopenia, cognitive impairment	46, XY, del(20)q
13	M	19	258+2T>C/183–184TA>CT+258+2T>C	1130	PI, FTT, recurrent infections, bone malformation, thrombocytopenia, anemia, cognitive impairment	46, XY, del(20)q
26	M	16	258+2T>C/183–184TA>CT	58	PI, FTT, bone malformation, thrombocytopenia	46, XY
33	F	8	258+2T>C/183–184TA>CT	1100	PI, FTT, bone malformation, thrombocytopenia	46, XX
35	F	15	258+2T>C/183–184TA>CT	970	PI, FTT, thrombocytopenia, anemia	46, XX
37	F	10	258+2T>C/183–184TA>CT	1280	PI, FTT, recurrent infections	46, XX, i(7)(q10)
43	M	22	258+2T>C/258+2T>C+533–549+403del	970	PI, FTT, bone malformation, thrombocytopenia, cognitive impairment	46, XY
47	M	12	258+2T>C/183–184TA>CT	550	PI, FTT, recurrent infections, bone malformation, thrombocytopenia	46, XY

Table 1. *Cont.*

UPN	Gender	Age	Genotype	PMN (Cell/mm^3)	Phenotype	Cytogenetics
52	M	10	258+2T>C/183–184TA>CT+258+2T>C	1070	PI, FTT, bone malformation	46, XY
56	F	15	258+2T>C/183–184TA>CT	1840	PI, FTT, recurrent infections, HbF > 2%, bone malformation, thrombocytopenia	46, XX
57	F	40	258+2T>C/G63C	500	PI, FTT, HbF > 2%, bone malformation, thrombocytopenia, cognitive impairment	46, XX
58	M	12	258+2T>C/183–184TA>CT	390	PI, FTT, HbF > 2%, bone malformation, thrombocytopenia, anemia	46, XY
63	M	15	258+2T>C/183–184TA>CT+258+2T>C	536	PI, FTT, recurrent infections, HbF > 2%, bone malformation, thrombocytopenia	46, XY
65	M	19	258+2T>C/258+2T>C	1390	PI, recurrent infections, bone malformation, thrombocytopenia, cognitive impairment	46, XY, del(20)q
66	M	23	258+2T>C/183–184TA>CT	1340	PI, bone malformation, thrombocytopenia, cognitive impairment	46, XY
67	M	8	258+2T>C/183–184TA>CT	500	PI, FTT, HbF > 2%, bone malformation	46, XY
68	M	22	258+2T>C/183–184TA>CT+258+2T>C	600	PI, FTT, recurrent infections, bone malformation, thrombocytopenia, cognitive impairment	46, XY
69	F	8	258+2T>C/183–184TA>CT	770	PI, FTT, HbF > 2%, anemia, cognitive impairment	46, XX
72	M	28	258+2T>C/183–184TA>CT	380	PI, FTT, recurrent infections, HbF > 2%, bone malformation, thrombocytopenia, anemia, cognitive impairment	46, XY
73	F	7	258+2T>C/183–184TA>CT	520	PI, FTT, HbF > 2%, thrombocytopenia	46, XX
74	M	9	258+2T>C/183–184TA>CT	1430	PI, FTT, HbF > 2%, cognitive impairment	46, XY
75	F	7	258+2T>C/183–184TA>CT	1000	PI, FTT, HbF > 2%, bone malformation, thrombocytopenia, cognitive impairment	46, XX
80	M	7	258+2T>C/183–184TA>CT	680	PI, FTT, recurrent infections, bone malformation, anemia, cognitive impairment	46, XY
82	M	16	258+2T>C/183–184TA>CT	300	PI, FTT, recurrent infections, bone malformation, thrombocytopenia, anemia, cognitive impairment	46, XY
87	M	18	258+2T>C/183–184TA>CT	880	PI, FTT, recurrent infections, bone malformation, cognitive impairment	46, XY
91	M	4	258+2T>C/183–184TA>CT+258+2T>C	1050	PI, FTT, bone malformation	46, XY
94	F	19	258+2T>C/352A>G	2420	PI, HbF > 2%, thrombocytopenia	46, XX
104	M	10	258+2T>C/183–184TA>CT	500	PI, FTT, recurrent infections, HbF > 2%, bone malformation, thrombocytopenia	46, XY
106	M	36	258+2T>C/183–184TA>CT	1210	PI, FTT, bone malformation, anemia	46, XY
108	M	17	258+2T>C/183–184TA>CT	970	PS, FTT, bone malformation, anemia	46, XY

UPN, unique patient number; PI, pancreas insufficiency; PS, pancreas sufficiency; FTT, failure to thrive; HbF, fetal hemoglobin.

2. Results

2.1. mTOR-STAT3 Pathway is Hyper-Activated Also in SDS Lymphocyte Subsets and Everolimus Can Reduce This Process In Vitro

The JAK/STAT3 pathway regulates T cell cytotoxic gene expression, proliferation, and survival. STAT3 inhibition in the myeloid compartment displays a remarkable induction of the T cell anti-tumor capabilities and promotes their expansion in vivo [16,17]. Thus, STAT3 activation in immune cells is associated with suppression of anti-tumor immunity. The protein kinase mTOR acts as an activator of the STAT3 pathway [24,25,31]. Accordingly, we recently reported that mTOR can activate STAT3 pathway in B cells, neutrophils and monocytes from SDS patients [26]. To assess the activation of mTOR-STAT3 pathway in SDS lymphoid compartment, we determined the phosphorylation of mTOR (S2448) and STAT3 (Y705 and S727) in CD4+ and CD8+ T cells, DN T cells, $\gamma\delta$T cells, and NK cells isolated from peripheral blood from seven patients with SDS. In SDS patients, all cell subsets except NK displayed significantly elevated levels of phospho-mTOR (Figure 1) and -STAT3 (Figure 2) compared to age-matched healthy donors. Then, we tested the effect of the clinically-approved rapamycin analog, everolimus (RAD001) on mTOR, and STAT3 phosphorylation in SDS patient-derived T cell subsets. Results show that everolimus restores normal levels of phosphorylation of mTOR (Figure 1) and STAT3 (Figure 2), confirming the existence of an mTOR-STAT3 axis activation in the lymphoid compartment of SDS patients. To determine whether upregulation *STAT3* gene expression eventually exists in SDS cells along with hyper-phosphorylation, we measured and correlated *STAT3* transcript expression with protein levels in LCL and primary PBMC isolated from several SDS patients with different genotypes, compared to age-matched healthy controls. Our studies showed that *STAT3* expression is elevated in lymphocytes obtained from SDS patients compared to control subjects (Figures 3 and 4, Figures S1 and S2).

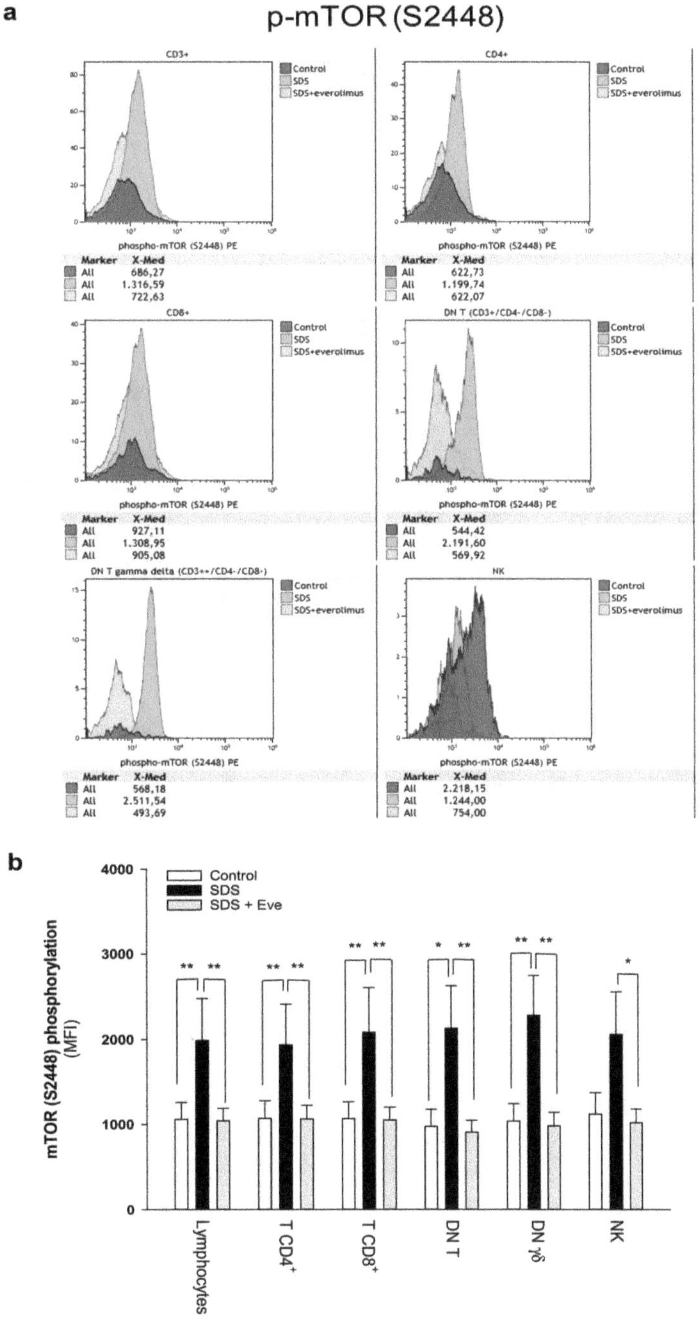

Figure 1. Phospho-flow analysis of mTOR S2448 in a panel of primary lymphocyte subsets. (**a**) Representative experiments conducted on peripheral blood samples obtained from patients with

SDS. Red histograms represent age-matched healthy control cells; green histograms represent SDS lymphocytes; blue histograms represent SDS lymphocytes upon everolimus (350 nM) treatment. (**b**) Data are mean ± SEM of seven experiments conducted on SDS lymphocytes obtained from seven SDS patients (UPN37, UPN58, UPN69, UPN73, UPN87, UPN106, UPN108). Statistics: normal distribution was tested by the Shapiro–Wilk test. Subsequently, the Mann–Whitney Rank Sum Test was calculated. * $p < 0.05$; ** $p < 0.01$.

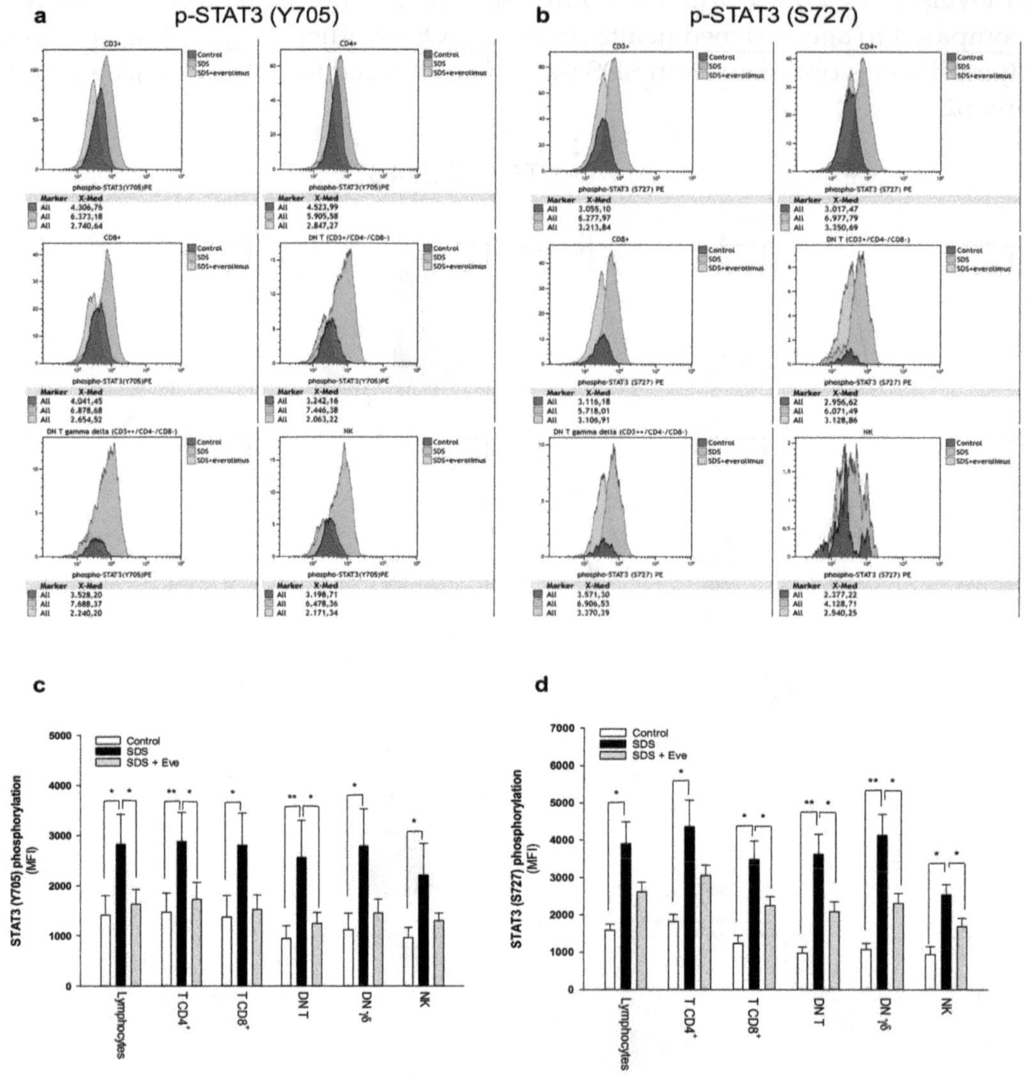

Figure 2. Phospho-flow analysis of STAT3 in a panel of primary lymphocyte subsets. (**a**) Representative analysis of phospho-STAT3 Y705 in lymphocyte subsets. (**b**) Representative analysis of phospho-STAT3 S727 in lymphocyte subsets. Red histograms represent age-matched healthy control cells; green histograms represent SDS patient-derived lymphocytes; blue histograms represent SDS lymphocytes upon everolimus (350nM) treatment. Data are mean ± SEM of seven experiments conducted on SDS patient-derived lymphocytes obtained from seven SDS patients (UPN37, UPN58, UPN69, UPN73, UPN87, UPN106, UPN108). Statistical Student's t test for paired data has been calculated. * $p < 0.05$; ** $p < 0.01$. (**c**) STAT3 (Y705) Median Fluorescence Intensity as measured by phospho-flow assays. (**d**) STAT3 (S727) Median Fluorescence Intensity as measured by phospho-flow assays. Data are mean ± SEM of seven experiments conducted on SDS lymphocytes obtained from seven SDS patients (UPN37, UPN58, UPN69, UPN73, UPN87, UPN106, UPN108). Statistics: Normal distribution was tested by the Shapiro–Wilk test. Subsequently, the Mann–Whitney Rank Sum Test was calculated. * $p < 0.05$; ** $p < 0.01$.

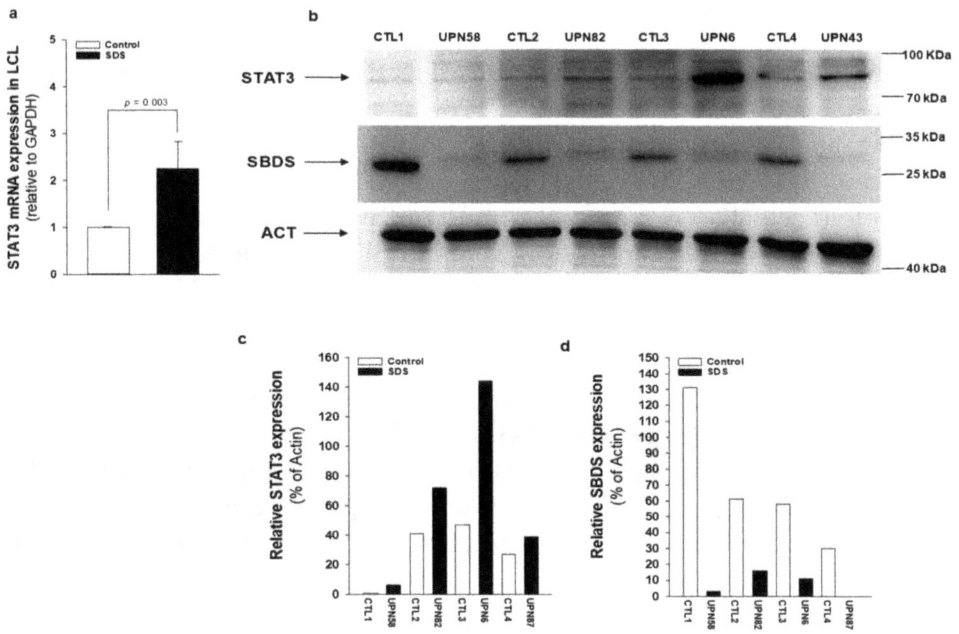

Figure 3. *STAT3* transcript and protein expression is upregulated in SDS patient-derived LCL. (**a**) *STAT3* mRNA expression in LCL isolated from UPN6, UPN43, UPN58, UPN82 (black bar), and from age-matched controls (white bar), measured by qRT-PCR. Data are mean ± SEM of four experiments performed in duplicate. (**b**) STAT3 protein level was measured in LCL (UPN6, UPN43, UPN58, UPN82) by Western blot analysis. (**c**,**d**) Densitometric analysis of Western blots showed in panel (**b**). Statistics: Normal distribution was tested by the Shapiro–Wilk test. Subsequently, the Mann–Whitney Rank Sum Test was calculated and reported within histograms.

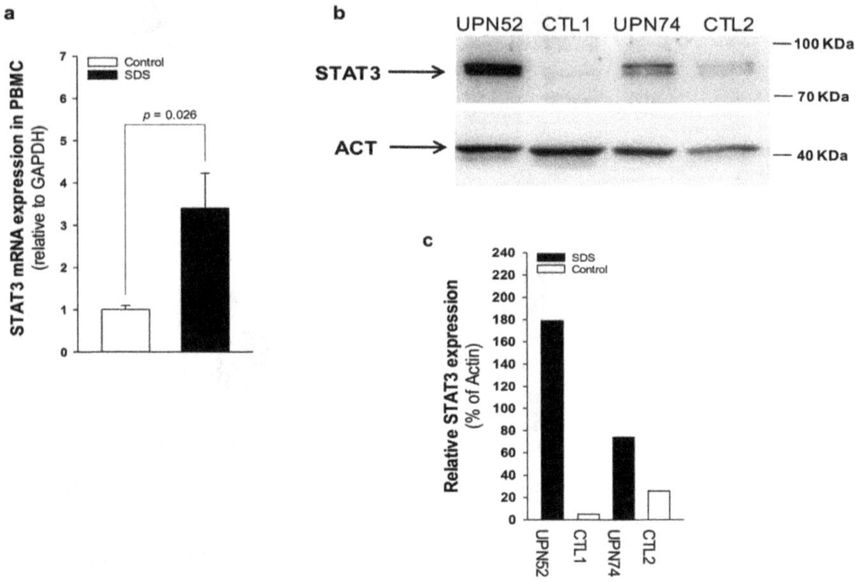

Figure 4. *STAT3* transcript and protein expression is upregulated in SDS patient-derived primary PBMC. (**a**) *STAT3* mRNA expression in primary PBMC isolated from UPN26, UPN69, UPN73, UPN87, UPN94, UPN106, and UPN108 (black bar), and from age-matched controls (white bar), measured by qRT-PCR. Data are mean ± SEM of seven experiments. (**b**) STAT3 protein level was measured in PBMC (UPN52 and UPN74) by Western blot analysis. (**c**) Densitometric analysis of Western blots showed in panel **b**. Statistics: Normal distribution was tested by the Shapiro–Wilk test. Subsequently, the Mann-Whitney Rank Sum Test was calculated and reported within histograms.

2.2. IL-6 Expression Is Upregulated in SDS

Plasma levels of IL-6 are generally close to the detection limit (1 pg/mL) in healthy individuals but significantly increase during the inflammatory process and cancers [32]. IL-6 is a major activator of JAK-STAT3 signaling, and *IL-6* transcript expression is upregulated by STAT3 activation, generating an autocrine/paracrine loop of activation [14]. Since mTOR-STAT3 axis is upregulated in SDS patient-derived myeloid cells [26], we sought to find out whether hyper-activation of this pathway is associated with *IL-6* over-expression in non-myeloid SDS cell models. Both LCL and BM-MSC obtained from patients with SDS displayed significantly upregulated IL-6 release into cell culture supernatants compared with age-matched healthy controls (Figure 5a,b). In particular, IL-6 levels were significantly elevated in primary SDS BM-MSC, which released ~8 ng/mL in 2×10^5 cells in our experimental conditions (Figure 5b). Plasma samples obtained from the peripheral blood collected from an expanded cohort of 21 patients with SDS showed significantly increased levels of IL-6 (mean 3.66 ± 4.58) compared to aged-matched controls (mean 1.19 ± 1.89), consistent with *in vitro* results (Figure 5c). Since SDS BM-MSC showed an impressive upregulation of *IL-6* expression, we sought to find out whether IL-6 were further concentrated in plasma derived from the bone marrow of patients. IL-6 levels in bone marrow plasma were even more elevated (mean 4.75 ± 3.82) than those found in peripheral blood (mean 3.04 ± 2.05) obtained in parallel, from the same patients (Figure 5d).

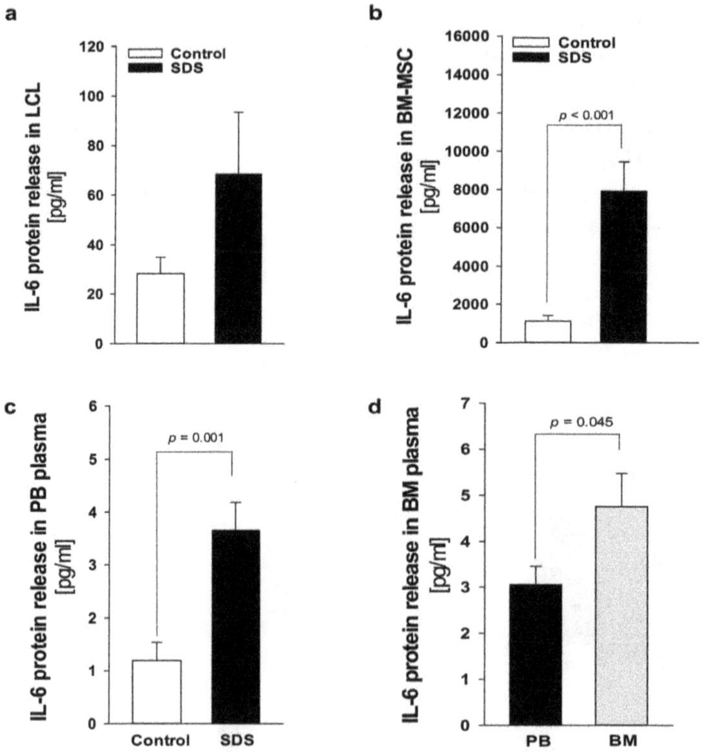

Figure 5. IL-6 release is elevated in SDS specimens compared to age-matched donor controls. (**a**) Measurement of IL-6 released in supernatants collected from 1×10^6 LCL after 48h. Data are mean ± SEM of 10 experiments conducted on LCL obtained from UPN24, UPN26, UPN58, UPN68, UPN75, and UPN106. (**b**) IL-6 released in supernatants collected from 2×10^5 primary BM-MSC after 48 h. (**c**) IL-6 concentration in peripheral blood (PB) plasma samples obtained from 21 patients with SDS (UPN1, 13, 26, 37, 47, 52, 56, 57, 58, 63, 65, 66, 69, 72, 73, 74, 87, 94, 104, 106, 108) compared with age-matched plasma controls. (**d**) IL-6 concentration in bone marrow (BM) plasma samples obtained from eight patients with SDS (UPN 47, 56, 65, 74, 87, 94, 106, 108) compared to PB plasma samples obtained from the same patients. Normal distribution was tested by the Shapiro–Wilk test. The Mann–Whitney Rank Sum Test was calculated and reported within histograms.

2.3. Patients with SDS ShowReduced Levels of Soluble IL-6 Receptor

IL-6 signaling cascade occurs by classical activation through mbIL-6R or trans-signaling via the soluble sIL-6R [19]. The latter mechanism allows cells that do not express mbIL-6R, but do express gp130, to be responsive to IL-6. Undifferentiated MSC lack the expression of IL-6 receptor although they normally express gp130 [33]. Thus, trans-signaling should be required to activate these cells upon IL-6 stimulation. Nevertheless, BM-MSC can constitutively release large quantities of IL-6 [34]. In this work, we quantified sIL-6R in plasma samples obtained from 21 patients with SDS. In healthy individuals, plasma levels of sIL-6R range between 50–70 ng/mL [35]. Of note, we found that the soluble receptor release is reduced in SDS patients (44.3 ± 15.4 pg/μL) compared to age-matched healthy donors (71.6 ± 14.7 pg/μL) (Figure 6a). To verify whether the bone marrow compartment, which showed increased levels of IL-6, exhibits also higher levels of sIL-6R, we compared sIL-6R expression in peripheral blood plasma with the expression found in bone marrow plasma obtained in parallel from the same patients. However, peripheral blood and bone marrow plasma showed comparable levels of sIL-6R expression (Figure 6b).

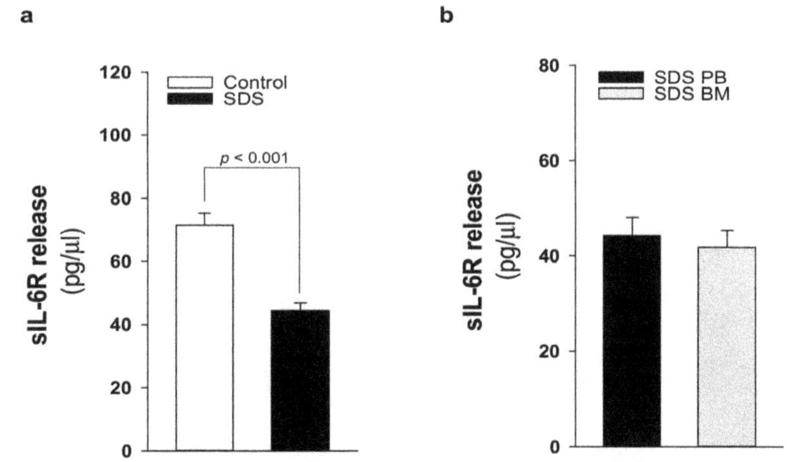

Figure 6. sIL6-R release is reduced in patients with SDS. (**a**) sIL-6R protein release was quantified by ELISA in PB plasma samples obtained from 21 patients with SDS (UPN 1, 13, 26, 37, 47, 52, 56, 57, 58, 63, 65, 66, 69, 72, 73, 74, 87, 94, 104, 106, 108) compared with age-matched plasma controls. (**b**) sIL-6R concentration in BM plasma samples obtained from eight patients with SDS (UPN 47, 56, 65, 74, 87, 94, 106, 108) compared with PB plasma samples obtained from the same patients. Normal distribution was tested by the Shapiro–Wilk test. The Mann–Whitney Rank Sum Test was calculated and reported within histograms.

2.4. Elevated IL-6 Gene Expressionin Hematopoietic CellsIs Primarily Driven by mTOR-STAT3 Pathway in SDS

Fifty to eighty percent of patients with AML present a constitutive activation of the mTOR pathway, showing significantly shorter disease-free and overall survival rates compared with patients without constitutive activation [36,37]. In the last decade, the development of new rapamycin (sirolimus) analogs showing improved pharmacokinetic profile, such as the clinically approved rapalog everolimus, have given rise to great interest for anti-cancer therapy [29]. We previously reported that rapamycin-dependent mTOR inhibition leads to normal levels of phosphorylation of STAT3 in SDS cells [26]. Here we show that everolimus (350 nM) and STAT3 inhibitor stattic (7.5 μM) significantly reduce *IL-6* mRNA expression in LCL and in primary BM-MNC obtained from patients with SDS (Figure 7a,c). Decreased *IL-6* mRNA expression is paralleled by a reduction of IL-6 release in culture supernatants upon everolimus and stattic treatments both in LCL (-46.6% and -68%, respectively) and BM-MNC (-34.6% and -66%, respectively) (Figure 7b,d). In order to validate these data, we also knocked-down the expression of *mTOR* and *STAT3* in SDS cells using a short interference (si)RNA strategy. To this aim, we transfected two different siRNA sequences for each target gene and a negative control sequence (scrambled), which was previously validated [29], using siRNA-specific liposomes as a vector. We significantly knocked-down *mTOR* and *STAT3* gene expression in SDS LCL (Figure 8a,b). Consistently with pharmacological inhibition, both *mTOR* and *STAT3* gene silencing lead to a strong inhibition of *IL-6* expression in SDS cells (Figure 8c,d). In particular, knock-down of the *STAT3* gene resulted in a statistically significant reduction of *IL-6* expression in terms of mRNA (-85%) and protein release (-51%). However, no effect of everolimus nor stattic on IL-6 release was reported in BM-MSC obtained from four SDS patients (Figure 7f), and *IL-6* mRNA transcription was surprisingly increased (two-fold increase) upon stattic treatment in these cells (Figure 7e).

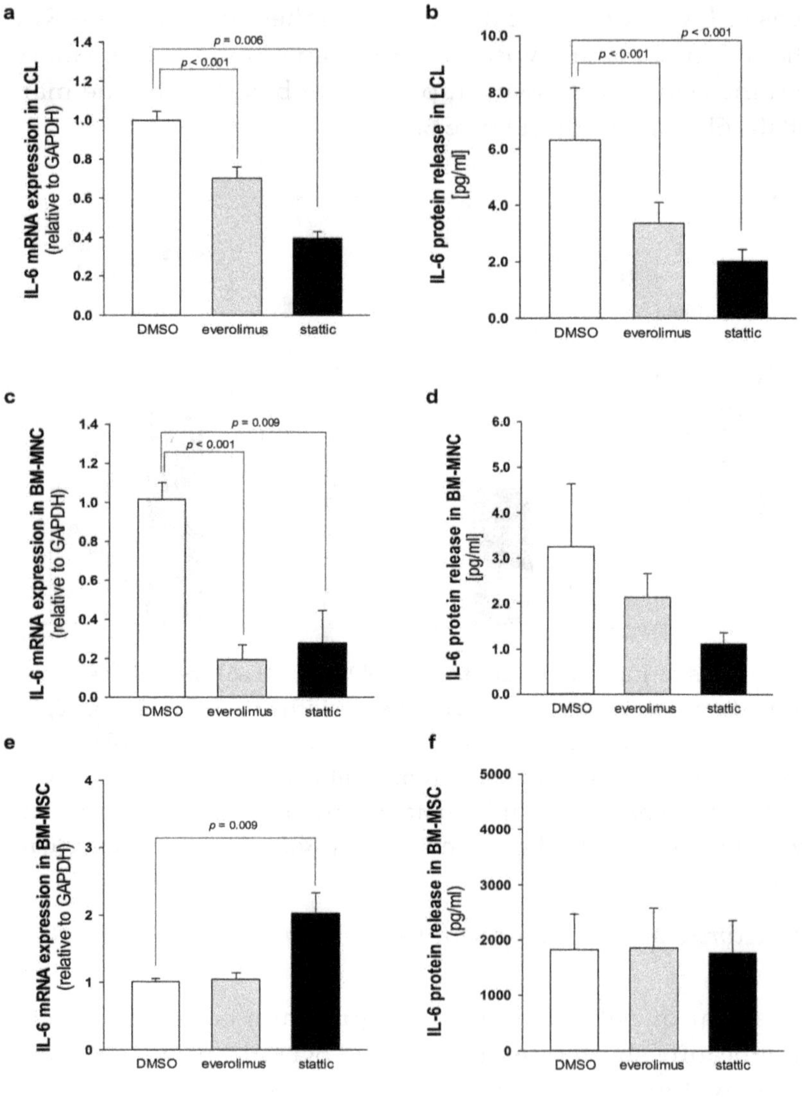

Figure 7. Everolimus and stattic inhibit *IL-6* expression in SDS patient-derived hematopoietic cells. (**a**) *IL-6* transcript expression in LCL incubated in the absence (DMSO) or in the presence of 350 nM everolimus or 7.5 μM stattic for 24 h was quantified by qRT-PCR. Data are mean ± SEM of six experiments performed in duplicate from three affected individuals (UPN58, UPN75, and UPN106). (**b**) IL-6 release in supernatants collected from LCL incubated in the absence (DMSO) or in the presence of 350 nM everolimus or 7.5 μM stattic for 24 h, as measured by Bio-plex assay. Data are mean ± SEM of six experiments conducted as reported in panel a. (**c**) *IL-6* mRNA expression in primary BM-MNC incubated in the absence (DMSO) or in the presence of 350 nM everolimus or 7.5 μM stattic for 24 h was

quantified by qRT-PCR. Data are mean ± SEM of four experiments performed in duplicate from four affected individuals (UPN74, UPN80, UPN94, and UPN106). (**d**) IL-6 release in supernatants collected from BM-MNC as measured by Bio-plex assay. Data are mean ± SEM of four experiments conducted as reported in panel **c**. (**e**) IL-6 transcript expression in BM-MSC incubated in the absence (DMSO) or in the presence of 350 nM everolimus or 7.5 μM stattic for 24 h was quantified by qRT-PCR. Data are mean ± SEM of four experiments performed in duplicate from four affected individuals (UPN33, UPN35, UPN67 and UPN91). (**f**) IL-6 release in supernatants collected from BM-MSC (UPN33, UPN35, UPN67 and UPN91) as measured by Bio-plex assay. Data are mean ± SEM of four experiments conducted as reported in panel **e**. Statistics: Normal distribution was tested by the Shapiro–Wilk test, and the Student's t test for paired data has been calculated and reported within histograms, accordingly.

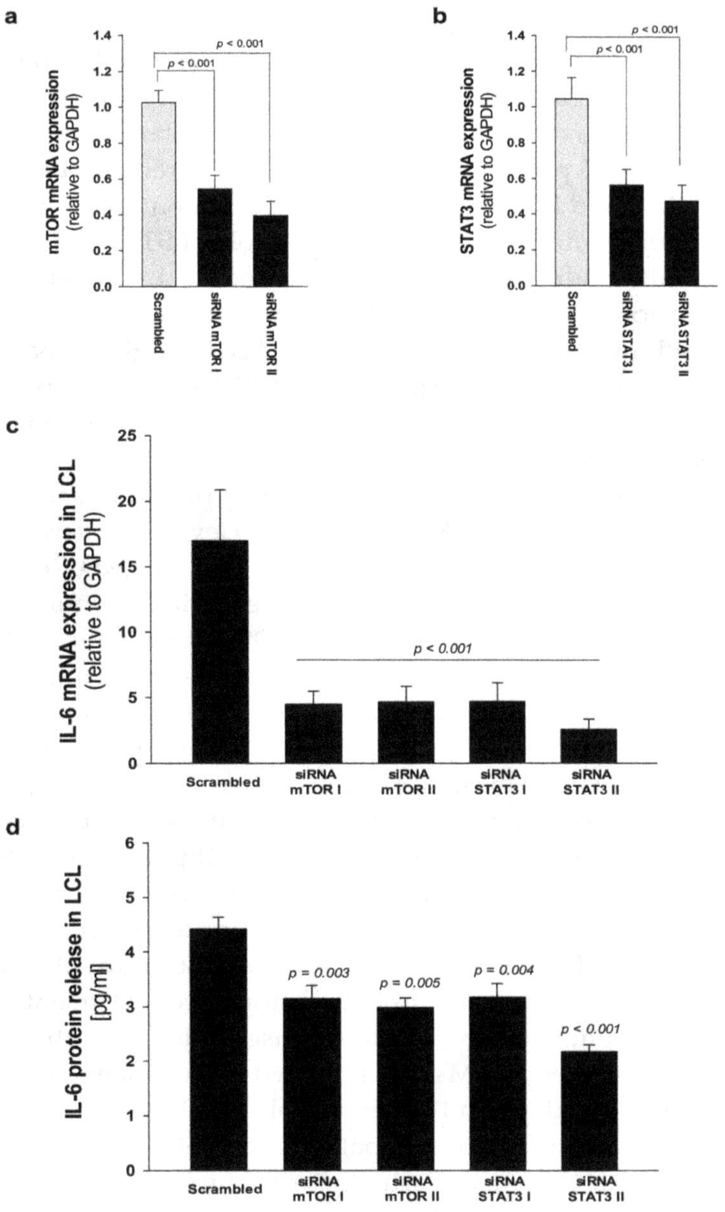

Figure 8. *STAT3* and *mTOR* gene silencing inhibit *IL-6* expression in LCL from SDS patients. (**a**) Reduced expression of *mTOR* mRNA after siRNA-mediated gene silencing. LCL obtained from UPN58, UPN75

and UPN106 were cultured with two different siRNA sequences against target genes, or scrambled sequence (negative control for 48 h). Data are mean ± SEM of four experiments performed in duplicate. (**b**) Reduced expression of STAT3 mRNA after siRNA-mediated gene silencing in LCL, as performed in panel a. Data are mean ± SEM of four experiments performed in duplicate. (**c**) *IL-6* mRNA expression in LCL treated as indicated in panels (**a**) and (**b**). (**d**) IL-6 release was measured by Bio-plex assay in supenatants obtained from LCL treated as indicated in **a**. Statistics: Normal distribution was tested by the Shapiro–Wilk test. Subsequently, the Mann–Whitney Rank Sum Test was calculated and reported within histograms.

3. Discussion

Although it has been widely reported that SDS mainly involves the neutrophil lineage, a number of patients suffer from anemia, thrombocytopenia or pancytopenia (Table 1). Because the bone marrow is often hypocellular, lymphoid and stromal cells may contribute to reduced blood cell formation and myelodysplasia. We recently described a severe deficit of T cells, in particular of DN T subpopulation in SDS patients [38]. Here we show that T cell subpopulations isolated from SDS patients display also hyper-activation of mTOR-STAT3 pathway. In addition, *STAT3* transcript and protein expression are markedly increased in PBMC and LCL obtained from SDS patients, confirming the involvement of the STAT3 pathway in lymphoid lineages. Given the key role of STAT3 in reducing T regulatory cell accumulation [39], our results could partially explain the reduced number of DN T cells observed in SDS patients [38]. Furthermore, STAT3 activation in lymphocytes is associated with T cell impaired functions [17] and reduced anti-tumor activity [16,40]. Accordingly, impaired functions of T cells have been previously described in SDS [41]. Thus, STAT3 pathway upregulation could lead to harmful consequences both on myeloid differentiation in the bone marrow and on innate and adaptive immunosurveillance mechanisms in SDS.

IL-6 is a major activator of STAT3, and *IL-6* transcript expression is itself a target of STAT3 [14]. Thus, we measured *IL-6* expression both at mRNA and protein levels in several SDS cell types. We found that *IL-6* expression is elevated in LCL, primary BM-MNC and BM-MSC, compared to age-matched healthy controls. BM-MSC produce large amounts of IL-6 even in unstimulated and undifferentiated conditions [34]. Accordingly, BM-MSC obtained from SDS patients released huge amounts of IL-6 protein into the supernatants (~8 ng/mL, from 2×10^5 cells), that is an amount comparable with the dose of IL-6 commonly used to stimulate in vitro these cells (10 ng/mL). Interestingly, that concentration of IL-6 has been also reported as the driving force leading to a further increase of mTOR-STAT3 activation in SDS leukocytes [26], suggesting the existence of a feedforward autocrine/paracrine feedback loop between STAT3 and IL-6 in SDS bone marrow. In order to verify this hypothesis in clinical samples, we measured peripheral blood plasma levels of IL-6 in a cohort of 21 patients with SDS. We found that IL-6 levels in peripheral blood plasma are elevated in SDS patients. Then, we measured IL-6 released into bone marrow plasma and we found even more elevated cytokine levels. Since BM-MNC do release very low amounts of IL-6 compared to BM-MSC, we speculate that IL-6 accumulation into the bone marrow environment is mainly due to the contribution from the stromal compartment. The MSC compartment is involved in AML development, contributing to disease initiation both in animal models and in patients [42–44]. For instance, studies on BM-MSC obtained from patients with MDS or AML reported altered expression of several cytokines and other soluble pro-inflammatory mediators [45]. Thus, increased interleukin-6 levels in bone marrow of patients with SDS, together with mTOR-STAT3 axis hyper-activation in hematopoietic cells, may clarify the reason why these subjects are prone to develop AML. We previously reported that rapamycin (sirolimus) treatment reduced the phosphorylation of both mTOR (S2448) and STAT3 (Y705 and S727) in cell lines as well as in primary neutrophils, monocytes and B cells isolated from patients with SDS. Here, we show that the clinically approved rapalog, everolimus, can restore normal level of activation of both mTOR and STAT3 in SDS lymphocyte populations. Importantly, in this study we show that everolimus and a commercially available STAT3 chemical inhibitor, namely stattic, can significantly inhibit IL-6 release in SDS LCL and BM-MNC. In the light of these findings, novel clinically approved inhibitors of mTOR and STAT3 might be helpful in reducing pro-leukemic pathways in SDS hematopoietic cells. STAT3 inhibitors are being evaluated as chemotherapeutic agents in leukemias, due to their strong pro-apoptotic activity [13]. Although we used 7.5 μM stattic, which was a previously reported dose that can inhibit STAT3 without affecting lymphoid cell viability in vitro [46], we observed a stattic-dependent induction of late apoptosis in LCL (Figure S3). We cannot exclude the fact that the effect observed in this case may partially be due to pro-apoptotic activity. However, since everolimus reduced both STAT3 activation and *IL-6* expression in SDS LCL and BM-MNC without inducing pro-apoptotic processes (Figure S3), we can assume that mTOR-STAT3 pathway inhibition might be useful in reducing the excessive cytokine release. Both everolimus and stattic did not reduce the huge release of IL-6 from undifferentiated SDS BM-MSC, which remains one of the main sources of IL-6 within the bone marrow compartment. This finding suggests other regulatory pathways of *IL-6* gene expression exist in BM-MSC.

IL-6 can trigger JAK-STAT3 signaling activation through the direct binding to mbIL-6R or via the soluble sIL-6R (IL-6 trans-signaling), as result of alternative splicing or protease cleavage of mbIL-6R [19]. In order to evaluate whether the mTOR-STAT3-IL6 loop is generated by increased IL-6 trans-signaling in SDS, we measured sIL-6R levels in plasma samples obtained from 21 patients. However, results indicate that sIL-6R expression is reduced in SDS patients compared to age-matched healthy donors, thus suggesting that this loop is mainly generated via the classical IL-6 signaling in hematopoietic cells. Interestingly, ex vivo undifferentiated BM-MSC lack the expression of mbIL-6R, even if they normally do express the gp130 protein [33]. Besides, only a few cell types do express mbIL-6R and therefore are able to respond to IL-6 generating the IL-6-STAT3-IL6 loop of activation. Among these cells, major players are represented by macrophages, neutrophils, T-cells, and hepatocytes. On the contrary, gp130 is generally ubiquitously expressed [47,48]. It has been suggested that the lack of mbIL-6R is a regulatory mechanism of BM-MSC that inhibits the IL-6-dependent chondrogenic/osteogenic differentiation, thus maintaining the stemness [33]. During osteogenic differentiation, the expression of IL-6R indeed increases in BM-MSC, allowing the autocrine/paracrine activation of the IL-6-STAT3 signaling pathway [49]. Thus, reduced sIL-6R levels in plasma suggest that the impaired ossification reported in SDS [50] might be partially due to decreased IL-6 trans-signaling in BM-MSC. The lack of IL-6 trans-signaling might therefore also partially explain why everolimus and stattic cannot reduce *IL-6* expression in BM-MSC. Our results indicate that BM-MSC are unable to activate the autocrine/paracrine feedforward loop of mTOR-STAT3-IL6.

4. Materials and Methods

4.1. Human Subjects

Human samples were obtained and analyzed in accordance with the declaration of Helsinki, after written consent. All protocols were approved by the Ethics Committee of San Gerardo Hospital (Monza, Italy), approval No. 504 4/9/2012, Azienda Ospedaliera Universitaria Integrata (Verona, Italy), approval No. 658 CESC, and Azienda Ospedaliero Universitaria Ospedali Riuniti (Ancona, Italy), approval No. CERM 2018-82.

4.2. Plasma Isolation

Bone marrow and peripheral blood specimens were collected into EDTA-containing tubes from patients. Specimens were collected during a bone marrow harvest from healthy donors serving as donors for a related HLA-matched transplant, as permitted by the local hospital ethics committee and after informed consent was obtained. Peripheral blood and bone marrow plasma samples were prepared by centrifugation for 10 min at 600× g at 4 °C. To obtain platelet-poor plasma fractions, another centrifugation was performed for 15 min at 1800× g. Plasma specimens were then stored at −80 °C until analysis.

4.3. Cell Cultures

LCL, PBMC, BM-MNC, and BM-MSC were obtained from peripheral blood or bone marrow samples from31 patients with Shwachman–Diamond Syndrome. All patients enrolled in this study were diagnosed as affected by SDS on the basis of clinical and genetics criteria, and they were excluded if MDS/AML were present (Table 1). None of the patients underwent granulocyte colony-stimulating factor (G-CSF, filgrastim) nor steroids therapies. PBMC and BM-MNC were separated by stratification on Ficoll-Paque PLUS (density 1.077 g/mL, GE Healthcare, Waukesha, WI, USA) gradient and washed twice with PBS (Euroclone, Milan, Italy). Cells were seeded in 6-wells cell culture plates at 1×10^6 BM-MNC/well in 1 mL RPMI-1640 Medium (Gibco, Waltham, MA, USA) supplemented with 10% FBS (Sigma-Aldrich, St. Louis, MO) and incubated in the presence or absence of 350 nM everolimus (Sigma-Aldrich) or 7.5 μM stattic (Selleckchem, Houston, TX, USA) at 37 °C in humidified atmosphere, with 5% CO_2 for 24 h. Treated cell pellets were collected by centrifugation and the supernatant isolated and stored at −80 °C. BM-MSC were isolated after seeding 1.6×10^5 BM-MNC cells/cm^2 from fresh bone marrow, in Dulbecco's Modified Eagle Medium low glucose (Gibco) supplemented with 10% of FBS, 1% L-glutamine and 1% penicillin/streptomycin. BM-MNC were incubated at 37 °C in humidified atmosphere, with 5% CO_2 for 24 h. Subsequently, non-adherent cells were removed by washing with PBS and culture medium was finally replaced.BM-MSC were seeded at 4×10^3 cell/cm^2 and incubated at 37 °C in humidified atmosphere, with 5% CO_2 since they reached 70–80% confluence. 1×10^5 MSC were seeded in a 6-well plate and incubated for 48 h at 37 °C in humidified atmosphere, with 5% CO_2. Eventually, treated cells were collected and the supernatant isolated and stored at −80 °C. To obtain LCL, B cells were seeded at a density of 3×10^6 cells in 12-well cell culture plates, incubated with 3 mL RPMI-1640 supplemented with 10% FBS, and infected for 18 h with Epstein–Barr virus (EBV) released from marmoset blood leukocytes B95.8 virus-producer cell lines as previously described [26]. 1×10^6 LCL were seeded in 6 well-cell culture plate in RPMI-1640 supplemented with 10% FBS and incubated for 48 h at 37 °C in humidified atmosphere, with 5% CO_2. Eventually,1×10^6 LCL were seeded in 6 well-cell culture plate in RPMI-1640 supplemented with 10% FBS, in the presence or absence of 350 nM everolimus or 7.5 μM stattic at 37 °C in humidified atmosphere, with 5% CO_2 for 24 h. Treated cell pellets were collected by centrifugation and the supernatant isolated and stored at −80 °C. LCL, BM-MNC and MSC cultured in medium containing DMSO (Sigma, dilution 1:10,000) were used as negative control.

4.4. Flow Cytometry

Plasma was separated from peripheral blood samples derived from SDS patients or healthy subjects by centrifugation at 1500× *g* for 10 min. Red blood cells were lysed in 40 mL of solution containing 0.89% (*w/v*) NH_4Cl, 0.10% (*w/v*) $KHCO_3$ and 200 μM EDTA. Leukocytes were cultured in 6-well plates containing RPMI-1640 supplemented with 10% freshly prepared, heat-inactivated human plasma, in the presence or absence (DMSO) of 350 nM everolimus for 1h. Fifty to one-hundred-100 microliter aliquots of blood cells were incubated for 15 min at room temperature with combinations of the following fluorochrome-conjugated monoclonal antibodies: CD3-APC750, CD4-PC7, CD8-ECD, CD16-Pacific Blue, CD19-Chrome-Orange, CD45-APC700, and CD56-PC5 (Beckman-Coulter). Cells were gently centrifuged (600× *g*) for 10 min, washed with ice-cold PBS, fixed, and permeabilized with Intracellular Fixation and Permeabilization Buffer Set (eBioscience, San Diego, CA, USA), following the manufacturer's recommendations. Subsequently, cells were washed with flow buffer and stained with anti-pS727-STAT3-PE, anti-Y705-STAT3-PE, anti-p-S2448-mTOR-PE or isotype control-PE conjugated antibodies for 30 min. Data acquisition was performed by a 10 color, three laser Navios flow cytometer (Beckman Coulter, Indianapolis, IN, USA). Analysis of the acquired data was performed by the "Navios" or Kaluza software, version 1.3 (Beckman Coulter, Indianapolis, IN, USA). In order to define different subsets of lymphocytes, gating strategy and data filtering were established as follows. Cell debris were excluded using a dot plot for morphological parameters (FS, SS). Lymphocytes were gated into the side scatter (SS low) and CD45 positive area. Within the lymphocyte compartment, $CD3^+$

events were further gated into $CD4^+$ and $CD8^+$ T cells. NK cells were instead identified as CD3/CD19 double-negative events ($CD56^+$ and/or $CD16^+$). Double-negative T cells were gated by plotting T cells ($CD3^+$) in CD4 versus CD8 dot plots. Since it has been reported that TCR $\gamma\delta^+$T cells are recognizable by the bright expression on their membrane of the CD3 molecule [51], CD3 bright DN T cells (DN $\gamma\delta$) were gated by plotting CD3 bright positive area into CD4 versus CD8 dot plots. Flow cytometry was performed on at least 25,000 events. The total white blood cell count (WBC) was determined by Hematology Analyzer XN- 9000 (Sysmex Europe GmbH, Norderstedt, Germany).

4.5. qRT-PCR

Total RNA from LCL and MNC was isolated using High Pure RNA Isolation Kit (Roche, Mannheim, Germany) following the manufacturer's instructions. RNA was eluted in 50 µL of RNAse free water. Final RNA concentration was determined using NanoDrop 2000 spectrophotometer (Thermo Fisher Scientific, Foster City, CA, USA) and then stored at −80 °C until use. A total amount of 500 ng of RNA was reverse transcribed to cDNA using a High Capacity cDNA Reverse Transcription Kit with random primers (Applied Biosystems by Thermo Fisher Scientific) for 120 min at 37 °C and 5 min at 85 °C in a final reaction volume of 20 µL. For real-time qPCR analysis, 5 Ml of cDNA were used for each reaction to quantify the relative gene expression. The cDNA was then amplified for 45 PCR cycles using Platinum SYBR Green qPCR Super Mix-UDG (Invitrogen by Thermo Fisher Scientific) in a 20 µL reaction using the Rotor-Gene 6000 cycler (Qiagen, Hilden, Germany). In order to perform the PCR reaction, QuantiTect Primer Assays (Qiagen) for IL-6 (Hs_IL6_1_SG, NM_000600), STAT3 (Hs_STAT3_1_SG, NM_003150), mTOR (Hs_MTOR_1_SG, NM_004958), and GAPDH (HS_GAPDH_1_SG, NM_001256799) were purchased. Each target gene expression was normalized with GAPDH gene expression and relative quantification was calculated by the ΔCt method.

4.6. IL-6 and sIL-6R Detection

Bio-Plex Pro Human IL-6 Assay (Bio-Rad Laboratories, Philadelphia, PA, USA) was used to measure IL-6 protein released into cell supernatants and plasma samples (sample volume 50 µL). The assays were performed using the Bio-Plex Suspension Array System, with the Bio-Rad 96-well plate reader. Data were analyzed by the Bio-Plex Manager software (Bio-Rad Laboratories). The in-vitro quantitative determination of soluble IL-6 Receptor (sIL-6R) in plasma was performed by using the Human IL-6R/CD126 ELISA Kit (Origene, Rockville, MD, USA), according to the manufacturer's instructions. Plasma samples were assayed after 1:200 dilution.

4.7. Western Blot

A total of 30 µg of cell extract was denatured for 5 min at 95 °C in 4× Laemmli Sample Buffer, (277.8 mM Tris-HCl, pH 6.8, 44.4% glycerol, 4.4% LDS, 0.02% bromophenol blue) (Bio-Rad Laboratoires) supplemented with 355 mM 2-mercaptoethanol. The samples were loaded on 7.5% polyacrylamide SDS-PAGE gel in Tris-glycine Buffer (25 mM Tris, 192 mM glycine, and 0.1% SDS) using tag protein ladder (Spectra Multicolor Broad Range Protein Ladder, Thermo Scientific, Waltham, MA, USA) to determine molecular weight. The electrotransfer was performed into nitrocellulose membrane (iBlot Gel Transfer Stacks Nitrocellulose, Thermo Fisher) at 20V using iBlot Dry Blotting System (Invitrogen, Waltham, MA, USA) for 10 min. The membranes were blocked in BSA 5% for 90 min at room temperature and washed with TBS (5 mM Tris-HCl pH 7.6, 150 mM NaCl) supplemented with 0.05% tween-20 (TBS/T) for 15 min. Subsequently, membranes were probed with primary anti-human STAT3 rabbit polyclonal antibody (SAB2104912 Sigma-Aldrich, Missouri, USA, dilution 1:500) in primary antibody dilution buffer (TBS/T with 5% BSA) and incubated overnight at 4 °C. After washes, membranes were incubated with mouse anti-rabbit IgG- Horseradish Peroxidase-Coupled secondary antibody (Sigma-Aldrich, dilution 1:2000) diluted TBS/T for 90 min at room temperature. Immunocomplexes were visualized using chemiluminescence (ECL Plus Western Blotting Substrate, Pierce, Thermo Scientific), incubating ECL for 5 min at room temperature. Band intensity was calculated by scanning video densitometry using the Chemi Doc imaging system (UVP, LCC, Upland, CA, USA). Blots were re-probed with monoclonal β-Actin-Peroxidase clone AC-15 antibody (Sigma-Aldrich, dilution1:5000) in TBS/T for 90 min.

4.8. Gene Silencing

Both *mTOR* and *STAT3* genes were knocked-down by siRNA. The lipid-based agent for reverse transfection, siPORT NeoFX (Thermo Fisher), was used according to the manufacturer's instructions. LCL derived from healthy donors and SDS patients were transiently transfected with two different specific siRNA sequences designed against mTOR and STAT3 genes or with a scrambled sequence as control. The siRNA molecules were complexed with cationic liposomes, siPORT Neo-FX (Thermo Fisher). Briefly, siPORT Neo-FX (1 μL/well) was complexed with siRNA or Scrambled oligos (40 nM each) in 250 μL RPMI-1640 medium supplemented with 0.5% FBS. 1×10^5 LCL were incubated in 24-wells plates for 48 h at 37 °C. Knock-down of *mTOR* and *STAT3* gene expression was determined by Real Time qRT-PCR.

4.9. Statistical Analysis

Normal distribution was tested in each experiment using the Shapiro–Wilk test. Based on that evaluation, independent group determination was tested using Mann–Whitney test, while Student's t-test was used in case of paired data. A p-value < 0.05 was considered statistically significant. The statistical software SigmaPlot (Systat Software Inc., San Jose, CA, USA) was used.

5. Conclusions

STAT3 acts as a double-edged sword in SDS cells, as this pathways controls myeloid progenitor growth and proliferation [52] and promotes leukemogenesis. Extending our findings on mTOR-STAT3 signaling dysregulation in myeloid lineage [26], we now show that constitutive activation of mTOR-STAT3 axis occurs in the lymphoid compartment of SDS patients. An autocrine or paracrine feedforward loop of STAT3-IL6 exists in hematopoietic SDS cells. Since the loss of *SBDS* expression in healthy donor-derived cells is sufficient to reproduce the hyper-activation of mTOR-STAT3 signaling [26], we assume that alteration of the mTOR-STAT3-IL6 axis could be used by SDS cells as a survival mechanism that induces cell proliferation and myeloid growth in bone marrow, thus trying to escape from incoming neutropenia/aplasia processes.

Author Contributions: Conceptualization, V.B. (Valentino Bezzerri), M.B. and M.C.; methodology, A.V. and V.B.(Valentino Bezzerri); formal analysis, A.V., E.D., M.B. and V.B. (Valentino Bezzerri); investigation, E.D., M.A. (Martina Api), M.A. (Marisole Allegri), G.B. (Giulia Breveglieri), A.G., E.M.B., T.C. and G.B. (Gloria Bedini).; data curation, V.B.(Valentino Bezzerri),A.V., S.J.C. and M.B.; writing—original draft preparation, V.B. (Valentino Bezzerri); writing—review and editing, S.J.C., M.B. and M.C.; supervision, V.B. (Valentino Bezzerri), M.B., M.C.; project administration, V.B. (Valentino Bezzerri), G.B. (Giulia Breveglieri). and M.B.; resources, B.F., V.B. (Vincenzo Bronte)., M.C., C.S., A.B., A.P., C.B. and G.D.; Funding acquisition, V.B. (Valentino Bezzerri), G.B. (Giulia Breveglieri), M.B and M.C. All authors have read and agreed to the published version of the manuscript.

Acknowledgments: The authors are grateful to Anastasia D'Antuono and Federica Masseria (Azienda Ospedaliero Universitaria Ospedali Riuniti, Ancona) for excellent patient care, to Emily Pintani for SDS Italian Registry data management, to Francesco Pasquali (Università degli Studi dell'Insubria) for cytogenetics of SDS patients, and to Laura Donnini and Silvia Duranti (Azienda Ospedaliero Universitaria Ospedali Riuniti, Ancona) for the admirable administrative support.

References

1. Dror, Y.; Donadieu, J.; Koglmeier, J.; Dodge, J.; Toiviainen-Salo, S.; Makitie, O.; Kerr, E.; Zeidler, C.; Shimamura, A.; Shah, N.; et al. Draft consensus guidelines for diagnosis and treatment of shwachman-diamond syndrome. *Ann. N Y Acad. Sci.* **2011**, *1242*, 40–55. [CrossRef]

2. Kargas, V.; Castro-Hartmann, P.; Escudero-Urquijo, N.; Dent, K.; Hilcenko, C.; Sailer, C.; Zisser, G.; Marques-Carvalho, M.J.; Pellegrino, S.; Wawiórka, L.; et al. Mechanism of completion of peptidyltransferase centre assembly in eukaryotes. *Elife* **2019**, *8*, e44904. [CrossRef] [PubMed]
3. Weis, F.; Giudice, E.; Churcher, M.; Jin, L.; Hilcenko, C.; Wong, C.C.; Traynor, D.; Kay, R.R.; Warren, A.J. Mechanism of eif6 release from the nascent 60s ribosomal subunit. *Nat. Struct. Mol. Biol.* **2015**, *22*, 914–919. [CrossRef]
4. Finch, A.J.; Hilcenko, C.; Basse, N.; Drynan, L.F.; Goyenechea, B.; Menne, T.F.; González Fernández, A.; Simpson, P.; D'Santos, C.S.; Arends, M.J.; et al. Uncoupling of gtp hydrolysis from eif6 release on the ribosome causes shwachman-diamond syndrome. *Genes Dev.* **2011**, *25*, 917–929. [CrossRef] [PubMed]
5. Corey, S.J.; Minden, M.D.; Barber, D.L.; Kantarjian, H.; Wang, J.C.; Schimmer, A.D. Myelodysplastic syndromes: The complexity of stem-cell diseases. *Nat. Rev. Cancer* **2007**, *7*, 118–129. [CrossRef] [PubMed]
6. Donadieu, J.; Leblanc, T.; Bader Meunier, B.; Barkaoui, M.; Fenneteau, O.; Bertrand, Y.; Maier-Redelsperger, M.; Micheau, M.; Stephan, J.L.; Phillipe, N.; et al. Analysis of risk factors for myelodysplasias, leukemias and death from infection among patients with congenital neutropenia. Experience of the french severe chronic neutropenia study group. *Haematologica* **2005**, *90*, 45–53. [PubMed]
7. Lindsley, R.C.; Saber, W.; Mar, B.G.; Redd, R.; Wang, T.; Haagenson, M.D.; Grauman, P.V.; Hu, Z.H.; Spellman, S.R.; Lee, S.J.; et al. Prognostic mutations in myelodysplastic syndrome after stem-cell transplantation. *N. Engl. J. Med.* **2017**, *376*, 536–547. [CrossRef]
8. Curran, E.K.; Godfrey, J.; Kline, J. Mechanisms of immune tolerance in leukemia and lymphoma. *Trends Immunol.* **2017**, *38*, 513–525. [CrossRef]
9. D'Acquisto, F.; Crompton, T. Cd3+cd4-cd8- (double negative) t cells: Saviours or villains of the immune response? *Biochem. Pharmacol.* **2011**, *82*, 333–340. [CrossRef]
10. Chen, B.; Lee, J.B.; Kang, H.; Minden, M.D.; Zhang, L. Targeting chemotherapy-resistant leukemia by combining dnt cellular therapy with conventional chemotherapy. *J. Exp. Clin. Cancer Res.* **2018**, *37*, 88. [CrossRef]
11. Lee, J.; Minden, M.D.; Chen, W.C.; Streck, E.; Chen, B.; Kang, H.; Arruda, A.; Ly, D.; Der, S.D.; Kang, S.; et al. Allogeneic human double negative t cells as a novel immunotherapy for acute myeloid leukemia and its underlying mechanisms. *Clin. Cancer Res.* **2018**, *24*, 370–382. [CrossRef] [PubMed]
12. Zhang, L.; Chen, X.; Liu, X.; Kline, D.E.; Teague, R.M.; Gajewski, T.F.; Kline, J. Cd40 ligation reverses t cell tolerance in acute myeloid leukemia. *J. Clin. Investig.* **2013**, *123*, 1999–2010. [CrossRef] [PubMed]
13. O'Shea, J.J.; Holland, S.M.; Staudt, L.M. Jaks and stats in immunity, immunodeficiency, and cancer. *N. Engl. J. Med.* **2013**, *368*, 161–170. [CrossRef] [PubMed]
14. Johnson, D.E.; O'Keefe, R.A.; Grandis, J.R. Targeting the il-6/jak/stat3 signalling axis in cancer. *Nat. Rev. Clin. Oncol.* **2018**, *15*, 234–248. [CrossRef] [PubMed]
15. Yu, H.; Kortylewski, M.; Pardoll, D. Crosstalk between cancer and immune cells: Role of stat3 in the tumour microenvironment. *Nat. Rev. Immunol.* **2007**, *7*, 41–51. [CrossRef] [PubMed]
16. Herrmann, A.; Kortylewski, M.; Kujawski, M.; Zhang, C.; Reckamp, K.; Armstrong, B.; Wang, L.; Kowolik, C.; Deng, J.; Figlin, R.; et al. Targeting stat3 in the myeloid compartment drastically improves the in vivo antitumor functions of adoptively transferred t cells. *Cancer Res.* **2010**, *70*, 7455–7464. [CrossRef]
17. Kujawski, M.; Zhang, C.; Herrmann, A.; Reckamp, K.; Scuto, A.; Jensen, M.; Deng, J.; Forman, S.; Figlin, R.; Yu, H. Targeting stat3 in adoptively transferred t cells promotes their in vivo expansion and antitumor effects. *Cancer Res.* **2010**, *70*, 9599–9610. [CrossRef]
18. Kumari, N.; Dwarakanath, B.S.; Das, A.; Bhatt, A.N. Role of interleukin-6 in cancer progression and therapeutic resistance. *Tumour Biol.* **2016**, *37*, 11553–11572. [CrossRef]
19. Chalaris, A.; Garbers, C.; Rabe, B.; Rose-John, S.; Scheller, J. The soluble interleukin 6 receptor: Generation and role in inflammation and cancer. *Eur. J. Cell Biol.* **2011**, *90*, 484–494. [CrossRef]
20. Audet, J.; Miller, C.L.; Rose-John, S.; Piret, J.M.; Eaves, C.J. Distinct role of gp130 activation in promoting self-renewal divisions by mitogenically stimulated murine hematopoietic stem cells. *Proc. Natl. Acad. Sci. USA* **2001**, *98*, 1757–1762. [CrossRef]
21. Walker, F.; Zhang, H.H.; Matthews, V.; Weinstock, J.; Nice, E.C.; Ernst, M.; Rose-John, S.; Burgess, A.W. Il6/sil6r complex contributes to emergency granulopoietic responses in g-csf- and gm-csf-deficient mice. *Blood* **2008**, *111*, 3978–3985. [CrossRef] [PubMed]

22. Han, Y.; Ye, A.; Bi, L.; Wu, J.; Yu, K.; Zhang, S. Th17 cells and interleukin-17 increase with poor prognosis in patients with acute myeloid leukemia. *Cancer Sci.* **2014**, *105*, 933–942. [CrossRef] [PubMed]
23. Sanchez-Correa, B.; Bergua, J.M.; Campos, C.; Gayoso, I.; Arcos, M.J.; Bañas, H.; Morgado, S.; Casado, J.G.; Solana, R.; Tarazona, R. Cytokine profiles in acute myeloid leukemia patients at diagnosis: Survival is inversely correlated with il-6 and directly correlated with il-10 levels. *Cytokine* **2013**, *61*, 885–891. [CrossRef] [PubMed]
24. Yokogami, K.; Wakisaka, S.; Avruch, J.; Reeves, S.A. Serine phosphorylation and maximal activation of stat3 during cntf signaling is mediated by the rapamycin target mtor. *Curr. Biol.* **2000**, *10*, 47–50. [CrossRef]
25. Dodd, K.M.; Yang, J.; Shen, M.H.; Sampson, J.R.; Tee, A.R. Mtorc1 drives hif-1α and vegf-a signalling via multiple mechanisms involving 4e-bp1, s6k1 and stat3. *Oncogene* **2015**, *34*, 2239–2250. [CrossRef]
26. Bezzerri, V.; Vella, A.; Calcaterra, E.; Finotti, A.; Gasparello, J.; Gambari, R.; Assael, B.M.; Cipolli, M.; Sorio, C. New insights into the shwachman-diamond syndrome-related haematological disorder: Hyper-activation of mtor and stat3 in leukocytes. *Sci. Rep.* **2016**, *6*, 33165. [CrossRef]
27. Chapuis, N.; Tamburini, J.; Green, A.S.; Willems, L.; Bardet, V.; Park, S.; Lacombe, C.; Mayeux, P.; Bouscary, D. Perspectives on inhibiting mtor as a future treatment strategy for hematological malignancies. *Leukemia* **2010**, *24*, 1686–1699. [CrossRef]
28. Hoshii, T.; Matsuda, S.; Hirao, A. Pleiotropic roles of mtor complexes in haemato-lymphopoiesis and leukemogenesis. *J. Biochem.* **2014**, *156*, 73–83. [CrossRef]
29. Porta, C.; Paglino, C.; Mosca, A. Targeting pi3k/akt/mtor signaling in cancer. *Front. Oncol.* **2014**, *4*, 64. [CrossRef]
30. Witzig, T.E.; Reeder, C.; Han, J.J.; LaPlant, B.; Stenson, M.; Tun, H.W.; Macon, W.; Ansell, S.M.; Habermann, T.M.; Inwards, D.J.; et al. The mtorc1 inhibitor everolimus has antitumor activity in vitro and produces tumor responses in patients with relapsed t-cell lymphoma. *Blood* **2015**, *126*, 328–335. [CrossRef]
31. Cai, Y.; Xue, F.; Qin, H.; Chen, X.; Liu, N.; Fleming, C.; Hu, X.; Zhang, H.G.; Chen, F.; Zheng, J.; et al. Differential roles of the mtor-stat3 signaling in dermal $\gamma\delta$ t cell effector function in skin inflammation. *Cell Rep.* **2019**, *27*, 3034–3048. [CrossRef] [PubMed]
32. Rose-John, S. Il-6 trans-signaling via the soluble il-6 receptor: Importance for the pro-inflammatory activities of il-6. *Int. J. Biol. Sci.* **2012**, *8*, 1237–1247. [CrossRef] [PubMed]
33. Erices, A.; Conget, P.; Rojas, C.; Minguell, J.J. Gp130 activation by soluble interleukin-6 receptor/interleukin-6 enhances osteoblastic differentiation of human bone marrow-derived mesenchymal stem cells. *Exp. Cell Res.* **2002**, *280*, 24–32. [CrossRef] [PubMed]
34. Pricola, K.L.; Kuhn, N.Z.; Haleem-Smith, H.; Song, Y.; Tuan, R.S. Interleukin-6 maintains bone marrow-derived mesenchymal stem cell stemness by an erk1/2-dependent mechanism. *J. Cell Biochem.* **2009**, *108*, 577–588. [CrossRef]
35. Calabrese, L.H.; Rose-John, S. Il-6 biology: Implications for clinical targeting in rheumatic disease. *Nat. Rev. Rheumatol.* **2014**, *10*, 720–727. [CrossRef]
36. Altman, J.K.; Sassano, A.; Platanias, L.C. Targeting mtor for the treatment of aml. New agents and new directions. *Oncotarget* **2011**, *2*, 510–517. [CrossRef]
37. Dinner, S.; Platanias, L.C. Targeting the mtor pathway in leukemia. *J. Cell Biochem.* **2016**, *117*, 1745–1752. [CrossRef]
38. Bezzerri, V.; Vella, A.; Gennaro, G.D.; Ortolani, R.; Nicolis, E.; Cesaro, S.; Fabrizzi, B.; Bronte, V.; Corey, S.J.; Cipolli, M. Peripheral blood immunophenotyping in a large cohort of patients with shwachman-diamond syndrome. *Pediatr. Blood Cancer* **2019**, *66*, e27597. [CrossRef]
39. Durant, L.; Watford, W.T.; Ramos, H.L.; Laurence, A.; Vahedi, G.; Wei, L.; Takahashi, H.; Sun, H.W.; Kanno, Y.; Powrie, F.; et al. Diverse targets of the transcription factor stat3 contribute to t cell pathogenicity and homeostasis. *Immunity* **2010**, *32*, 605–615. [CrossRef]
40. Hossain, D.M.; Dos Santos, C.; Zhang, Q.; Kozlowska, A.; Liu, H.; Gao, C.; Moreira, D.; Swiderski, P.; Jozwiak, A.; Kline, J.; et al. Leukemia cell-targeted stat3 silencing and tlr9 triggering generate systemic antitumor immunity. *Blood* **2014**, *123*, 15–25. [CrossRef]
41. Dror, Y.; Ginzberg, H.; Dalal, I.; Cherepanov, V.; Downey, G.; Durie, P.; Roifman, C.M.; Freedman, M.H. Immune function in patients with shwachman-diamond syndrome. *Br. J. Haematol.* **2001**, *114*, 712–717. [CrossRef]

42. Raaijmakers, M.H.; Mukherjee, S.; Guo, S.; Zhang, S.; Kobayashi, T.; Schoonmaker, J.A.; Ebert, B.L.; Al-Shahrour, F.; Hasserjian, R.P.; Scadden, E.O.; et al. Bone progenitor dysfunction induces myelodysplasia and secondary leukaemia. *Nature* **2010**, *464*, 852–857. [CrossRef]
43. Santamaría, C.; Muntión, S.; Rosón, B.; Blanco, B.; López-Villar, O.; Carrancio, S.; Sánchez-Guijo, F.M.; Díez-Campelo, M.; Alvarez-Fernández, S.; Sarasquete, M.E.; et al. Impaired expression of dicer, drosha, sbds and some micrornas in mesenchymal stromal cells from myelodysplastic syndrome patients. *Haematologica* **2012**, *97*, 1218–1224. [CrossRef] [PubMed]
44. Geyh, S.; Oz, S.; Cadeddu, R.P.; Fröbel, J.; Brückner, B.; Kündgen, A.; Fenk, R.; Bruns, I.; Zilkens, C.; Hermsen, D.; et al. Insufficient stromal support in mds results from molecular and functional deficits of mesenchymal stromal cells. *Leukemia* **2013**, *27*, 1841–1851. [CrossRef]
45. Schroeder, T.; Geyh, S.; Germing, U.; Haas, R. Mesenchymal stromal cells in myeloid malignancies. *Blood Res.* **2016**, *51*, 225–232. [CrossRef] [PubMed]
46. Severin, F.; Frezzato, F.; Visentin, A.; Martini, V.; Trimarco, V.; Carraro, S.; Tibaldi, E.; Brunati, A.M.; Piazza, F.; Semenzato, G.; et al. In chronic lymphocytic leukemia the jak2/stat3 pathway is constitutively activated and its inhibition leads to cll cell death unaffected by the protective bone marrow microenvironment. *Cancers* **2019**, *11*, 1939. [CrossRef] [PubMed]
47. Oberg, H.H.; Wesch, D.; Grüssel, S.; Rose-John, S.; Kabelitz, D. Differential expression of cd126 and cd130 mediates different stat-3 phosphorylation in cd4+cd25- and cd25high regulatory t cells. *Int. Immunol.* **2006**, *18*, 555–563. [CrossRef]
48. Scheller, J.; Rose-John, S. Interleukin-6 and its receptor: From bench to bedside. *Med. Microbiol. Immunol.* **2006**, *195*, 173–183. [CrossRef]
49. Kondo, M.; Yamaoka, K.; Sakata, K.; Sonomoto, K.; Lin, L.; Nakano, K.; Tanaka, Y. Contribution of the interleukin-6/stat-3 signaling pathway to chondrogenic differentiation of human mesenchymal stem cells. *Arthritis Rheumatol.* **2015**, *67*, 1250–1260. [CrossRef]
50. Bezzerri, V.; Cipolli, M. Shwachman-diamond syndrome: Molecular mechanisms and current perspectives. *Mol. Diagn. Ther.* **2018**, *23*, 281–290. [CrossRef]
51. Lambert, C.; Genin, C. Cd3 bright lymphocyte population reveal gammadelta t cells. *Cytom. B Clin. Cytom.* **2004**, *61*, 45–53. [CrossRef] [PubMed]
52. Zhang, H.; Nguyen-Jackson, H.; Panopoulos, A.D.; Li, H.S.; Murray, P.J.; Watowich, S.S. Stat3 controls myeloid progenitor growth during emergency granulopoiesis. *Blood* **2010**, *116*, 2462–2471. [CrossRef] [PubMed]

10

Pharmacological Inhibition of Oncogenic STAT3 and STAT5 Signaling in Hematopoietic Cancers

Marie Brachet-Botineau [1], Marion Polomski [2], Heidi A. Neubauer [3], Ludovic Juen [2], Damien Hédou [2], Marie-Claude Viaud-Massuard [2], Gildas Prié [2] and Fabrice Gouilleux [1,*]

1 Leukemic Niche and Oxidative metabolism (LNOx), CNRS ERL 7001, University of Tours, 37000 Tours, France; marie.brachet@univ-tours.fr
2 Innovation Moléculaire et Thérapeutique (IMT), EA 7501, University of Tours, 37000 Tours, France; marion.polomski@etu.univ-tours.fr (M.P.); ludovic.juen@mcsaf.fr (L.J.); damien.hedou@univ-tours.fr (D.H.); marie-claude.viaud-massuard@univ-tours.fr (M.-C.V.-M.); gildas.prie@univ-tours.fr (G.P.)
3 Institute of Animal Breeding and Genetics, University of Veterinary Medicine Vienna, A-1210 Vienna, Austria; heidi.neubauer@vetmeduni.ac.at
* Correspondence: fabrice.gouilleux@univ-tours.fr

Abstract: Signal Transducer and Activator of Transcription (STAT) 3 and 5 are important effectors of cellular transformation, and aberrant STAT3 and STAT5 signaling have been demonstrated in hematopoietic cancers. STAT3 and STAT5 are common targets for different tyrosine kinase oncogenes (TKOs). In addition, STAT3 and STAT5 proteins were shown to contain activating mutations in some rare but aggressive leukemias/lymphomas. Both proteins also contribute to drug resistance in hematopoietic malignancies and are now well recognized as major targets in cancer treatment. The development of inhibitors targeting STAT3 and STAT5 has been the subject of intense investigations during the last decade. This review summarizes the current knowledge of oncogenic STAT3 and STAT5 functions in hematopoietic cancers as well as advances in preclinical and clinical development of pharmacological inhibitors.

Keywords: STAT3; STAT5; hematopoietic cancers; therapeutic targeting; pharmacological inhibitors

1. Introduction

Signal Transducer and Activator of Transcription (STAT) proteins are a seven-member family of cytoplasmic transcription factors that relay signals emanating from cell-surface cytokine and growth factor receptors to the nucleus [1,2]. STAT proteins control fundamental cellular processes, including survival, proliferation, differentiation, and immune responses [3]. It is now well-established that three of these members, STAT3 and the closely related STAT5A and STAT5B proteins (Figure 1) are also important effectors of cellular transformation. Aberrant STAT3, STAT5A, and STAT5B signaling have been described in different solid tumors such as prostate, breast, colon, gliomas, head and neck cancer, melanoma, and in hematopoietic malignancies [4–7] (see also [8] in this issue). Historically, persistent activation of these transcription factors was frequently found in many tumor cells as a consequence of deregulated tyrosine kinase activity. STAT5A/5B and/or STAT3 are downstream effectors of various tyrosine kinase oncogenes (TKOs) such as TEL-JAK2, JAK2^{V617F}, SRC, TEL-ABL, BCR-ABL, TEL-SYK, NPM-ALK, TEL-PDGFR, and mutated forms of FLT3 and KIT receptors [9–19].

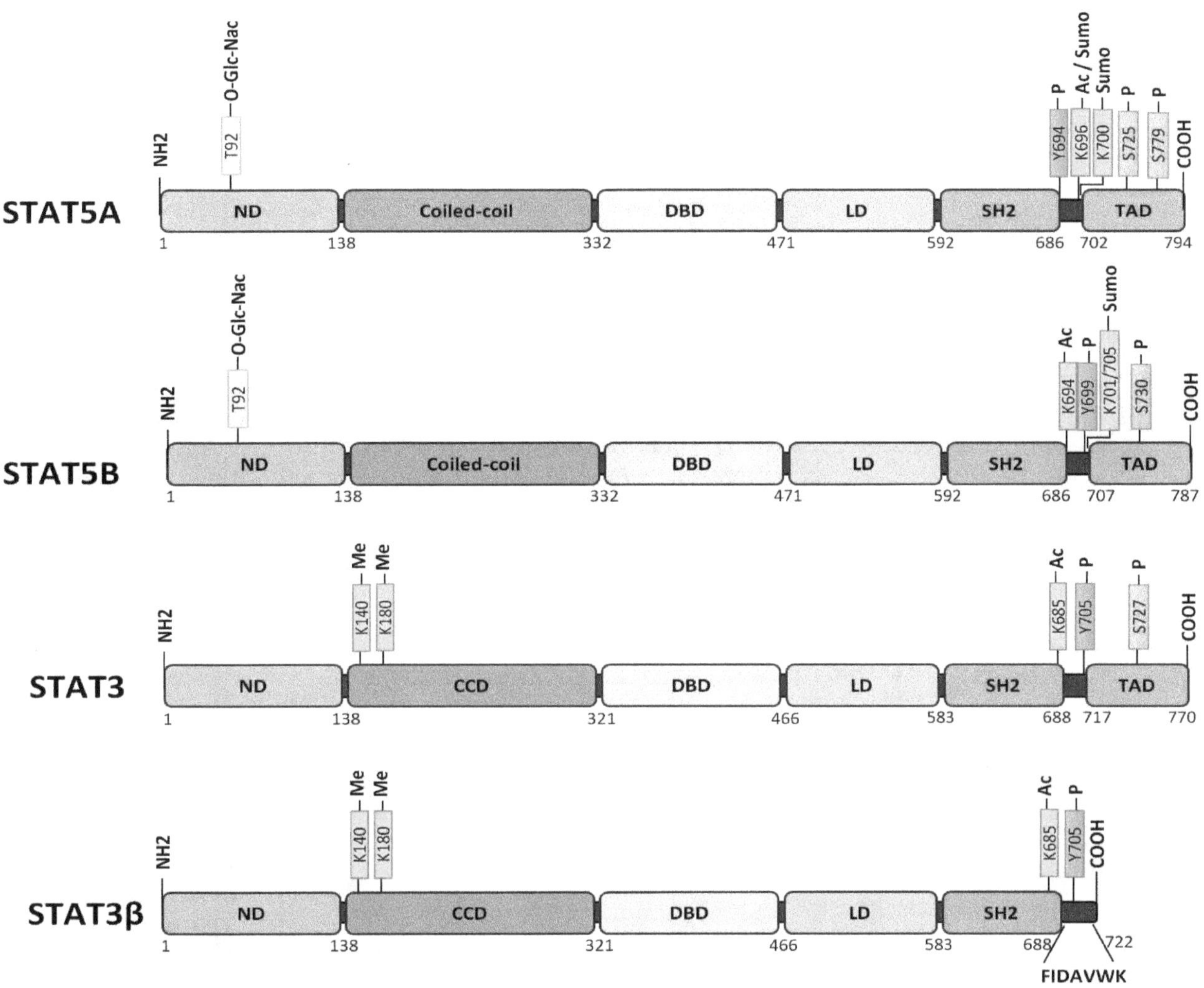

Figure 1. Structure of Signal Transducer and Activator of Transcription (STAT)5A, STAT5B, STAT3, and the spliced isoform STAT3β proteins. Functional domains: ND, NH2-terminal domain; CCD, coiled-coil domain; DBD, DNA binding domain; LD, linker domain; SH2, Src homology 2 domain; TAD, transactivation domain. Post-translational modifications: Ac, acetylation; Me, methylation; O-Glc-Nac, O-GlcNacylation; P, phosphorylation; Sumo, sumoylation.

Inhibition of STAT3/STAT5 signaling has a negative impact on the transforming potential of these tyrosine kinases in vitro and in vivo. Evidence for a direct role of STAT3 or STAT5A/5B in cell transformation was provided by the use of constitutively active variants. These proteins, designated STAT3C, STAT51*6 or $cS5^F$, are able to induce cell transformation in vitro and solid tumors or leukemias in vivo [20–23]. More recently, gain-of-function (GOF) mutations in STAT5B and STAT3 have been found in patients with leukemias and lymphomas (Figure 2) [24–27]. Occurring primarily within the SH2 domain, these mutations confer persistent and prolonged signaling and have been linked to poorer prognosis and relapse in patients [28–30]. Collectively, these data would undoubtedly define STAT3 and STAT5A/5B as important therapeutic targets in hematologic cancers. Nevertheless, STAT3 and STAT5 also behave as tumor suppressors in other tissues and regulate the antitumoral response of immune cells [31–37]. Thus, the respective and specific functions of STAT3 and STAT5, as well as their interactions in hematopoietic cancers, still need to be refined to develop therapeutic strategies that selectively block STAT3 and/or STAT5 activity in these diseases. In this review, we will first summarize the respective contribution of STAT3 and STAT5 in hematologic cancers as well as the canonical and non-canonical oncogenic properties of STAT3/STAT5. Finally, we will describe the different strategies used to target STAT3 or STAT5 and will discuss the potential future development of single or combined therapies to block STAT3 and/or STAT5A/5B activity/expression in hematologic cancers.

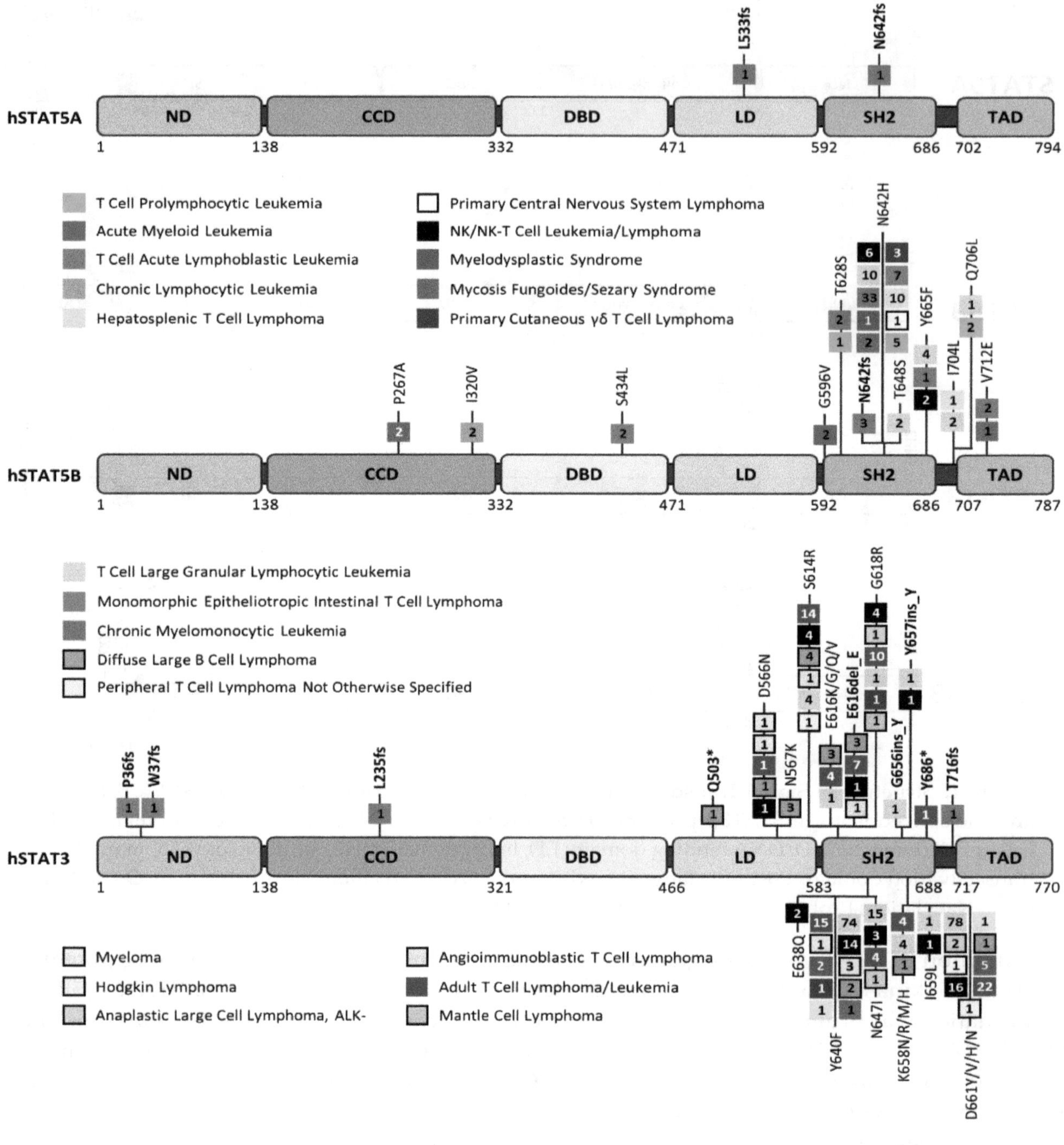

Figure 2. Map of somatic mutations detected in human STAT5A, STAT5B, and STAT3 in patients with hematologic cancers. Individual missense mutations found in at least two patients, as well as all reported nonsense (*) and frameshift (fs) mutations (bold), are depicted. The numbers in each box represent the number of cases reported for each mutation. Data were mined from the Catalogue of Somatic Mutations in Cancer (COSMIC) database. ND, NH2-terminal domain; CCD, coiled-coil domain; DBD, DNA binding domain; LD, linker domain; SH2, Src homology 2 domain; TAD, transactivation domain.

2. STAT3 and STAT5A/5B in Hematopoietic Cancers

2.1. STAT3/STAT5 in Hematopoietic Cancers: An Amazing 23-Year-Old Story

In 1996, pioneering works demonstrated that STAT3 and/or STAT5 are constitutively activated in leukemic cells from patients with acute myeloid leukemia (AML) or acute lymphoblastic leukemia (ALL) [38]. Surprising was the high constitutive STAT5 DNA binding activity detected in leukemic cells from Philadelphia chromosome-positive ALL (Ph^+ ALL) patients. These original findings already pointed to STAT5 as a potential effector of the BCR-ABL tyrosine kinase fusion protein, the transforming agent in Ph^+ ALL and chronic myeloid leukemia (CML), and were confirmed a few months later by other groups [15,16]. 23 years later with more than 4000 publications in the field, it is now clearly established that STAT3/STAT5A/STAT5B are essential and/or contribute to the development of hematopoietic malignancies, affecting both myeloid and lymphoid compartments (Table 1). Data also showed that deregulated STAT3 and STAT5 activity promotes drug resistance in leukemias/lymphomas/myelomas, highlighting the crucial interest to develop pharmacological molecules that selectively target STAT3 and/or STAT5 in hematologic cancers.

Table 1. STAT3 and STAT5 in hematologic cancers.

Hematopoietic Lineage	Hematologic Malignancies		Contribution of STAT3 and/or STAT5A/5B		
Myeloid compartment	Ph^+ MPN	**CML (BCR-ABL)**	STAT3	STAT5A	STAT5B
	Ph^- MPN	PV $JAK2^{V617F}$	STAT3	STAT5	
		ET $JAK2^{V617F}$			
		PMF $JAK2^{V617F}$			
		SM KIT^{D816V}			
	AML	Flt3-ITD		STAT5	
		CBF-AML Kit^{D816V}	STAT3	STAT5	
		APL	STAT3	STAT5	
		t (15;17)			
Lymphoid compartment	ALL	Pre-B-ALL	STAT3	STAT5	
		B-ALL			
		T-ALL	STAT3	STAT5A	STAT5B
	T-LGL Leukemias		STAT3	STAT5A	STAT5B
	CLL		STAT3		
	Lymphomas	HL	STAT3	STAT5	
		DLBCL	STAT3	STAT5	
		ALCL	STAT3		STAT5B
		γδ-T cell Lymphomas	STAT3		STAT5B
		NK-T cell Lymphomas	STAT3		STAT5B
	Multiple Myelomas		STAT3		

Changes in font size and bold text refer to the level of contribution of each STAT protein in the disease. ALCL (anaplastic large cell lymphoma), AML (acute myeloid leukemia), APL (acute promyelocytic leukemia), B- or T-ALL (B- or T- acute lymphoblastic leukemia), CBF-AML (core binding factor-acute myeloid leukemia), CLL (chronic lymphocytic leukemia), CML (chronic myeloid leukemia), DLBCL (diffuse large B cell lymphoma), ET (essential thrombocythemia), Flt3-ITD (FMS-like tyrosine kinase 3-internal tandem duplication), HL (Hodgkin lymphoma), Ph^+MPN and Ph^-MPN (Philadelphia chromosome-positive and Philadelphia chromosome-negative myeloproliferative neoplasm), NK (natural killer cell), PMF (primary myelofibrosis), PV (polycythemia vera), SM (systemic mastocytosis), and T-LGL (T cell large granular lymphocytic) leukemia.

2.2. STAT3/5 in Myeloproliferative Neoplasms (MPNs)

Myeloproliferative neoplasms (MPNs) are hematologic diseases characterized by abnormal proliferation and accumulation of mature myeloid cells in the bone marrow and peripheral blood [39]. An increased risk of developing acute myeloid leukemia is also associated with MPNs. MPNs are classified as *BCR-ABL*-positive (or Ph^+) and *BCR-ABL*-negative (Ph^-) MPNs [40]. Both Ph^+ and Ph^- MPNs are clonal disorders that result from the transformation of hematopoietic stem cells (HSCs). While the BCR-ABL fusion protein is the transforming agent in CML, driver mutations in *JAK2*, *CALR*, and *MPL* genes are variably present and are mostly mutually exclusive in Ph^-MPNs, which include essential thrombocythemia (ET), polycythemia vera (PV), and primary myelofibrosis (MF) [41]. The JAK2 GOF mutation ($JAK2^{V617F}$) has been identified in 95% to 97% of PV patients [42,43]. This mutation, located in the pseudokinase domain of the JAK2 protein, constitutively activates the kinase. JAK2, MPL, and CALR mutants have been functionally validated and are sufficient to induce MPNs in mice [41]. Systemic mastocytosis (SM), a subcategory of MPNs, is a heterogeneous clonal disorder characterized by an accumulation of mast cells in various organs [44]. The GOF mutation in KIT (KIT^{D816V}) causing activation of the KIT receptor tyrosine kinase was found in 80–95% of patients with SM. Studies with transgenic mice suggested that this mutation alone is sufficient to cause SM [45]. The KIT^{D816V} mutant has also been detected in leukemic cells from AML patients [46]. The presence of KIT^{D816V} in AML is highly associated with co-existing SM [47]. Activation of STAT3 and/or STAT5 by BCR-ABL, $JAK2^{V617F}$, and KIT^{D816V} has been abundantly documented in the literature. However, conflicting results (cell lines vs. primary cells and/or human vs. murine leukemic cells) have emerged from these studies. For instance, tyrosine phosphorylation of STAT3 (Y^{705}) was observed in murine BCR-ABL^+ cells but barely detected in human BCR-ABL^+ cells [16,48]. Using *stat3*- or *stat5a/5b*-deficient mice, previous studies demonstrated that while STAT3 and STAT5 are required for the initial step of BCR-ABL-dependent cell transformation, only STAT5 is necessary for the maintenance of BCR-ABL-induced leukemia [49]. The effective role of STAT3 and STAT5 in maintenance, self-renewal, or transformation of normal HSCs might explain why both proteins are required in the BCR-ABL-dependent leukemia-initiating stem cell population [50–53]. More recently, dissection of the respective contributions of STAT5A and STAT5B in BCR-ABL-dependent transformation revealed that STAT5B, but not STAT5A, is a critical effector of BCR-ABL-driven leukemia development [54]. Besides survival- and growth-promoting effects, STAT5B facilitates leukemogenesis by suppressing IFN-α and IFN-γ signaling in these animal models. The development of tyrosine kinase inhibitors (TKIs) targeting BCR-ABL, such as imatinib mesylate (IM), has revolutionized the treatment of CML. However, imatinib mesylate (IM) is not totally curative and approximately 50% of patients remain therapy-free after IM discontinuation. The inability of IM to completely eradicate quiescent leukemic stem cells (LSCs) is probably responsible for the relapse of CML patients [55]. Moreover, the occurrence of BCR-ABL mutations in progressive or relapsed diseases promotes IM resistance of CML cells [56]. Studies indicated that high levels of phosphorylated STAT5 enhance the resistance of CML cells to TKIs but also triggers BCR-ABL mutations by inducing the production of reactive oxygen species (ROS) responsible for DNA damage [57–59]. Moreover, STAT5 was shown to play a key role in the maintenance of IM-resistant LSCs from CML patients [60,61].

The contribution of STAT3 in IM resistance was also demonstrated using an in vitro model of cocultures mimicking the bone marrow microenvironment of CML cells. In this model, bone marrow stromal cells stimulated the phosphorylation of STAT3 in CML cells in a BCR-ABL-independent manner and promoted the resistance of these leukemic cells to IM [62]. Interestingly, combined targeting of STAT3 and STAT5 has been proposed to overcome drug resistance in CML cells [63].

While STAT5 plays a critical role in $JAK2^{V617F}$-driven mouse models of MPN, studies have shown that STAT3 is not required for myeloid expansion induced by this TKO [64–67]. Moreover, activation of STAT3 negatively regulates $JAK2^{V617F}$-driven MPN in mice by enhancing thrombocytosis and shortening overall survival [67]. Further studies are therefore required to determine if STAT3 plays a pathogenic role in human MPNs.

STAT3, STAT5, and also STAT1 are activated by the KIT^{D816V} mutant in transformed mast cells, but only STAT5 seems to be transcriptionally active in these cells [68]. Phosphorylation of STAT5 has been detected in mast cells from SM patients. shRNA-mediated knockdown, dominant-negative mutant and pharmacological molecules targeting STAT5 all abrogate the growth of human neoplastic mast cells in vitro, indicating that STAT5 is a critical effector of KIT^{D816V} in human neoplastic mast cells [12].

Collectively, all data indicate that STAT5A/5B proteins are particularly relevant therapeutic targets in Ph^+ MPN and Ph^- MPN and in the resistance to TKIs. Although the pathogenic role of STAT3 in human MPN still remains questionable, the contribution of STAT3 to TKI resistance elicited by the leukemic microenvironment would suggest that combination therapy or dual molecules targeting STAT3 and STAT5 might help to eradicate resistant leukemic cells in their "niche."

2.3. STAT3/5 in Acute Myeloid Leukemia (AML)

AML is a heterogeneous clonal disorder characterized by immature myeloid cell proliferation and bone marrow failure. A two-hit model has been suggested as the probable mechanism in the pathogenesis of AML [69]. In this model, gene mutations that give a growth advantage and block normal hematopoietic differentiation are responsible for AML development. For instance, activating mutations in FLT3 (FMS-related tyrosine kinase 3) and KIT receptors promote proliferation and survival, while mutations affecting the transcription factor CEBPα inhibit myeloid differentiation. However, there are several other gene classes such as those involved in epigenetic regulation or metabolism that are mutated in AML [70].

Analyses of primary peripheral blood and bone marrow specimens have demonstrated constitutive activation of STAT3 and/or STAT5 in AML [6,37]. Importantly, constitutive activation of STAT3 and STAT5 has been linked with disease outcomes in AML. Bone marrow evaluation of AML patients revealed that the activation of STAT3 was significantly associated with poor overall survival and reduced progression-free survival [71]. In sharp contrast, the spliced STAT3β isoform was shown to have a suppressor function in AML [33]. A higher STAT3β/α mRNA ratio was found in AML cells and correlated with a favorable prognosis and increased overall survival. Stat3β expression in mouse models of AML resulted in decelerated disease progression and extended survival. It is, however, unclear whether tyrosine phosphorylation and dimerization are required for the tumor suppressor activity of STAT3β in these animal models. The contribution of STAT3 in AML may not only depend on the STAT3β/α ratio but also on the subtype of AML cells or acquired mutations in this disease. For instance, KIT^{D816V}, which is frequently found in core-binding factor (CBF)-AML leukemias, stimulates autophagy through activation of STAT3 [72]. Inhibition of STAT3 blocked autophagy and reduced tumor growth in mouse xenograft models. Importantly, inhibition of STAT3 also stimulates the antitumoral immune response in animal models of AML, indicating that targeting STAT3 would not only block the growth and survival of AML but also AML-induced immune evasion [73,74].

Mutations in FLT3, either involving internal tandem duplications (FLT3-ITD) or point mutations in the activating loop of the tyrosine kinase domain (FLT3-KD), were observed in approximately 30% of AML patients and are associated with poor prognosis [75]. Although FLT3 mutants activate both STAT3 and STAT5 in cell lines, results showed that tyrosine-phosphorylated STAT5 is selectively associated with expression of FLT3-ITD in primary blasts from AML patients. Activation of STAT5 by FLT3-ITD is required to induce primary cell survival in vitro and leukemia in vivo [76–80]. Pharmacological inhibition of STAT5 also blocks FLT3-ITD-driven leukemias in mouse xenograft models [81]. FLT3-ITD promotes genomic instability by increasing ROS production via activation and association of STAT5 with the GTPase Rac1, which is an essential component of certain NADPH oxidases such as Nox2 [82]. Conversely, ROS production and p22phox, a membrane subunit of NADPH oxidase, are required for FLT3-ITD-induced STAT5 phosphorylation in the endoplasmic reticulum (ER) [83]. Importantly, recent works demonstrated that FLT3-ITD-independent activation of STAT5 induced by the leukemic microenvironment promotes resistance of $FLT3\text{-}ITD^+$ AML cells to quizartinib, an FLT3 inhibitor that is now in phase 3 clinical trial [84].

Lastly, we should also mention the identification of a fusion between *STAT5B* and *Retinoic Acid Receptor (RAR)α* resulting from an interstitial deletion on chromosome 17 in acute promyelocytic leukemia (APL) [85]. The corresponding fusion protein enhances STAT3 signaling and blocks myeloid maturation by inhibiting RARα/retinoid X receptor (RXR)α transcriptional activity [86].

2.4. *STAT3/5 in Acute Lymphoblastic Leukemia (ALL)*

ALL is the most common form of cancer in children and predominantly arises from the transformation of B cell progenitors (80–85% of cases) [87]. Mouse studies suggest that STAT5 is functionally important in certain types of B-ALL [88]. Transgenic overexpression of a constitutively active STAT5A mutant ($cS5^F$) cooperates with p53 deficiency to promote B-ALL in mice [89]. Genetic or pharmacological targeting of STAT5 suppresses human Ph+ ALL cell growth and leukemia development in mouse xenograft models [90]. Deregulation of precursor B cell antigen receptor (pre-BCR) signaling has been shown to be important in the development of B-ALL, and constitutive activation of STAT5B cooperates with defects in pre-BCR signaling components to initiate B-ALL [91]. Similarly, haploinsufficiency of B cell-specific transcription factors such as EBF1 or PAX5 synergizes with activated STAT5 in ALL [92]. Despite strong evidence for the oncogenic activity of STAT5 in TKO-driven B-ALL, the role of STAT5 appears to be context-dependent. For example, the deletion of STAT5 accelerates the development of B-ALL induced by c-myc in mouse models [93]. Activating mutations in *STAT5B* have been found in T-ALL [24,28]. The amino acid substitution N642H in the phosphotyrosine binding pocket of the SH2 domain promotes the constitutive activation of STAT5B and the capacity to induce T cell neoplasia in transgenic mice [29,30]. The role of STAT3 in ALL is poorly documented. However, data indicated that blockade of STAT3 signaling compromises the growth of B-ALL cells overexpressing the high mobility group A1 (HMGA1)-STAT3 pathway [94]. Unlike STAT5B, there are no recurrent STAT3 mutations detected in T-ALL and, in fact, only single frameshift mutations are reported (Figure 2).

2.5. *STAT3/5 in T Cell Large Granular Lymphocytic (T-LGL) Leukemia*

Activating mutations in the SH2 domain of STAT3 (Y640F, D661Y/V) and STAT5B (N642H) were also described in T-LGL leukemia which is a chronic lymphoproliferative disorder characterized by the expansion of some cytotoxic T cell or NK cell populations (Figure 2) [95–97]. *STAT3* mutations have been described in 30–40% of T-LGL leukemia patients while *STAT5B* mutations were found in rare but typical $CD4^+$ T-LGL leukemia cases. However, *STAT5B* mutations were more frequently detected in patients with a severe clinical course. In all cases, mutations were shown to increase the transcriptional activity of both STAT3 and STAT5B proteins, but only the $STAT5B^{N642H}$ mutation was demonstrated to drive T-LGL leukemias in mouse models [98,99].

2.6. *STAT3/5 in Chronic Lymphocytic Leukemias (CLL)*

CLL is characterized by the accumulation of mature clonal B cells in peripheral blood, bone marrow, and lymphoid tissues. These cells are characterized by an extended lifespan due to intrinsic defects in apoptosis [100]. Increasing STAT3 phosphorylation on S^{727} but not on Y^{705} is believed to be a hallmark of CLL progression [101]. Phosphorylation of S^{727} regulates the transcriptional activity of the STAT3 protein but it is also involved in the mitochondrial localization of STAT3 in primary cells from CLL patients [102]. Cytokines such as interleukin (IL)-15 secreted by the microenvironment contribute to the survival of CLL cells through JAK-mediated tyrosine phosphorylation of STAT5 [103].

2.7. *STAT3/5 in Lymphomas*

Lymphomas are cancers of the lymphatic system. They are divided into two categories: Hodgkin lymphoma (HL) and non-Hodgkin lymphoma (NHL). Data from the literature underscored the important contribution of STAT3 and STAT5 in the proliferation and/or survival of HL cells [104–106]. Most NHLs are B cell lymphomas. Diffuse large B cell lymphoma (DLBCL) is the most common subtype of NHL and consists of at least two phenotypic subtypes: the germinal center B cell-like (GCB-DLBCL) and the activated B cell-like (ABC-DLBCL) groups. High-level STAT3 expression and activation are preferentially detected in ABC-DLBCL, which is associated with poor outcomes [107,108]. Inhibition of STAT3 expression/activity in ABC-DLBCL cells abrogates lymphoma cell growth and triggers apoptosis. Moreover, STAT3 coordinates migration to facilitate the dissemination of DLBCL [109]. Among NHL subtypes, peripheral T cell lymphoma (PTCL) and natural killer (NK)/T cell lymphoma (NKTL) represent a heterogeneous group of diseases with varied clinical features, prognosis and response to treatment. PTCL has been categorized into several subtypes including PTCL-not otherwise specified (PTCL-NOS), angioimmunoblastic TCL (AITL), anaplastic large cell lymphoma (ALCL), and the predominant subsets of cutaneous TCL (CTCL) [110]. Deregulation of STAT3/STAT5 activity was shown to be important for CTCL pathogenesis and cancer progression [111]. ALCL can be divided into anaplastic lymphoma kinase (ALK) positive and ALK negative subgroups, based on ALK gene rearrangements. The nucleophosmin-anaplastic lymphoma kinase (NPM-ALK) fusion protein is the major oncogenic driver in ALK+ ALCL and it activates STAT3 and STAT5 [17,18]. While STAT3 is required for NPM-ALK-induced cell transformation and B cell lymphomagenesis, the contribution of STAT5 is still unclear [112]. Studies indicated that STAT5A, but not STAT5B, is epigenetically silenced in NPM-ALK tumors and behaves as a tumor suppressor when reactivated to suppress NPM-ALK expression [34]. Activating mutations of STAT3 and STAT5B have been found in NKTLs and $\gamma\delta$-T cell lymphomas [25,27]. Recurrent hot spot mutations Y640F and D661Y/V/H/N in the SH2 domain of STAT3 have been identified most commonly in T-LGL leukemia, NK, NK/T, and adult T cell leukemia/lymphoma (ATLL) patients, while the aggressive STAT5B^{N642H} mutation has been predominantly found in T-ALL and $\gamma\delta$-T cell lymphomas, such as hepatosplenic TCL, monomorphic epitheliotropic intestinal TCL, and primary CTCL (Figure 2) [113]. These mutations were shown to be associated with increased phosphorylated STAT3 and STAT5B proteins and to confer a growth advantage to transduced cell lines or normal NK cells [27]. Activating STAT3 mutations E616G and E616K in the SH2 domain found in NK/T cell lymphoma patients were shown to increase the phosphorylation and transcriptional activity of STAT3 [114]. Interestingly, programmed cell death-ligand 1 (PD-L1) was overexpressed in NK/T cells harboring hotspot STAT3 mutations, and overexpression of a STAT3^{E616K} or STAT3^{E616G} mutant was sufficient to enhance PD-L1 expression. These data are consistent with the role of STAT3 as an important regulator of tumor immune evasion [115]. In summary, STAT3 and STAT5B, but not STAT5A, are relevant therapeutic targets in the treatment of lymphomas. All recurrent STAT3 and STAT5B mutations in hematopoietic cancers are summarized in Figure 2.

2.8. STAT3/5 in Myelomas

Multiple myeloma (MM) is a B cell malignancy characterized by the proliferation of clonal plasma cells in the bone marrow accompanied by secretion of monoclonal immunoglobulin [116]. Constitutive tyrosine phosphorylation of STAT3 has been evidenced in MM cell lines and primary CD138^{+} cells from MM patients [117,118]. Patients with STAT3 activation were found to have significantly shorter progression-free and overall survival [119]. STAT3 activation has been reported to contribute to MM progression both directly, by upregulating survival and anti-apoptotic target genes, as well as indirectly by activating myeloid-derived suppressor-cells (MDSCs) in the bone marrow microenvironment, which facilitates tumor progression [120,121]. Myeloma cells rely on the pleiotropic cytokine IL-6, which activates the JAK/STAT3 pathway. IL-6 signaling is tightly regulated by tyrosine phosphatases SHP-1 and SHP-2, and suppressor of cytokine signaling 1 (SOCS1). The disruption of this negative feedback loop results in the constitutive activation of STAT3 [122]. Accordingly, *SHP-1* and *SOCS1* genes were found to be silenced by hypermethylation in MM patients [123]. In addition to the induction of anti-apoptotic gene expression in MM cells, STAT3 also regulates the expression of microRNA-21 with strong anti-apoptotic potential, suggesting that noncoding RNAs have an impact on the pathogenesis of human MM [124]. Moreover, the expression of five long noncoding RNAs (lncRNAs) was found to be regulated by STAT3 in MM cells [125]. However, the role of these lncRNAs in the progression of MM remains to be elucidated. The role of STAT5 seems to be marginal and restricted to immunoglobulin production in MM [126].

3. Canonical and Non-Canonical Roles of STAT3/STAT5 in Hematopoietic Cancers

It was presumed that most, if not all, of the oncogenic activities of STAT3/STAT5 are due to canonical functions. However, STAT3/5 proteins are also active through alternate non-canonical pathways impacting cell transformation. Post-translational modifications affecting canonical and non-canonical roles of STAT3 and/or STAT5 have been extensively reviewed in the literature. To be in line with this review, we will focus on canonical and non-canonical activities of STAT3/5 that have been described in cell transformation (Figure 3).

3.1. Canonical Function of STAT3/STAT5

In the canonical model, STAT3, STAT5A, and STAT5B are primarily activated by phosphorylation on tyrosine residues Y705, Y694, and Y699, respectively (Figure 1). In a physiological situation, STAT3 and STAT5 are transiently phosphorylated by JAK in response to cytokine receptor signaling, and contribute to hematopoietic cell proliferation and differentiation. In contrast, persistent tyrosine phosphorylation of STAT3/5 induced by TKOs promotes hematopoietic cell transformation [6,7]. Phosphorylation of additional tyrosine residues in STAT5A and STAT5B proteins in TKO-transformed cells has also been described in the literature, but the functional and physiological meaning of such phosphorylations remains unclear [127,128]. P-Y^{705}-STAT3 (herein referred to in this review as P-Y-STAT3), P-Y^{694}-STAT5A (P-Y-STAT5A), and P-Y^{699}-STAT5B (P-Y-STAT5B) proteins dimerize through reciprocal interactions between the SH2 domain of one monomer and the phospho-tyrosine residue of the other. P-Y-STAT3, P-Y-STAT5A, and P-Y-STAT5B form homodimers but P-Y-STAT5A can also dimerize with P-Y-STAT5B. The formation of STAT5A/5B heterodimers is not only dependent on the relative abundance of each protein but can also be differentially affected by upstream signaling events. For instance, BCR-ABL is less efficient in inducing STAT5A/5B heterodimerization than IL-3 [128]. Heterodimers between P-Y-STAT3 and P-Y-STAT5A or P-Y-STAT5B have never been reported despite the capacity of some TKOs to simultaneously activate these proteins. After dimerization, STAT3, STAT5A, and STAT5B are translocated into the nucleus. Nuclear import is dependent on a nuclear localization signal present in the coiled-coil domain [129]. The nuclear import of STAT3/STAT5 occurs independently of their tyrosine phosphorylation and is mediated by the importin-$\alpha 3/\beta 1$ system coupled to Ras-related nuclear (Ran) proteins bound to GDP or GTP [129]. Molecules other than importins and Ran also participate in the regulation of the nuclear translocation of STAT3/5. The small GTPase Rac1 and the GTPase-activating protein MgcRacGAP form a ternary complex with P-Y-STAT3 or P-Y-STAT5 to induce their translocation via the importin α/β pathway [130]. The MgcRacGAP/Rac1 complex also regulates the tyrosine phosphorylation of STAT3/5 induced by cytokines [131,132]. Interestingly, p21-activated kinases (PAK1 and PAK2), which are important effectors of Rac1, were shown to induce phosphorylation of the S779 residue and nuclear translocation of STAT5A in BCR-ABL-expressing cells [133]. In a similar vein, activation of focal adhesion kinase (FAK) by FLT3-ITD or KIT^{D816V} in AML cells was demonstrated to induce the nuclear translocation of STAT5 via the Rac1/PAK1 pathway [134]. In sharp contrast, several reports indicated that P-Y-STAT5 is abundantly detected in the cytoplasm of BCR-ABL^+ cells and that Src kinases might be responsible for the cytoplasmic retention of P-Y-STAT5A [135,136]. The reasons for these apparent controversial data remain unclear but might be related to the cellular context.

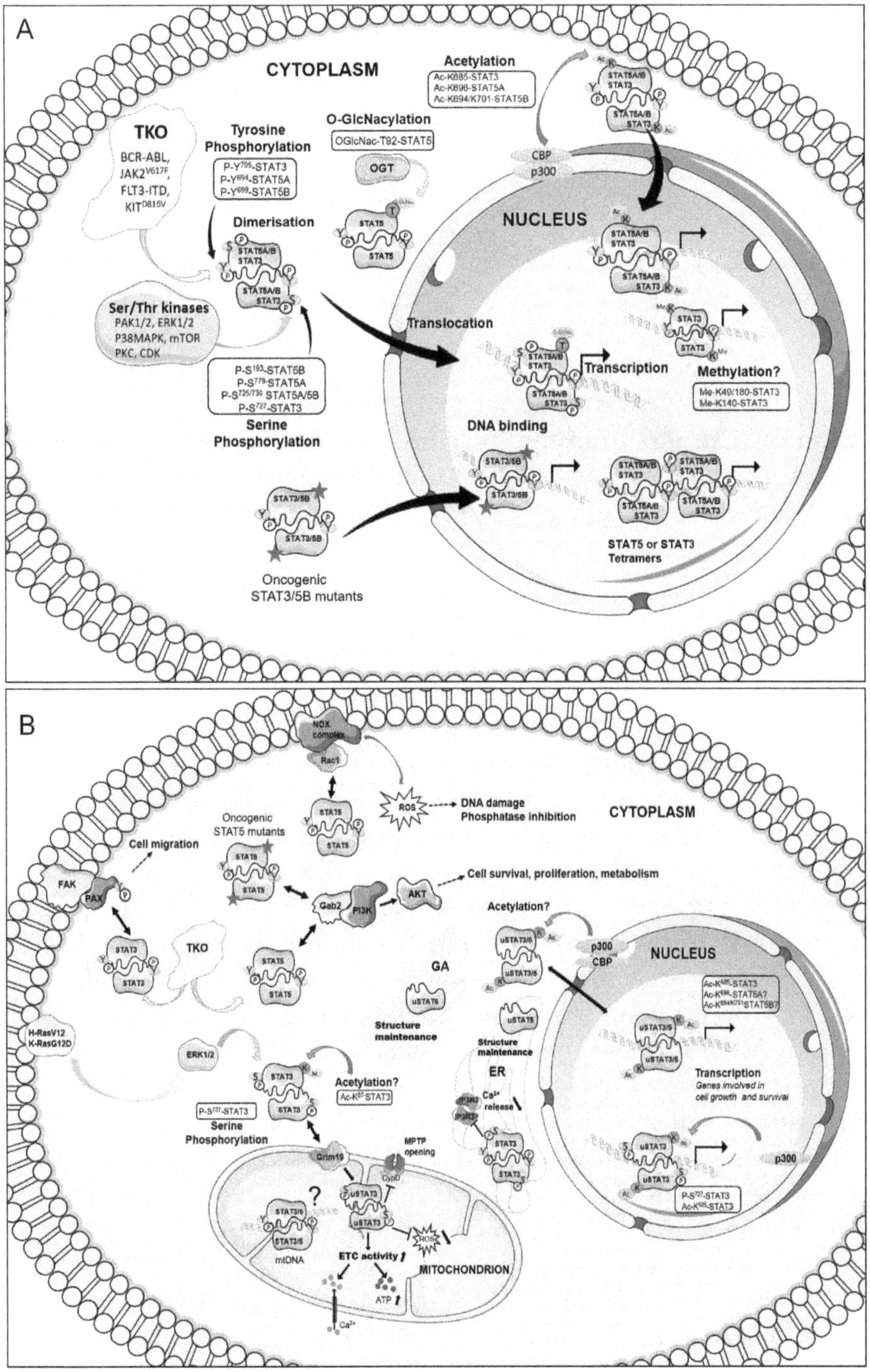

Figure 3. STAT3 and STAT5 signaling in hematologic malignancies. (**A**) Canonical function. In the canonical model, TKOs, or GOF mutations induce persistent activation (e.g., tyrosine phosphorylation) and nuclear translocation of STAT3 and STAT5, which then bind as dimers or tetramers to specific promoter sequences to regulate transcription of genes involved in cell growth, apoptosis, angiogenesis and immune response. Other post-translational modifications, as indicated, regulate the oncogenic activity of STAT3/STAT5. OGT (O-GlcNAc transferase). (**B**) Non-canonical function. uSTAT3 and uSTAT5 (non-tyrosine phosphorylated STAT3/5) proteins also have an active role in the nucleus, mitochondrion, ER (endoplasmic reticulum), GA (golgi apparatus), ETC (electron transport chain). Besides their transcriptional activity, P-Y-STAT3 and P-Y-STAT5 interact with different signaling pathways in the cytoplasm and ER.

Tyrosine phosphorylated STAT3/STAT5 dimers bind to specific DNA elements found in promoters, enhancers, and the first intron of target genes. These binding sites are characterized by clusters of conserved motifs with an interferon gamma-activated site (GAS)-like core sequence (TTCT/CNA/GGAA). STAT5 was shown to bind as a tetramer on adjacent GAS sequences with high or weak affinity in target gene promoters [137,138]. The tetramerization domain located in the NH2-terminal region of STAT5 was found to promote constitutively active, STAT5A mutant (cS5^{F})-induced leukemia in mice [23]. The NH2-terminal domain of STAT3 is also required for oligomerization and IL-6-dependent transcriptional regulation [139]. This domain is involved in STAT3-mediated survival of solid tumor cells but its role in hematologic cancers is currently unknown [140].

TKO-mediated activation of STAT3 and/or STAT5 regulates expression of common as well as specific genes involved in hematopoietic cell survival, proliferation, metabolism, hypoxia, autophagy, migration and tumor immune evasion. BCL2-family members, cell cycle-regulated genes, proto-oncogenes such as PIM1, c-MYC, BCL6, and genes involved in cytokine receptor signaling or immune response are often targets of STAT3/STAT5-mediated transcription in hematopoietic cancers [141]. Transcription of these genes is also dependent on P-Y-STAT3/5 levels in the nucleus. For instance, previous works demonstrated that nuclear P-Y-STAT5 proteins at low, intermediate, and high levels differentially affect self-renewal, proliferation, and differentiation of leukemic stem cells by regulating the expression of distinct sets of genes [142]. Accordingly, P-Y-STAT5 regulates D-type cyclins and c-MYC expression at intermediate levels to promote proliferation but upregulates expression of the cyclin-dependent kinase (CDK) inhibitor p21waf1 at high levels to induce growth arrest and differentiation. Similarly, while constitutively active STAT3 was shown to protect CLL cells from apoptosis, at high levels it induced apoptosis and caspase-3 expression [143]. These data emphasize that increasing nuclear levels of P-Y-STAT3/5 might confer a growth disadvantage to leukemic cells [143,144]. Protein–protein interactions also drive the regulation of STAT3/STAT5-dependent gene expression. STAT3 and STAT5 interact with many transcription factors, co-factors, and/or chromatin remodeling proteins such as enhancer of zeste homolog 2 (EZH2), CREB-binding protein (CBP)/p300, ten-eleven translocation 2 (TET2), and DNA methyltransferase 1 (DNMT1), which largely impact hematopoiesis or leukemogenesis. Blocking these proteins also interferes with STAT3/STAT5-mediated transcription and cancer cell growth. Most of these interacting partners were exhaustively described in a recent review and will not be discussed here [141]. In addition, serine phosphorylation plays an important role in STAT3/5-dependent cell transformation. Constitutive phosphorylation of S727, which is located in the COOH-terminal transactivation domain of STAT3 (Figure 1), was shown to increase the transcriptional activity of this protein in CLL cells [101]. The oncogenic activity of STAT5A is highly dependent on S725 and S779 phosphorylation, and both residues (S726 and S780 in humans) were found to be constitutively phosphorylated in AML, ALL, and CML cells [145]. Importantly, PAK-dependent S779 phosphorylation of STAT5A is required for BCR-ABL-induced leukemogenesis but is not affected by BCR-ABL tyrosine kinase inhibitor treatment, indicating that oncogenic activation of STAT5A is mediated by independent pathways in BCR-ABL-expressing cells [133]. Phosphorylation of S779/780 on STAT5A also promotes the expansion and transformation of human hematopoietic stem/progenitor cells (HSCs/HPCs) [53]. Constitutive phosphorylation of S193 on STAT5B has been detected in various lymphoid tumor cell lines as well as in primary cells from leukemia or lymphoma patients [146]. Phosphorylation of S193 was found to be sensitive to inhibitors of mammalian targets of rapamycin (mTOR) and to play an important role in the DNA binding and transcriptional activity of STAT5B. Other post-translational modifications such as glycosylation affect the oncogenic functions of STAT5. O-GlcNAcylation at threonine 92 in STAT5A and STAT5B was shown to regulate tyrosine phosphorylation and transcriptional activity of oncogenic STAT5 in leukemic cells [147].

STAT3 and STAT5 also undergo acetylation on multiple lysine residues by the CBP/p300 histone acetyltransferase [148]. Acetylation of K685 located in the SH2 domain of STAT3 can enhance transcriptional activity by increasing dimer stability. Mutation of K685 affects dimerization but not tyrosine or serine phosphorylation of STAT3, and acetylation of STAT3 can occur in the absence of tyrosine phosphorylation [149,150]. AcK^{685}-STAT3 also recruits DNMT1 to induce the epigenetic silencing of tumor suppressor genes [151]. Acetylation of STAT5A and STAT5B at K696 and K694/K701 residues, respectively, is also required for STAT5 dimerization [152]. The contribution of acetylation in the canonical functions of oncogenic STAT3/5 on one hand, and in hematopoietic cancers, on the other hand, remains very unclear. Previous work indicated that acetylation of K685 does not play an essential role in the expression of the great majority of P-Y-STAT3-dependent genes, suggesting that acetylation might be more important for the non-canonical functions of STAT3/5 [153].

Methylation of nuclear STAT3 on K140 or K49/K180 by the histone-modifying enzymes SET domain-containing lysine methyltransferase 9 (SET9) or EZH2, respectively, have been reported to mediate opposing effects on STAT3-dependent transcriptional activity [154–156]. STAT5 also interacts with EZH2 in B cells to repress the Igκ locus [157]. The methylation status of STAT3 and STAT5 in hematopoietic cancers has yet to be investigated. Similarly, other post-translational modifications such as oxidation, glutathionylation, or sumoylation were shown to negatively regulate STAT3/5 activity, but their impact on STAT3/5-driven hematologic malignancies are still unknown [158–162].

The concomitant activation of STAT3/5 by TKOs is probably the most intriguing event in leukemogenesis as both proteins in cancer cells have compensatory and/or opposing effects on gene expression and cell fate [163]. For instance, the competitive binding of STAT3 and STAT5 to the regulatory loci of BCL6 and IL-17 has been shown to modify gene expression and cell phenotype [164,165]. The differential recruitment of co-activators and/or co-repressors might explain these opposing effects on gene expression. Activation status of both STAT5 and STAT3 might, therefore, provide important diagnostic and prognostic information in hematologic cancers. Notably, $JAK2^{V617F}$ activates both STAT3 and STAT5, but only STAT3 negatively regulates $JAK2^{V617F}$-dependent MPN development [66,67].

3.2. Non-Canonical Functions of STAT3/STAT5

Subcellular localization of P-Y-STAT3/5 is not restricted to the nucleus and previously published data highlighted important functions of P-Y-STAT3/5 in the cytoplasm and/or other subcellular organelles of cancer cells. Cytoplasmic localization of P-Y-STAT5 was abundantly found in leukemic cells expressing BCR-ABL, $JAK2^{V617F}$, or KIT^{D816V}. Here, it was shown to interact with the scaffolding adapter Gab2 to favor the activation of the phosphatidylinositol-3-kinase (PI3K)/AKT pathway and leukemic cell survival [12,135,166]. The deletion of Gab2 attenuated the transforming potential of an oncogenic STAT5 mutant in mouse models [167]. Interaction of P-Y-STAT5 with Rac1, which is also an important component of certain membrane-bound NADPH-oxidase (NOX) enzymatic complexes, promotes ROS production in FLT3-ITD-expressing AML cells thereby increasing cell growth concomitantly with DNA damage [82]. BCR-ABL- and $JAK2^{V617F}$-induced ROS generation is mediated by P-Y-STAT5, and binding of P-Y-STAT5 to the NOX2 complex in $BCR\text{-}ABL^{+}$ cells also requires active Rac1 (unpublished data) [58,59,168]. P-Y-STAT3 has been detected in focal adhesions of cancer cells where it interacts with phosphorylated paxillin and FAK, thereby regulating cell migration [169]. Constitutively active STAT3 also controls $Ca2^{+}$ release in the ER by interacting with inositol 1,4,5-triphosphate receptors (IP3R3), facilitating its proteasomal degradation [170]. Accordingly, the binding of STAT3 to IP3R3 protects cells from oxidative/ER stress and apoptosis. Finally, the mitochondrial localization and the potential role of P-Y-STAT3 and P-Y-STAT5 in regulating the mitochondrial genome have also been reported [171–173]. Collectively, these data suggest that P-Y-STAT3/5 distribution in subcellular organelles or cytoplasmic structures may directly impact cancer cell growth and survival independently of their transcriptional activity in the nucleus. It would then be relevant to explore in further detail the subcellular localization and function of P-Y-STAT3/5 in hematologic malignancies.

Constitutive phosphorylation of STAT3 on residue S727, but not on Y705, plays a role in the pathogenesis of CLL by regulating STAT3-dependent expression of genes associated with cell growth and survival [101]. DNA binding and transcriptional activity of $P\text{-}S^{727}$-STAT3 probably require acetylation of K685 because (1) persistent acetylation of K685 on STAT3 has been observed in CLL cells, (2) acetylation of K685 promotes dimer formation, DNA binding and activation of transcription, and (3) K685 acetylation provides CLL cells with survival advantages [174]. In addition to its nuclear localization, P-S^{727}-STAT3 was also found in mitochondria where it contributes to the viability of CLL cells and protects against oxidative stress [102]. The oncogenic activity of mitochondrial STAT3 was first demonstrated in H-Ras^{V12}-transformed cells where it promotes anchorage-independent cell growth and tumor induction in mice [175]. Phosphorylation of S727 is carried out by the MAPK/ERK kinase (MEK)/Extracellular signal-regulated kinases (ERK) pathway and is necessary for Ras-mediated cell transformation [176]. Mitochondrial P-S^{727}-STAT3 also supports K-Ras^{G12D}-driven hematologic neoplasms [177]. Retinoid interferon-induced cell mortality (GRIM)19, a component of the electron transport chain (ETC) complex 1 and binding partner of STAT3, was shown to mediate STAT3 uptake into mitochondria [178]. However, other proteins such as the chaperone TOM20 have also been suggested as mitochondrial carriers of STAT3 [179]. Phosphorylation of S727 is important for mitochondrial STAT3 to upregulate ETC activity and ATP production, and mutation of S727 decreases the mitochondrial translocation of STAT3 [178]. In addition, CBP-dependent acetylation of K87 was also found to promote the mitochondrial localization of STAT3 [180]. Hematopoietic cell-targeted deletion of the *stat3* gene demonstrated the critical role of mitochondrial STAT3 in HSC/HPC function. *stat3*$^{-/-}$ mice have a blood phenotype with similarities to human diseases of myelodysplastic syndrome (MDS) and MPN, supporting, as mentioned above, a negative regulatory role of STAT3 in MPN development [181]. These data also suggest that targeting mitochondrial STAT3 in K-Ras-induced hematopoietic malignancies might have adverse effects on normal HSCs. In addition to mitochondrial functions, non-tyrosine phosphorylated STAT3 and STAT5 proteins (uSTAT3/5) also have important roles in the nucleus, ER, and Golgi apparatus (GA). The knockdown of STAT5A/5B in human pulmonary arterial endothelial and smooth muscle cells resulted in a loss of uSTAT5A and uSTAT5B associated with the GA, leading to a dramatic destabilization of the ER and GA. This was accompanied by the activation of ER stress pathways, indicating a potential role of uSTAT5 in maintaining the structure and function of these organelles [182]. Inhibition of uSTAT5 activity/expression induced ROS production and apoptosis in pre-B leukemic cells, indicating that uSTAT5 may also provide protection against oxidative stress [183] (unpublished data).

uSTAT3/5 are also involved in the regulation of transcription and chromatin remodeling as preformed anti-parallel dimers [129,141]. The NH2-terminal domain mediates the dimerization of uSTAT3 and is essential for its nuclear accumulation, DNA binding, chromatin remodeling, and regulation of gene expression [184,185]. Acetylation of K685 was shown to be crucial for uSTAT3 to form stable dimers and regulate gene transcription [149]. Accordingly, the majority of uSTAT3-mediated gene expression depends on the ability of K685 to become acetylated [153]. Interestingly, nuclear uSTAT3 regulates the expression of well-known genes in cancer, suggesting that uSTAT3 might contribute to oncogenesis [186]. The role of uSTAT5 is less documented but data from the literature indicates that nuclear uSTAT5 behaves as a partial antagonist of P-Y-STAT5 and acts as a repressor to maintain self-renewal of hematopoietic cells and to block differentiation [187]. The role of uSTAT3/5 in the initiation, emergence and/or progression of hematologic cancers has yet to be determined. However, previous studies suggested that uSTAT5 provides preferential and critical cell survival signals in lymphoid tumor cells, indicating that uSTAT3/5 should also be considered as therapeutic targets in certain hematologic malignancies [188]. Our understanding of the molecular mechanisms involved in non-canonical functions of STAT3/5 is still incomplete. Most of these activities contribute to the known roles of STAT3/5 in hematopoiesis and hematopoietic neoplasms, and this knowledge complicates the already difficult task of targeting STAT3/5 for therapeutic purposes.

4. Pharmacological Inhibitors of STAT3 and STAT5

4.1. Introduction

Modalities for targeting STAT3 and STAT5 in hematologic cancers can be classified into direct and indirect approaches. Compounds directly targeting STAT3 or STAT5 canonical functions may either inhibit dimerization, DNA binding, or transcriptional activity. Indirect approaches include preventing ligands binding to growth factor or cytokine receptors, inhibiting upstream tyrosine kinases such as TKOs, targeting the nucleocytoplasmic shuttling of STAT3/5, or activating negative regulators of STAT3/5 such as the tyrosine phosphatases SOCS or PIAS [189–191]. Both direct or indirect approaches might also be applied to non-canonical functions. Molecules targeting STAT3/5 protein expression/stability or disrupting interactions with partners that play critical roles in oncogenic STAT3/5 activity should also be considered.

Historically, abnormal activation of STAT3 in solid tumors was first appreciated before STAT5 was implicated, and as such, various therapeutics to target STAT3 were first developed based on one or more of these strategies described above. Pioneering works using phospho-Y-peptides to compete with P-Y^{705} for binding to the SH2 domain of STAT3 and peptidomimetic derivatives provided the proof of concept that disrupting the P-Y^{705}/SH2 domain interaction could be an efficient strategy to block oncogenic STAT3 signaling [192,193]. However, peptides and peptidomimetics have certain limitations in line with their in vivo instability and poor membrane permeability [194]. Nevertheless, these limitations provided the necessary impetus in many programs to design small molecules with greater and more specific inhibitory effects. Blocking DNA binding using decoy oligonucleotides (Duplex ODN), G-quartet oligonucleotides (GQ-ODN), or DBD peptide aptamers was another strategy to suppress canonical functions of STAT3 and STAT5. Treatment with STAT3 or STAT5 ODNs inhibits cell growth and/or induces apoptosis by preventing nuclear translocation of STAT3/5 in cancer cells [195]. RNA interference (RNAi) or antisense oligonucleotides (ASO) were also employed to target STAT3/5 mRNA in leukemia or lymphoma cells [195]. Some of these tools have been promising in modulating STAT3/5 signaling and inhibiting tumor cell growth. However, the therapeutic success of these different approaches relies on the effective entry and stability of the oligonucleotides in the cells and therefore requires chemical modifications for this purpose. Finally, cell-based screening with chemical compound libraries has allowed the identification of natural, synthetic, or clinically used molecules that inhibit STAT3/5 transcriptional activity [196,197]. However, hits derived from these assays have indirect effects that are challenging to determine. Due to an ever-growing list of STAT3/5 inhibitors and reviews in the field, we will focus on those that have been tested in hematopoietic cancers. We will discuss the limitations of STAT3/5 inhibitors in the treatment of these diseases but also promising outcomes when combined with other pharmacological compounds.

4.2. Indirect Inhibitors of STAT3 and STAT5 Signaling in Hematopoietic Cancers

4.2.1. Targeting Upstream Tyrosine Kinases

Most agents that are described as indirect STAT3/5 inhibitors actually target upstream kinases such as JAK, Src, BCR-ABL, FLT3, or KIT receptors. Activation of STAT3 and/or STAT5 is dependent on the leukemic cell type in which the kinase is active. The development of IM and related BCR-ABL kinase inhibitors such as nilotinib, dasatinib, bosutinib, and ponatinib has made a major breakthrough in targeted cancer therapy, CML and Ph^{+}ALL treatment [198–204]. IM leads to complete inhibition of BCR-ABL-dependent STAT5 activation and this is likely an important part of the effectiveness of this molecule [49,90]. However, some patients cannot tolerate the side effects of IM and related TKIs. In addition, the development of resistance to TKIs is a significant clinical problem, and is due, in part, to acquired point mutations in BCR-ABL such as T315I [205]. Although second- and third-generation TKIs were found to be effective against some BCR-ABL mutants, TKIs alone do not eradicate LSCs and de novo resistance of CML cells [55,56,62]. Targeting the kinase activity of FLT3 mutants has also been adopted to inhibit aberrant signaling in AML [206]. Among the agents that have been or are being evaluated in preclinical studies and/or clinical trials are multi-kinase inhibitors midostaurin (PKC412), sorafenib (BAY 43-9006), lestaurtinib (CEP701), KW-2449 (which also targets T315I-mutated BCR-ABL), sunitinib (SU11248), tandutinib (MLN518), and quizartinib (AC220), as well as compounds more specific for FLT3 mutants, including crenolanib (CP-868596) and gilteritinib (ASP2215) (Table 2) [207–219]. Few of them were approved for the treatment of FLT3-mutated AML as a single-agent or in combination with other therapeutic drugs. Although the effectiveness of these inhibitors was shown in preclinical studies, mixed results have been observed in clinical trials. Clinical activity of some of these molecules was evidenced in patients with FLT3-mutated AML but was often transient and relapse eventually occurred [206]. Secondary mutations in FLT3 and/or mutations associated with epigenetic regulators or transcription factors are responsible for the loss of therapeutic response to FLT3 inhibitors [206,220]. In addition, the bone marrow microenvironment provides protection against FLT3-TKIs [221]. Midostaurin, crenolanib, and tandutinib were also employed to inhibit KIT^{D816V} activity in mast cells from SM patients, while lestaurtinib was tested in two phase 2 trials for MPN treatment (Table 2) [222–226]. Other potent and selective KIT^{D816V} inhibitors including avapritinib (BLU-285) and DCC-2618 have entered clinical trials with promising results for the treatment of SM [222].

A breakthrough in understanding myeloproliferative diseases occurred after the discovery of GOF mutations in JAK2, leading to the development of small-molecule inhibitors of JAK2 for the treatment of MPNs [227]. In clinical trials, responses obtained with JAK inhibitors are independent of the driver mutations. However, treatment with JAK inhibitors was shown to have some limitations, partly because the targeted pathway is required for normal hematopoiesis and because specific inhibitors targeting the $JAK2^{V617F}$ mutant are yet to be developed. Consequently, JAK2 inhibitors have been disappointing in their ability to induce molecular remissions in MPN patients, indicating that JAK2 inhibitors do not preferentially target MPN cells over normal cells [228]. In addition, some of the JAK inhibitors that entered clinical trials were discontinued due to their adverse effects. A crystal structure and biochemical properties of the pseudokinase domain of JAK2 will certainly assist in developing $JAK2^{V617F}$-specific inhibitors in the future. Selective JAK and pan-JAK inhibitors that variably affect P-Y-STAT3/5 levels in hematologic neoplasms are presented in Table 2. Ruxolitinib is the first clinically-approved JAK1/2 inhibitor for PV and MF treatment and is also in clinical trials either alone or in combination with other pharmacological agents or TKIs for HL, MM, or CML treatment (Table 2) [227–232]. Ruxolitinib and most of the JAK inhibitors that are in clinical trials are type I inhibitors, which means that they block the ATP-binding site of JAKs under the active conformation of the kinase domain [227,233–237]. Type II inhibitors bind to the ATP-binding pocket of the JAK2 kinase domain in the inactive conformation, while allosteric inhibitors interact with other sites in the JAK2 protein [227]. Importantly, JAK2 target inhibition in MPN can be improved with the type II inhibitor NVP-CHZ868 offering increased therapeutic efficacy [238]. NVP-CHZ868 was also shown to act synergistically with dexamethasone in suppressing the growth and survival of human B-ALL cells in PDX models [239]. JAK2-specific inhibitors such as pacritinib are in the final stages of clinical trials for primary and secondary MF, and display increased potency compared to currently available JAK inhibitors [233]. Gandotinib, which is in a phase 2 study for $JAK2^{V617F}$-mutated MPN treatment, showed an increased potency for the $JAK2^{V617F}$ mutant [240]. Another example of a JAK2 inhibitor, fedratinib, which was previously burdened with a clinical hold in 2013, was recently FDA-approved for the treatment of MPN patients who have failed therapy with ruxolitinib [241]. Fedratinib also blocks the growth of HL and Mediastinal Large B-cell Lymphoma (MLBCL) in vitro and in vivo as demonstrated in preclinical studies [242]. Many JAK inhibitors with different selectivity and/or mechanisms of action have been tested in leukemias, lymphomas and MM (Table 2) [243–268]. For instance, AZD1480, INCB20, INCB16562, NS-018 and momelotinib (CYT-387) showed promising in vitro and/or in vivo efficacy against MM cells (Table 2) [246–248,251,261]. In all cases, P-Y-STAT3 was markedly reduced in

MM cells treated with these inhibitors. Collectively, these data strongly support that JAK inhibition has significant potential as a therapeutic strategy in MM.

4.2.2. Natural and Synthetic Molecules

Historically, natural compounds have been successfully used in the management of various human diseases. Natural products may also serve as a basis for the synthesis of derivatives aiming to increase their efficacy. There are several compounds that are known to exert anti-tumor effects through their indirect or direct action on STAT3 and/or STAT5 signaling. These natural molecules have a low toxicity profile and can act synergistically with other pharmacological agents to reverse chemoresistance. A number of plant-derived molecules such as avicin D, curcubitacin I, butein, honokiol, capsaicin, celestrol, and piperlongumine have been reported to inhibit growth or survival of leukemia, lymphoma, or myeloma cells in preclinical studies (Table 2) [269–277]. However, the mechanistic basis of their effects on STAT3/STAT5 signaling is still unknown. Inhibition of upstream kinases JAK1/2 and Src, and upregulation and/or activation of SHP1 or other protein tyrosine phosphatases appear to be a common feature of these compounds. Curcumin, a naturally derived phytochemical from plants such as turmeric (*Curcuma longa*), has been extensively investigated for its anti-tumor effects [278]. Curcumin was shown to block STAT3 and/or STAT5 phosphorylation in leukemia, lymphoma, and myeloma cells (Table 2) [279–283]. Although the administration of curcumin has been shown to be safe in humans, its clinical utility is somewhat limited due to the poor bioavailability and target selectivity. Therefore, efforts were made to design and synthesize novel curcumin analogs. FLLL32, one of these analogs, was shown to inhibit P-Y-STAT3 and growth of MM cells with greater efficacy but, again, target selectivity and mechanisms of action remained poorly defined [284]. The synthetic triterpenoid, CDDO-Imidazolide, which acts as an anti-inflammatory and anti-cancer drug, was demonstrated to suppress STAT3 and STAT5 phosphorylation and to induce apoptosis in MM cells [285]. EC804, a synthetic derivative of indirubin, an active component in a traditional Chinese medicine formulation, was reported to inhibit STAT3 and/or STAT5 phosphorylation as well as the growth of sensitive or TKI-resistant CML and AML cells [286,287]. This compound also blocks STAT3 activity in solid tumors [288]. Natural and synthetic compounds such as sulforaphane and the BET inhibitor JQ1 were found to inhibit STAT5-mediated transcription in CML and T-ALL cells, probably via epigenetic mechanisms [289,290]. Naphthoquinone (NPQ)-based derivatives could be also mentioned as indirect inhibitors of STAT5 probably through their multikinase modulatory effects in leukemias [291]. Research in our laboratory is focused on the synthesis and development of small-molecule inhibitors of STAT5. We recently identified 17f as a compound that selectively inhibits STAT5 phosphorylation and expression in AML and CML cells [292]. Moreover, we found that 17f overcomes the resistance of CML and AML cells to IM and Ara-C, a conventional therapeutic agent used in AML treatment [293]. We also found that 17f, when associated with IM or Ara-C, inhibits expression of STAT5B but not STAT5A in resistant CML and AML cells via translational or post-translational mechanisms. The mechanistic basis of this inhibitory effect is currently under investigation.

4.2.3. Drug Repositioning

Cell-based assays for high-throughput screening were employed to identify compounds that specifically block the transcriptional activity of STAT3/5 [196,197]. This type of strategy utilizes cells that are stably transfected with a construct containing the luciferase reporter gene under the control of a specific and high-affinity STAT3 or STAT5 responsive promoter. Chemical libraries of compounds biased toward bioactives and drugs known to be safe in humans were used in the screening. Using this approach, nifuroxazide, niclosamide, and pyrimethamine were identified as specific inhibitors of STAT3, while pimozide was found to inhibit STAT5 activity (Table 2). Nifuroxazide, an antidiarrheic agent, was shown to decrease STAT3 tyrosine phosphorylation, most probably via inhibition of JAK kinase activity, and to reduce the viability of MM cells [294]. Niclosamide, an antiparasitic drug, blocks P-Y-STAT3 and myeloma cell growth via unknown mechanisms. Niclosamide lacks selectivity because it also inhibits NFkB activity in MM cells [295]. The antiparasitic and antifolate drug pyrimethamine also displays significant activity in vitro against MM cell lines harboring P-Y-STAT3 [196]. Pyrimethamine inhibited P-Y-STAT3 and transcriptional activity without affecting its upstream kinase JAK2. Three-dimensional modeling studies indicated that pyrimethamine binds to the SH2 domain, suggesting that it might be a direct inhibitor of STAT3, but this interaction needs to be biophysically demonstrated. Pyrimethamine was also found to be a potent inducer of apoptosis in AML cells [296]. However, it is still unclear whether the antitumor activity of pyrimethamine is due to its inhibitory effect on STAT3 and/or on folic acid metabolism [297]. Pyrimethamine is currently in a phase 1 clinical trial for the treatment of high-risk MDS and in phase $\frac{1}{2}$ trials for CLL and small lymphocytic lymphoma (SLL) treatment.

Pimozide was identified by high-throughput drug screening as a potent inhibitor of STAT5. Pimozide was shown to inhibit P-Y-STAT5 and survival of CML and MPN cells without affecting the kinase activity of BCR-ABL, JAK2 or Src [298,299]. Pimozide shows synergistic effects with IM/nilotinib in killing CML cells and overcomes TKI resistance in BCR-ABL^{T315I} mutant cells [298]. The effects of pimozide are not limited to CML and MPN cells, and the efficacy of this drug was also demonstrated in AML. Pimozide can also inhibit P-Y-STAT5 and STAT5-dependent gene expression in AML cells expressing FLT3-ITD, and it acts synergistically with FLT3 inhibitors to induce apoptosis in these leukemic cells [79]. The mechanisms involved in pimozide-mediated inhibition of P-Y-STAT5 is currently not known. Pimozide also inhibits P-Y-STAT3 in myeloma cells indicating that it is not a selective STAT5 inhibitor [196]. Antidiabetic drugs such as pioglitazone and rosiglitazone were shown to have antileukemic activity [300]. Both synthetic compounds belong to the thiazolidinedione (TZD) class of ligands that bind to the nuclear receptor PPARγ. Activation of PPARγ by pioglitazone not only inhibits the growth of CML cells but also reduces the expression of *STAT5* genes [60]. Quiescent CML stem cells, which are known to be resistant to TKI treatment, are highly sensitive to pioglitazone when combined with IM. This suggests that besides phosphorylation, targeting STAT5 expression might be important for eradicating resistant CML stem cells. Whether PPARγ directly regulates STAT5A and STAT5B gene promoter activity remains to be investigated. Pioglitazone combined with IM are now in a phase 2 trial to evaluate the impact of this combination therapy on CML residual disease [301]. Activation of PPARγ was also shown to induce apoptosis in Ph^{+} ALL cells [302]. Finally, PPARγ agonists were found to inhibit the transcriptional activity of STAT3 in MM cells. In that case, it has been proposed that PPARγ and STAT3 may compete for binding to nuclear co-factors [303].

Table 2. Indirect inhibitors of STAT3 and STAT5 that have been tested in hematologic cancers.

Drug	Target(s) and Mode of Inhibition	Hematologic Malignancy	Stage of Clinical Development	References
		Upstream Kinases		
Atiprimod	JAK2/JAK3 Inhibits kinase activity	AML JAK2^{V617F}-MPN	Preclinical (cell lines, primary cells) Preclinical (cell lines, primary cells)	[249,250]
AZ960	JAK2/JAK3 Type I inhibitor	T cell leukemia/ lymphoma	Preclinical (cell lines)	[251]
AZD1480	JAK2/JAK1/Aurora/FGFR1/FLT4 Type I inhibitor	MPN B-ALL Lymphomas MM	Phase 1 (terminated) Preclinical (xenografts) Preclinical (cell lines) Preclinical (cell lines, xenografts)	[243–246]
Bosutinib (SKI-606)	BCR-ABL/Src Type I inhibitor	CML	Market authorization (Bosulif®, Pfizer)	[201,202]
Crenolanib (CP-868596)	FLT3/PDGFR/KIT Type I inhibitor	FLT3-mutated AML SM, AML	Phase 2 Phase 3 (CT) Preclinical (cell lines, primary cells)	[218,223]
Dasatinib (BMS-354825)	BCR-ABL/Src. Inhibits kinase activity Type I inhibitor	CML, Ph$^+$ ALL	Market authorization (Sprycel®, BMS)	[200]
Fedratinib (TG101348)	JAK2/FLT3/BRD4 Type I inhibitor	MPN HL, MLBCL	Market authorization (Inrebic®, Celgene) Preclinical (cell lines, xenografts)	[241,242]
Gandotinib (LY2784544)	JAK2/JAK1/JAK3 Type I inhibitor	JAK2^{V617F}-MPN	Phase 2	[240]
Gilteritinib (ASP2215)	FLT3/AXL Type I inhibitor	FLT3-mutated AML	Market authorization (Xospata®, Astellas)	[219]
Imatinib (STI-571)	BCR-ABL. Type II inhibitor	CML, Ph$^+$ALL	Market authorization (Glivec®, Novartis)	[198,199]
INCB20	Pan JAK	MM	Preclinical (cell lines, primary cells, mouse models)	[247]
INCB16562	JAK1/JAK2. Type I inhibitor	MM	Preclinical (cell lines, primary cells)	[248]
Itacitinib (INCB-039110)	JAK1 Type I inhibitor	MF B-cell lymphoma	Phase 2 Phase 1/2	[252,253]

Table 2. *Cont.*

Drug	Target(s) and Mode of Inhibition	Hematologic Malignancy	Stage of Clinical Development	References
KW-2449	FLT3/BCR-ABL/ BCR-ABL315I/Aurora Inhibits FLT3 and BCR-ABL phosphorylation	CML, AML MDS, AML, ALL, CML	Preclinical (cell lines, primary cells, xenografts) Phase 1	[212,213]
Lestaurtinib (CEP-701)	JAK2/FLT3/TrkA/ Aurora kinase Type I inhibitor	AML MF PV, ET	Phase 2 Phase 2 Phase 2	[211,225,226]
LS104 (AG490 analog)	JAK2/BCR-ABL Allosteric inhibitor	$JAK2^{V617F}$-MPN AML Ph^+ and Ph^- ALL	Preclinical (cell lines, primary cells) Preclinical (cell lines, primary cells) Preclinical (cell lines, primary cells, xenografts)	[262–264]
Midostaurin (PKC412)	FLT3-ITD/KIT Type I inhibitor	FLT3 mutated AML, SM	Market authorization (Rydapt®, Novartis)	[207,222]
Momelotinib (CYT-387)	JAK2/JAK1/JAK3/ALK2 Type I inhibitor	MPN MM	2 Phase 3 Preclinical (cell lines, primary cells)	[259,260]
Nilotinib (AMN107)	BCR-ABL Type II inhibitor	CML, Ph^+ ALL	Market authorization (Tasigna®, Novartis)	[203]
NS-018	JAK2/Src Type I inhibitor	MPN MM	Phase 1-2 Preclinical (cell lines, mouse model)	[254,255]
NVP-BSK805	JAK2 ($JAK2^{V617F}$) Type I inhibitor	$JAK2^{V617F}$-MPN	Preclinical (cell lines, mouse model)	[256]
NVP-BVB808	JAK2 Type I inhibitor	MPN	Preclinical (cell lines)	[257]
NVP-CHZ868	JAK2/TYK2/KIT/PDGF/VEGFR Type II inhibitor	MPN B-ALL	Preclinical (cell lines, primary cells, mouse models) Preclinical (cell lines, xenografts, PDX)	[238,239]
ON044580	JAK2/BCR-ABL Allosteric inhibitor	$JAK2^{V617F}$-MPN $BCR\text{-}ABL^{T315I}$-CML	Preclinical (cell lines, primary cells)	[258]
Pacritinib (SB1518)	JAK2/FLT3/IRAK1 Type I inhibitor	MF Lymphomas, CLL, LPD	Phase 3 Phase 1	[233,234]
Ponatinib (AP24534)	BCR-ABL and BCR-ABL mutants FGFR/PDGFR/Src/RET/KIT/FLT1 Type II inhibitor	CML, Ph+ ALL	Market authorization (Iclusig®, Ariad)	[204]

Table 2. *Cont.*

Drug	Target(s) and Mode of Inhibition	Hematologic Malignancy	Stage of Clinical Development	References
Ruxolitinib (INCB18424)	JAK2/JAK1/JAK3 Type I Inhibitor	MPN CML HL MM AML, ALL	Market authorization (Jakavi®, Novartis) Phase 1 (CT) Phase 2 Phase 1 (CT) Phase1/2 (terminated)	[227–232]
Quizartinib (AC220)	FLT3 Type II inhibitor	FLT3-mutated AML	Market authorization (Japan) (Vanflyta®, Daiichi-Sankyo)	[217,218]
Sorafenib (BAY43-9006)	FLT3/PDGFR/ VEGFR/c-KIT/RAF Type II inhibitor	FLT3-mutated AML ALL, AML, MDS	Phase 1 Phase $\frac{1}{2}$ (CT) Phase 1	[208–210]
Sunitinib (SU11248)	PDGFR/ VEGFR/c-KIT/FLT3 Type I Inhibitor	FLT3-mutated AML	Phase 1 Phase $\frac{1}{2}$ (CT)	[214,215]
Tandutinib (MLN518)	FLT3/PDGFR/KIT Inhibits type III receptor kinases	AML/MDS AML SM	Phase $\frac{1}{2}$ Phase $\frac{1}{2}$ (CT) Preclinical (cell lines)	[216,224]
TG101209	JAK2/FLT3 Type I inhibitor	MPN MM T-ALL	Preclinical (cell lines, primary cells, mouse models) Preclinical (cell lines) Preclinical (cell lines, primary cells)	[235–237]
WP1066 and WP1034 (AG490 analogs)	JAK2 Phosphorylation and/or degradation	JAK2^{V617F}-mutated MPN AML	Preclinical (cell lines, primary cells) Preclinical (cell lines, primary cells)	[265–267]
XL019	Pan-JAK. Type I inhibitor	MPN	Phase 1 (terminated)	[268]
		Natural compounds		
Avicin D	JAK/STAT3 Inhibits P-Y/JAK1/2 phosphorylation Unknown mechanism	MM CTCL	Preclinical (cell lines) Preclinical (cell lines, primary cells)	[269,270]
Butein	STAT3. Inhibits P-Y Activates tyrosine phosphatase SHP1 Unknown mechanism	MM	Preclinical (cell lines)	[271]

Table 2. *Cont.*

Drug	Target(s) and Mode of Inhibition	Hematologic Malignancy	Stage of Clinical Development	References
Capsaicin	JAK1/Src/ STAT3. Inhibits P-Y/ Inhbits kinase activity Unknown mechanism	MM	Preclinical (cell lines, xenografts)	[272]
Celastrol	JAK/Src/STAT3. Inhibits P-Y/Kinase activity Unknown mechanism	MM	Preclinical (cell lines)	[273]
Curcubitacin I (JSI-124)	STAT3. Inhibits P-Y/P-S of STAT3 Unknown mechanism	B-ALL, CLL, lymphomas	Preclinical (cell lines, primary cells)	[274,275]
Curcumin diferuloylmethane	JAK/STAT3/STAT5. Inhibits P-Y/kinase activity Unknown mechanism	CML T-cell leukemia T-cell lymphoma HL MM	Preclinical (cell lines) Preclinical (cell lines) Preclinical (cell lines, primary cells) Preclinical (cell lines) Preclinical (cell lines)	[279–283]
Honokiol (HNK)	STAT3. Inhibits P-Y Activate SHP1 Unknown mechanism	AML	Preclinical (cell lines, primary cells)	[276]
Piperlongumine	STAT3. Inhibits P-Y and other pathways Binds to cysteine 712 of STAT3 Unknown mechanism	MM	Preclinical (cell lines, xenografts)	[277]
Sulforaphane	STAT5. Inhibits transcriptional activity. Deacetylase inhibitor Unknown mechanism	CML	Preclinical (cell lines)	[289]
		Synthetic molecules		
CDDO (Imidazolide)	STAT3/ STAT5. Inhibits P-Y Unknown mechanism	MM	Preclinical (cell lines)	[285]
E804 (Indirubin derivative)	SFK/STAT5 Inhibits P-Y/ Inhibits kinase activity Unknown mechanism	CML AML	Preclinical (cell lines, primary cells) Preclinical (cell lines, xenografts)	[286,287]
FLLL3 (Curcumin derivative)	JAK/STAT3. Inhibits P-Y Unknown mechanism	MM	Preclinical (cell lines)	[284]

Table 2. *Cont.*

Drug	Target(s) and Mode of Inhibition	Hematologic Malignancy	Stage of Clinical Development	References
JQ1 (Thienodiazepine derivative)	STAT5. Inhibits transcriptional activity.BET inhibitor. Unknown mechanism	T-ALL	Preclinical (cell lines, primary cells)	[290]
17f	STAT5. Inhibits P-Y and expression. Unknown Mechanism	AML, CML	Preclinical (cell lines)	[292]
Naphtoquinone (NPQ) derivative	STAT5. Inhibits P-Y/ Unknown mechanism	CML	Preclinical (cell lines)	[291]
		Drug repositionning		
Niclosamide (Antiparasitic)	STAT3. Inhibits P-Y Unknown mechanism	MM	Preclinical (cell lines, primary cells)	[295]
Nifuroxazide (Antidiarrheic)	JAK/STAT3. Inhibits P-Y Unknown mechanism	MM	Preclinical (cell lines, primary cells)	[294]
Pimozide (Antipsychotic)	STAT3/STAT5. Inhibits P-Y Unknown mechanism	CML MPN FLT3-mutated AML MM	Preclinical (cell lines, primary cells) Preclinical (cell lines) (CT) Preclinical (cell lines, mouse model) Preclinical (cell lines)	[79,196,298, 299]
PPARγ ligands (Antidiabetic) (Thiazolinediones)	STAT5. Inhibits *STAT5A* and *STAT5B* gene expression STAT3. Inhibits transcriptional activity	CM, Ph^+ ALL MM	Preclinical (cell lines, primary cells) Phase 2 (CT) Preclinical (cell lines, primary cells)	[60,301–303]
Pyrimethamine (Antiparasitic) (Antifolate)	STAT3. Inhibits P-Y Unknown mechanism	AML CLL/SLL MM	Preclinical (cell lines, primary cells, xenografts) Phase $\frac{1}{2}$ Preclinical (cell lines)	[196,296]

AML (acute myeloid leukemia), B- and T-ALL (B- and T-acute lymphoblastic leukemia), CLL (chronic lymphocytic leukemia), CTCL (cutaneous T cell lymphoma), CML (chronic myeloid leukemia), CTCL (cutaneous T cell lymphoma), ET (essential thrombocythemia), HL (Hodgkin's lymphoma), LPD (lymphoproliferative disorders), myelodysplastic syndromes (MDS), MF (myelofibrosis), MCL (mantle cell lymphoma), MLBCL (mediastinal large B cell lymphoma), MM (multiple myeloma), MPN (myeloproliferative neoplasm), Ph^+ and Ph^- ALL (Philadelphia chromosome-positive and -negative acute lymphoblastic leukemia); PV (polycythemia vera), SLL (small lymphocytic lymphoma), SM (systemic mastocytosis), CT (combination therapy).

4.3. Direct Inhibitors of STAT3 and STAT5 in Hematopoietic Cancers (Table 3)

The mechanisms of direct STAT3/5 inhibition include disruption of tyrosine phosphorylation, dimerization, nuclear translocation and/or DNA binding. Targeting the SH2 domain was, therefore, the main focus for the design and identification of selective inhibitors.

4.3.1. Inhibitors Targeting the SH2 Domain

Dimerization is an essential step in the canonical functions of STAT3 and STAT5. Blocking this process seems to be an optimal solution to directly inhibit aberrant STAT3/5 signaling in hematopoietic cancers. The SH2 domain is not only required for dimer formation but also in the recruitment of STAT3/5 to tyrosine-phosphorylated receptor complexes. Therefore, initial strategies aiming to identify phosphopeptide (P-Y-peptide) inhibitors able to block SH2/P-Y interactions were pursued by several teams. This approach was first attempted by Turkson et al. who demonstrated that the peptide PY*LKTK (where Y* represents P-Y^{705}) inhibits STAT3 dimerization and tumor cell growth [192]. P-Y-peptides derived from STAT3 docking sites of gp130 and other cytokine or growth factors receptors were also used to identify the P-Y-peptide YLPQTV as a potent blocker of STAT3 dimerization and DNA binding [304,305]. Although highly specific, peptides usually have poor membrane permeability, low stability, and consequently low biological activities. This prompted investigators to develop peptidomimetics using the tripeptide PY*L, the minimum P-Y-peptide sequence required for STAT3 inhibition. One of these peptidomimetics, ISS610 was shown to be more potent in disrupting STAT3 dimerization and DNA binding but still had poor membrane permeability [193]. Structural and computational analysis of the interaction between ISS610 and the STAT3 SH2 domain led to the development of the peptidomimetic molecule, S3I-M2001, having increased membrane permeability but similar capacity in inhibiting STAT3 DNA binding [306]. Despite hard work in developing peptides or peptidomimetics with potent STAT3-inhibitory activity, poor permeability and metabolic stability have precluded their clinical testing. For the same reason, no peptide/peptidomimetic inhibitors of STAT5 have been developed. Based on the proof of concept that the STAT3-SH2/pY-peptide interaction was amenable to targeting, small nonpeptidic compounds that could specifically bind to the SH2 domains of STAT3 or STAT5 have become attractive candidates. Here, we will discuss the SH2 domain inhibitors that have been used in preclinical studies and clinical trials for the treatment of hematologic malignancies.

Most of the small molecules targeting the SH2 domain of STAT3 were identified by structure-based high-throughput virtual screening. The available X-ray crystallographic data of both the monomer and dimerized STAT3 bound to DNA were essential for these in silico studies [307]. Among these molecules, S3I-201, STA-21, C188, and STX-0119 were shown to block the phosphorylation, dimerization, DNA binding, and transcriptional activity of STAT3 as well as growth and survival of cancer cells [308–311]. In modeling studies, S3I-201 was shown to bind to the STAT3-SH2 domain through its salicylic acid moiety, which is known as a P-Y mimetic [308]. However, Genetic Optimization for Ligand Docking (GOLD) studies indicated that binding of S3I-201 was suboptimal, and structure–activity relationship (SAR) analyses based on this "hit" compound were conducted to derive analogs with improved STAT3 inhibitory activity. One of these analogs, S3I-1757, bound to the SH2 domain of STAT3 but still had a modest potency for decreasing STAT3 DNA binding [312]. However, S3I-1757 embedded in nanoparticles that had been conjugated with monoclonal antibodies against CD38 (denoted as CD38-S3I-NP) was demonstrated to have some increased efficacy in suppressing P-Y-STAT3 and tumor growth in xenograft models of MM. Data describing this compound are presented in this Cancers issue [313]. Among various S3I-201 derivatives, SF-1-066 was found to be the most potent. SF-1-066 was shown to directly bind STAT3 and to inhibit P-Y-STAT3 and growth of AML and MM cell lines with greater potency than S3I-201 [314,315]. A family of sixteen sulfonamide analogs of SF-1-066 was synthetized and BP-1-102 (17o) was proved to be the most active compound in this series. BP-1-102 exhibited improved inhibition of P-Y-STAT3 in MM cell lines [316]. In addition, BP-1-102 blocked P-Y-STAT3 in B-ALL cells overexpressing HMGA1 chromatin remodeling proteins and suppressed their growth [94]. Growth inhibition induced by this salicylic acid-based inhibitor was also reported for T-ALL cell lines [94]. N-alkylated derivatives of both SF-1-066 and BP-1-102, such as 16i and 21h molecules, were demonstrated to inhibit STAT3 function and to induce MM cell apoptosis, but these molecules also lost selectivity toward STAT3 [317]. BP-5-087, another analog of SF-1-066, was identified by SAR-based drug design and compound library screening. BP-5-087 was shown to be a more potent salicylic acid-based STAT3 inhibitor than SF-1-066 and to overcome IM resistance in CML stem cells [318].

A library of phosphopeptide mimicking salicylic acid-based molecules targeting the STAT3-SH2 domain was also employed to identify potent and STAT5-selective inhibitors. Among compounds that were selected in the screening via fluorescence polarization (FP) assays, BP-1-108 was found to be the more potent STAT5 inhibitor [319]. BP-1-108 significantly reduced P-Y-STAT5 and exhibited growth inhibitory properties in CML and AML cell lines. In silico-based studies using BP-1-108 as a scaffold identified compound 13a (or AC-4-130) as a selective and first nanomolar inhibitor of the STAT5B-SH2 domain [320]. AC-4-130 did not exhibit off-target kinase activity and suppressed P-Y-STAT5 at low micromolar concentrations in CML and AML cell lines, as well as in primary AML cells [81]. AC-4-130 inhibited the growth of AML cells in vitro and tumor formation in xenograft models of AML. Furthermore, AC-4-130 was also shown to synergistically increase the cytotoxic effect of the JAK1/2 inhibitor ruxolitinib and the p300/pCAF inhibitor garcinol in FLT3-ITD^{+} AML cells [81]. IST5-002, another STAT5 inhibitor, was identified using in silico structure-based screening and medicinal chemistry. IST5-002 was demonstrated to bind to the SH2 domain of STAT5B, to inhibit P-Y-STAT5 and to reduce the growth and viability of sensitive and IM-resistant CML and Ph^{+}ALL cell lines [321]. Patient-derived CML or newly diagnosed and relapsed/TKI-resistant Ph^{+}ALL cells were also sensitive to IST5-002 treatment. Moreover, IST5-002 was found to reduce leukemia development in PDX models of Ph^{+}ALL [90]. A chromone-derived acyl hydrazine compound, identified through a high-throughput FP screen, was the first inhibitor to directly target the SH2 domain of STAT5B. However, high concentrations of this compound was required to disrupt STAT5 DNA binding and P-Y-STAT5 in CML and Burkitt's lymphoma (BL) cell lines, respectively [322]. The same group next identified catechol-biphosphate as a selective inhibitor of the STAT5B SH2 domain. One of the catechol-biphosphate derivatives, stafib-1, was shown to be a nanomolar inhibitor of the STAT5B SH2 domain with more than 50-fold selectivity over the STAT5A-SH2 domain [323]. SAR studies based on the Stafib-1 compound and rational design led to the development of Stafib-2 with improved and high selective activity against STAT5B in CML cells [324].

Structure-based high-throughput virtual screening of four different chemical libraries followed by cell-based reporter assays identified STA-21 (or deoxytetrangomycin) as a new SH2 domain-targeting inhibitor of STAT3 [309]. STA-21 induced apoptosis of primary cells from chronic lymphoproliferative disorders of natural killer cells (CLPD-NK) and T-LGL patients and DLBCL cells when associated with YM155, a survivin suppressant [325,326]. A structural analog of STA-21, LLL-3, was developed to improve the cellular permeability of STA-21. LLL-3 was shown to promote apoptosis of CML cell lines and to synergistically act with IM to suppress CML cell growth and survival [327]. LLL-3 was further optimized by replacing its acetyl group with a sulfonamide to produce LLL-12, which prevented the phosphorylation of STAT3 and induced apoptosis in MM cell lines and primary MM cells in vitro, even in samples from patients with relapsed/refractory MM. LLL-12 also suppressed tumor formation in xenograft models of MM [328]. However, LLL-12 was required in high concentrations for in vivo studies due to low bioavailability. C188 compound was identified by virtual ligand screening of eight chemical libraries [310]. C188 was shown to inhibit P-Y-peptide binding to the SH2 domain of STAT3 by surface plasmon resonance (SPR) assays. Using C188 as the scaffold, C188-9 was identified in a hit-to-lead program and was shown to bind to the STAT3-SH2 domain with greater potency than the C188 compound. C188-9 effectively suppressed G-CSF-induced P-Y-STAT3 in AML cell lines and patient samples without affecting JAK or Src kinases. C188-9 also induced apoptosis and blocked the clonogenic growth of AML cells [329]. In silico screening studies followed by STAT3-dependent luciferase reporter gene assays led to the identification of STX-0119, another STAT3 inhibitor impacting the growth of lymphoma cell lines. Oral administration of STX-0119 was shown to reduce STAT3 target gene expression and to induce apoptosis in a xenograft model of lymphoma [311]. The STAT3-SH2 domain inhibitor STATTIC was identified by screening chemical libraries in an FP-based binding assay. This compound was demonstrated to selectively inhibit the dimerization and transcriptional activity of STAT3, although a report indicated that it also blocks STAT1 phosphorylation [330]. STATTIC reduced P-Y-STAT3 and viability of MM cells in three-dimensional (3D) culture and sensitized them to bortezomib, a proteasome inhibitor that is clinically used in MM treatment (this issue) [331]. Moreover, STATTIC suppressed growth and survival of NK/T cell lymphoma cell lines expressing the STAT3^{Y640F} mutant [332]. Among the STAT3-SH2 domain targeting inhibitors, OPB-31121 and OPB-51602 were the only compounds to have reached early phase clinical trials for treatment of hematopoietic malignancies [333]. OPB-31121 was shown to inhibit P-Y-STAT3 and P-Y-STAT5 without affecting upstream kinase activity in various myeloid leukemia cell lines expressing BCR-ABL, FLT3-ITD, or JAK2^{V617F}, as well as in BL and MM cell lines [334]. Treatment with OPB-31121 also induced growth inhibition of these hematopoietic malignant cells and reduced tumor formation in PDX models of ALL, CML, and AML. Computational docking and molecular dynamics simulations (MDS) demonstrated that OPB-31121 bound with high affinity to the SH2 domain of STAT3 [335]. In the same way, OPB-51602 was also shown to interact with high affinity with the STAT3 SH2 domain to inhibit P-Y-STAT3 and to be effective against MM, BL, and AML cells in preclinical in vitro and in vivo studies [336]. Although promising data were obtained from these preclinical studies, clinical trials were terminated for both of these inhibitors due to poor pharmacokinetic properties, significant toxicity and lack of antitumor activity. Very recently, SD-36, a proteolysis targeting chimera (PROTAC) that selectively degrades STAT3 protein has been developed. This molecule consists of the cell-permeable STAT3-SH2 domain-targeting inhibitor, SI-109, conjugated to a ligand of Cereblon, an important component of the E3 ubiquitin ligase complex that is involved in ubiquitination and degradation of multiple cellular proteins. The resulting PROTAC binds to and recruits both STAT3 and the E3 ligase to form a productive ternary complex for ubiquitination and degradation. SD-36 was shown to induce STAT3 degradation, cell growth arrest, and apoptosis in lymphoma and leukemia cells lines as well as tumor regression in multiple xenograft mouse models of leukemias and lymphomas [337]. Lastly, several plant-derived molecules were also demonstrated to be direct inhibitors of STAT3. Using computational modeling and docking simulations, Withaferin A, a natural product isolated from the medicinal plant *Withania somnifera*, was found to interact with the SH2 domain of STAT3. Consequently, Withaferin A prevented IL-6–mediated or persistently activated P-Y-STAT3 and induced apoptosis in MM cells [338]. In a similar vein, YL064, a derivative of Sinomenine, a plant component that has been used to treat rheumatic diseases, was shown to target the SH2 domain of STAT3 and to induce MM cell death [339].

4.3.2. Inhibitors Targeting the DNA Binding Domain (DBD)

STAT3/5 bind to specific DNA response elements within promoters to mediate transcriptional activation of target genes. Concerted efforts were therefore made to identify specific inhibitors targeting the DBD of STAT proteins. Most of the DBD inhibitors presented in this section were historically used against STAT3 and proved to be effective in some hematopoietic cancers.

Platinum (IV) compounds such as CPA-1, CPA-7, and platinum (IV) tetrachloride, were shown to inhibit STAT3 DNA binding activity [340]. IS3 295, a member of this class of molecules, was identified by screening of the NCI 2000 diversity set using electrophoretic mobility shift assays (EMSA). IS3 295 was demonstrated to irreversibly bind to the DBD of STAT3 in its active and inactive form and to prevent its interaction with specific response elements. This compound inhibited the constitutive DNA binding of STAT3 and induced apoptosis in MM cells [341]. Peptide aptamers as inhibitors of STAT3 represent one of the effective approaches to disrupt STAT3 DNA binding. The DBD-directed peptide aptamer DBD-1 was identified in yeast two-hybrid screenings, and its cell-penetrating form DBD-1-9R was shown to inhibit STAT3 DNA binding in EMSAs and to induce growth inhibition and apoptosis in MM cells [342].

Another class of STAT DBD inhibitors being used are duplex ODN (decoy oligonucleotides or dODN) and GQ-ODN. dODNs targeting STAT proteins are double-stranded DNA molecules mimicking the consensus STAT DNA binding sequence. These duplex ODNs act by competitively inhibiting the DNA binding of STAT proteins to their endogenous promoter elements thereby preventing their nuclear function [195]. For instance, a STAT5 dODN was shown to block DNA binding and transcriptional activity of STAT5 and to induce growth arrest and apoptosis in CML cell lines [343]. STAT3 dODN, and the second-generation cyclic STAT3 decoy (CS3D) with a longer half-life, were only tested in solid tumors and will not be discussed here. However, a dual-function molecule CpG-STAT3 dODN, which consists of a STAT3 dODN fused to the TLR9 agonist cytosine guanine dinucleotide (CpG), was reported to induce growth-inhibitory and immune-mediated effects against AML and DLBCL [74,344]. These compounds are rapidly internalized by $TLR9^{+}$ immune and malignant cells to block the oncogenic activity of STAT3 and to promote an antitumor immune response. Both TLR9 stimulation and concurrent STAT3 inhibition were critical for immune-mediated therapeutic effects. GQ-ODN are G-rich oligodeoxynucleotides that form intra- and inter-molecular four-stranded structures [195]. For example, the GQ-ODN T40214 was shown to inhibit IL-6–induced P-Y-STAT3 and STAT3-mediated transcription [345]. Computer-simulated docking studies indicated that GQ-ODNs interact mainly with the SH2 domain of STAT3 and are able to disrupt STAT3 dimers bound to DNA. Blocking STAT3 with the GQ-ODN T40214 loaded into nanoparticles was shown to be effective in a mouse model of T-ALL [346]. However, the use of GQ-ODN remains problematic due to the large size and potassium-dependence of this molecular probe.

Table 3. Direct inhibitors of STAT3 and STAT5 that have been tested in hematologic cancers.

Target	Drugs	Hematologic Malignancy	Stage of Clinical Development	References
		Direct STAT3 inhibitors		
mRNA	AZD9150 (IONIS-STAT3Rx)	AML, MDS DLBCL, HL, NHL	Preclinical (cell lines, primary cells PDX) Phase 1	[347–349]
mRNA	CpG-STAT3-siRNA	AML, MM	Preclinical (cell lines, primary cells, mouse model of AML, xenografts)	[73,350]
SH2 domain	BP-5-087, BP-1-102/17o (S3I-201 derivatives)	CML B- ALL, T-ALL MM	Preclinical (cell lines, primary cells) Preclinical (cell lines, xenografts) Preclinical (cell lines)	[94,316–318]
SH2 domain	CD38-S3I-NP (S3I-1757)	MM	Preclinical (cell lines, xenografts)	[313]
SH2 domain	C188-9	AML	Preclinical (cell lines, primary cells)	[329]
SH2 domain	LLL-3	CML	Preclinical (cell lines)	[327]
SH2 domain	LLL-12	MM	Preclinical (cell lines, primary cells, xenografts)	[328]
SH2 domain	OPB-51602	MM, NHL, AML, CML	Phase 1 (terminated)	[336]
SH2 domain	SF-1-066 (S3I-201 derivative)	AML, MM	Preclinical (cell lines)	[314,315]
SH2 domain	STA-21	CLPD-NKs T-LGL DLBCL	Preclinical (primary cells) Preclinical (primary cells) Preclinical (cell lines)	[325,326]
SH2 domain	STATTIC	MM NK lymphomas T-cell lymphomas	Preclinical (cell lines) Preclinical (primary cells) Preclinical (primary cells)	[331,332]
SH2 domain	STX-0119	NHL	Preclinical (cell lines, xenografts)	[311]
SH2 domain	SD-36	AML, ALCL	Preclinical (cell lines, xenografts)	[337]
SH2 domain	YL064	MM	Preclinical (cell lines, primary cells)	[339]
SH2 domain	Withaferin A	MM	Preclinical (cell lines)	[338]

Table 3. *Cont.*

Target	Drugs	Hematologic Malignancy	Stage of Clinical Development	References
DBD	CpG-STAT3dODN (decoy oligonucleotides)	AML DLBCL	Preclinical (cell lines, xenografts, mouse models) Preclinical (cell lines, xenografts, mouse models)	[74,344]
	DBD-1-9R (peptide aptamer)	MM	Preclinical (cell lines)	[342]
	IS3-295 (Platinum IV)	MM	Preclinical (cell lines)	[341]
	T40214 (GQ-ODN)	T-ALL	Preclinical (mouse model)	[346]
Direct STAT5 inhibitors				
SH2 domain	AC-4-130	AML, CML	Preclinical (cell lines, primary cells, xenografts)	[81,320]
	BP-1-107, BP-1-108	AML, CML	Preclinical (cell lines)	[319]
	IST5-002	CML Ph+ ALL	Preclinical (cell lines, primary cells) Preclinical (cell lines, primary cells, PDX)	[90,321]
	Stafib-1	CML	Preclinical (cell lines)	[323]
	Stafib-2	CML	Preclinical (cell lines)	[324]
DBD	STAT5 dODN	CML	Preclinical (cell lines)	[343]
Dual STAT3/STAT5 inhibitors				
SH2 domain	OPB-31121	CML, AML, ALL, BL, MM, NHL, MM	Preclinical (cell lines, xenografts, PDX) Phase 1 (terminated)	[334]

AML (acute myeloid leukemia), ALCL (anaplastic large cell lymphoma), B-ALL/T-ALL (B- or T- acute lymphoblastic leukemia), Ph^+ ALL (Philadelphia chromosome-positive acute lymphoblastic leukemia), BL (Burkitt's lymphoma), CLPD-NKs (chronic lymphoproliferative disorders–natural killer cells), CML (chronic myeloid leukemia), DLBCL (diffuse large B cell lymphoma), HL (Hodgkin's lymphoma), MDS (myelodysplastic syndrome), MM (multiple myeloma), NHL (non-Hodgkin's lymphoma), T-LGL (T cell large granular lymphocytic) leukemia.

4.3.3. Inhibitors Targeting STAT3/5 mRNAs

Oligonucleotide-based inhibition of STAT3 and STAT5 has also been achieved by using antisense oligonucleotides (ASO) or mRNA knockdown (siRNA) [195]. Although siRNAs and ASOs were widely used to illuminate the oncogenic activity of STAT3 and STAT5 in hematopoietic malignant cells, few of these molecules were developed as anticancer therapeutic agents. We will discuss here two examples of ASOs and siRNAs that gave promising results in preclinical studies and clinical trials. The first example is the chemically modified ASO, AZD9150, which specifically targets STAT3 mRNA and is now in phase 1/2 clinical trials. AZD9150, without any delivery agent, was shown to inhibit STAT3 expression in primary AML/MDS leukemia stem cells and to inhibit leukemic cell growth in vitro and in vivo using PDX models of AML/MDS [347]. Inhibitory effects of AZD9150 on STAT3 expression and cell growth were also demonstrated in preclinical models of lymphomas [348]. Furthermore, AZD9150 was well tolerated and demonstrated efficacy in patients with highly treatment-refractory lymphoma [349]. The second example is a STAT3 siRNA conjugated to the TLR9 ligand CpG described above. STAT3 silencing mediated by this CpG-siRNA inhibited tumor growth in xenograft models of AML or MM [350]. In an elegant study, Hossain et al. demonstrated that the CpG-STAT3 siRNA conjugate stimulates systemic antitumor immunity and antigen-specific activation of $CD8^+$ T cells in a mouse model of AML [73]. Intravenous administration of CpG-STAT3 siRNA showed a direct immunogenic effect on leukemic cells indicating that targeted STAT3 inhibition and TLR9 triggering blocks leukemia cell growth by promoting antitumor immunity rather than direct tumor cell killing. To date, no inhibitors targeting STAT5A/STAT5B mRNAs were developed for therapeutic purposes.

5. STAT3/5 Inhibitors in Hematopoietic Cancers: Inherent Limitations and Future Challenges

Despite intensive efforts made during the last 15 years, clinically applicable STAT3/5 inhibitors still remain elusive. Toxicity, poor tumor targeting, lack of cell penetrance and rapid degradation are some examples of the problems to be considered and addressed. There are also several other reasons that could hinder the development of such inhibitors. First, STAT3 and STAT5 display multifaceted activity due to their canonical and non-canonical functions in cancer cells. The definition of constitutively active STAT3/5 only based on P-Y-STAT3 and/or P-Y-STAT5 levels as described in chapter 3 may not be broadly representative. For example, the use of decoy ODNs or SH2 domain-targeting inhibitors would not be effective in CLL cells, in which a predominant mitochondrial localization and function of P-S^{727}-STAT3 has been evidenced [102]. Furthermore, it has been shown that nuclear translocation and DNA binding of STAT3 or STAT5 can occur independently of their P-Y status. These observations indicate that SH2 domain-targeting inhibitors may not be sufficient to fully abrogate STAT3/5 oncogenic functions, which may contribute to the limited success of these compounds. Second, extensive crosstalk and alternative signaling pathways present in TKO-driven hematopoietic malignancies probably render single-agent STAT3/5 inhibition less effective. Third, adverse effects of STAT3/5 inhibitors may be associated with the physiological function of these proteins in normal tissues including hematopoietic and lymphoid tissues, resulting in STAT3- or STAT5-specific toxicity. Previous studies indicated that STAT5-deficient NK cells induce angiogenesis and promote tumor progression, highlighting the potentially detrimental effects of STAT5 inhibitors on NK cell-mediated tumor immune surveillance [37]. Moreover, STAT3 and STAT5 were also shown to activate tumor suppressor pathways, strengthening the possible adverse effects of STAT3/5 inhibitors. Most of these tumor suppressor functions were described in different reviews including one published as part of this special issue [8,32]. One of the tumor suppressor activities of STAT3 is linked to the level of STAT3β expression, which could explain why different AML patient subsets, based on the STAT3β/STAT3α mRNA ratio, are more or less sensitive to STAT3 inhibitors [34]. In such a case, STAT3 direct inhibitors would also target STAT3β which is unwanted for AML treatment. Despite these limitations, optimization of currently available STAT3/5 inhibitors and discovery of new compounds that specifically target other functional domains such as the NH2-terminal domain, domains involved in the subcellular localization of STAT3/5 or STAT3/5/Rac1 interacting domains, as well as drugs activating negative regulators of STAT3/5 such as SOCS and PIAS proteins, still have to be considered in the future. Although most of the direct inhibitors have shown a moderate efficacy in treating hematologic cancers as single agents, they act synergistically with clinically used chemotherapeutic drugs or TKIs. Importantly, targeting STAT3 and/or STAT5 signaling was demonstrated to overcome drug resistance in hematopoietic malignant cells, suggesting that combination therapy using STAT3/5 inhibitors is the most attractive approach to fight against relapsed hematopoietic malignancies. For example, STAT3 and STAT5 were both recognized as important effectors of de novo and acquired IM-resistance in CML cells [57,60,62]. In fact, activation of STAT3 was demonstrated to be an important positive autocrine-paracrine feedback loop in the therapeutic treatment of oncogene-addicted cancer cells [351]. Activation of STAT3/5 elicited by bone marrow microenvironment-derived signals was found to mediate resistance of CML cells to IM treatment, and inhibition of STAT3/5 suppressed IM-resistant CML cells in the niche. In both cases, extrinsic activation of STAT3 or STAT5 was induced by JAK kinases, and the JAK1/2 inhibitor ruxolitinib was shown to act synergistically with IM in killing resistant CML cells [62,229]. In a similar vein, the STAT5-SH2 domain inhibitor AC-4-130 increased the cytotoxicity of ruxolitinib in AML cells [81]. Combined targeting of STAT3 and STAT5 is another approach to overcome TKI resistance in CML cells. For example, the triterpenoid CDDO-Me (bardoxolone methyl) was shown to act synergistically with IM to kill resistant BCR-ABL-expressing cells [63]. In addition, blocking STAT3 and/or STAT5 with natural compounds or drugs that have been proven to be safe in humans might help to reduce the side effects of combination therapies. For example, piperlongumine (Table 2) enhanced the anti-MM effect of the proteasome inhibitor bortezomib, while pimozide increased the cytotoxic effects of TKIs in myeloid leukemias [277,298]. Besides its oncogenic activity, STAT3 is a central immune checkpoint regulator in cancer cells and tumor-associated immune cells. Oncogenic STAT3 drives PD-L1 expression in lymphoma and AML cells and may promote tumor immune evasion by inducing an immunosuppressive microenvironment [73,114]. In this context, the combination of STAT3 inhibitors with anti-checkpoint antibodies blocking PD-L1/programmed cell death 1 (PD-1) interactions might be a promising therapeutic approach for lymphomas and leukemias. Collectively, these data indicate that combination therapies using inhibitors that indirectly or directly target STAT3 and/or STAT5 might provide new therapeutic opportunities for relapsed or refractory hematopoietic malignancies.

6. Conclusions

The design and development of STAT3/5 inhibitors evolved rapidly during the last 10 years. Computer simulations and other in silico studies such as high-throughput virtual screening, cell-based assays, biophysical and biochemical approaches led to the identification of selective STAT3/5 inhibitors with potent anticancer effects. STAT3/5 inhibitors are likely to become a valuable addition to the expanding arsenal of drugs against hematopoietic cancers and solid tumors. However, there are some limitations to this relative success story that must be taken into account. Optimization of STAT3/STAT5 inhibitors by chemical modifications and drug delivery systems are both required for resolving issues linked to stability, cell permeability, and targeted delivery to tumor cells. The development of antibodies conjugated to STAT3/5 inhibitors or antibody-conjugated nanoparticles harboring STAT3/5 inhibitors might be promising but remains challenging. Identification of new inhibitors targeting the non-canonical functions of STAT3/5 is also desirable in treating some hematologic malignancies. Last but not least, the optimization of combination therapies using STAT3/5 inhibitors with molecules targeting tyrosine kinases or other key players in cancer will be required for finding the right combination that safely unlocks drug resistance in hematologic cancers.

Author Contributions: M.B.-B., M.P., H.A.N., L.J. and D.H. performed the literature search, designed the draft of the review and drew all figures and tables. M.-C.V.-M., G.P. and F.G. contributed to the editing and critical reading of the manuscript and provided the final version of the manuscript. All authors have read and agreed to the published version of the manuscript.

Acknowledgments: Open Access Funding by the Austrian Science Fund (FWF).

References

1. Leonard, W.J.; O'Shea, J.J. JAKs and STATs: Biological implications. *Annu. Rev. Immunol.* **1998**, *16*, 293–322. [CrossRef]
2. Leaman, D.W.; Leung, S.; Li, X.; Stark, G.R. Regulation of STAT-dependent pathways by growth factors and cytokines. *FASEB J.* **1996**, *10*, 1578–1588. [CrossRef] [PubMed]
3. Kiu, H.; Nicholson, S.E. Biology and significance of the JAK/STAT signalling pathways. *Growth Factors* **2012**, *30*, 88–106. [CrossRef] [PubMed]
4. Yu, H.; Jove, R. The STATs of cancer—New molecular targets come of age. *Nat. Rev. Cancer* **2004**, *4*, 97–105. [CrossRef] [PubMed]
5. Liao, Z.; Nevalainen, M.T. Targeting transcription factor Stat5a/b as a therapeutic strategy for prostate cancer. *Am. J. Transl. Res.* **2011**, *3*, 133–138.
6. Benekli, M.; Baer, M.R.; Baumann, H.; Wetzler, M. Signal transducer and activator of transcription proteins in leukemias. *Blood* **2003**, *101*, 2940–2954. [CrossRef]
7. Dorritie, K.A.; McCubrey, J.A.; Johnson, D.E. STAT transcription factors in hematopoiesis and leukemogenesis: Opportunities for therapeutic intervention. *Leukemia* **2014**, *28*, 248–257. [CrossRef]
8. Igelmann, S.; Neubauer, H.A.; Ferbeyre, G. STAT3 and STAT5 activation in solid cancers. *Cancers* **2019**, *11*, 1428. [CrossRef]

9. Vainchenker, W.; Constantinescu, S.N. JAK/STAT signaling in hematological malignancies. *Oncogene* **2013**, *32*, 2601–2613. [CrossRef]
10. Yu, C.L.; Meyer, D.J.; Campbell, G.S.; Larner, A.C.; Carter-Su, C.; Schwartz, J.; Jove, R. Enhance DNA-binding activity of a Stat3-related protein in cells transformed by the Src oncoprotein. *Science* **1995**, *269*, 81–83. [CrossRef]
11. Lacronique, V.; Boureux, A.; Monni, R.; Dumon, S.; Mauchauffe, M.; Mayeux, P.; Gouilleux, F.; Berger, R.; Gisselbrecht, S.; Ghysdael, J.; et al. Transforming properties of chimeric TEL-JAK proteins in Ba/F3 cells. *Blood* **2000**, *95*, 2076–2083. [CrossRef] [PubMed]
12. Harir, N.; Boudot, C.; Friedbichler, K.; Sonneck, K.; Kondo, R.; Martin-Lanneree, S.; Kenner, L.; Kerenyi, M.; Yahiaoui, S.; Gouilleux-Gruart, V.; et al. Oncogenic Kit controls neoplastic mast cell growth through a Stat5/PI3-kinase signaling cascade. *Blood* **2008**, *112*, 2463–2473. [CrossRef] [PubMed]
13. Kanie, T.; Abe, A.; Matsuda, T.; Kuno, Y.; Towatari, M.; Yamamoto, T.; Saito, H.; Emi, N.; Naoe, T. TEL-Syk fusion constitutively activates PI3-K/Akt, MAPK and JAK2-independent STAT5 signal pathways. *Leukemia* **2004**, *18*, 548–555. [CrossRef] [PubMed]
14. Mizuki, M.; Fenski, R.; Halfter, H.; Matsumura, I.; Schmidt, R.; Müller, C.; Grüning, W.; Kratz-Albers, K.; Serve, S.; Steur, C.; et al. Flt3 mutations from patients with acute myeloid leukemia induce transformation of 32D cells mediated by the Ras and STAT5 pathways. *Blood* **2000**, *96*, 3907–3914. [CrossRef]
15. Shuai, K.; Halpern, J.; ten Hoeve, J.; Rao, X.; Sawyers, C.L. Constitutive activation of STAT5 by the BCR-ABL oncogene in chronic myelogenous leukemia. *Oncogene* **1996**, *13*, 247–254.
16. Carlesso, N.; Frank, D.A.; Griffin, J.D. Tyrosyl phosphorylation and DNA binding activity of signal transducers and activators of transcription (STAT) proteins in hematopoietic cell lines transformed by Bcr-Abl. *J. Exp. Med.* **1996**, *183*, 811–820. [CrossRef]
17. Nieborowska-Skorska, M.; Slupianek, A.; Xue, L.; Zhang, Q.; Raghunath, P.N.; Hoser, G.; Wasik, M.A.; Morris, S.W.; Skorski, T. Role of signal transducer and activator of transcription 5 in nucleophosmin/anaplastic lymphoma kinase-mediated malignant transformation of lymphoid cells. *Cancer Res.* **2001**, *61*, 6517–6523.
18. Zamo, A.; Chiarle, R.; Piva, R.; Howes, J.; Fan, Y.; Chilosi, M.; Levy, D.E.; Inghirami, G. Anaplastic lymphoma kinase (ALK) activates Stat3 and protects hematopoietic cells from cell death. *Oncogene* **2002**, *21*, 1038–1047. [CrossRef]
19. Wilbanks, A.M.; Mahajan, S.; Frank, D.A.; Druker, B.J.; Gilliland, D.G.; Carroll, M. TEL/PDGFbetaR fusion protein activates STAT1 and STAT5: A common mechanism for transformation by tyrosine kinase fusion proteins. *Exp. Hematol.* **2000**, *28*, 584–593. [CrossRef]
20. Bromberg, J.F.; Wrzeszczynska, M.H.; Devgan, G.; Zhao, Y.; Pestell, R.G.; Albanese, C.; Darnell, J.E., Jr. Stat3 as an oncogene. *Cell* **1999**, *98*, 295–303. [CrossRef]
21. Schwaller, J.; Parganas, E.; Wang, D.; Cain, D.; Aster, J.C.; Williams, I.R.; Lee, C.K.; Gerthner, R.; Kitamura, T.; Frantsve, J.; et al. Stat5 is essential for the myelo- and lymphoproliferative disease induced by TEL/JAK2. *Mol. Cell* **2000**, *6*, 693–704. [CrossRef]
22. Onishi, M.; Nosaka, T.; Misawa, K.; Mui, A.L.F.; Gorman, D.; McMahon, M.; Miyajima, A.; Kitamura, T. Identification and Characterization of a Constitutively Active STAT5 Mutant That Promotes Cell Proliferation. *Mol. Cell. Biol.* **1998**, *18*, 3871–3879. [CrossRef] [PubMed]
23. Moriggl, R.; Sexl, V.; Kenner, L.; Duntsch, C.; Stangl, K.; Gingras, S.; Hoffmeyer, A.; Bauer, A.; Piekorz, R.; Wang, D.; et al. Stat5 tetramer formation is associated with leukemogenesis. *Cancer Cell* **2005**, *7*, 87–99. [CrossRef] [PubMed]
24. Kontro, M.; Kuusanmäki, H.; Eldfors, S.; Burmeister, T.; Andersson, E.I.; Bruserud, O.; Brümmendorf, T.H.; Edgren, H.; Gjertsen, B.T.; Itälä-Remes, M.; et al. Novel activating STAT5B mutations as putative drivers of T-cell acute lymphoblastic leukemia. *Leukemia* **2014**, *28*, 1738–1742. [CrossRef] [PubMed]
25. Nicolae, A.; Xi, L.; Pittaluga, S.; Abdullaev, Z.; Pack, S.D.; Chen, J.; Waldmann, T.A.; Jaffe, E.S.; Raffeld, M. Frequent STAT5B mutations in $\gamma\delta$ hepatosplenic T-cell lymphomas. *Leukemia* **2014**, *28*, 2244–2248. [CrossRef] [PubMed]
26. Koskela, H.L.; Eldfors, S.; Ellonen, P.; van Adrichem, A.J.; Kuusanmaki, H.; Andersson, E.I.; Lagström, S.; Clemente, M.J.; Olson, T.; Jalkanen, S.E.; et al. Somatic STAT3 mutations in large granular lymphocytic leukemia. *N. Engl. J. Med.* **2012**, *366*, 1905–1913. [CrossRef]

27. Küçük, C.; Jiang, B.; Hu, X.; Zhang, W.; Chan, J.K.; Xiao, W.; Lack, N.; Alkan, C.; Williams, J.C.; Avery, K.N.; et al. Activating mutations of STAT5B and STAT3 in lymphomas derived from $\gamma\delta$-T or NK cells. *Nat. Commun.* **2015**, *6*, 6025. [CrossRef]
28. Bandapalli, O.R.; Schuessele, S.; Kunz, J.B.; Rausch, T.; Stütz, A.M.; Tal, N.; Geron, I.; Gershman, N.; Izraeli, S.; Eilers, J.; et al. The activating STAT5B N642H mutation is a common abnormality in pediatric T-cell acute lymphoblastic leukemia and confers a higher risk of relapse. *Haematologica* **2014**, *99*, 188–192. [CrossRef]
29. Pham, H.T.T.; Maurer, B.; Prchal-Murphy, M.; Grausenburger, R.; Grundschober, E.; Javaheri, T.; Nivarthi, H.; Boersma, A.; Kolbe, T.; Elabd, M.; et al. STAT5BN642H is a driver mutation for T cell neoplasia. *J. Clin. Investig.* **2018**, *128*, 387–401. [CrossRef]
30. de Araujo, E.D.; Erdogan, F.; Neubauer, H.A.; Meneksedag-Erol, D.; Manaswiyoungkul, P.; Eram, M.S.; Seo, H.S.; Qadree, A.K.; Israelian, J.; Orlova, A.; et al. Structural and functional consequences of the STAT5BN642H driver mutation. *Nat. Commun.* **2019**, *10*, 2517. [CrossRef]
31. Hosui, A.; Kimura, A.; Yamaji, D.; Zhu, B.M.; Na, R.; Hennighausen, L. Loss of STAT5 causes liver fibrosis and cancer development through increased TGF-beta and STAT3 activation. *J. Exp. Med.* **2009**, *206*, 819–831. [CrossRef] [PubMed]
32. Ferbeyre, G.; Moriggl, R. The role of Stat5 transcription factors as tumor suppressors or oncogenes. *Biochim. Biophys. Acta* **2011**, *1815*, 104–114. [CrossRef] [PubMed]
33. Zhang, H.F.; Lai, R. STAT3 in Cancer—Friend or Foe. *Cancers* **2014**, *6*, 1408–1440. [CrossRef] [PubMed]
34. Aigner, P.; Mizutani, T.; Horvath, J.; Eder, T.; Heber, S.; Lind, K.; Just, V.; Moll, H.P.; Yeroslaviz, A.; Fischer, M.J.M.; et al. STAT3β is a tumor suppressor in acute myeloid leukemia. *Blood Adv.* **2019**, *3*, 1989–2002. [CrossRef] [PubMed]
35. Zhang, Q.; Wang, H.Y.; Liu, X.; Wasik, M.A. STAT5A is epigenetically silenced by the tyrosine kinase NPM1-ALK and acts as a tumor suppressor by reciprocally inhibiting NPM1-ALK expression. *Nat. Med.* **2007**, *13*, 1341–1348. [CrossRef] [PubMed]
36. Yu, H.; Kortylewski, M.; Pardoll, D. Crosstalk between cancer and immune cells: Role of STAT3 in the tumour microenvironment. *Nat. Rev. Immunol.* **2007**, *7*, 41–51. [CrossRef]
37. Gotthardt, D.; Putz, E.M.; Grundschober, E.; Prchal-Murphy, M.; Straka, E.; Kudweis, P.; Heller, G.; Bago-Horvath, Z.; Witalisz-Siepracka, A.; Cumaraswamy, A.A.; et al. STAT5 Is a Key Regulator in NK Cells and Acts as a Molecular Switch from Tumor Surveillance to Tumor Promotion. *Cancer Discov.* **2016**, *6*, 414–429. [CrossRef]
38. Gouilleux-Gruart, V.; Gouilleux, F.; Desaint, C.; Claisse, J.F.; Capiod, J.C.; Delobel, J.; Weber-Nordt, R.; Dusanter-Fourt, I.; Dreyfus, F.; Groner, B.; et al. STAT related transcription factors are constitutively activated in peripheral blood cells from acute leukemia patients. *Blood* **1996**, *87*, 1692–1697. [CrossRef]
39. Spivak, J.L. The chronic myeloproliferative disorders: Clonality and clinical heterogeneity. *Semin. Hematol.* **2004**, *41*, 1–5. [CrossRef]
40. Van Etten, R.A.; Koschmieder, S.; Delhommeau, F.; Perrotti, D.; Holyoake, T.; Pardanani, A.; Mesa, R.; Green, T.; Ibrahim, A.R.; Mughal, T.; et al. The Ph-positive and Ph-negative myeloproliferative neoplasms: Some topical pre-clinical and clinical issues. *Haematologica* **2011**, *96*, 590–601. [CrossRef]
41. Vainchenker, W.; Kralovics, R. Genetic basis and molecular pathophysiology of classical myeloproliferative neoplasms. *Blood* **2017**, *129*, 667–679. [CrossRef] [PubMed]
42. James, C.; Ugo, V.; Le Couedic, J.P.; Staerk, J.; Delhommeau, F.; Lacout, C.; Garçon, L.; Raslova, H.; Berger, R.; Bennaceur-Griscelli, A.; et al. A unique clonal JAK2 mutation leading to constitutive signalling causes polycythaemia vera. *Nature* **2005**, *434*, 1144–1148. [CrossRef] [PubMed]
43. Kralovics, R.; Passamonti, F.; Busers, A.S.; Teo, S.S.; Tiedt, R.; Passweg, J.R.; Tichelli, A.; Cazzola, M.; Skoda, R.C. A gain-of-function mutation of JAK2 in myeloproliferative disorders. *N. Engl. J. Med.* **2005**, *352*, 1779–1790. [CrossRef] [PubMed]
44. Valent, P.; Horny, H.P.; Escribano, L.; Longley, B.J.; Li, C.Y.; Schwartz, L.B.; Marone, G.; Nuñez, R.; Akin, C.; Sotlar, K.; et al. Diagnostic criteria and classification of mastocytosis: A consensus proposal. *Leuk. Res.* **2001**, *25*, 603–625. [CrossRef]
45. Lim, K.H.; Pardanani, A.; Tefferi, A. KIT and mastocytosis. *Acta Haematol.* **2008**, *119*, 194–198. [CrossRef]
46. Goemans, B.F.; Zwaan, C.M.; Miller, M.; Zimmermann, M.; Harlow, A.; Meshinchi, S.; Loonen, A.H.; Hählen, K.; Reinhardt, D.; Creutzig, U.; et al. Mutations in KIT and RAS are frequent events in pediatric core-binding factor acute myeloid leukemia. *Leukemia* **2005**, *19*, 1536–1542. [CrossRef]

47. Fritsche-Polanz, R.; Fritz, M.; Huber, A.; Sotlar, K.; Sperr, W.R.; Mannhalter, C.; Födinger, M.; Valent, P. High frequency of concomitant mastocytosis in patients with acute myeloid leukemia exhibiting the transforming KIT mutation D816V. *Mol. Oncol.* **2010**, *4*, 335–346. [CrossRef]
48. Coppo, P.; Flamant, S.; De Mas, V.; Jarrier, P.; Guillier, M.; Bonnet, M.L.; Lacout, C.; Guilhot, F.; Vainchenker, W.; Turhan, A.G. BCR-ABL activates STAT3 via JAK and MEK pathways in human cells. *Br. J. Haematol.* **2006**, *134*, 171–179. [CrossRef]
49. Hoelbl, A.; Schuster, C.; Kovacic, B.; Zhu, B.; Wickre, M.; Hoelzl, M.A.; Fajmann, S.; Grebien, F.; Warsch, W.; Stengl, G.; et al. Stat5 is indispensable for the maintenance of Bcr/abl-positive leukemia. *EMBO Mol. Med.* **2010**, *2*, 98–110. [CrossRef]
50. Chung, Y.J.; Park, B.B.; Kang, Y.J.; Kim, T.M.; Eaves, C.J.; Oh, I.H. Unique effects of Stat3 on the early phase of hematopoietic stem cell regeneration. *Blood* **2006**, *108*, 1208–1215. [CrossRef]
51. Schepers, H.; van Gosliga, D.; Wierenga, A.T.; Eggen, B.J.; Schuringa, J.J.; Vellenga, E. STAT5 is required for long-term maintenance of normal and leukemic human stem/progenitor cells. *Blood* **2007**, *110*, 2880–2888. [CrossRef] [PubMed]
52. Kato, Y.; Iwama, A.; Tadokoro, Y.; Shimoda, K.; Minoguchi, M.; Akira, S.; Tanaka, M.; Miyajima, A.; Kitamura, T.; Nakauchi, H. Selective activation of STAT5 unveils its role in stem cell self-renewal in normal and leukemic hematopoiesis. *J. Exp. Med.* **2005**, *202*, 169–179. [CrossRef] [PubMed]
53. Ghanem, S.; Friedbichler, K.; Boudot, C.; Bourgeais, J.; Gouilleux-Gruart, V.; Régnier, A.; Herault, O.; Moriggl, R.; Gouilleux, F. STAT5A/5B-specific expansion and transformation of hematopoietic stem cells. *Blood Cancer J.* **2017**, *7*, 514. [CrossRef] [PubMed]
54. Kollmann, S.; Grundschober, E.; Maurer, B.; Warsch, W.; Grausenburger, R.; Edlinger, L.; Huuhtanen, J.; Lagger, S.; Hennighausen, L.; Valent, P.; et al. Twins with different personalities: STAT5B-but not STAT5A-has a key role in BCR/ABL-induced leukemia. *Leukemia* **2019**, *33*, 1583–1597. [CrossRef] [PubMed]
55. Chomel, J.C.; Bonnet, M.L.; Sorel, N.; Sloma, I.; Bennaceur-Griscelli, A.; Rea, D.; Legros, L.; Marfaing-Koka, A.; Bourhis, J.H.; Ame, S.; et al. Leukemic stem cell persistence in chronic myeloid leukemia patients in deep molecular response induced by tyrosine kinase inhibitors and the impact of therapy discontinuation. *Oncotarget* **2016**, *7*, 35293–35301. [CrossRef] [PubMed]
56. Gambacorti-Passerini, C.B.; Gunby, R.H.; Piazza, R.; Galietta, A.; Rostagno, R.; Scapozza, L. Molecular mechanisms of resistance to imatinib in Philadelphia-chromosome-positive leukaemias. *Lancet Oncol.* **2003**, *4*, 75–85. [CrossRef]
57. Warsch, W.; Kollmann, K.; Eckelhart, E.; Fajmann, S.; Cerny-Reiterer, S.; Hölbl, A.; Gleixner, K.V.; Dworzak, M.; Mayerhofer, M.; Hoermann, G.; et al. High STAT5 levels mediate imatinib resistance and indicate disease progression in chronic myeloid leukemia. *Blood* **2011**, *117*, 3409–3420. [CrossRef]
58. Warsch, W.; Grundschober, E.; Berger, A.; Gille, L.; Cerny-Reiterer, S.; Tigan, A.S.; Hoelbl-Kovacic, A.; Valent, P.; Moriggl, R.; Sexl, V. STAT5 triggers BCR-ABL1 mutation by mediating ROS production in chronic myeloid leukemia. *Oncotarget* **2012**, *3*, 1669–1687. [CrossRef]
59. Bourgeais, J.; Ishac, N.; Medrzycki, M.; Brachet-Botineau, M.; Desbourdes, L.; Gouilleux-Gruart, V.; Pecnard, E.; Rouleux-Bonnin, F.; Gyan, E.; Domenech, J.; et al. Oncogenic STAT5 signaling promotes oxidative stress in chronic myeloid leukemia cells by repressing antioxidant defenses. *Oncotarget* **2017**, *8*, 41876–41889. [CrossRef]
60. Prost, S.; Relouzat, F.; Spentchian, M.; Ouzegdouh, Y.; Saliba, J.; Massonnet, G.; Beressi, J.P.; Verhoeyen, E.; Raggueneau, V.; Maneglier, B.; et al. Erosion of the chronic myeloid leukemia stem cell pool by PPARγ agonists. *Nature* **2015**, *525*, 380–383. [CrossRef]
61. Casetti, L.; Martin-Lannerée, S.; Najjar, I.; Plo, I.; Augé, S.; Roy, L.; Chomel, J.C.; Lauret, E.; Turhan, A.G.; Dusanter-Fourt, I. Differential contributions of STAT5A and STAT5B to stress protection and tyrosine kinase inhibitor resistance of chronic myeloid leukemia stem/progenitor cells. *Cancer Res.* **2013**, *73*, 2052–2058. [CrossRef] [PubMed]
62. Bewry, N.N.; Nair, R.R.; Emmons, M.; Boulware, D.; Pinilla-Ibarz, J.; Hazlehurst, L.A. Stat3 contributes to resistance toward BCR-ABL inhibitors in a bone marrow microenvironment model of drug resistance. *Mol. Cancer Ther.* **2008**, *7*, 3169–3175. [CrossRef] [PubMed]

63. Gleixner, K.V.; Schneeweiss, M.; Eisenwort, G.; Berger, D.; Herrmann, H.; Blatt, K.; Greiner, G.; Byrgazov, K.; Hoermann, G.; Konopleva, M.; et al. Combined targeting of STAT3 and STAT5: A novel approach to overcome drug resistance in chronic myeloid leukemia. *Haematologica* **2017**, *102*, 1519–1529. [CrossRef] [PubMed]
64. Walz, C.; Ahmed, W.; Lazarides, K.; Betancur, M.; Patel, N.; Hennighausen, L.; Zaleskas, V.M.; Van Etten, R.A. Essential role for Stat5a/b in myeloproliferative neoplasms induced by BCR-ABL1 and JAK2(V617F) in mice. *Blood* **2012**, *119*, 3550–3560. [CrossRef]
65. Yan, D.; Hutchison, R.E.; Mohi, G. Critical requirement for Stat5 in a mouse model of polycythemia vera. *Blood* **2012**, *119*, 3539–3549. [CrossRef]
66. Yan, D.; Jobe, F.; Hutchison, R.E.; Mohi, G. Deletion of Stat3 enhances myeloid cell expansion and increases the severity of myeloproliferative neoplasms in Jak2V617F knock-in mice. *Leukemia* **2015**, *29*, 2050–2061. [CrossRef]
67. Grisouard, J.; Shimizu, T.; Duek, A.; Kubovcakova, L.; Hao-Shen, H.; Dirnhofer, S.; Skoda, R.C. Deletion of Stat3 in hematopoietic cells enhances thrombocytosis and shortens survival in a JAK2-V617F mouse model of MPN. *Blood* **2015**, *125*, 2131–2140. [CrossRef]
68. Chaix, A.; Lopez, S.; Voisset, E.; Gros, L.; Dubreuil, P.; De Sepulveda, P. Mechanisms of STAT protein activation by oncogenic KIT mutants in neoplastic mast cells. *J. Biol. Chem.* **2011**, *286*, 5956–5966. [CrossRef]
69. Gilliland, D.G. Molecular genetics of human leukemias: New insights into therapy. *Semin. Hematol.* **2002**, *39*, 6–11. [CrossRef]
70. Martignoles, J.A.; Delhommeau, F.; Hirsch, P. Genetic Hierarchy of Acute Myeloid Leukemia: From Clonal Hematopoiesis to Molecular Residual Disease. *Int. J. Mol. Sci.* **2018**, *19*, 3850. [CrossRef]
71. Benekli, M.; Xia, Z.; Donohue, K.A.; Ford, L.A.; Pixley, L.A.; Baer, M.R.; Baumann, H.; Wetzler, M. Constitutive activity of signal transducer and activator of transcription 3 protein in acute myeloid leukemia blasts is associated with short disease-free survival. *Blood* **2002**, *99*, 252–257. [CrossRef] [PubMed]
72. Larrue, C.; Heydt, Q.; Saland, E.; Boutzen, H.; Kaoma, T.; Sarry, J.E.; Joffre, C.; Récher, C. Oncogenic KIT mutations induce STAT3-dependent autophagy to support cell proliferation in acute myeloid leukemia. *Oncogenesis* **2019**, *8*, 39. [CrossRef] [PubMed]
73. Hossain, D.M.; Dos Santos, C.; Zhang, Q.; Kozlowska, A.; Liu, H.; Gao, C.; Moreira, D.; Swiderski, P.; Jozwiak, A.; Kline, J.; et al. Leukemia cell-targeted STAT3 silencing and TLR9 triggering generate systemic antitumor immunity. *Blood* **2014**, *123*, 15–25. [CrossRef] [PubMed]
74. Zhang, Q.; Hossain, D.M.; Duttagupta, P.; Moreira, D.; Zhao, X.; Won, H.; Buettner, R.; Nechaev, S.; Majka, M.; Zhang, B.; et al. Serum-resistant CpG-STAT3 decoy for targeting survival and immune checkpoint signaling in acute myeloid leukemia. *Blood* **2016**, *127*, 1687–1700. [CrossRef]
75. Kiyoi, H.; Naoe, T. FLT3 in human hematologic malignancies. *Leuk. Lymphoma* **2002**, *43*, 1541–1547. [CrossRef]
76. Spiekermann, K.; Bagrintseva, K.; Schwab, R.; Schmieja, K.; Hiddemann, W. Overexpression and constitutive activation of FLT3 induces STAT5 activation in primary acute myeloid leukemia blast cells. *Clin. Cancer Res.* **2003**, *9*, 2140–2150.
77. Obermann, E.C.; Arber, C.; Jotterand, M.; Tichelli, A.; Hirschmann, P.; Tzankov, A. Expression of pSTAT5 predicts FLT3 internal tandem duplications in acute myeloid leukemia. *Ann. Hematol.* **2010**, *89*, 663–669. [CrossRef]
78. Rocnik, J.L.; Okabe, R.; Yu, J.C.; Lee, B.H.; Giese, N.; Schenkein, D.P.; Gilliland, D.G. Roles of tyrosine 589 and 591 in STAT5 activation and transformation mediated by FLT3-ITD. *Blood* **2006**, *108*, 339–345. [CrossRef]
79. Nelson, E.A.; Walker, S.R.; Xiang, M.; Weisberg, E.; Bar-Natan, M.; Barrett, R.; Liu, S.; Kharbanda, S.; Christie, A.L.; Nicolais, M.; et al. The STAT5 Inhibitor Pimozide Displays Efficacy in Models of Acute Myelogenous Leukemia Driven by FLT3 Mutations. *Genes Cancer* **2012**, *3*, 503–511. [CrossRef]
80. Müller, T.A.; Grundler, R.; Istvanffy, R.; Rudelius, M.; Hennighausen, L.; Illert, A.L.; Duyster, J. Lineage-specific STAT5 target gene activation in hematopoietic progenitor cells predicts the FLT3(+) -mediated leukemic phenotype. *Leukemia* **2016**, *30*, 1725–1733. [CrossRef]
81. Wingelhofer, B.; Maurer, B.; Heyes, E.C.; Cumaraswamy, A.A.; Berger-Becvar, A.; de Araujo, E.D.; Orlova, A.; Freund, P.; Ruge, F.; Park, J.; et al. Pharmacologic inhibition of STAT5 in acute myeloid leukemia. *Leukemia* **2018**, *32*, 1135–1146. [CrossRef] [PubMed]

82. Sallmyr, A.; Fan, J.; Datta, K.; Kim, K.T.; Grosu, D.; Shapiro, P.; Small, D.; Rassool, F. Internal tandem duplication of FLT3 (FLT3/ITD) induces increased ROS production, DNA damage, and misrepair: Implications for poor prognosis in AML. *Blood* **2008**, *111*, 3173–3182. [CrossRef] [PubMed]
83. Woolley, J.F.; Naughton, R.; Stanicka, J.; Gough, D.R.; Bhatt, L.; Dickinson, B.C.; Chang, C.J.; Cotter, T.G. H_2O_2 production downstream of FLT3 is mediated by p22phox in the endoplasmic reticulum and is required for STAT5 signalling. *PLoS ONE* **2012**, *7*, e34050. [CrossRef] [PubMed]
84. Dumas, P.Y.; Naudin, C.; Martin-Lannerée, S.; Izac, B.; Casetti, L.; Mansier, O.; Rousseau, B.; Artus, A.; Dufossée, M.; Giese, A.; et al. Hematopoietic niche drives FLT3-ITD acute myeloid leukemia resistance to quizartinib via STAT5-and hypoxia-dependent upregulation of AXL. *Haematologica* **2019**, *104*, 2017–2027. [CrossRef] [PubMed]
85. Arnould, C.; Philippe, C.; Bourdon, V.; Gregoire, M.J.; Berger, R.; Jonveaux, P. The signal transducer and activator of transcription STAT5b gene is a new partner of retinoic acid receptor alpha in acute promyelocytic-like leukaemia. *Hum. Mol. Genet.* **1999**, *8*, 1741–1749. [CrossRef] [PubMed]
86. Dong, S.; Tweardy, D.J. Interactions of STAT5b-RARalpha, a novel acute promyelocytic leukemia fusion protein, with retinoic acid receptor and STAT3 signaling pathways. *Blood* **2002**, *99*, 2637–2646. [CrossRef] [PubMed]
87. Pui, C.H.; Gajjar, A.J.; Kane, J.R.; Qaddoumi, I.A.; Pappo, A.S. Challenging issues in pediatric oncology. *Nat. Rev. Clin. Oncol.* **2011**, *8*, 540–549. [CrossRef]
88. Hoelbl, A.; Kovacic, B.; Kerenyi, M.A.; Simma, O.; Warsch, W.; Cui, Y.; Beug, H.; Hennighausen, L.; Moriggl, R.; Sexl, V. Clarifying the role of Stat5 in lymphoid development and Abelson induced transformation. *Blood* **2006**, *107*, 4898–4906. [CrossRef]
89. Joliot, V.; Cormier, F.; Medyouf, H.; Alcalde, H.; Ghysdael, J. Constitutive STAT5 activation specifically cooperates with the loss of p53 function in B-cell lymphomagenesis. *Oncogene* **2006**, *25*, 4573–4584. [CrossRef]
90. Minieri, V.; De Dominici, M.; Porazzi, P.; Mariani, S.A.; Spinelli, O.; Rambaldi, A.; Peterson, L.F.; Porcu, P.; Nevalainen, M.T.; Calabretta, B. Targeting STAT5 or STAT5-Regulated Pathways Suppresses Leukemogenesis of Ph+ Acute Lymphoblastic Leukemia. *Cancer Res.* **2018**, *78*, 5793–5807. [CrossRef]
91. Katerndahl, C.D.S.; Heltemes-Harris, L.M.; Willette, M.J.L.; Henzler, C.M.; Frietze, S.; Yang, R.; Schjerven, H.; Silverstein, K.A.T.; Ramsey, L.B.; Hubbard, G.; et al. Antagonism of B cell enhancer networks by STAT5 drives leukemia and poor patient survival. *Nat. Immunol.* **2017**, *18*, 694–704. [CrossRef] [PubMed]
92. Heltemes-Harris, L.M.; Willette, M.J.; Ramsey, L.B.; Qiu, Y.H.; Neeley, E.S.; Zhang, N.; Thomas, D.A.; Koeuth, T.; Baechler, E.C.; Kornblau, S.M.; et al. Ebf1 or Pax5 haploinsufficiency synergizes with STAT5 activation to initiate acute lymphoblastic leukemia. *J. Exp. Med.* **2011**, *208*, 1135–1149. [CrossRef] [PubMed]
93. Wang, Z.; Medrzycki, M.; Bunting, S.T.; Bunting, K.D. Stat5-deficient hematopoiesis is permissive for Myc-induced B-cell leukemogenesis. *Oncotarget* **2015**, *6*, 28961–28972. [CrossRef] [PubMed]
94. Belton, A.; Xian, L.; Huso, T.; Koo, M.; Luo, L.Z.; Turkson, J.; Page, B.D.; Gunning, P.T.; Liu, G.; Huso, D.L.; et al. STAT3 inhibitor has potent antitumor activity in B-lineage acute lymphoblastic leukemia cells overexpressing the high mobility group A1 (HMGA1)-STAT3 pathway. *Leuk. Lymphoma* **2016**, *57*, 2681–2684. [CrossRef]
95. Fasan, A.; Kern, W.; Grossmann, V.; Haferlach, C.; Haferlach, T.; Schnittger, S. STAT3 mutations are highly specific for large granular lymphocytic leukemia. *Leukemia* **2013**, *27*, 1598–1600. [CrossRef]
96. Rajala, H.L.; Porkka, K.; Maciejewski, J.P.; Loughran, T.P., Jr.; Mustjoki, S. Uncovering the pathogenesis of large granular lymphocytic leukemia-novel STAT3 and STAT5b mutations. *Ann. Med.* **2014**, *46*, 114–122. [CrossRef]
97. Rajala, H.L.; Eldfors, S.; Kuusanmäki, H.; van Adrichem, A.J.; Olson, T.; Lagström, S.; Andersson, E.I.; Jerez, A.; Clemente, M.J.; Yan, Y.; et al. Discovery of somatic STAT5b mutations in large granular lymphocytic leukemia. *Blood* **2013**, *121*, 4541–4550. [CrossRef]
98. Dutta, A.; Yan, D.; Hutchison, R.E.; Mohi, G. STAT3 mutations are not sufficient to induce large granular lymphocytic leukaemia in mice. *Br. J. Haematol.* **2018**, *180*, 911–915. [CrossRef]
99. Klein, K.; Witalisz-Siepracka, A.; Maurer, B.; Prinz, D.; Heller, G.; Leidenfrost, N.; Prchal-Murphy, M.; Suske, T.; Moriggl, R.; Sexl, V. STAT5BN642H drives transformation of NKT cells: A novel mouse model for CD56+ T-LGL leukemia. *Leukemia* **2019**, *33*, 2336–2340. [CrossRef]
100. Chiorazzi, N.; Rai, K.R.; Ferrarini, M. Chronic lymphocytic leukemia. *N. Engl. J. Med.* **2005**, *352*, 804–815. [CrossRef]

101. Hazan-Halevy, I.; Harris, D.; Liu, Z.; Liu, J.; Li, P.; Chen, X.; Shanker, S.; Ferrajoli, A.; Keating, M.J.; Estrov, Z. STAT3 is constitutively phosphorylated on serine 727 residues, binds DNA, and activates transcription in CLL cells. *Blood* **2010**, *115*, 2852–2863. [CrossRef] [PubMed]
102. Capron, C.; Jondeau, K.; Casetti, L.; Jalbert, V.; Costa, C.; Verhoeyen, E.; Massé, J.M.; Coppo, P.; Béné, M.C.; Bourdoncle, P.; et al. Viability and stress protection of chronic lymphoid leukemia cells involves overactivation of mitochondrial phosphoSTAT3Ser727. *Cell Death Dis.* **2014**, *5*, e1451. [CrossRef] [PubMed]
103. de Totero, D.; Meazza, R.; Capaia, M.; Fabbi, M.; Azzarone, B.; Balleari, E.; Gobbi, M.; Cutrona, G.; Ferrarini, M.; Ferrini, S. The opposite effects of IL-15 and IL-21 on CLL B cells correlate with differential activation of the JAK/STAT and ERK1/2 pathways. *Blood* **2008**, *111*, 517–524. [CrossRef] [PubMed]
104. Kube, D.; Holtick, U.; Vockerodt, M.; Ahmadi, T.; Haier, B.; Behrmann, I.; Heinrich, P.C.; Diehl, V.; Tesch, H. STAT3 is constitutively activated in Hodgkin cell lines. *Blood* **2001**, *98*, 762–770. [CrossRef]
105. Holtick, U.; Vockerodt, M.; Pinkert, D.; Schoof, N.; Stürzenhofecker, B.; Kussebi, N.; Lauber, K.; Wesselborg, S.; Löffler, D.; Horn, F.; et al. STAT3 is essential for Hodgkin lymphoma cell proliferation and is a target of tyrphostin AG17 which confers sensitization for apoptosis. *Leukemia* **2005**, *19*, 936–944. [CrossRef]
106. Scheeren, F.A.; Diehl, S.A.; Smit, L.A.; Beaumont, T.; Naspetti, M.; Bende, R.J.; Blom, B.; Karube, K.; Ohshima, K.; van Noesel, C.J.; et al. IL-21 is expressed in Hodgkin lymphoma and activates STAT5: Evidence that activated STAT5 is required for Hodgkin lymphomagenesis. *Blood* **2008**, *111*, 4706–4715. [CrossRef]
107. Ding, B.B.; Yu, J.J.; Yu, R.Y.; Mendez, L.M.; Shaknovich, R.; Zhang, Y.; Cattoretti, G.; Ye, B.H. Constitutively activated STAT3 promotes cell proliferation and survival in the activated B-cell subtype of diffuse large B-cell lymphomas. *Blood* **2008**, *111*, 1515–1523. [CrossRef]
108. Lam, L.T.; Wright, G.; Davis, R.E.; Lenz, G.; Farinha, P.; Dang, L.; Chan, J.W.; Rosenwald, A.; Gascoyne, R.D.; Staudt, L.M. Cooperative signaling through the signal transducer and activator of transcription 3 and nuclear factor-{kappa}B pathways in subtypes of diffuse large B-cell lymphoma. *Blood* **2008**, *111*, 3701–3713. [CrossRef]
109. Pan, Y.R.; Chen, C.C.; Chan, Y.T.; Wang, H.J.; Chien, F.T.; Chen, Y.L.; Liu, J.L.; Yang, M.H. STAT3-coordinated migration facilitates the dissemination of diffuse large B-cell lymphomas. *Nat. Commun.* **2018**, *9*, 3696. [CrossRef]
110. Skarbnik, A.P.; Burki, M.; Pro, B. Peripheral T-cell lymphomas: A review of current approaches and hopes for the future. *Front Oncol.* **2013**, *3*, 138. [CrossRef]
111. Netchiporouk, E.; Litvinov, I.V.; Moreau, L.; Gilbert, M.; Sasseville, D.; Duvic, M. Deregulation in STAT signaling is important for cutaneous T-cell lymphoma (CTCL) pathogenesis and cancer progression. *Cell Cycle* **2014**, *13*, 3331–3335. [CrossRef] [PubMed]
112. Chiarle, R.; Simmons, W.J.; Cai, H.; Dhall, G.; Zamo, A.; Raz, R.; Karras, J.G.; Levy, D.E.; Inghirami, G. Stat3 is required for ALK-mediated lymphomagenesis and provides a possible therapeutic target. *Nat. Med.* **2005**, *11*, 623–629. [CrossRef] [PubMed]
113. Waldmann, T.A.; Chen, J. Disorders of the JAK/STAT Pathway in T Cell Lymphoma Pathogenesis: Implications for Immunotherapy. *Annu. Rev. Immunol.* **2017**, *35*, 533–550. [CrossRef] [PubMed]
114. Song, T.L.; Nairismägi, M.L.; Laurensia, Y.; Lim, J.Q.; Tan, J.; Li, Z.M.; Pang, W.L.; Kizhakeyil, A.; Wijaya, G.C.; Huang, D.C.; et al. Oncogenic activation of the STAT3 pathway drives PD-L1 expression in natural killer/T-cell lymphoma. *Blood* **2018**, *132*, 1146–1158. [CrossRef] [PubMed]
115. Kortylewski, M.; Yu, H. Role of Stat3 in suppressing anti-tumor immunity. *Curr. Opin. Immunol.* **2008**, *20*, 228–233. [CrossRef]
116. Roodman, G.D. Pathogenesis of myeloma bone disease. *Leukemia* **2009**, *23*, 435–441. [CrossRef]
117. Catlett-Falcone, R.; Landowski, T.H.; Oshiro, M.M.; Turkson, J.; Levitzki, A.; Savino, R.; Ciliberto, G.; Moscinski, L.; Fernández-Luna, J.L.; Núñez, G.; et al. Constitutive Activation of Stat3 Signaling Confers Resistance to Apoptosis in Human U266 Myeloma Cells. *Immunity* **1999**, *10*, 105–115. [CrossRef]
118. Bharti, A.C.; Shishodia, S.; Reuben, J.M.; Weber, D.; Alexanian, R.; Raj-Vadhan, S.; Estrov, Z.; Talpaz, M.; Aggarwal, B.B. Nuclear factor-kappaB and STAT3 are constitutively active in CD138+ cells derived from multiple myeloma patients, and suppression of these transcription factors leads to apoptosis. *Blood* **2004**, *103*, 3175–3184. [CrossRef]
119. Jung, S.H.; Ahn, S.Y.; Choi, H.W.; Shin, M.G.; Lee, S.S.; Yang, D.H.; Ahn, J.S.; Kim, Y.K.; Kim, H.J.; Lee, J.J. STAT3 expression is associated with poor survival in non-elderly adult patients with newly diagnosed multiple myeloma. *Blood Res.* **2017**, *52*, 293–299. [CrossRef]

120. Chong, P.S.Y.; Chng, W.J.; de Mel, S. STAT3: A Promising Therapeutic Target in Multiple Myeloma. *Cancers* **2019**, *11*, 731. [CrossRef]
121. Wang, J.; De Veirman, K.; Faict, S.; Frassanito Maria, A.; Ribatti, D.; Vacca, A.; Menu, E. Multiple myeloma exosomes establish a favourable bone marrow microenvironment with enhanced angiogenesis and immunosuppression. *J. Pathol.* **2016**, *239*, 162–173. [CrossRef] [PubMed]
122. Beldi-Ferchiou, A.; Skouri, N.; Ben Ali, C.; Safra, I.; Abdelkefi, A.; Ladeb, S.; Mrad, K.; Ben Othman, T.; Ben Ahmed, M. Abnormal repression of SHP-1, SHP-2 and SOCS-1 transcription sustains the activation of the JAK/STAT3 pathway and the progression of the disease in multiple myeloma. *PLoS ONE* **2017**, *12*, e0174835. [CrossRef] [PubMed]
123. Chim, C.S.; Fung, T.K.; Cheung, W.C.; Liang, R.; Kwong, Y.L. SOCS1 and SHP1 hypermethylation in multiple myeloma: Implications for epigenetic activation of the Jak/STAT pathway. *Blood* **2004**, *103*, 4630–4635. [CrossRef] [PubMed]
124. Löffler, D.; Brocke-Heidrich, K.; Pfeifer, G.; Stocsits, C.; Hackermüller, J.; Kretzschmar, A.K.; Burger, R.; Gramatzki, M.; Blumert, C.; Bauer, K.; et al. Interleukin-6 dependent survival of multiple myeloma cells involves the Stat3-mediated induction of microRNA-21 through a highly conserved enhancer. *Blood* **2007**, *110*, 1330–1333. [CrossRef] [PubMed]
125. Binder, S.; Hösler, N.; Riedel, D.; Zipfel, I.; Buschmann, T.; Kämpf, C.; Reiche, K.; Burger, R.; Gramatzki, M.; Hackermüller, J.; et al. STAT3-induced long noncoding RNAs in multiple myeloma cells display different properties in cancer. *Sci. Rep.* **2017**, *7*, 7976. [CrossRef] [PubMed]
126. Brown, R.; Yang, S.; Weatherburn, C.; Gibson, J.; Ho, P.J.; Suen, H.; Hart, D.; Joshua, D. Phospho-flow detection of constitutive and cytokine-induced pSTAT3/5, pAKT and pERK expression highlights novel prognostic biomarkers for patients with multiple myeloma. *Leukemia* **2015**, *29*, 483–490. [CrossRef]
127. Kloth, M.T.; Catling, A.D.; Silva, C.M. Novel activation of STAT5b in response to epidermal growth factor. *J. Biol. Chem.* **2002**, *277*, 8693–8701. [CrossRef]
128. Schaller-Schönitz, M.; Barzan, D.; Williamson, A.J.; Griffiths, J.R.; Dallmann, I.; Battmer, K.; Ganser, A.; Whetton, A.D.; Scherr, M.; Eder, M. BCR-ABL affects STAT5A and STAT5B differentially. *PLoS ONE* **2014**, *9*, e97243. [CrossRef]
129. Reich, N.C. STATs get their move on. *JAK-STAT* **2013**, *2*, e27080. [CrossRef]
130. Kawashima, T.; Bao, Y.C.; Nomura, Y.; Moon, Y.; Tonozuka, Y.; Minoshima, Y.; Hatori, T.; Tsuchiya, A.; Kiyono, M.; Nosaka, T.; et al. Rac1 and a GTPase-activating protein, MgcRacGAP, are required for nuclear translocation of STAT transcription factors. *J. Cell Biol.* **2006**, *175*, 937–946. [CrossRef]
131. Kawashima, T.; Bao, Y.C.; Minoshima, Y.; Nomura, Y.; Hatori, T.; Hori, T.; Fukagawa, T.; Fukada, T.; Takahashi, N.; Nosaka, T.; et al. A Rac GTPase-activating protein, MgcRacGAP, is a nuclear localizing signal-containing nuclear chaperone in the activation of STAT transcription factors. *Mol. Cell. Biol.* **2009**, *29*, 1796–1813. [CrossRef] [PubMed]
132. Tonozuka, Y.; Minoshima, Y.; Bao, Y.C.; Moon, Y.; Tsubono, Y.; Hatori, T.; Nakajima, H.; Nosaka, T.; Kawashima, T.; Kitamura, T. A GTPase-activating protein binds STAT3 and is required for IL-6-induced STAT3 activation and for differentiation of a leukemic cell line. *Blood* **2004**, *104*, 3550–3557. [CrossRef] [PubMed]
133. Berger, A.; Hoelbl-Kovacic, A.; Bourgeais, J.; Hoefling, L.; Warsch, W.; Grundschober, E.; Uras, I.Z.; Menzl, I.; Putz, E.M.; Hoermann, G.; et al. Pak-dependent stat5 serine phosphorylation is required for bcr-abl-induced leukemogenesis. *Leukemia* **2014**, *28*, 629–641. [CrossRef] [PubMed]
134. Chatterjee, A.; Ghosh, J.; Ramdas, B.; Mali, R.S.; Martin, H.; Kobayashi, M.; Vemula, S.; Canela, V.H.; Waskow, E.R.; Visconte, V.; et al. Regulation of Stat5 by FAK and PAK1 in Oncogenic FLT3- and KIT-Driven Leukemogenesis. *Cell Rep.* **2014**, *9*, 1333–1348. [CrossRef]
135. Harir, N.; Pecquet, C.; Kerenyi, M.; Sonneck, K.; Kovacic, B.; Nyga, R.; Brevet, M.; Dhennin, I.; Gouilleux-Gruart, V.; Beug, H.; et al. Constitutive activation of Stat5 promotes its cytoplasmic localization and association with PI3-kinase in myeloid leukemias. *Blood* **2007**, *109*, 1678–1686. [CrossRef]
136. Chatain, N.; Ziegler, P.; Fahrenkamp, D.; Jost, E.; Moriggl, R.; Schmitz-Van de Leur, H.; Muller-Newen, G. Src family kinases mediate cytoplasmic retention of activated stat5 in bcr-abl-positive cells. *Oncogene* **2013**, *32*, 3587–3597. [CrossRef]

137. Meyer, W.K.; Reichenbach, P.; Schindler, U.; Soldaini, E.; Nabholz, M. Interaction of STAT5 dimers on two low affinity binding sites mediates interleukin 2 (IL-2) stimulation of IL-2 receptor alpha gene transcription. *J. Biol. Chem.* **1997**, *272*, 31821–31828. [CrossRef]
138. Soldaini, E.; John, S.; Moro, S.; Bollenbacher, J.; Schindler, U.; Leonard, W.J. DNA binding site selection of dimeric and tetrameric Stat5 proteins reveals a large repertoire of divergent tetrameric Stat5a binding sites. *Mol. Cell. Biol.* **2000**, *20*, 389–401. [CrossRef]
139. Hu, T.; Yeh, J.E.; Pinello, L.; Jacob, J.; Chakravarthy, S.; Yuan, G.C.; Chopra, R.; Frank, D.A. Impact of the N-Terminal Domain of STAT3 in STAT3-Dependent Transcriptional Activity. *Mol. Cell. Biol.* **2015**, *35*, 3284–3300. [CrossRef]
140. Timofeeva, O.A.; Tarasova, N.I.; Zhang, X.; Chasovskikh, S.; Cheema, A.K.; Wang, H.; Brown, M.L.; Dritschilo, A. STAT3 suppresses transcription of proapoptotic genes in cancer cells with the involvement of its N-terminal domain. *Proc. Natl. Acad. Sci. USA* **2013**, *110*, 1267–1272. [CrossRef]
141. Wingelhofer, B.; Neubauer, H.A.; Valent, P.; Han, X.; Constantinescu, S.N.; Gunning, P.T.; Müller, M.; Moriggl, R. Implications of STAT3 and STAT5 signaling on gene regulation and chromatin remodeling in hematopoietic cancer. *Leukemia* **2018**, *32*, 1713–1726. [CrossRef] [PubMed]
142. Wierenga, A.T.; Vellenga, E.; Schuringa, J.J. Maximal STAT5-induced proliferation and self-renewal at intermediate STAT5 activity levels. *Mol. Cell. Biol.* **2008**, *28*, 6668–6680. [CrossRef] [PubMed]
143. Rozovski, U.; Harris, D.M.; Li, P.; Liu, Z.; Wu, J.Y.; Grgurevic, S.; Faderl, S.; Ferrajoli, A.; Wierda, W.G.; Martinez, M.; et al. At High Levels, Constitutively Activated STAT3 Induces Apoptosis of Chronic Lymphocytic Leukemia Cells. *J. Immunol.* **2016**, *196*, 4400–4409. [CrossRef] [PubMed]
144. Nosaka, T.; Kawashima, T.; Misawa, K.; Ikuta, K.; Mui, A.L.; Kitamura, T. STAT5 as a molecular regulator of proliferation, differentiation and apoptosis in hematopoietic cells. *EMBO J.* **1999**, *18*, 4754–4765. [CrossRef] [PubMed]
145. Friedbichler, K.; Kerenyi, M.A.; Kovacic, B.; Li, G.; Hoelbl, A.; Yahiaoui, S.; Sexl, V.; Mullner, E.W.; Fajmann, S.; Cerny-Reiterer, S.; et al. Stat5a serine 725 and 779 phosphorylation is a prerequisite for hematopoietic transformation. *Blood* **2010**, *116*, 1548–1558. [CrossRef]
146. Mitra, A.; Ross, J.A.; Rodriguez, G.; Nagy, Z.S.; Wilson, H.L.; Kirken, R.A. Signal transducer and activator of transcription 5b (Stat5b) serine 193 is a novel cytokine-induced phospho-regulatory site that is constitutively activated in primary hematopoietic malignancies. *J. Biol. Chem.* **2012**, *287*, 16596–16608. [CrossRef]
147. Freund, P.; Kerenyi, M.A.; Hager, M.; Wagner, T.; Wingelhofer, B.; Pham, H.T.T.; Elabd, M.; Han, X.; Valent, P.; Gouilleux, F.; et al. O-GlcNAcylation of STAT5 controls tyrosine phosphorylation and oncogenic transcription in STAT5-dependent malignancies. *Leukemia* **2017**, *31*, 2132–2142. [CrossRef]
148. Zhuang, S. Regulation of STAT Signaling by Acetylation. *Cell Signal.* **2013**, *25*, 1924–1931. [CrossRef]
149. Yuan, Z.L.; Guan, Y.J.; Chatterjee, D.; Chin, Y.E. Stat3 dimerization regulated by reversible acetylation of a single lysine residue. *Science* **2005**, *307*, 269–273. [CrossRef]
150. Wang, R.; Cherukuri, P.; Luo, J. Activation of Stat3 sequence-specific DNA binding and transcription by p300/CREB-binding protein-mediated acetylation. *J. Biol. Chem.* **2005**, *280*, 11528–11534. [CrossRef]
151. Lee, H.; Zhang, P.; Herrmann, A.; Yang, C.; Xin, H.; Wang, Z.; Hoon, D.S.; Forman, S.J.; Jove, R.; Riggs, A.D. Acetylated STAT3 is crucial for methylation of tumor-suppressor gene promoters and inhibition by resveratrol results in demethylation. *Proc. Natl. Acad. Sci. USA* **2012**, *109*, 7765–7769. [CrossRef] [PubMed]
152. Kosan, C.; Ginter, T.; Heinzel, T.; Krämer, O.H. STAT5 acetylation: Mechanisms and consequences for immunological control and leukemogenesis. *JAKSTAT* **2013**, *2*, e26102. [CrossRef] [PubMed]
153. Dasgupta, M.; Unal, H.; Willard, B.; Yang, J.; Karnik, S.S.; Stark, G.R. Critical role for lysine 685 in gene expression mediated by transcription factor unphosphorylated STAT3. *J. Biol. Chem.* **2014**, *289*, 30763–30771. [CrossRef] [PubMed]
154. Yang, J.; Huang, J.; Dasgupta, M.; Sears, N.; Miyagi, M.; Wang, B.; Chance, M.R.; Chen, X.; Du, Y.; Wang, Y.; et al. Reversible methylation of promoter-bound stat3 by histone-modifying enzymes. *Proc. Natl. Acad. Sci. USA* **2010**, *107*, 21499–21504. [CrossRef]
155. Kim, E.; Kim, M.; Woo, D.H.; Shin, Y.; Shin, J.; Chang, N.; Oh, Y.T.; Kim, H.; Rheey, J.; Nakano, I.; et al. Phosphorylation of ezh2 activates stat3 signaling via stat3 methylation and promotes tumorigenicity of glioblastoma stem-like cells. *Cancer Cell* **2013**, *23*, 839–852. [CrossRef]
156. Dasgupta, M.; Dermawan, J.K.; Willard, B.; Stark, G.R. STAT3-driven transcription depends upon the dimethylation of K49 by EZH2. *Proc. Natl. Acad. Sci. USA* **2015**, *112*, 3985–3990. [CrossRef]

157. Mandal, M.; Powers, S.E.; Maienschein-Cline, M.; Bartom, E.T.; Hamel, K.M.; Kee, B.L.; Dinner, A.R.; Clark, M.R. Epigenetic repression of the Igk locus by STAT5-mediated recruitment of the histone methyltransferase Ezh2. *Nat. Immunol.* **2011**, *12*, 1212–1220. [CrossRef]
158. Butturini, E.; Darra, E.; Chiavegato, G.; Cellini, B.; Cozzolino, F.; Monti, M.; Pucci, P.; Dell'Orco, D.; Mariotto, S. S-Glutathionylation at Cys328 and Cys542 impairs STAT3 phosphorylation. *ACS Chem. Biol.* **2014**, *9*, 1885–1893. [CrossRef]
159. Li, L.; Cheung, S.H.; Evans, E.L.; Shaw, P.E. Modulation of gene expression and tumor cell growth by redox modification of STAT3. *Cancer Res.* **2010**, *70*, 8222–8232. [CrossRef]
160. Sebastián, C.; Herrero, C.; Serra, M.; Lloberas, J.; Blasco, M.A.; Celada, A. Telomere shortening and oxidative stress in aged macrophages results in impaired STAT5a phosphorylation. *J. Immunol.* **2009**, *183*, 2356–2364. [CrossRef]
161. Krämer, O.H.; Moriggl, R. Acetylation and sumoylation control STAT5 activation antagonistically. *JAKSTAT* **2012**, *1*, 203–207. [CrossRef] [PubMed]
162. Zhou, Z.; Wang, M.; Li, J.; Xiao, M.; Chin, Y.E.; Cheng, J.; Yeh, E.T.; Yang, J.; Yi, J. SUMOylation and SENP3 regulate STAT3 activation in head and neck cancer. *Oncogene* **2016**, *35*, 5826–5838. [CrossRef] [PubMed]
163. Walker, S.R.; Xiang, M.; Frank, D.A. Distinct roles of STAT3 and STAT5 in the pathogenesis and targeted therapy of breast cancer. *Mol. Cell. Endocrinol.* **2014**, *382*, 616–621. [CrossRef] [PubMed]
164. Walker, S.R.; Nelson, E.A.; Yeh, J.E.; Pinello, L.; Yuan, G.C.; Frank, D.A. STAT5 outcompetes STAT3 to regulate the expression of the oncogenic transcriptional modulator BCL6. *Mol. Cell. Biol.* **2013**, *33*, 2879–2890. [CrossRef] [PubMed]
165. Yang, X.P.; Ghoreschi, K.; Steward-Tharp, S.M.; Rodriguez-Canales, J.; Zhu, J.; Grainger, J.R.; Hirahara, K.; Sun, H.W.; Wie, L.; Vahedi, G.; et al. Opposing regulation of the locus encoding IL-17 through direct, reciprocal actions of STAT3 and STAT5. *Nat. Immunol.* **2011**, *12*, 247–254. [CrossRef] [PubMed]
166. Weber, A.; Borghouts, C.; Brendel, C.; Moriggl, R.; Delis, N.; Brill, B.; Vafaizadeh, V.; Groner, B. Stat5 Exerts Distinct, Vital Functions in the Cytoplasm and Nucleus of Bcr-Abl+ K562 and Jak2(V617F)+ HEL Leukemia Cells. *Cancers* **2015**, *7*, 503–537. [CrossRef]
167. Li, G.; Miskimen, K.L.; Wang, Z.; Xie, X.Y.; Tse, W.; Gouilleux, F.; Moriggl, R.; Bunting, K.D. Effective targeting of stat5-mediated survival in myeloproliferative neoplasms using abt-737 combined with rapamycin. *Leukemia* **2010**, *24*, 1397–1405. [CrossRef]
168. Marty, C.; Lacout, C.; Droin, N.; Le Couédic, J.P.; Ribrag, V.; Solary, E.; Vainchenker, W.; Villeval, J.L.; Plo, I. A role for reactive oxygen species in JAK2 V617F myeloproliferative neoplasm progression. *Leukemia* **2013**, *27*, 2187–2195. [CrossRef]
169. Silver, D.L.; Naora, H.; Liu, J.; Cheng, W.; Montell, D.J. Activated signal transducer and activator of transcription (stat) 3: Localization in focal adhesions and function in ovarian cancer cell motility. *Cancer Res.* **2004**, *64*, 3550–3558. [CrossRef]
170. Avalle, L.; Camporeale, A.; Morciano, G.; Caroccia, N.; Ghetti, E.; Orecchia, V.; Viavattene, D.; Giorgi, C.; Pinton, P.; Poli, V. Stat3 localizes to the ER, acting as a gatekeeper for er-mitochondrion ca (2+) fluxes and apoptotic responses. *Cell Death Differ.* **2019**, *26*, 932–942. [CrossRef]
171. Yang, R.; Lirussi, D.; Thornton, T.M.; Jelley-Gibbs, D.M.; Diehl, S.A.; Case, L.K.; Madesh, M.; Taatjes, D.J.; Teuscher, C.; Haynes, L. Mitochondrial Ca^{2+} and membrane potential, an alternative pathway for Interleukin 6 to regulate CD4 cell effector function. *Elife* **2015**, *4*, e06376. [CrossRef] [PubMed]
172. Carbognin, E.; Betto, R.M.; Soriano, M.E.; Smith, A.G.; Martello, G. Stat3 promotes mitochondrial transcription and oxidative respiration during maintenance and induction of naive pluripotency. *EMBO J.* **2016**, *35*, 618–634. [CrossRef] [PubMed]
173. Chueh, F.Y.; Leong, K.F.; Yu, C.L. Mitochondrial translocation of signal transducer and activator of transcription 5 (STAT5) in leukemic T cells and cytokine-stimulated cells. *Biochem. Biophys. Res. Commun.* **2010**, *402*, 778–783. [CrossRef] [PubMed]
174. Rozovski, U.; Harris, D.M.; Li, P.; Liu, Z.; Jain, P.; Ferrajoli, A.; Burger, J.; Thompson, P.; Jain, N.; Wierda, W.; et al. STAT3 is constitutively acetylated on lysine 685 residues in chronic lymphocytic leukemia cells. *Oncotarget* **2018**, *9*, 33710–33718. [CrossRef] [PubMed]
175. Gough, D.J.; Corlett, A.; Schlessinger, K.; Wegrzyn, J.; Larner, A.C.; Levy, D.E. Mitochondrial STAT3 supports Ras-dependent oncogenic transformation. *Science* **2009**, *324*, 1713–1716. [CrossRef] [PubMed]

176. Gough, D.J.; Koetz, L.; Levy, D.E. The MEK-ERK pathway is necessary for serine phosphorylation of mitochondrial STAT3 and Ras-mediated transformation. *PLoS ONE* **2013**, *8*, e83395. [CrossRef]
177. Gough, D.J.; Marié, I.J.; Lobry, C.; Aifantis, I.; Levy, D.E. STAT3 supports experimental K-RasG12D-induced murine myeloproliferative neoplasms dependent on serine phosphorylation. *Blood* **2014**, *124*, 2252–2261. [CrossRef]
178. Tammineni, P.; Anugula, C.; Mohammed, F.; Anjaneyulu, M.; Larner, A.C.; Sepuri, N.B. The import of the transcription factor stat3 into mitochondria depends on grim-19, a component of the electron transport chain. *J. Biol. Chem.* **2013**, *288*, 4723–4732. [CrossRef]
179. Avalle, L.; Poli, V. Nucleus, Mitochondrion, or Reticulum? STAT3 à La Carte. *Int. J. Mol. Sci.* **2018**, *19*, 2820. [CrossRef]
180. Xu, Y.S.; Liang, J.J.; Wang, Y.; Zhao, X.J.; Xu, L.; Xu, Y.Y.; Zou, Q.C.; Zhang, J.M.; Tu, C.E.; Cui, Y.G.; et al. STAT3 Undergoes Acetylation-dependent Mitochondrial Translocation to Regulate Pyruvate Metabolism. *Sci. Rep.* **2016**, *6*, 39517. [CrossRef]
181. Mantel, C.; Messina-Graham, S.; Moh, A.; Cooper, S.; Hangoc, G.; Fu, X.Y.; Broxmeyer, H.E. Mouse hematopoietic cell-targeted STAT3 deletion: Stem/progenitor cell defects, mitochondrial dysfunction, ROS overproduction, and a rapid aging-like phenotype. *Blood* **2012**, *120*, 2589–2599. [CrossRef] [PubMed]
182. Lee, J.E.; Yang, Y.M.; Liang, F.X.; Gough, D.J.; Levy, D.E.; Sehgal, P.B. Nongenomic stat5-dependent effects on golgi apparatus and endoplasmic reticulum structure and function. *Am. J. Physiol. Cell Physiol.* **2012**, *302*, 804–820. [CrossRef] [PubMed]
183. Cholez, E.; Debuysscher, V.; Bourgeais, J.; Boudot, C.; Leprince, J.; Tron, F.; Brassart, B.; Regnier, A.; Bissac, E.; Pecnard, E.; et al. Evidence for a protective role of the STAT5 transcription factor against oxidative stress in human leukemic pre-B cells. *Leukemia* **2012**, *26*, 2390–2397. [CrossRef] [PubMed]
184. Vogt, M.; Domoszlai, T.; Kleshchanok, D.; Lehmann, S.; Schmitt, A.; Poli, V.; Richtering, W.; Muller-Newen, G. The role of the N-terminal domain in dimerization and nucleocytoplasmic shuttling of latent STAT3. *J. Cell Sci.* **2011**, *124*, 900–909. [CrossRef] [PubMed]
185. Timofeeva, O.A.; Chasovskikh, S.; Lonskaya, I.; Tarasova, N.I.; Khavrutskii, L.; Tarasov, S.G.; Zhang, X.; Korostyshevskiy, V.R.; Cheema, A.; Zhang, L.; et al. Mechanisms of unphosphorylated STAT3 transcription factor binding to DNA. *J. Biol. Chem.* **2012**, *287*, 14192–14200. [CrossRef] [PubMed]
186. Yang, J.; Chatterjee-Kishore, M.; Staugaitis, S.M.; Nguyen, H.; Schlessinger, K.; Levy, D.E.; Stark, G.R. Novel roles of unphosphorylated STAT3 in oncogenesis and transcriptional regulation. *Cancer Res.* **2005**, *65*, 939–947.
187. Park, H.J.; Li, J.; Hannah, R.; Biddie, S.; Leal-Cervantes, A.I.; Kirschner, K.; Flores Santa Cruz, D.; Sexl, V.; Göttgens, B.; Green, A.R. Cytokine-induced megakaryocytic differentiation is regulated by genome-wide loss of a uSTAT transcriptional program. *EMBO J.* **2016**, *35*, 580–594. [CrossRef]
188. Nagy, Z.S.; Rui, H.; Stepkowski, S.M.; Karras, J.; Kirken, R.A. A preferential role for STAT5, not constitutively active STAT3, in promoting survival of a human lymphoid tumor. *J. Immunol.* **2006**, *177*, 5032–5040. [CrossRef]
189. Xu, D.; Qu, C.K. Protein tyrosine phosphatases in the jak/stat pathway. *Front. Biosci.* **2008**, *13*, 4925–4932. [CrossRef]
190. Krebs, D.L.; Hilton, D.J. Socs: Physiological suppressors of cytokine signaling. *J. Cell Sci.* **2000**, *113*, 2813–2819.
191. Shuai, K. Modulation of stat signaling by stat-interacting proteins. *Oncogene* **2000**, *19*, 2638–2644. [CrossRef] [PubMed]
192. Turkson, J.; Ryan, D.; Kim, J.S.; Zhang, Y.; Chen, Z.; Haura, E.; Laudano, A.; Sebti, S.; Hamilton, A.D.; Jove, R. Phosphotyrosyl peptides block Stat3-mediated DNA binding activity, gene regulation, and cell transformation. *J. Biol. Chem.* **2001**, *276*, 45443–45455. [CrossRef] [PubMed]
193. Turkson, J.; Kim, J.S.; Zhang, S.; Yuan, J.; Huang, M.; Glenn, M.; Haura, E.; Sebti, S.; Hamilton, A.D.; Jove, R. Novel peptidomimetic inhibitors of signal transducer and activator of transcription 3 dimerization and biological activity. *Mol. Cancer Ther.* **2004**, *3*, 261–269. [PubMed]
194. Yue, P.; Turkson, J. Targeting STAT3 in cancer: How successful are we? *Expert Opin. Investig. Drugs* **2009**, *18*, 45–56. [CrossRef]
195. Sen, M.; Grandis, J.R. Nucleic acid-based approaches to STAT inhibition. *JAKSTAT* **2012**, *1*, 285–291. [CrossRef] [PubMed]

196. Nelson, E.A.; Sharma, S.V.; Settleman, J.; Frank, D.A. A chemical biology approach to developing STAT inhibitors: Molecular strategies for accelerating clinical translation. *Oncotarget* **2011**, *2*, 518–524. [CrossRef] [PubMed]
197. Walker, S.R.; Frank, D.A. Screening approaches to generating STAT inhibitors: Allowing the hits to identify the targets. *JAKSTAT* **2012**, *1*, 292–299. [CrossRef]
198. Druker, B.J.; Talpaz, M.; Resta, D.J.; Peng, B.; Buchdunger, E.; Ford, J.M.; Lydon, N.B.; Kantarjian, H.; Capdeville, R.; Ohno-Jones, S.; et al. Efficacy and safety of a specific inhibitor of the BCR-ABL tyrosine kinase in chronic myeloid leukemia. *N. Engl. J. Med.* **2001**, *344*, 1031–1037. [CrossRef]
199. Druker, B.J.; Sawyers, C.L.; Kantarjian, H.; Resta, D.J.; Reese, S.F.; Ford, J.M.; Capdeville, R.; Talpaz, M. Activity of a specific inhibitor of the BCR-ABL tyrosine kinase in the blast crisis of chronic myeloid leukemia and acute lymphoblastic leukemia with the Philadelphia chromosome. *N. Engl. J. Med.* **2001**, *344*, 1038–1042. [CrossRef]
200. Talpaz, M.; Shah, N.P.; Kantarjian, H.; Donato, N.; Nicoll, J.; Paquette, R.; Cortes, J.; O'Brien, S.; Nicaise, C.; Bleickardt, E.; et al. Dasatinib in imatinib-resistant Philadelphia chromosome-positive leukemias. *N. Engl. J. Med.* **2006**, *354*, 2531–2541. [CrossRef]
201. Golas, J.M.; Arndt, K.; Etienne, C.; Lucas, J.; Nardin, D.; Gibbons, J.; Frost, P.; Ye, F.; Boschelli, D.H.; Boschelli, F. SKI-606, a 4-anilino-3-quinolinecarbonitrile dual inhibitor of Src and Abl kinases, is a potent antiproliferative agent against chronic myelogenous leukemia cells in culture and causes regression of K562 xenografts in nude mice. *Cancer Res.* **2003**, *63*, 375–381. [PubMed]
202. Konig, H.; Holyoake, T.L.; Bhatia, R. Effective and selective inhibition of chronic myeloid leukemia primitive hematopoietic progenitors by the dual Src/Abl kinase inhibitor SKI-606. *Blood* **2008**, *111*, 2329–2338. [CrossRef] [PubMed]
203. Weisberg, E.; Manley, P.W.; Breitenstein, W.; Brüggen, J.; Cowan-Jacob, S.W.; Ray, A.; Huntly, B.; Fabbro, D.; Fendrich, G.; Hall-Meyers, E.; et al. Characterization of AMN107, a selective inhibitor of native and mutant Bcr-Abl. *Cancer Cell* **2005**, *7*, 129–141. [CrossRef] [PubMed]
204. O'Hare, T.; Shakespeare, W.C.; Zhu, X.; Eide, C.A.; Rivera, V.M.; Wang, F.; Adrian, L.T.; Zhou, T.; Huang, W.S.; Xu, Q.; et al. AP24534, a pan-BCR-ABL inhibitor for chronic myeloid leukemia, potently inhibits the T315I mutant and overcomes mutation-based resistance. *Cancer Cell* **2009**, *16*, 401–412. [CrossRef]
205. Shah, N.P.; Nicoll, J.M.; Nagar, B.; Gorre, M.E.; Paquette, R.L.; Kuriyan, J.; Sawyers, C.L. Multiple BCR-ABL kinase domain mutations confer polyclonal resistance to the tyrosine kinase inhibitor imatinib (STI571) in chronic phase and blast crisis chronic myeloid leukemia. *Cancer Cell* **2002**, *2*, 117–125. [CrossRef]
206. Daver, N.; Schlenk, R.F.; Russell, N.H.; Levis, M.J. Targeting FLT3 mutations in AML: Review of current knowledge and evidence. *Leukemia* **2019**, *33*, 299–312. [CrossRef]
207. Levis, M. Midostaurin approved for FLT3-mutated AML. *Blood* **2017**, *129*, 3403–3406. [CrossRef]
208. Crump, M.; Hedley, D.; Kamel-Reid, S.; Leber, B.; Wells, R.; Brandwein, J.; Buckstein, R.; Kassis, J.; Minden, M.; Matthews, J.; et al. A randomized phase I clinical and biologic study of two schedules of sorafenib in patients with myelodysplastic syndrome or acute myeloid leukemia: A NCIC (National Cancer Institute of Canada) Clinical Trials Group Study. *Leuk. Lymphoma* **2010**, *51*, 252–260. [CrossRef]
209. Borthakur, G.; Kantarjian, H.; Ravandi, F.; Zhang, W.; Konopleva, M.; Wright, J.J.; Faderl, S.; Verstovsek, S.; Mathews, S.; Andreeff, M.; et al. Phase I study of sorafenib in patients with refractory or relapsed acute leukemias. *Haematologica* **2011**, *96*, 62–68. [CrossRef]
210. Ravandi, F.; Cortes, J.E.; Jones, D.; Faderl, S.; Garcia-Manero, G.; Konopleva, M.Y.; O'Brien, S.; Estrov, Z.; Borthakur, G.; Thomas, D.; et al. Phase I/II study of combination therapy with sorafenib, idarubicin, and cytarabine in younger patients with acute myeloid leukemia. *J. Clin. Oncol.* **2010**, *28*, 1856–1862. [CrossRef]
211. Knapper, S.; Burnett, A.K.; Littlewood, T.; Kell, W.J.; Agrawal, S.; Chopra, R.; Clark, R.; Levis, M.J.; Small, D. A phase 2 trial of the FLT3 inhibitor lestaurtinib (CEP701) as first-line treatment for older patients with acute myeloid leukemia not considered fit for intensive chemotherapy. *Blood* **2006**, *108*, 3262–3270. [CrossRef] [PubMed]
212. Shiotsu, Y.; Kiyoi, H.; Ishikawa, Y.; Tanizaki, R.; Shimizu, M.; Umehara, H.; Ishii, K.; Mori, Y.; Ozeki, K.; Minami, Y.; et al. KW-2449, a novel multikinase inhibitor, suppresses the growth of leukemia cells with FLT3 mutations or T315I-mutated BCR/ABL translocation. *Blood* **2009**, *114*, 1607–1617. [CrossRef] [PubMed]

213. Pratz, K.W.; Cortes, J.; Roboz, G.J.; Rao, N.; Arowojolu, O.; Stine, A.; Shiotsu, Y.; Shudo, A.; Akinaga, S.; Small, D.; et al. A pharmacodynamic study of the FLT3 inhibitor KW-2449 yields insight into the basis for clinical response. *Blood* **2009**, *113*, 3938–3946. [CrossRef] [PubMed]
214. O'Farrell, A.M.; Foran, J.M.; Fiedler, W.; Serve, H.; Paquette, R.L.; Cooper, M.A.; Yuen, H.A.; Louie, S.G.; Kim, H.; Nicholas, S.; et al. An innovative phase I clinical study demonstrates inhibition of FLT3 phosphorylation by SU11248 in acute myeloid leukemia patients. *Clin. Cancer Res.* **2003**, *9*, 5465–5476. [PubMed]
215. Fiedler, W.; Kayser, S.; Kebenko, M.; Janning, M.; Krauter, J.; Schittenhelm, M.; Götze, K.; Weber, D.; Göhring, G.; Teleanu, V.; et al. A phase I/II study of sunitinib and intensive chemotherapy in patients over 60 years of age with acute myeloid leukaemia and activating FLT3 mutations. *Br. J. Haematol.* **2015**, *169*, 694–700. [CrossRef] [PubMed]
216. Cheng, Y.; Paz, K. Tandutinib, an oral, small-molecule inhibitor of FLT3 for the treatment of AML and other cancer indications. *IDrugs* **2008**, *11*, 46–56. [PubMed]
217. Zhou, F.; Ge, Z.; Chen, B. Quizartinib (AC220): A promising option for acute myeloid leukemia. *Drug Des. Dev. Ther.* **2019**, *13*, 1117–1125. [CrossRef]
218. Sutamtewagul, G.; Vigil, C.E. Clinical use of FLT3 inhibitors in acute myeloid leukemia. *Onco Targets Ther.* **2018**, *11*, 7041–7052. [CrossRef]
219. Zhao, J.; Song, Y.; Liu, D. Gilteritinib: A novel FLT3 inhibitor for acute myeloid leukemia. *Biomark Res.* **2019**, *7*, 19. [CrossRef]
220. Zhang, H.; Savage, S.; Schultz, A.R.; Bottomly, D.; White, L.; Segerdell, E.; Wilmot, B.; McWeeney, S.K.; Eide, C.A.; Nechiporuk, T.; et al. Clinical resistance to crenolanib in acute myeloid leukemia due to diverse molecular mechanisms. *Nat. Commun.* **2019**, *10*, 244. [CrossRef]
221. Ghiaur, G.; Levis, M. Mechanisms of Resistance to FLT3 Inhibitors and the Role of the Bone Marrow Microenvironment. *Hematol. Oncol. Clin. N. Am.* **2017**, *31*, 681–692. [CrossRef] [PubMed]
222. Baird, J.H.; Gotlib, J. Clinical Validation of KIT Inhibition in Advanced Systemic Mastocytosis. *Curr. Hematol. Malig. Rep.* **2018**, *13*, 407–416. [CrossRef] [PubMed]
223. Kampa-Schittenhelm, K.M.; Frey, J.; Haeusser, L.A.; Illing, B.; Pavlovsky, A.A.; Blumenstock, G.; Schittenhelm, M.M. Crenolanib is a type I tyrosine kinase inhibitor that inhibits mutant KIT D816 isoforms prevalent in systemic mastocytosis and core binding factor leukemia. *Oncotarget* **2017**, *8*, 82897–82909. [CrossRef] [PubMed]
224. Corbin, A.S.; Griswold, I.J.; La Rosée, P.; Yee, K.W.; Heinrich, M.C.; Reimer, C.L.; Druker, B.J.; Deininger, M.W. Sensitivity of oncogenic KIT mutants to the kinase inhibitors MLN518 and PD180970. *Blood* **2004**, *104*, 3754–3757. [CrossRef] [PubMed]
225. Mascarenhas, J.; Baer, M.R.; Kessler, C.; Hexner, E.; Tremblay, D.; Price, L.; Sandy, L.; Weinberg, R.; Pahl, H.; Silverman, L.R.; et al. Phase II trial of Lestaurtinib, a JAK2 inhibitor, in patients with myelofibrosis. *Leuk. Lymphoma* **2019**, *60*, 1343–1345. [CrossRef] [PubMed]
226. Hexner, E.; Roboz, G.; Hoffman, R.; Luger, S.; Mascarenhas, J.; Carroll, M.; Clementi, R.; Bensen-Kennedy, D.; Moliterno, A. Open-label study of oral CEP-701 (lestaurtinib) in patients with polycythaemia vera or essential thrombocythaemia with JAK2-V617F mutation. *Br. J. Haematol.* **2014**, *164*, 83–93. [CrossRef]
227. Vainchenker, W.; Leroy, E.; Gilles, L.; Marty, C.; Plo, I.; Constantinescu, S.N. JAK inhibitors for the treatment of myeloproliferative neoplasms and other disorders. *F1000 Res.* **2018**, *7*, 82. [CrossRef]
228. Deininger, M.; Radich, J.; Burn, T.C.; Huber, R.; Paranagama, D.; Verstovsek, S. The effect of long-term ruxolitinib treatment on JAK2p.V617F allele burden in patients with myelofibrosis. *Blood* **2015**, *126*, 1551–1554. [CrossRef]
229. Gallipoli, P.; Cook, A.; Rhodes, S.; Hopcroft, L.; Wheadon, H.; Whetton, A.D.; Jørgensen, H.G.; Bhatia, R.; Holyoake, T.L. JAK2/STAT5 inhibition by nilotinib with ruxolitinib contributes to the elimination of CML CD34+ cells in vitro and in vivo. *Blood* **2014**, *124*, 1492–1501. [CrossRef]
230. Van Den Neste, E.; André, M.; Gastinne, T.; Stamatoullas, A.; Haioun, C.; Belhabri, A.; Reman, O.; Casasnovas, O.; Ghesquieres, H.; Verhoef, G.; et al. A phase II study of the oral JAK1/JAK2 inhibitor ruxolitinib in advanced relapsed/refractory Hodgkin lymphoma. *Haematologica* **2018**, *103*, 840–848. [CrossRef]
231. Ghermezi, M.; Spektor, T.M.; Berenson, J.R. The role of JAK inhibitors in multiple myeloma. *Clin. Adv. Hematol. Oncol.* **2019**, *17*, 500–505. [PubMed]

232. Pemmaraju, N.; Kantarjian, H.; Kadia, T.; Cortes, J.; Borthakur, G.; Newberry, K.; Garcia-Manero, G.; Ravandi, F.; Jabbour, E.; Dellasala, S.; et al. A phase I/II study of the Janus kinase (JAK)1 and 2 inhibitor ruxolitinib in patients with relapsed or refractory acute myeloid leukemia. *Clin. Lymphoma Myeloma Leuk.* **2015**, *15*, 171–176. [CrossRef] [PubMed]
233. Diaz, A.E.; Mesa, R.A. Pacritinib and its use in the treatment of patients with myelofibrosis who have thrombocytopenia. *Future Oncol.* **2018**, *14*, 797–807. [CrossRef] [PubMed]
234. Younes, A.; Romaguera, J.; Fanale, M.; McLaughlin, P.; Hagemeister, F.; Copeland, A.; Neelapu, S.; Kwak, L.; Shah, J.; de Castro Faria, S.; et al. Phase I study of a novel oral Janus kinase 2 inhibitor, SB1518, in patients with relapsed lymphoma: Evidence of clinical and biologic activity in multiple lymphoma subtypes. *J. Clin. Oncol.* **2012**, *30*, 4161–4167. [CrossRef]
235. Pardanani, A.; Hood, J.; Lasho, T.; Levine, R.L.; Martin, M.B.; Noronha, G.; Finke, C.; Mak, C.C.; Mesa, R.; Zhu, H.; et al. TG101209, a small molecule JAK2-selective kinase inhibitor potently inhibits myeloproliferative disorder-associated JAK2V617F and MPLW515L/K mutations. *Leukemia* **2007**, *21*, 1658–1668. [CrossRef]
236. Ramakrishnan, V.; Kimlinger, T.; Haug, J.; Timm, M.; Wellik, L.; Halling, T.; Pardanani, A.; Tefferi, A.; Rajkumar, S.V.; Kumar, S. TG101209, a novel JAK2 inhibitor, has significant in vitro activity in multiple myeloma and displays preferential cytotoxicity for CD45+ myeloma cells. *Am. J. Hematol.* **2010**, *85*, 675–686. [CrossRef]
237. Cheng, Z.; Yi, Y.; Xie, S.; Yu, H.; Peng, H.; Zhang, G. The effect of the JAK2 inhibitor TG101209 against T cell acute lymphoblastic leukemia (T-ALL) is mediated by inhibition of JAK-STAT signaling and activation of the crosstalk between apoptosis and autophagy signaling. *Oncotarget* **2017**, *8*, 106753–106763. [CrossRef]
238. Meyer, S.C.; Keller, M.D.; Chiu, S.; Koppikar, P.; Guryanova, O.A.; Rapaport, F.; Xu, K.; Manova, K.; Pankov, D.; O'Reilly, R.J.; et al. CHZ868, a Type II JAK2 Inhibitor, Reverses Type I JAK Inhibitor Persistence and Demonstrates Efficacy in Myeloproliferative Neoplasms. *Cancer Cell* **2015**, *28*, 15–28. [CrossRef]
239. Wu, S.C.; Li, L.S.; Kopp, N.; Montero, J.; Chapuy, B.; Yoda, A.; Christie, A.L.; Liu, H.; Christodoulou, A.; van Bodegom, D.; et al. Activity of the Type II JAK2 Inhibitor CHZ868 in B Cell Acute Lymphoblastic Leukemia. *Cancer Cell* **2015**, *28*, 29–41. [CrossRef]
240. Berdeja, J.; Palandri, F.; Baer, M.R.; Quick, D.; Kiladjian, J.J.; Martinelli, G.; Verma, A.; Hamid, O.; Walgren, R.; Pitou, C.; et al. Phase 2 study of gandotinib (LY2784544) in patients with myeloproliferative neoplasms. *Leuk. Res.* **2018**, *71*, 82–88. [CrossRef]
241. Blair, H.A. Fedratinib: First Approval. *Drugs* **2019**, *79*, 1719–1725. [CrossRef] [PubMed]
242. Hao, Y.; Chapuy, B.; Monti, S.; Sun, H.H.; Rodig, S.J.; Shipp, M.A. Selective JAK2 inhibition specifically decreases Hodgkin lymphoma and mediastinal large B-cell lymphoma growth in vitro and in vivo. *Clin. Cancer Res.* **2014**, *20*, 2674–2683. [CrossRef] [PubMed]
243. Verstovsek, S.; Hoffman, R.; Mascarenhas, J.; Soria, J.C.; Bahleda, R.; McCoon, P.; Tang, W.; Cortes, J.; Kantarjian, H.; Ribrag, V. A phase I, open-label, multi-center study of the JAK2 inhibitor AZD1480 in patients with myelofibrosis. *Leuk. Res.* **2015**, *39*, 157–163. [CrossRef] [PubMed]
244. Suryani, S.; Bracken, L.S.; Harvey, R.C.; Sia, K.C.; Carol, H.; Chen, I.M.; Evans, K.; Dietrich, P.A.; Roberts, K.G.; Kurmasheva, R.T.; et al. Evaluation of the in vitro and in vivo efficacy of the JAK inhibitor AZD1480 against JAK-mutated acute lymphoblastic leukemia. *Mol. Cancer Ther.* **2015**, *14*, 364–374. [CrossRef]
245. Derenzini, E.; Lemoine, M.; Buglio, D.; Katayama, H.; Ji, Y.; Davis, R.E.; Sen, S.; Younes, A. The JAK inhibitor AZD1480 regulates proliferation and immunity in Hodgkin lymphoma. *Blood Cancer J.* **2011**, *1*, e46. [CrossRef]
246. Scuto, A.; Krejci, P.; Popplewell, L.; Wu, J.; Wang, Y.; Kujawski, M.; Kowolik, C.; Xin, H.; Chen, L.; Wang, Y.; et al. The novel JAK inhibitor AZD1480 blocks STAT3 and FGFR3 signaling, resulting in suppression of human myeloma cell growth and survival. *Leukemia* **2011**, *25*, 538–550. [CrossRef]
247. Burger, R.; Le Gouill, S.; Tai, Y.T.; Shringarpure, R.; Tassone, P.; Neri, P.; Podar, K.; Catley, L.; Hideshima, T.; Chauhan, D.; et al. Janus kinase inhibitor INCB20 has antiproliferative and apoptotic effects on human myeloma cells in vitro and in vivo. *Mol. Cancer Ther.* **2009**, *8*, 26–35. [CrossRef]
248. Li, J.; Favata, M.; Kelley, J.A.; Caulder, E.; Thomas, B.; Wen, X.; Sparks, R.B.; Arvanitis, A.; Rogers, J.D.; Combs, A.P.; et al. INCB16562, a JAK1/2 selective inhibitor, is efficacious against multiple myeloma cells and reverses the protective effects of cytokine and stromal cell support. *Neoplasia* **2010**, *12*, 28–38. [CrossRef]

249. Faderl, S.; Ferrajoli, A.; Harris, D.; Van, Q.; Kantarjian, H.M.; Estrov, Z. Atiprimod blocks phosphorylation of JAK-STAT and inhibits proliferation of acute myeloid leukemia (AML) cells. *Leuk. Res.* **2007**, *31*, 91–95. [CrossRef]
250. Quintás-Cardama, A.; Manshouri, T.; Estrov, Z.; Harris, D.; Zhang, Y.; Gaikwad, A.; Kantarjian, H.M.; Verstovsek, S. Preclinical characterization of atiprimod, a novel JAK2 AND JAK3 inhibitor. *Investig. New Drugs* **2011**, *29*, 818–826. [CrossRef]
251. Yang, J.; Ikezoe, T.; Nishioka, C.; Furihata, M.; Yokoyama, A. AZ960, a novel Jak2 inhibitor, induces growth arrest and apoptosis in adult T-cell leukemia cells. *Mol. Cancer Ther.* **2010**, *9*, 3386–3395. [CrossRef] [PubMed]
252. Mascarenhas, J.O.; Talpaz, M.; Gupta, V.; Foltz, L.M.; Savona, M.R.; Paquette, R.; Turner, A.R.; Coughlin, P.; Winton, E.; Burn, T.C.; et al. Primary analysis of a phase II open-label trial of INCB039110, a selective JAK1 inhibitor, in patients with myelofibrosis. *Haematologica* **2017**, *102*, 327–335. [CrossRef] [PubMed]
253. Phillips, T.J.; Forero-Torres, A.; Sher, T.; Diefenbach, C.S.; Johnston, P.; Talpaz, M.; Pulini, J.; Zhou, L.; Scherle, P.; Chen, X.; et al. Phase 1 study of the PI3Kδ inhibitor INCB040093 ± JAK1 inhibitor itacitinib in relapsed/refractory B-cell lymphoma. *Blood* **2018**, *132*, 293–306. [CrossRef] [PubMed]
254. Verstovsek, S.; Talpaz, M.; Ritchie, E.; Wadleigh, M.; Odenike, O.; Jamieson, C.; Stein, B.; Uno, T.; Mesa, R.A. A phase I, open-label, dose-escalation, multicenter study of the JAK2 inhibitor NS-018 in patients with myelofibrosis. *Leukemia* **2017**, *31*, 393–402. [CrossRef] [PubMed]
255. Honda, A.; Kuramoto, K.; Niwa, T.; Naito, H. NS-018 reduces myeloma cell proliferation and suppresses osteolysis through inhibition of the JAK2 and Src signaling pathways. *Blood Cancer J.* **2018**, *8*, 62. [CrossRef] [PubMed]
256. Baffert, F.; Régnier, C.H.; De Pover, A.; Pissot-Soldermann, C.; Tavares, G.A.; Blasco, F.; Brueggen, J.; Chène, P.; Drueckes, P.; Erdmann, D.; et al. Potent and selective inhibition of polycythemia by the quinoxaline JAK2 inhibitor NVP-BSK805. *Mol. Cancer Ther.* **2010**, *9*, 1945–1955. [CrossRef] [PubMed]
257. Ringel, F.; Kaeda, J.; Schwarz, M.; Oberender, C.; Grille, P.; Dörken, B.; Marque, F.; Manley, P.W.; Radimerski, T.; le Coutre, P. Effects of Jak2 type 1 inhibitors NVP-BSK805 and NVP-BVB808 on Jak2 mutation-positive and Bcr-Abl-positive cell lines. *Acta Haematol.* **2014**, *132*, 75–86. [CrossRef]
258. Jatiani, S.S.; Cosenza, S.C.; Reddy, M.V.; Ha, J.H.; Baker, S.J.; Samanta, A.K.; Olnes, M.J.; Pfannes, L.; Sloand, E.M.; Arlinghaus, R.B.; et al. A Non-ATP-Competitive Dual Inhibitor of JAK2 and BCR-ABL Kinases: Elucidation of a Novel Therapeutic Spectrum Based on Substrate Competitive Inhibition. *Genes Cancer.* **2010**, *1*, 331–345. [CrossRef]
259. Mesa, R.A.; Kiladjian, J.J.; Catalano, J.V.; Devos, T.; Egyed, M.; Hellmann, A.; McLornan, D.; Shimoda, K.; Winton, E.F.; Deng, W.; et al. SIMPLIFY-1: A Phase III Randomized Trial of Momelotinib Versus Ruxolitinib in Janus Kinase Inhibitor-Naïve Patients With Myelofibrosis. *J. Clin. Oncol.* **2017**, *35*, 3844–3850. [CrossRef]
260. Harrison, C.N.; Vannucchi, A.M.; Platzbecker, U.; Cervantes, F.; Gupta, V.; Lavie, D.; Passamonti, F.; Winton, E.F.; Dong, H.; Kawashima, J.; et al. Momelotinib versus best available therapy in patients with myelofibrosis previously treated with ruxolitinib (SIMPLIFY 2): A randomised, open-label, phase 3 trial. *Lancet Haematol.* **2018**, *5*, e73–e81. [CrossRef]
261. Monaghan, K.A.; Khong, T.; Burns, C.J.; Spencer, A. The novel JAK inhibitor CYT387 suppresses multiple signalling pathways, prevents proliferation and induces apoptosis in phenotypically diverse myeloma cells. *Leukemia* **2011**, *25*, 1891–1899. [CrossRef] [PubMed]
262. Lipka, D.B.; Hoffmann, L.S.; Heidel, F.; Markova, B.; Blum, M.C.; Breitenbuecher, F.; Kasper, S.; Kindler, T.; Levine, R.L.; Huber, C.; et al. LS104, a non-ATP-competitive small-molecule inhibitor of JAK2, is potently inducing apoptosis in JAK2V617F-positive cells. *Mol. Cancer Ther.* **2008**, *7*, 1176–1184. [CrossRef] [PubMed]
263. Kasper, S.; Breitenbuecher, F.; Hoehn, Y.; Heidel, F.; Lipka, D.B.; Markova, B.; Huber, C.; Kindler, T.; Fischer, T. The kinase inhibitor LS104 induces apoptosis, enhances cytotoxic effects of chemotherapeutic drugs and is targeting the receptor tyrosine kinase FLT3 in acute myeloid leukemia. *Leuk. Res.* **2008**, *32*, 1698–1708. [CrossRef] [PubMed]
264. Grunberger, T.; Demin, P.; Rounova, O.; Sharfe, N.; Cimpean, L.; Dadi, H.; Freywald, A.; Estrov, Z.; Roifman, C.M. Inhibition of acute lymphoblastic and myeloid leukemias by a novel kinase inhibitor. *Blood* **2003**, *102*, 4153–4158. [CrossRef]
265. Verstovsek, S.; Manshouri, T.; Quintás-Cardama, A.; Harris, D.; Cortes, J.; Giles, F.J.; Kantarjian, H.; Priebe, W.; Estrov, Z. WP1066, a novel JAK2 inhibitor, suppresses proliferation and induces apoptosis in erythroid human cells carrying the JAK2 V617F mutation. *Clin. Cancer Res.* **2008**, *14*, 788–796. [CrossRef]

266. Ferrajoli, A.; Faderl, S.; Van, Q.; Koch, P.; Harris, D.; Liu, Z.; Hazan-Halevy, I.; Wang, Y.; Kantarjian, H.M.; Priebe, W.; et al. WP1066 disrupts Janus kinase-2 and induces caspase-dependent apoptosis in acute myelogenous leukemia cells. *Cancer Res.* **2007**, *67*, 11291–11299. [CrossRef]
267. Faderl, S.; Ferrajoli, A.; Harris, D.; Van, Q.; Priebe, W.; Estrov, Z. WP-1034, A Novel JAK-STAT Inhibitor, with Proapoptotic and Antileukemic Activity in Acute Myeloid Leukemia (AML). *Anticancer Res.* **2005**, *25*, 1841–1850.
268. Verstovsek, S.; Tam, C.S.; Wadleigh, M.; Sokol, L.; Smith, C.C.; Bui, L.A.; Song, C.; Clary, D.O.; Olszynski, P.; Cortes, J.; et al. Phase I evaluation of XL019, an oral, potent, and selective JAK2 inhibitor. *Leuk. Res.* **2014**, *38*, 316–322. [CrossRef]
269. Haridas, V.; Nishimura, G.; Xu, Z.X.; Connolly, F.; Hanausek, M.; Walaszek, Z.; Zoltaszek, R.; Gutterman, J.U. Avicin D: A protein reactive plant isoprenoid dephosphorylates Stat 3 by regulating both kinase and phosphatase activities. *PLoS ONE* **2009**, *4*, e5578. [CrossRef]
270. Zhang, C.; Li, B.; Gaikwad, A.S.; Haridas, V.; Xu, Z.; Gutterman, J.U.; Duvic, M. Avicin D selectively induces apoptosis and downregulates p-STAT-3, bcl-2, and survivin in cutaneous T-cell lymphoma cells. *J. Investig. Dermatol.* **2008**, *128*, 2728–2735. [CrossRef]
271. Pandey, M.K.; Sung, B.; Ahn, K.S.; Aggarwal, B.B. Butein suppresses constitutive and inducible signal transducer and activator of transcription (STAT) 3 activation and STAT3-regulated gene products through the induction of a protein tyrosine phosphatase SHP-1. *Mol. Pharmacol.* **2009**, *75*, 525–533. [CrossRef] [PubMed]
272. Bhutani, M.; Pathak, A.K.; Nair, A.S.; Kunnumakkara, A.B.; Guha, S.; Sethi, G.; Aggarwal, B.B. Capsaicin is a novel blocker of constitutive and interleukin-6-inducible STAT3 activation. *Clin. Cancer Res.* **2007**, *13*, 3024–3032. [CrossRef] [PubMed]
273. Kannaiyan, R.; Hay, H.S.; Rajendran, P.; Li, F.; Shanmugam, M.K.; Vali, S.; Abbasi, T.; Kapoor, S.; Sharma, A.; Kumar, A.P.; et al. Celastrol inhibits proliferation and induces chemosensitization through down-regulation of NF-κB and STAT3 regulated gene products in multiple myeloma cells. *Br. J. Pharmacol.* **2011**, *164*, 1506–1521. [CrossRef] [PubMed]
274. Ishdorj, G.; Johnston, J.B.; Gibson, S.B. Inhibition of constitutive activation of STAT3 by curcurbitacin-I (JSI-124) sensitized human B-leukemia cells to apoptosis. *Mol. Cancer Ther.* **2010**, *9*, 3302–3314. [CrossRef]
275. van Kester, M.S.; Out-Luiting, J.J.; von dem Borne, P.A.; Willemze, R.; Tensen, C.P.; Vermeer, M.H. Cucurbitacin I inhibits Stat3 and induces apoptosis in Sézary cells. *J. Investig. Dermatol.* **2008**, *128*, 1691–1695. [CrossRef]
276. Bi, L.; Yu, Z.; Wu, J.; Yu, K.; Hong, G.; Lu, Z.; Gao, S. Honokiol Inhibits Constitutive and Inducible STAT3 Signaling via PU.1-Induced SHP1 Expression in Acute Myeloid Leukemia Cells. *Tohoku J. Exp. Med.* **2015**, *237*, 163–172. [CrossRef]
277. Yao, Y.; Sun, Y.; Shi, M.; Xia, D.; Zhao, K.; Zeng, L.; Yao, R.; Zhang, Y.; Li, Z.; Niu, M.; et al. Piperlongumine induces apoptosis and reduces bortezomib resistance by inhibiting STAT3 in multiple myeloma cells. *Oncotarget* **2016**, *7*, 73497–73508. [CrossRef]
278. Kuttan, G.; Kumar, K.B.; Guruvayoorappan, C.; Kuttan, R. Antitumor, anti-invasion, and antimetastatic effects of curcumin. *Adv. Exp. Med. Biol.* **2007**, *595*, 173–184.
279. Blasius, R.; Reuter, S.; Henry, E.; Dicato, M.; Diederich, M. Curcumin regulates signal transducer and activator of transcription (STAT) expression in K562 cells. *Biochem. Pharmacol.* **2006**, *72*, 1547–1554. [CrossRef]
280. Rajasingh, J.; Raikwar, H.P.; Muthian, G.; Johnson, C.; Bright, J.J. Curcumin induces growth-arrest and apoptosis in association with the inhibition of constitutively active JAK-STAT pathway in T cell leukemia. *Biochem. Biophys. Res. Commun.* **2006**, *340*, 359–368. [CrossRef]
281. Uddin, S.; Hussain, A.R.; Manogaran, P.S.; Al-Hussein, K.; Platanias, L.C.; Gutierrez, M.I.; Bhatia, K.G. Curcumin suppresses growth and induces apoptosis in primary effusion lymphoma. *Oncogene* **2005**, *24*, 7022–7030. [CrossRef] [PubMed]
282. Mackenzie, G.G.; Queisser, N.; Wolfson, M.L.; Fraga, C.G.; Adamo, A.M.; Oteiza, P.I. Curcumin induces cell-arrest and apoptosis in association with the inhibition of constitutively active NF-kappaB and STAT3 pathways in Hodgkin's lymphoma cells. *Int. J. Cancer.* **2008**, *123*, 56–65. [CrossRef] [PubMed]
283. Bharti, A.C.; Donato, N.; Aggarwal, B.B. Curcumin (diferuloylmethane) inhibits constitutive and IL-6-inducible STAT3 phosphorylation in human multiple myeloma cells. *J. Immunol.* **2003**, *171*, 3863–3871. [CrossRef] [PubMed]

284. Lin, L.; Deangelis, S.; Foust, E.; Fuchs, J.; Li, C.; Li, P.K.; Schwartz, E.B.; Lesinski, G.B.; Benson, D.; Lü, J.; et al. A novel small molecule inhibits STAT3 phosphorylation and DNA binding activity and exhibits potent growth suppressive activity in human cancer cells. *Mol. Cancer.* **2010**, *9*, 217. [CrossRef]
285. Liby, K.; Voong, N.; Williams, C.R.; Risingsong, R.; Royce, D.B.; Honda, T.; Gribble, G.W.; Sporn, M.B.; Letterio, J.J. The synthetic triterpenoid CDDO-Imidazolide suppresses STAT phosphorylation and induces apoptosis in myeloma and lung cancer cells. *Clin. Cancer Res.* **2006**, *12*, 4288–4293. [CrossRef]
286. Nam, S.; Scuto, A.; Yang, F.; Chen, W.; Park, S.; Yoo, H.S.; Konig, H.; Bhatia, R.; Cheng, X.; Merz, K.H.; et al. Indirubin derivatives induce apoptosis of chronic myelogenous leukemia cells involving inhibition of Stat5 signaling. *Mol. Oncol.* **2012**, *6*, 276–283. [CrossRef]
287. Zhou, J.; Bi, C.; Janakakumara, J.V.; Liu, S.C.; Chng, W.J.; Tay, K.G.; Poon, L.F.; Xie, Z.; Palaniyandi, S.; Yu, H.; et al. Enhanced activation of STAT pathways and overexpression of survivin confer resistance to FLT3 inhibitors and could be therapeutic targets in AML. *Blood* **2009**, *113*, 4052–4062. [CrossRef]
288. Nam, S.; Buettner, R.; Turkson, J.; Kim, D.; Cheng, J.Q.; Muehlbeyer, S.; Hippe, F.; Vatter, S.; Merz, K.H.; Eisenbrand, G.; et al. Indirubin derivatives inhibit Stat3 signaling and induce apoptosis in human cancer cells. *Proc. Natl. Acad. Sci. USA* **2005**, *102*, 5998–6003. [CrossRef]
289. Pinz, S.; Unser, S.; Rascle, A. The natural chemopreventive agent sulforaphane inhibits STAT5 activity. *PLoS ONE* **2014**, *9*, e99391. [CrossRef]
290. Liu, S.; Walker, S.R.; Nelson, E.A.; Cerulli, R.; Xiang, M.; Toniolo, P.A.; Qi, J.; Stone, R.M.; Wadleigh, M.; Bradner, J.E.; et al. Targeting STAT5 in hematologic malignancies through inhibition of the bromodomain and extra-terminal (BET) bromodomain protein BRD2. *Mol. Cancer Ther.* **2014**, *13*, 1194–1205. [CrossRef]
291. Guerra, B.; Martín-Rodríguez, P.; Díaz-Chico, J.C.; McNaughton-Smith, G.; Jiménez-Alonso, S.; Hueso-Falcón, I.; Montero, J.C.; Blanco, R.; León, J.; Rodríguez-González, G.; et al. CM363, a novel naphthoquinone derivative which acts as multikinase modulator and overcomes imatinib resistance in chronic myelogenous leukemia. *Oncotarget* **2017**, *8*, 29679–29698. [CrossRef] [PubMed]
292. Juen, L.; Brachet-Botineau, M.; Parmenon, C.; Bourgeais, J.; Hérault, O.; Gouilleux, F.; Viaud-Massuard, M.C.; Prié, G. New inhibitor targeting signal transducer and activator of transcription 5 (STAT5) signaling in myeloid leukemias. *J. Med. Chem.* **2017**, *60*, 6119–6136. [CrossRef] [PubMed]
293. Brachet-Botineau, M.; Deynoux, M.; Vallet, N.; Polomski, M.; Juen, L.; Hérault, O.; Mazurier, F.; Viaud-Massuard, M.C.; Prié, G.; Gouilleux, F. A Novel Inhibitor of STAT5 Signaling Overcomes Chemotherapy Resistance in Myeloid Leukemia Cells. *Cancers* **2019**, *11*, 2043. [CrossRef] [PubMed]
294. Nelson, E.A.; Walker, S.R.; Kepich, A.; Gashin, L.B.; Hideshima, T.; Ikeda, H.; Chauhan, D.; Anderson, K.C.; Frank, D.A. Nifuroxazide inhibits survival of multiple myeloma cells by directly inhibiting STAT3. *Blood* **2008**, *112*, 5095–5102. [CrossRef]
295. Khanim, F.L.; Merrick, B.A.; Giles, H.V.; Jankute, M.; Jackson, J.B.; Giles, L.J.; Birtwistle, J.; Bunce, C.M.; Drayson, M.T. Redeployment-based drug screening identifies the anti-helminthic niclosamide as anti-myeloma therapy that also reduces free light chain production. *Blood Cancer J.* **2011**, *1*, e39. [CrossRef]
296. Sharma, A.; Jyotsana, N.; Lai, C.K.; Chaturvedi, A.; Gabdoulline, R.; Görlich, K.; Murphy, C.; Blanchard, J.E.; Ganser, A.; Brown, E.; et al. Pyrimethamine as a Potent and Selective Inhibitor of Acute Myeloid Leukemia Identified by High-throughput Drug Screening. *Curr. Cancer Drug Targets* **2016**, *16*, 818–828. [CrossRef]
297. Liu, H.; Qin, Y.; Zhai, D.; Zhang, Q.; Gu, J.; Tang, Y.; Yang, J.; Li, K.; Yang, L.; Chen, S.; et al. Antimalarial Drug Pyrimethamine Plays a Dual Role in Antitumor Proliferation and Metastasis through Targeting DHFR and TP. *Mol. Cancer Ther.* **2019**, *18*, 541–555. [CrossRef]
298. Nelson, E.A.; Walker, S.R.; Weisberg, E.; Bar-Natan, M.; Barrett, R.; Gashin, L.B.; Terrell, S.; Klitgaard, J.L.; Santo, L.; Addorio, M.R.; et al. The STAT5 inhibitor pimozide decreases survival of chronic myelogenous leukemia cells resistant to kinase inhibitors. *Blood* **2011**, *117*, 3421–3429. [CrossRef]
299. Bar-Natan, M.; Nelson, E.A.; Walker, S.R.; Kuang, Y.; Distel, R.J.; Frank, D.A. Dual inhibition of Jak2 and STAT5 enhances killing of myeloproliferative neoplasia cells. *Leukemia* **2012**, *26*, 1407–1410. [CrossRef]
300. Tabe, Y.; Konopleva, M.; Andreeff, M.; Ohsaka, A. Effects of PPARγ Ligands on Leukemia. *PPAR Res.* **2012**, *2012*. [CrossRef]
301. Rousselot, P.; Prost, S.; Guilhot, J.; Roy, L.; Etienne, G.; Legros, L.; Charbonnier, A.; Coiteux, V.; Cony-Makhoul, P.; Huguet, F.; et al. Pioglitazone together with imatinib in chronic myeloid leukemia: A proof of concept study. *Cancer* **2017**, *123*, 1791–1799. [CrossRef] [PubMed]

302. Liu, H.; Zang, C.; Fenner, M.H.; Liu, D.; Possinger, K.; Koeffler, H.P.; Elstner, E. Growth inhibition and apoptosis in human Philadelphia chromosome-positive lymphoblastic leukemia cell lines by treatment with the dual PPARα/γ ligand TZD18. *Blood* **2006**, *107*, 3683–3692. [CrossRef] [PubMed]
303. Wang, L.H.; Yang, X.Y.; Zhang, X.; Huang, J.; Hou, J.; Li, J.; Xiong, H.; Mihalic, K.; Zhu, H.; Xiao, W.; et al. Transcriptional inactivation of STAT3 by PPARgamma suppresses IL-6-responsive multiple myeloma cells. *Immunity* **2004**, *20*, 205–218. [CrossRef]
304. Ren, Z.; Cabell, L.A.; Schaefer, T.S.; McMurray, J.S. Identification of a high-affinity phosphopeptide inhibitor of Stat3. *Bioorg. Med. Chem. Lett.* **2003**, *13*, 633–636. [CrossRef]
305. Coleman, D.R., IV; Ren, Z.; Mandal, P.K.; Cameron, A.G.; Dyer, G.A.; Muranjan, S.; Campbell, M.; Chen, X.; McMurray, J.S. Investigation of the binding determinants of phosphopeptides targeted to the SRC homology 2 domain of the signal transducer and activator of transcription 3. Development of a high-affinity peptide inhibitor. *J. Med. Chem.* **2005**, *48*, 6661–6670. [CrossRef]
306. Siddiquee, K.A.; Gunning, P.T.; Glenn, M.; Katt, W.P.; Zhang, S.; Schrock, C.; Sebti, S.M.; Jove, R.; Hamilton, A.D.; Turkson, J. An oxazole-based small-molecule Stat3 inhibitor modulates Stat3 stability and processing and induces antitumor cell effects. *ACS Chem. Biol.* **2007**, *2*, 787–798. [CrossRef]
307. Becker, S.; Groner, B.; Müller, C.W. Three-dimensional structure of the Stat3beta homodimer bound to DNA. *Nature* **1998**, *394*, 145–151. [CrossRef]
308. Siddiquee, K.; Zhang, S.; Guida, W.C.; Blaskovich, M.A.; Greedy, B.; Lawrence, H.R.; Yip, M.L.; Jove, R.; McLaughlin, M.M.; Lawrence, N.J.; et al. Selective chemical probe inhibitor of Stat3, identified through structure-based virtual screening, induces antitumor activity. *Proc. Natl. Acad. Sci. USA* **2007**, *104*, 7391–7396. [CrossRef]
309. Song, H.; Wang, R.; Wang, S.; Lin, J. A low-molecular-weight compound discovered through virtual database screening inhibits Stat3 function in breast cancer cells. *Proc. Natl. Acad. Sci. USA* **2005**, *102*, 4700–4705. [CrossRef]
310. Xu, X.; Kasembeli, M.; Jiang, X.; Tweardy, B.J.; Tweardy, D.J. Chemical probes that competitively and selectively inhibit Stat3 activation. *PLoS ONE* **2009**, *4*, e4783. [CrossRef]
311. Matsuno, K.; Masuda, Y.; Uehara, Y.; Sato, H.; Muroya, A.; Takahashi, O.; Yokotagawa, T.; Furuya, T.; Okawara, T.; Otsuka, M.; et al. Identification of a New Series of STAT3 Inhibitors by Virtual Screening. *ACS Med. Chem. Lett.* **2010**, *1*, 371–375. [CrossRef] [PubMed]
312. Zhang, X.; Sun, Y.; Pireddu, R.; Yang, H.; Urlam, M.K.; Lawrence, H.R.; Guida, W.C.; Lawrence, N.J.; Sebti, S.M. A novel inhibitor of STAT3 homodimerization selectively suppresses STAT3 activity and malignant transformation. *Cancer Res.* **2013**, *73*, 1922–1933. [CrossRef] [PubMed]
313. Huang, Y.H.; Vakili, M.R.; Molavi, O.; Morrissey, Y.; Wu, C.; Paiva, I.; Soleimani, A.H.; Sanaee, F.; Lavasanifar, A.; Lai, R. Decoration of Anti-CD38 on Nanoparticles Carrying a STAT3 Inhibitor Can Improve the Therapeutic Efficacy Against Myeloma. *Cancers* **2019**, *11*, 248. [CrossRef] [PubMed]
314. Fletcher, S.; Singh, J.; Zhang, X.; Yue, P.; Page, B.D.; Sharmeen, S.; Shahani, V.M.; Zhao, W.; Schimmer, A.D.; Turkson, J.; et al. Disruption of transcriptionally active Stat3 dimers with non-phosphorylated, salicylic acid-based small molecules: Potent in vitro and tumor cell activities. *Chembiochem* **2009**, *10*, 1959–1964. [CrossRef]
315. Fletcher, S.; Page, B.D.; Zhang, X.; Yue, P.; Li, Z.H.; Sharmeen, S.; Singh, J.; Zhao, W.; Schimmer, A.D.; Trudel, S.; et al. Antagonism of the Stat3-Stat3 protein dimer with salicylic acid based small molecules. *Chem. Med. Chem.* **2011**, *6*, 1459–1470. [CrossRef]
316. Page, B.D.; Fletcher, S.; Yue, P.; Li, Z.; Zhang, X.; Sharmeen, S.; Datti, A.; Wrana, J.L.; Trudel, S.; Schimmer, A.D.; et al. Identification of a non-phosphorylated, cell permeable, small molecule ligand for the Stat3 SH2 domain. *Bioorg. Med. Chem. Lett.* **2011**, *21*, 5605–5609. [CrossRef]
317. Page, B.D.; Croucher, D.C.; Li, Z.H.; Haftchenary, S.; Jimenez-Zepeda, V.H.; Atkinson, J.; Spagnuolo, P.A.; Wong, Y.L.; Colaguori, R.; Lewis, A.M.; et al. Inhibiting aberrant signal transducer and activator of transcription protein activation with tetrapodal, small molecule Src homology 2 domain binders: Promising agents against multiple myeloma. *J. Med. Chem.* **2013**, *56*, 7190–7200. [CrossRef]
318. Eiring, A.M.; Page, B.D.G.; Kraft, I.L.; Mason, C.C.; Vellore, N.A.; Resetca, D.; Zabriskie, M.S.; Zhang, T.Y.; Khorashad, J.S.; Engar, A.J.; et al. Combined STAT3 and BCR-ABL1 inhibition induces synthetic lethality in therapy-resistant chronic myeloid leukemia. *Leukemia* **2015**, *29*, 586–597. [CrossRef]

319. Page, B.D.; Khoury, H.; Laister, R.C.; Fletcher, S.; Vellozo, M.; Manzoli, A.; Yue, P.; Turkson, J.; Minden, M.D.; Gunning, P.T. Small molecule STAT5-SH2 domain inhibitors exhibit potent antileukemia activity. *J. Med. Chem.* **2012**, *55*, 1047–1055. [CrossRef]
320. Cumaraswamy, A.A.; Lewis, A.M.; Geletu, M.; Todic, A.; Diaz, D.B.; Cheng, X.R.; Brown, C.E.; Laister, R.C.; Muench, D.; Kerman, K.; et al. Nanomolar-potency small molecule inhibitor of STAT5 protein. *ACS Med. Chem. Lett.* **2014**, *5*, 1202–1206. [CrossRef]
321. Liao, Z.; Gu, L.; Vergalli, J.; Mariani, S.A.; De Dominici, M.; Lokareddy, R.K.; Dagvadorj, A.; Purushottamachar, P.; McCue, P.A.; Trabulsi, E.; et al. Structure-Based Screen Identifies a Potent Small Molecule Inhibitor of Stat5a/b with Therapeutic Potential for Prostate Cancer and Chronic Myeloid Leukemia. *Mol. Cancer Ther.* **2015**, *14*, 1777–1793. [CrossRef] [PubMed]
322. Müller, J.; Sperl, B.; Reindl, W.; Kiessling, A.; Berg, T. Discovery of chromone-based inhibitors of the transcription factor STAT5. *Chembiochem* **2008**, *9*, 723–727. [CrossRef] [PubMed]
323. Elumalai, N.; Berg, A.; Natarajan, K.; Scharow, A.; Berg, T. Nanomolar inhibitors of the transcription factor STAT5b with high selectivity over STAT5a. *Angew. Chem. Int. Ed. Engl.* **2015**, *54*, 4758–4763. [CrossRef] [PubMed]
324. Elumalai, N.; Berg, A.; Rubner, S.; Blechschmidt, L.; Song, C.; Natarajan, K.; Matysik, J.; Berg, T. Rational development of Stafib-2: A selective, nanomolar inhibitor of the transcription factor STAT5b. *Sci. Rep.* **2017**, *7*, 819. [CrossRef]
325. Jerez, A.; Clemente, M.J.; Makishima, H.; Koskela, H.; Leblanc, F.; Peng Ng, K.; Olson, T.; Przychodzen, B.; Afable, M.; Gomez-Segui, I.; et al. STAT3 mutations unify the pathogenesis of chronic lymphoproliferative disorders of NK cells and T-cell large granular lymphocyte leukemia. *Blood* **2012**, *120*, 3048–3057. [CrossRef]
326. Kaneko, N.; Kita, A.; Yamanaka, K.; Mori, M. Combination of YM155, a survivin suppressant with a STAT3 inhibitor: A new strategy to treat diffuse large B-cell lymphoma. *Leuk. Res.* **2013**, *37*, 1156–1161. [CrossRef]
327. Mencalha, A.L.; Du Rocher, B.; Salles, D.; Binato, R.; Abdelhay, E. LLL-3, a STAT3 inhibitor, represses BCR-ABL-positive cell proliferation, activates apoptosis and improves the effects of Imatinib mesylate. *Cancer Chemother. Pharmacol.* **2010**, *65*, 1039–1046. [CrossRef]
328. Lin, L.; Benson, D.M., Jr.; DeAngelis, S.; Bakan, C.E.; Li, P.K.; Li, C.; Lin, J. A small molecule, LLL12 inhibits constitutive STAT3 and IL-6-induced STAT3 signaling and exhibits potent growth suppressive activity in human multiple myeloma cells. *Int. J. Cancer* **2012**, *130*, 1459–1469. [CrossRef]
329. Redell, M.S.; Ruiz, M.J.; Alonzo, T.A.; Gerbing, R.B.; Tweardy, D.J. Stat3 signaling in acute myeloid leukemia: Ligand-dependent and -independent activation and induction of apoptosis by a novel small-molecule Stat3 inhibitor. *Blood* **2011**, *117*, 5701–5709. [CrossRef]
330. Schust, J.; Sperl, B.; Hollis, A.; Mayer, T.U.; Berg, T. Stattic: A small-molecule inhibitor of STAT3 activation and dimerization. *Chem. Biol.* **2006**, *13*, 1235–1242. [CrossRef]
331. Huang, Y.H.; Molavi, O.; Alshareef, A.; Haque, M.; Wang, Q.; Chu, M.P.; Venner, C.P.; Sandhu, I.; Peters, A.C.; Lavasanifar, A.; et al. Constitutive Activation of STAT3 in Myeloma Cells Cultured in a Three-Dimensional, Reconstructed Bone Marrow Model. *Cancers* **2018**, *10*, 206. [CrossRef] [PubMed]
332. Liu, J.; Liang, L.; Li, D.; Nong, L.; Zheng, Y.; Huang, S.; Zhang, B.; Li, T. JAK3/STAT3 oncogenic pathway and PRDM1 expression stratify clinicopathologic features of extranodal NK/T-cell lymphoma, nasal type. *Oncol. Rep.* **2019**, *41*, 3219–3232. [CrossRef] [PubMed]
333. Beebe, J.D.; Liu, J.Y.; Zhang, J.T. Two decades of research in discovery of anticancer drugs targeting STAT3, how close are we? *Pharmacol. Ther.* **2018**, *191*, 74–91. [CrossRef] [PubMed]
334. Hayakawa, F.; Sugimoto, K.; Harada, Y.; Hashimoto, N.; Ohi, N.; Kurahashi, S.; Naoe, T. A novel STAT inhibitor, OPB-31121, has a significant antitumor effect on leukemia with STAT-addictive oncokinases. *Blood Cancer J.* **2013**, *3*, e166. [CrossRef]
335. Brambilla, L.; Genini, D.; Laurini, E.; Merulla, J.; Perez, L.; Fermeglia, M.; Carbone, G.M.; Pricl, S.; Catapano, C.V. Hitting the right spot: Mechanism of action of OPB-31121, a novel and potent inhibitor of the Signal Transducer and Activator of Transcription 3 (STAT3). *Mol. Oncol.* **2015**, *9*, 1194–1206. [CrossRef]
336. Ogura, M.; Uchida, T.; Terui, Y.; Hayakawa, F.; Kobayashi, Y.; Taniwaki, M.; Takamatsu, Y.; Naoe, T.; Tobinai, K.; Munakata, W.; et al. Phase I study of OPB-51602, an oral inhibitor of signal transducer and activator of transcription 3, in patients with relapsed/refractory hematological malignancies. *Cancer Sci.* **2015**, *106*, 896–901. [CrossRef]

337. Bai, L.; Zhou, H.; Xu, R.; Zhao, Y.; Chinnaswamy, K.; McEachern, D.; Chen, J.; Yang, C.Y.; Liu, Z.; Wang, M.; et al. A Potent and Selective Small-Molecule Degrader of STAT3 Achieves Complete Tumor Regression In Vivo. *Cancer Cell* **2019**, *36*, 498–511. [CrossRef]
338. Yco, L.P.; Mocz, G.; Opoku-Ansah, J.; Bachmann, A.S. Withaferin A Inhibits STAT3 and Induces Tumor Cell Death in Neuroblastoma and Multiple Myeloma. *Biochem. Insights* **2014**, *7*, 1–13. [CrossRef]
339. Wang, Y.; Wu, L.; Cai, H.; Lei, H.; Ma, C.M.; Yang, L.; Xu, H.; Zhu, Q.; Yao, Z.; Wu, Y. Sinomenine derivative YL064: A novel STAT3 inhibitor with promising anti-myeloma activity. *Cell Death Dis.* **2018**, *9*, 1093. [CrossRef]
340. Turkson, J.; Zhang, S.; Palmer, J.; Kay, H.; Stanko, J.; Mora, L.B.; Sebti, S.; Yu, H.; Jove, R. Inhibition of constitutive signal transducer and activator of transcription 3 activation by novel platinum complexes with potent antitumor activity. *Mol. Cancer Ther.* **2004**, *3*, 1533–1542.
341. Turkson, J.; Zhang, S.; Mora, L.B.; Burns, A.; Sebti, S.; Jove, R. A novel platinum compound inhibits constitutive Stat3 signaling and induces cell cycle arrest and apoptosis of malignant cells. *J. Biol. Chem.* **2005**, *280*, 32979–32988. [CrossRef] [PubMed]
342. Nagel-Wolfrum, K.; Buerger, C.; Wittig, I.; Butz, K.; Hoppe-Seyler, F.; Groner, B. The interaction of specific peptide aptamers with the DNA binding domain and the dimerization domain of the transcription factor Stat3 inhibits transactivation and induces apoptosis in tumor cells. *Mol. Cancer Res.* **2004**, *2*, 170–182. [PubMed]
343. Wang, X.; Zeng, J.; Shi, M.; Zhao, S.; Bai, W.; Cao, W.; Tu, Z.; Huang, Z.; Feng, W. Targeted blockage of signal transducer and activator of transcription 5 signaling pathway with decoy oligodeoxynucleotides suppresses leukemic k562 cell growth. *DNA Cell Biol.* **2011**, *30*, 71–78. [CrossRef] [PubMed]
344. Zhao, X.; Zhang, Z.; Moreira, D.; Su, Y.L.; Won, H.; Adamus, T.; Dong, Z.; Liang, Y.; Yin, H.H.; Swiderski, P.; et al. B Cell Lymphoma Immunotherapy Using TLR9-Targeted Oligonucleotide STAT3 Inhibitors. *Mol. Ther.* **2018**, *26*, 695–707. [CrossRef]
345. Jing, N.; Li, Y.; Xiong, W.; Sha, W.; Jing, L.; Tweardy, D.J. G-quartet oligonucleotides: A new class of signal transducer and activator of transcription 3 inhibitors that suppresses growth of prostate and breast tumors through induction of apoptosis. *Cancer Res.* **2004**, *64*, 6603–6609. [CrossRef]
346. Hillion, J.; Belton, A.M.; Shah, S.N.; Turkson, J.; Jing, N.; Tweardy, D.J.; di Cello, F.; Huso, D.L.; Resar, L.M. Nanoparticle delivery of inhibitory signal transducer and activator of transcription 3 G-quartet oligonucleotides blocks tumor growth in HMGA1 transgenic model of T-cell leukemia. *Leuk. Lymphoma* **2014**, *55*, 1194–1197. [CrossRef]
347. Shastri, A.; Choudhary, G.; Teixeira, M.; Gordon-Mitchell, S.; Ramachandra, N.; Bernard, L.; Bhattacharyya, S.; Lopez, R.; Pradhan, K.; Giricz, O.; et al. Antisense STAT3 inhibitor decreases viability of myelodysplastic and leukemic stem cells. *J. Clin. Investig.* **2018**, *128*, 5479–5488. [CrossRef]
348. Hong, D.; Kurzrock, R.; Kim, Y.; Woessner, R.; Younes, A.; Nemunaitis, J.; Fowler, N.; Zhou, T.; Schmidt, J.; Jo, M.; et al. AZD9150, a next-generation antisense oligonucleotide inhibitor of STAT3 with early evidence of clinical activity in lymphoma and lung cancer. *Sci. Transl. Med.* **2015**, *7*, 314. [CrossRef]
349. Reilley, M.J.; McCoon, P.; Cook, C.; Lyne, P.; Kurzrock, R.; Kim, Y.; Woessner, R.; Younes, A.; Nemunaitis, J.; Fowler, N.; et al. STAT3 antisense oligonucleotide AZD9150 in a subset of patients with heavily pretreated lymphoma: Results of a phase 1b trial. *J. Immunother. Cancer.* **2018**, *6*, 119. [CrossRef]
350. Zhang, Q.; Hossain, D.M.; Nechaev, S.; Kozlowska, A.; Zhang, W.; Liu, Y.; Kowolik, C.M.; Swiderski, P.; Rossi, J.J.; Forman, S.; et al. TLR9-mediated siRNA delivery for targeting of normal and malignant human hematopoietic cells in vivo. *Blood* **2013**, *121*, 1304–1315. [CrossRef]
351. Von Manstein, V.; Yang, C.M.; Richter, D.; Delis, N.; Vafaizadeh, V.; Groner, B. Resistance of cancer cells to targeted therapies through the activation of compensating signaling loops. *Curr. Signal. Transduct. Ther.* **2013**, *8*, 193–202. [CrossRef] [PubMed]

11

Companion Animals as Models for Inhibition of STAT3 and STAT5

Matthias Kieslinger *, Alexander Swoboda, Nina Kramer, Barbara Pratscher, Birgitt Wolfesberger and Iwan A. Burgener

Division of Small Animal Internal Medicine, Department of Companion Animals and Horses, University of Veterinary Medicine Vienna, 1210 Vienna, Austria; Alexander.Swoboda@vetmeduni.ac.at (A.S.); Nina.Kramer@vetmeduni.ac.at (N.K.); Barbara.Pratscher@vetmeduni.ac.at (B.P.); Birgitt.Wolfesberger@vetmeduni.ac.at (B.W.); Iwan.Burgener@vetmeduni.ac.at (I.A.B.)

* Correspondence: Matthias.Kieslinger@vetmeduni.ac.at

Abstract: The use of transgenic mouse models has revolutionized the study of many human diseases. However, murine models are limited in their representation of spontaneously arising tumors and often lack key clinical signs and pathological changes. Thus, a closer representation of complex human diseases is of high therapeutic relevance. Given the high failure rate of drugs at the clinical trial phase (i.e., around 90%), there is a critical need for additional clinically relevant animal models. Companion animals like cats and dogs display chronic inflammatory or neoplastic diseases that closely resemble the human counterpart. Cat and dog patients can also be treated with clinically approved inhibitors or, if ethics and drug safety studies allow, pilot studies can be conducted using, e.g., inhibitors of the evolutionary conserved JAK-STAT pathway. The incidence by which different types of cancers occur in companion animals as well as mechanisms of disease are unique between humans and companion animals, where one can learn from each other. Taking advantage of this situation, existing inhibitors of known oncogenic STAT3/5 or JAK kinase signaling pathways can be studied in the context of rare human diseases, benefitting both, the development of drugs for human use and their application in veterinary medicine.

Keywords: cancer models; companion animals; STAT3; STAT5; comparative oncology

---

1. Introduction

Almost half of all households in the United States have at least one companion animal. This means that approximately 77 million dogs and 58 million cats share a common environment with their human owners and are largely exposed to the same health risk factors [1]. In the absence of significant cardiovascular disease, cancer is the number one cause of death for dogs, killing between 40% and 50% of individuals older than 10 years, and between 20% and 25% regardless of age [2–4]. Numbers are less detailed for cats, but the overall tumor incidence ranges between 30% and 35% [5]. The prevalence of cancer in companion animals has increased in the last decades, which may be the result of a real increase in cancer incidence, an increase in the population of companion animals at risk or the awareness and willingness of the animal owners to pursue diagnostic and treatment options [6]. While the full spectrum of tumor types seen in humans also occurs in cats and dogs, the rates for individual tumor types are often different. Canine osteosarcoma, soft tissue sarcoma and feline non-Hodgkin's lymphoma for example are significantly higher than in humans (Table 1), whereas other tumor entities like lung, prostate and colon tumors are rare in companion animals. Cancer in companion animals, particularly in dogs, resembles cancer in humans in many ways, including spontaneous disease occurring without an isogenic background or genetic engineering and shared environmental and societal status with owners. Further similarities in chronology of the disease adapted to lifespan, organization into various well-characterized breeds that show different incidences of tumor types, and shared environmental and societal status with owners makes them attractive objects for comparison.

Table 1. Incidence rates of various tumor types from human, dog and cat.

Tissue	Human	Dog	Cat
Mammary	127.5 [7]	250 [8]	13–25 [9]
Melanoma	22.2 [7]	19.8 [10]	ND
Testes	5.9 [7]	16.7 [11]	ND
Connective Tissue	3.5 [7]	40.1 [10]	17.0 [1]
Skin	98.85 [12]	103.3 [10]	34.7 [1]
Oral	11.3 [7]	20.4 [1]	11.6 [1]
NHL/Leukemia	33.7 [7]	76.3 [13]	41 [14]
Bone	1.0 [7]	27.2 [15]	3.1–4.9 [16]

Numbers represent cases per 100,000. ND = not determined.

The mouse has been an extremely useful tool to gain cellular and mechanistic understanding into the development of cancer [17]. The concept of oncogenes and tumor suppressors balancing cellular proliferation of multicellular organisms has been proven in vivo in genetically modified mice [18,19]. The use of mice deleted for a single gene has allowed us to determine the involvement of signaling pathways, genetic regulators like transcription factors, epigenetic factors, etc. in the development and sustained growth of cancer. Overall, experiments in mouse models have been invaluable in understanding the mechanistic basis of cancer biology. However, while offering critical insights into basic concepts, murine models underrepresent the heterogeneity and complex interplay between human immune and cancer cells [20,21].

The Janus kinase (JAK)—signal transducer and activator of transcription (STAT) signaling pathway—provides a fast and efficient way for relaying signals from the extracellular space to the nucleus and modifying gene expression [22]. Main targets of this pathway represent regulators of cell division and apoptosis as broadly discussed in several publications in this special issue [23–25]. As these are central aspects in the development of cancer, it is not surprising that members of this pathway are over-activated in many cancers [26,27]. The kinases of this pathway, JAK 1-3 and TYK2 are constantly activated in several different tumor types and are subject to pharmacological inhibition. Recently, also STAT proteins have come into focus of cancer researchers. Particularly STAT3 and STAT5 are activated in over 70% of all human cancer types and constitute a critical node in the signaling networks of tumor cells [28].

This review will highlight, why companion animals and particularly the dog represent an attractive link between murine models, addressing basic mechanistic aspects and human diseases specifically in the context of JAK-STAT signaling.

2. Preclinical Models

The vast majority of cancer models currently are represented by mice, and their fundamental importance for preclinical research is clearly established [20,29]. These animals come as various strains that have been inbred over many generations and thereby are genetically highly homogenous [30]. When used for experiments, they are matched for age, sex and size, receive the same sterile diets and are housed under specific pathogen-free (SPF) conditions. All of these factors are controlled for in order to generate standardized conditions that allow us to draw scientifically sound conclusions based on the variation of one single parameter [31]. Furthermore, this reduction of variables and thereby "noise" allows us to reduce the number of animals necessary to reach statistically significant results. While this concept has proven its scientific merits and is logical within itself, it is questionable, whether young, sex-matched and inbred mice on sterile nutrition are a good representation of the typically older, obese and genetically diverse human cancer patients [32].

The controlled environment also affects the outcome of cancer-related experiments. For example, it has become clear recently, that the microbiome influences the response to cancer treatment [33,34]. Accordingly, mice raised under SPF conditions in various research institutions show differences in the composition of their gut microbiome, affecting tumor growth rates [35]. Again, while differences in the gut microbiome are likely to exist in human cancer patients, the single standardized composition under which these experiments are carried out in laboratory mice are very likely no close representation of the much broader spectrum in humans. Additionally, rodents have adopted to most ecosystems metabolically, but in regard to drug metabolism, due to growth-hormone regulated p450 cytochrome components, pharmacodynamic and pharmacokinetic properties, canines are by far superior and therefore used as key models for FDA-drug approved testing [36].

Another important aspect is the genetic background and its modification particularly with respect to the immune system. Mice used for tumor experiments usually are highly inbred, reducing genetic variability [37]. Experiments regarding the consequences of a narrow genetic background have demonstrated potential phenotypic tilting, which may result in unrepresentative biased phenotypes [38]. Furthermore, the buffering of genetic variation, including disease-causing mutations, is impaired under these circumstances [39]. One effort to overcome limitations due to a narrow genetic background is to establish new reference populations derived from the crossing of several different mouse strains, as exemplified by the Collaborative Cross project [40]. By broadening and defining the genetic basis, it offers the perspective of enhancing genetic stability and reproducibility, thereby also representing a new and potentially better resource of murine models for human diseases.

Additionally, and probably of higher practical relevance is the fact that development and proliferation of tumor cells happens in a complex interplay with cells of the immune system [41]. However, cancer-studies in mice are often performed in the absence of a fully functional immune system, using immunocompromised mice as hosts for transplantations of human cell lines, patient-derived xenografts or human tumor-derived immune cell xenografts [42]. The immune system constitutes the major player in the counter-selection against tumor cells, thereby necessitating evasion or adaption strategies on the side of the tumor cells. Since this aspect is missing in such mouse models, the results from experiments thereof likely reflect only partial aspects of tumor biology and challenge their biological relevance.

One way to overcome these problems are humanized mice, which express human instead of murine components of the immune system like major histocompatibility complex, allowing the transplantation of tumor cells in an immunologically at least partially competent environment [43]. These models clearly represent a step forward, however, they are costly, technically complicated and still do not represent all components of a functional and homogenous immune system [44]. Moreover, human patients that have developed cancer are not living under special pathogen-free conditions and are very likely to have chronic viral infections, like the Epstein–Barr virus, cytomegalovirus or herpes simplex, which are present in up to 90% of the total population [45–47]. Such chronic infections exert a constant pressure and shape the immune system, which is not the case in mice under SPF conditions. Co-housing of "dirty" outbred immunocompetent mice could be a way for improvement, still mammalian viral species barriers exist, that make companion animals superior in these infectious aspects.

Finally, genetically engineered mice harboring the deletion of a tumor suppressor gene or the ectopic expression of an oncogene or combinations thereof can be used [48]. In this case, the tumors develop in the presence of a competent immune system and problems related to cross-species compatibility do not arise. This setting also enables the introduction of defined mutations that occur in human tumors. However, this comes at the price of costly and time-consuming development, often requiring years of work before availability. Furthermore, whereas the targeted insertion of defined mutations reflecting the human situation is a clear advantage, cancer development is a multi-step process, and the heterogeneity of these further steps is often different between engineered mice and human patients [49,50].

Overall, the mouse has been highly instructive in determining genes involved and their mechanistic contribution to the origin and development of cancer [51]. However, the limitations of this model, in particular the differences in environment and microbiome, life span, tumor etiology and genetic status may be the reason why certain aspects are not reflected closely in this model, resulting in only 11% of oncology drugs that work in mice being approved for human use [52]. Therefore, it is desirable from a translational perspective to add another layer that closer reflects human biology and cancer development.

3. Advantages and Disadvantages of Canine Tumor Models

Roughly 4,000,000 dogs and a similar number of cats are diagnosed with cancer each year in the U.S. [53]. Although exact epidemiological data are not available for companion animals, this translates into approximately 5300 cases per 100,000 dogs, which is around ten times higher than in humans with 500 cases per 100,000 persons. This large number of pets provides the opportunity to study spontaneous cancers that are highly similar to those occurring in humans, especially since most pet owners are highly motivated to seek out novel treatments for their companion animals.

There are several advantages of using companion animals as models for human cancer. Among them is the fact that tumors arise spontaneously, just as in humans, and that tumor initiation and progression are influenced by similar factors like age, nutrition, sex, reproductive status and environmental exposure [54]. The risk for developing cancer of the nasal cavity for example is increased up to 60% in animals that are kept by smokers in comparison to pets of non-smoking owners [55,56]. Although there are differences in the diet of humans and companion animals, many components such as meat, vegetables and carbohydrates are derived from the same sources and are consumed non-sterile as opposed to mice under SPF conditions [57]. Furthermore, studies have shown that the contact between owner and pet leads to a large overlap in the microbiome, the importance of which for human tumor development has come into focus recently [58].

Pets, and in particular dogs are large and relatively outbred in comparison to laboratory mice. In fact, the genetic variation across dog breeds or in mixed breeds is similar to the variation in humans on the basis of single nucleotide polymorphisms [59]. In individual pure breeds however, the level of genetic diversity is more restricted [60]. The canine genome has been sequenced with a coverage of 99%, revealing that the approximately 19,000 genes identified in the dog match to homologous or orthologous genes in humans [59]. Actually, for many gene families, particularly for those associated with cancer, the homologies are significantly closer than the relationship between human and mouse [61]. Accordingly, most oncogenes and tumor suppressors that are known from human cancers have been shown to contribute to canine cancers [62].

Dogs and cats of all breeds develop cancer, and the spectrum of cancers seen in companion animals is as diverse as that seen in human patients [63]. The dog is the species in which comparative oncology has shown the most growth, and where it is best characterized [64]. Interestingly, there are breed-specific differences as to the cancer subtypes, reflecting the underlying genetics of the various breeds (Table 2). Mast cell tumors and gliomas for example are over-represented in Boxers, Staffordshire, Weimaraner and Golden Retriever, osteosarcoma in Rottweilers, Greyhounds and Golden Retrievers, bladder cancer in Scottish Terriers, histiocytic sarcomas in Flat-Coated Retrievers and Bernese Mountain Dogs and melanoma and gastric carcinoma in Chow-Chows [65–67].

Table 2. Oncological disposition of various dog breeds.

Breeds	Most Frequent Tumor Types
Bernese Mountain Dog	Histiocytic sarcoma [68], Lymphoma [68,69], Osteosarcoma [68]
Boxer	Glioma [67,69], Mast cell tumor [10,67]
Flat-Coated Retriever	Soft tissue sarcoma, Histiocytic sarcoma, Hemangiosarcoma [70]
Golden Retriever	Mast cell tumor, Lymphoma, Oral Melanoma, Fibrosarcoma [67]
Magyar Viszla	Mast cell tumor, Hemangiosarcoma, Lymphoma [10,71]
Giant Schnauzer	Epidermal tumor, Hair follicle tumor, Melanocytic tumor [10]
Airedale Terrier	Melanoma [72], Lymphoma [67], Prostatic carcinoma [67]
Bullmastiff	Mast cell tumor, Lymphoma [67]
St. Bernard	Lymphoma [67], Osteosarcoma [73]
Irish Wolfhound	Osteosarcoma [67,74], Lymphoma [75]

The most frequent tumor types of dog breeds with high tumor incidence.

The global expression pattern of canine and human osteosarcoma for example shows a strong similarity, and cluster analysis of orthologous gene signatures does not segregate human and canine tumors [73]. Finally, many chemotherapy protocols used for the treatment of canine cancers have been adopted from human medicine. The same chemotherapeutics used in human lymphoma for example are also active in canine lymphoma (e.g., vincristine, cyclophosphamide, doxorubicin, mitoxantrone, cytarabine and methotrexate), and drugs that are ineffective in human lymphomas are also inactive in canines (i.e., gemcitabine, cisplatin and carboplatin) [5].

One of the biggest advantages of companion animals as models for human cancer is the spontaneous development of tumors in the presence of an intact immune system. Immune cells pose a significant barrier to the development of cancer, and undergo changes themselves, as cancer cells co-opt the immune response [76,77]. As a result, the tumor influences innate as well as adaptive immune cells to become regulatory, rather than tumoricidal [78,79]. This interplay results in the selection of tumor cells that are invisible to anti-tumor T-cell-mediated destruction, and is central for tumor editing and immune evasion. Additionally, similar to humans, dogs with advanced cancer exhibit intrinsic T-cell defects as well as T-cell exhaustion [80]. Immune cell interplay with cancer cells is a JAK-STAT3/5 affair, as detailed in the special issue in several articles [24,25,81].

Like all other model systems, companion animals have strengths as well as weaknesses, both practically and conceptually. The biggest hurdle, when working with cats and dogs in research is the paucity of investigational tools. Many antibodies and recombinant products that are available for humans and mice do not show cross-species reactivity. However, the sequencing of the canine genome and the development of genome editing via the CRISPR/Cas9 technology has relieved many of the restrictions [82]. It is now possible to introduced genetic alterations with high efficiency into any desired locus, facilitating for example the visualization of specific cell types via the expression of marker proteins, or targeted deletion of single genes [83,84]. Conceptually, the lack of standardized housing conditions, making it difficult to control for variables, has been held against non-rodent models. However, as pointed out here, this represents an advantage when it comes to closely mimicking the translational aspects in tumor biology.

The potential of companion animals as biological models between the basic mechanistic work that is possible in the mouse and translation to humans is demonstrated by 1. The high conservation at the genomic level, 2. The involvement of similar genetic and environmental risk factors, 3. The successful use of canine cancer as biological models for the early development of bone marrow transplantation protocols and 4. canine trials for the development of drug level and exposure durations. This sets an ideal stage to combine new perspectives of targeted therapies and specific molecular inhibitors in the field of comparative oncology for the benefit of human as well as veterinary medicine (Figure 1). Furthermore, this approach is not limited to cancer but applicable to any comparative condition including infectious or inflammatory diseases like inflammatory bowel disease or pre-malignant conditions like adenoma formation or clonal hematopoiesis.

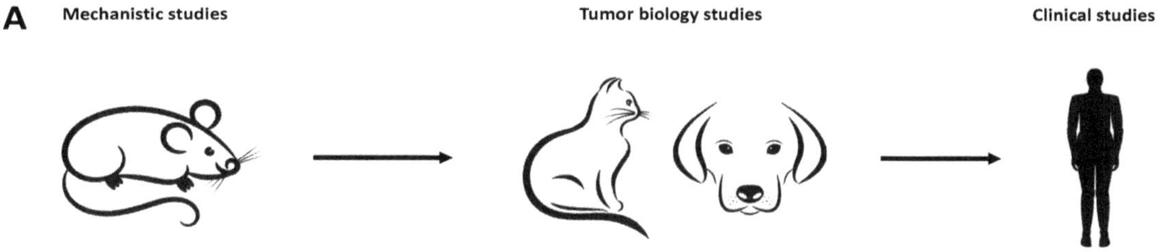

B **Advantages:**

- Controlled environment
- Defined genetic alterations
- Good reproducibility
- Mechanistic studies possible

- Shared environment and risk factors with humans
- Spontaneous tumor development
- Immunocompetent / chronic infections
- Microbiome comparable to human owners

Disadvantages:

- Environment not representative
- Limited spontaneous tumors
- Often immunocompromised for xenografting
- High failure rate in human patients

- Variables and environment difficult to control
- Heterogenous cancer biology / immune response
- Increased variability
- Limited reagents available

Figure 1. Advantages and disadvantages of different models during drug discovery. (**A**) Companion animals can be used as an intermediate step between the mechanistic work in murine models and clinical studies in humans, particularly with regard to comparative aspects of tumor biology. (**B**) Advantages and disadvantages of the individual models for translation into human clinical studies.

4. Relevance and Conservation of the JAK-STAT Signaling Pathway

JAK-STAT proteins constitute an evolutionary conserved signaling pathway [85]. Ligand binding of receptors leads to the activation of JAK kinases and STAT proteins, inducing transcription. The family of JAK kinases consists of four members, JAK1-3 and TYK2, and there are seven highly homologous STAT proteins, STAT1–4, STAT5a and STAT5b and STAT6 [22,86]. For details on the mechanism see also other reviews in this issue [87,88]. As such, this pathway provides a remarkably elegant and straightforward mechanism to transduce signals from receptors to the nucleus.

All family members show the same structural organization, i.e., an N-terminal domain required for oligomerization of dimers into tetramers, a coiled-coil domain, a DNA-binding domain, a linker domain, a Src homology 2 (Sh2) domain for dimerization and a C-terminal transactivation domain (Figure 2). Functionally, STAT2, STAT4 and STAT6 regulate specific immune cell responses, whereas STAT1, STAT3 and STAT5 have diverse physiological roles. STAT1 is mostly involved in immunity, host defense against pathogens and cell death, stimulating the transcription of pro-inflammatory and anti-proliferative genes like caspases, *NOS2*, *MDM2*, *CDKN1A* and *CDKN1B*. On the contrary, STAT3 and STAT5 are mostly involved in cell proliferation and prevention of apoptosis, activating the transcription of genes like *CCND1*, *BIRC5*, *c-MYC*, *VEGF*, *MCL1*, *BCL2L1* and *BCL2*. Additionally, STAT3 can also be found in mitochondria, where it supports RAS-dependent malignant transformation via sustained altered glycolytic and oxidative phosphorylation [89,90]. Given their roles in the stimulation of cellular proliferation, the prevention of apoptosis and the stimulation of metabolism, STAT5, and even more so STAT3, are activated in nearly 70% of solid and hematological human tumors [91–93].

Silencing or inhibition of STAT3 or STAT5 signaling impairs tumor growth and survival in murine and human studies, while only slightly affecting normal differentiated cells [94–97]. These findings lead to the concept of STAT3 and STAT5 constituting a "signaling bottleneck" situation for tumor cells, making them attractive targets for inhibition [98]. However, caution has to be exerted with regard to tissue-specificity, as tumor-suppressive functions have been ascribed to STAT3 in neuronal, hepatic and colorectal tumors and to STAT5 in breast cancer [99,100].

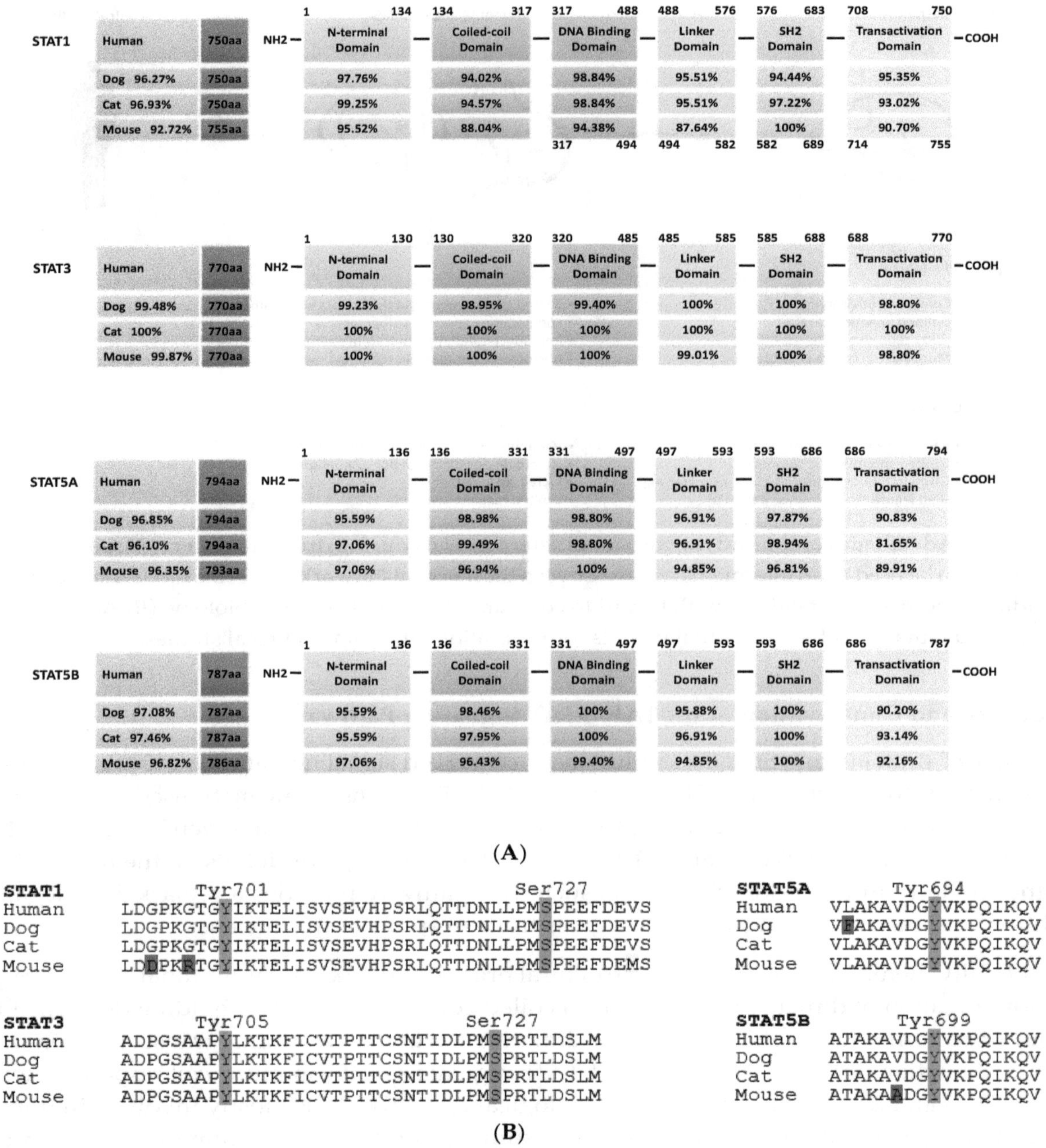

Figure 2. Cross-species conservation of STAT protein domains. (**A**) STAT1, STAT3, STAT5a and STAT5b from dog, cat and mouse are analyzed for their overall homology compared to the respective human protein (grey boxes, left). In the schematic representation of STAT protein domains, the amino acid positions are indicated above. All proteins share the same domain positions, except for murine STAT1, which has a five amino acid insertion in the DNA binding domain (numbers below the scheme indicate the aa position in this case). Percentages in the domain boxes of dog, cat and mouse STAT proteins show the homology of each domain to the human counterpart. Analyses were carried out using ClustalX. (**B**) Comparison of key phosphorylation sites in the transactivation domain of STAT1, STAT3, STAT5a and STAT5b from dog, cat and mouse to the human sequence. Amino acid sequence is shown, with phosphorylation sites in green and position indicated; positive amino acid exchanges (conserving protein function) are indicated in yellow, other exchanges in red. (STAT1: human NP_009330.1, dog XP_848353.1, cat XP_006935505.1, mouse NP_001192242.1; STAT3: human NP_644805.1, dog XP_005624514.1, cat XP_003996930.1, mouse NP_998824.1; STAT5a: human NP_001275647.1, dog XP_548091.2, cat XP_023099834.1, mouse NP_001157534.1; STAT5b: human NP_036580.2, dog XP_548092.1, cat XP_023100377.1, mouse NP_035619.3).

Several different ways of inhibiting STAT signaling are possible. Upstream of STAT proteins, JAK kinases are mutated in a broad range of diseases from severe combined immunodeficiency to various forms of cancer, including JAK1 in acute myeloid leukemia, JAK2 in myeloproliferative diseases and JAK3 in different leukemias and lymphomas, and inhibitors against JAK kinases are already approved by the US Food and Drug Administration (FDA) for clinical use [27]. Interestingly, different layers of negative regulators of JAK-STAT signaling are present such as suppressor of cytokine signaling (SOCS), protein inhibitor of activated STAT (PIAS) and protein tyrosine phosphatases, arguing for the necessity of a tightly controlled down-regulation of this signaling pathway [101].

Due to the broad activation, minor side-effects and the overall importance, major efforts by many laboratories and pharmaceutical companies are ongoing to develop inhibitors against STAT3 and STAT5. In both cases, all current inhibitors target one of three STAT motifs: the SH2 domain necessary for the interaction of phosphorylated monomers to form dimers, the N-terminal domain mediating the formation of tetramers from activated STAT dimers and the DNA-binding domain [102]. STAT3 and STAT5 from companion animals show more than 96% homology at the overall protein level to their human counterparts, with a particular high level of conservation of 98% to complete alignment in these three domains (Figure 2). This high level of conservation opens up the possibility to use pet animals as models for diseases in which the JAK-STAT signaling pathway is over-activated.

A good example for such a successful application is already established. Cytokine dysregulation has been implicated in allergic skin disease, particularly in atopic dermatitis in humans. T-helper cells type 2 (Th2) produce increased levels of IL4, IL5, IL10, IL13 and IL31, in addition to elevated production of IFNγ by T-helper cells type 1 (Th1), signals that all converge on the JAK-STAT signaling pathway [27,103–105]. Dermatological problems are the second most common reason for dogs to present to veterinary practices, frequently including allergic skin diseases like atopic dermatitis [106,107]. In the skin of atopic dogs, a cytokine profile can be found that resembles the human condition, and in an experimental model of canine allergic dermatitis, elevated transcripts of IL6, IL13 and IL18 and IFNγ were detected, supporting the idea that cytokine dysregulation plays a role in allergic skin disease [108–110]. The novel JAK inhibitor oclacitinib is most potent against JAK1, but also affects JAK2 and JAK3 at reduced efficiency, inhibiting the function of JAK1-dependent inflammatory cytokines [111]. Treatment of dogs suffering from atopic dermatitis results in a reduction of associated skin lesions and oclacitinib recently has been approved in the US and Europe for the treatment of allergic/atopic dermatitis [112–114]. Overall, inhibition of JAK1-dependent cytokines is an effective and novel way to treat canine allergic skin disease, proving the high similarity and cross-species conservation of JAK-STAT signaling.

5. Inhibition of STAT3 and STAT5 in Companion Animals: Current Status/Future Perspectives

Currently, STAT3 and STAT5 are studied in several cancer types of companion animals. Canine mammary cancer cells and diffuse large B cell lymphoma as well as feline oral squamous cell carcinoma and mammary tumors show activation of STAT3 or STAT5. First results indicate reduced proliferation and increased apoptosis upon inhibition of JAK1/2 or STAT3 respectively [115–121]. Two areas exemplify the potential of research and development concerning STAT proteins and human—canine comparative oncology best [122,123]. Osteosarcoma is the most frequent form of malignant bone disease in dogs, however, it is relatively rare in humans. The estimated incidence rate is at least 13.9/100,000 in canines and 1.02/100,000 in humans, affecting primarily children [124,125]. Such a high incidence rate provides a good opportunity to study a rare human disease using dogs as a preclinical model. Moreover, canine and human osteosarcoma share many key features like tumor location, early metastasis, development of chemotherapy-resistant metastases and altered expression or activation of several proteins [126].

Consistent activation of STAT3 occurs in a large subset of human and canine osteosarcoma and osteosarcoma cell lines, but not in normal osteoblasts. Down-regulation of STAT3 expression or activity reduces proliferation and induces apoptosis in human and canine cell lines [127–130]. Additionally, human and canine osteosarcoma possess overlapping transcriptional profiles, further supporting the concept that these diseases are similar at the molecular level [131]. Spontaneous canine osteosarcoma has been used for the development of novel therapeutics such as muramyl tripeptide, IGF-1R inhibitors and rapamycin [132]. However, despite aggressive treatment, 30–40% of children and >90% of dogs are still dying from disease, demonstrating the need for new therapeutic options. The similarities between human and canine osteosarcoma, together with the dependency on JAK-STAT signaling, particularly STAT3, make osteosarcoma one of the prime areas for comparative oncology studies, especially since a large number of STAT3 inhibitors are currently being developed.

Mast cell tumors arise from the uncontrolled proliferation of transformed mast cells mostly in skin, spreading primarily to spleen, liver and bone marrow [133]. While mast cell tumors are relatively rare in humans, they are frequent in dogs, accounting for 7–21% of cutaneous tumors [134,135]. Metastasized tumors in humans as well as in canines have a poor prognosis and short survival times [133,134,136]. Stem cell factor and its receptor c-KIT are essential for mast cell survival and inhibition of apoptosis, and gain-of-function mutations are present in human and canine mast cell tumors [137–142]. Accordingly, two inhibitors of c-KIT, masitinib and toceranib have been approved for use in c-KIT driven mast cell tumors in dogs. Both drugs are able to suppress tumor growth temporarily, however, relapses are high, indicating the need for the identification of further targets and therapeutic options [143,144].

STAT1, STAT3 and STAT5 are activated down-stream of mutant c-KIT, but only STAT5 is transcriptionally active in neoplastic mast cells [145,146]. Inhibition of JAK2 and STAT5 was recently discovered to inhibit proliferation and survival of canine mastocytoma cell lines, identifying JAK2/STAT5 signaling as a new potential target in mast cell tumors [147]. Therefore, an attractive option is to use the frequent occurrence of canine mast cell tumors to determine if inhibitors of STAT5 can be used alone or in combination with inhibitors of c-kit or other kinases as a new therapeutic option. Indeed, the potential of STAT5 inhibitors to overcome resistance to a multi-kinase inhibitor in neoplastic mast cells has been demonstrated [148]. Consequently, the next step is to determine the potential of new STAT5 inhibitors in vivo [149]. Both of these examples, osteosarcoma as well as mast cell tumors offer thereby the possibility of using new pharmaceutical in companion animals as a closer biological mimic of the human situation.

6. Conclusions

Taken together, cancer in dogs resembles cancer in humans in many ways, like its latency, clinical manifestation and metastatic behavior, its pathobiological characteristics like tumor heterogeneity, its genomic instability and pharmacogenomic signatures including chemoresistance and last but not least its multifactorial nature, including genetic and environmental risk factors [150]. The inability of murine cancer models to recapitulate certain aspects of human tumors is increasingly recognized and illuminates the huge potential of spontaneous canine and, to a lesser extent, feline cancer [151–153]. The JAK-STAT pathway is activated in the vast majority of solid and hematological tumors, and is necessary for tumor growth and prevention of apoptosis. The major efforts that are ongoing to develop inhibitors specifically for STAT3 and STAT5 can be extended to studies in companion animals, an option that is particularly attractive for rare human diseases occurring more frequently in dogs and/or cats.

Acknowledgments: Open Access Funding by the University of Veterinary Medicine Vienna.

References

1. Vail, D.M.; MacEwen, E.G. Spontaneously occurring tumors of companion animals as models for human cancer. *Cancer Invest.* **2000**, *18*, 781–792. [CrossRef]

2. Bonnett, B.N.; Egenvall, A.; Hedhammar, A.; Olson, P. Mortality in over 350,000 insured Swedish dogs from 1995-2000: I. Breed-, gender-, age- and cause-specific rates. *Acta Vet. Scand.* **2005**, *46*, 105–120. [CrossRef]
3. Egenvall, A.; Bonnett, B.N.; Hedhammar, A.; Olson, P. Mortality in over 350,000 insured Swedish dogs from 1995-2000: II. Breed-specific age and survival patterns and relative risk for causes of death. *Acta Vet. Scand.* **2005**, *46*, 121–136. [CrossRef] [PubMed]
4. Vascellari, M.; Baioni, E.; Ru, G.; Carminato, A.; Mutinelli, F. Animal tumour registry of two provinces in northern Italy: Incidence of spontaneous tumours in dogs and cats. *BMC Vet. Res.* **2009**, *5*, 39. [CrossRef] [PubMed]
5. Withrow, S.J.; Vail, D.M. *Withrow & MacEwen's Small Animal Clinical Oncology*, 5th ed.; Saunders Elsevier: Philadelphia, PA, USA, 2013.
6. Paoloni, M.; Khanna, C. Translation of new cancer treatments from pet dogs to humans. *Nat. Rev. Cancer* **2008**, *8*, 147–156. [CrossRef] [PubMed]
7. National Cancer Institute. Surveillance, Epidemiology and End Results (SEER) Program 2012–2016 Data. Available online: www.seer.cancer.gov/statfacts (accessed on 6 November 2019).
8. Vascellari, M.; Capello, K.; Carminato, A.; Zanardello, C.; Baioni, E.; Mutinelli, F. Incidence of mammary tumors in the canine population living in the Veneto region (Northeastern Italy): Risk factors and similarities to human breast cancer. *Prev. Vet. Med.* **2016**, *126*, 183–189. [CrossRef]
9. Hassan, B.B.; Elshafae, S.M.; Supsavhad, W.; Simmons, J.K.; Dirksen, W.P.; Sokkar, S.M.; Rosol, T.J. Feline Mammary Cancer. *Vet. Pathol.* **2017**, *54*, 32–43. [CrossRef]
10. Graf, R.; Pospischil, A.; Guscetti, F.; Meier, D.; Welle, M.; Dettwiler, M. Cutaneous tumors in Swiss dogs: Retrospective data from the Swiss canine cancer registry, 2008–2013. *Vet. Pathol.* **2018**, *55*, 809–820. [CrossRef]
11. Merlo, D.F.; Rossi, L.; Pellegrino, C.; Ceppi, M.; Cardellino, U.; Capurro, C.; Ratto, A.; Sambucco, P.L.; Sestito, V.; Tanara, G.; et al. Cancer incidence in pet dogs: Findings of the animal tumor registry of Genoa, Italy. *J. Vet. Intern. Med.* **2008**, *22*, 976–984. [CrossRef]
12. Lomas, A.; Leonardi-Bee, J.; Bath-Hextall, F. A systematic review of worldwide incidence of nonmelanoma skin cancer. *Br. J. Dermatol.* **2012**, *166*, 1069–1080. [CrossRef]
13. Pinello, K.C.; Niza-Ribeiro, J.; Fonseca, L.; de Matos, A.J. Incidence, characteristics and geographical distributions of canine and human non-Hodgkin's lymphoma in the Porto region (North West Portugal). *Vet. J.* **2019**, *245*, 70–76. [CrossRef] [PubMed]
14. Grover, S. Gastronintestinal lymphoma in cats. *Oncol. Compend.* **2005**, *27*, 741–751.
15. Egenvall, A.; Nodtvedt, A.; von Euler, H. Bone tumors in a population of 400 000 insured Swedish dogs up to 10 y of age: Incidence and survival. *Can. J. Vet. Res.* **2007**, *71*, 292–299. [PubMed]
16. Dimopoulou, M.; Kirpensteijn, J.; Moens, H.; Kik, M. Histologic prognosticators in feline osteosarcoma: A comparison with phenotypically similar canine osteosarcoma. *Vet. Surg.* **2008**, *37*, 466–471. [CrossRef] [PubMed]
17. Cheon, D.J.; Orsulic, S. Mouse models of cancer. *Annu. Rev. Pathol* **2011**, *6*, 95–119. [CrossRef] [PubMed]
18. Ghebranious, N.; Donehower, L.A. Mouse models in tumor suppression. *Oncogene* **1998**, *17*, 3385–3400. [CrossRef] [PubMed]
19. Fowlis, D.J.; Balmain, A. Oncogenes and tumour suppressor genes in transgenic mouse models of neoplasia. *Eur. J. Cancer* **1993**, *29A*, 638–645. [CrossRef]
20. Murphy, W.J. Of mice and men. *Biol. Blood Marrow Transplant.* **2013**, *19*, 1140–1141. [CrossRef]
21. Cekanova, M.; Rathore, K. Animal models and therapeutic molecular targets of cancer: Utility and limitations. *Drug Des. Devel. Ther.* **2014**, *8*, 1911–1921. [CrossRef]
22. Stark, G.R.; Darnell, J.E., Jr. The JAK-STAT pathway at twenty. *Immunity* **2012**, *36*, 503–514. [CrossRef]
23. Igelmann, S.; Neubauer, H.A.; Ferbeyre, G. STAT3 and STAT5 Activation in Solid Cancers. *Cancers* **2019**, *11*, 1428. [CrossRef] [PubMed]
24. Logotheti, S.; Putzer, B.M. STAT3 and STAT5 targeting for simultaneous management of melanoma and autoimmune diseases. *Cancers* **2019**, *11*, 1448. [CrossRef] [PubMed]
25. Rebe, C.; Ghiringhelli, F. STAT3, a master regulator of anti-tumor immune response. *Cancers* **2019**, *11*, 1280. [CrossRef] [PubMed]
26. Bromberg, J. Stat proteins and oncogenesis. *J. Clin. Invest.* **2002**, *109*, 1139–1142. [CrossRef]
27. O'Shea, J.J.; Schwartz, D.M.; Villarino, A.V.; Gadina, M.; McInnes, I.B.; Laurence, A. The JAK-STAT pathway: Impact on human disease and therapeutic intervention. *Annu. Rev. Med.* **2015**, *66*, 311–328. [CrossRef]

28. Fagard, R.; Metelev, V.; Souissi, I.; Baran-Marszak, F. STAT3 inhibitors for cancer therapy: Have all roads been explored? *JAKSTAT* **2013**, *2*, e22882. [CrossRef]
29. Rangarajan, A.; Weinberg, R.A. Opinion: Comparative biology of mouse versus human cells: Modelling human cancer in mice. *Nat. Rev. Cancer* **2003**, *3*, 952–959. [CrossRef]
30. Dobrowolski, P.; Fischer, M.; Naumann, R. Novel insights into the genetic background of genetically modified mice. *Transgenic Res.* **2018**, *27*, 265–275. [CrossRef]
31. Kilkenny, C.; Browne, W.J.; Cuthill, I.C.; Emerson, M.; Altman, D.G. Improving bioscience research reporting: The ARRIVE guidelines for reporting animal research. *PLoS Biol.* **2010**, *8*, e1000412. [CrossRef]
32. Bouchlaka, M.N.; Murphy, W.J. Impact of aging in cancer immunotherapy: The importance of using accurate preclinical models. *Oncoimmunology* **2013**, *2*, e27186. [CrossRef]
33. Hooper, L.V.; Littman, D.R.; Macpherson, A.J. Interactions between the microbiota and the immune system. *Science* **2012**, *336*, 1268–1273. [CrossRef] [PubMed]
34. Ivanov, I.I.; Honda, K. Intestinal commensal microbes as immune modulators. *Cell Host Microbe* **2012**, *12*, 496–508. [CrossRef] [PubMed]
35. Sivan, A.; Corrales, L.; Hubert, N.; Williams, J.B.; Aquino-Michaels, K.; Earley, Z.M.; Benyamin, F.W.; Lei, Y.M.; Jabri, B.; Alegre, M.L.; et al. Commensal bifidobacterium promotes antitumor immunity and facilitates anti-PD-L1 efficacy. *Science* **2015**, *350*, 1084–1089. [CrossRef] [PubMed]
36. Antonovic, L.; Martinez, M. Role of the cytochrome P450 enzyme system in veterinary pharmacokinetics: Where are we now? Where are we going? *Future Med. Chem.* **2011**, *3*, 855–879. [CrossRef] [PubMed]
37. Brekke, T.D.; Steele, K.A.; Mulley, J.F. Inbred or outbred? Genetic diversity in laboratory rodent colonies. *G3 (Bethesda)* **2018**, *8*, 679–686. [CrossRef] [PubMed]
38. Cutler, G.; Kassner, P.D. Copy number variation in the mouse genome: Implications for the mouse as a model organism for human disease. *Cytogenet. Genome Res.* **2008**, *123*, 297–306. [CrossRef]
39. Hartman, J.L.t.; Garvik, B.; Hartwell, L. Principles for the buffering of genetic variation. *Science* **2001**, *291*, 1001–1004. [CrossRef]
40. Srivastava, A.; Morgan, A.P.; Najarian, M.L.; Sarsani, V.K.; Sigmon, J.S.; Shorter, J.R.; Kashfeen, A.; McMullan, R.C.; Williams, L.H.; Giusti-Rodriguez, P.; et al. Genomes of the mouse collaborative cross. *Genetics* **2017**, *206*, 537–556. [CrossRef]
41. Dunn, G.P.; Bruce, A.T.; Ikeda, H.; Old, L.J.; Schreiber, R.D. Cancer immunoediting: From immunosurveillance to tumor escape. *Nat. Immunol.* **2002**, *3*, 991–998. [CrossRef]
42. Cespedes, M.V.; Casanova, I.; Parreno, M.; Mangues, R. Mouse models in oncogenesis and cancer therapy. *Clin. Transl. Oncol.* **2006**, *8*, 318–329. [CrossRef]
43. Holzapfel, B.M.; Wagner, F.; Thibaudeau, L.; Levesque, J.P.; Hutmacher, D.W. Concise review: Humanized models of tumor immunology in the 21st century: Convergence of cancer Research and tissue engineering. *Stem Cells* **2015**, *33*, 1696–1704. [CrossRef] [PubMed]
44. Richmond, A.; Su, Y. Mouse xenograft models vs. GEM models for human cancer therapeutics. *Dis Model. Mech.* **2008**, *1*, 78–82. [CrossRef] [PubMed]
45. Wald, A.; Corey, L. Persistence in the population: Epidemiology, transmission. In *Human Herpesviruses: Biology, Therapy, and Immunoprophylaxis*; Arvin, A., Campadelli-Fiume, G., Mocarski, E., Moore, P.S., Roizman, B., Whitley, R., Yamanishi, K., Eds.; Cambridge University Press: Cambridge, UK, 2007.
46. Smatti, M.K.; Al-Sadeq, D.W.; Ali, N.H.; Pintus, G.; Abou-Saleh, H.; Nasrallah, G.K. Epstein-Barr virus epidemiology, serology, and genetic variability of LMP-1 oncogene among healthy population: An update. *Front. Oncol.* **2018**, *8*, 211. [CrossRef] [PubMed]
47. Colugnati, F.A.; Staras, S.A.; Dollard, S.C.; Cannon, M.J. Incidence of cytomegalovirus infection among the general population and pregnant women in the United States. *BMC Infect. Dis* **2007**, *7*, 71. [CrossRef] [PubMed]
48. Day, C.P.; Merlino, G.; Van Dyke, T. Preclinical mouse cancer models: A maze of opportunities and challenges. *Cell* **2015**, *163*, 39–53. [CrossRef] [PubMed]
49. Hanahan, D.; Weinberg, R.A. Hallmarks of cancer: The next generation. *Cell* **2011**, *144*, 646–674. [CrossRef]
50. Willyard, C. The mice with human tumours: Growing pains for a popular cancer model. *Nature* **2018**, *560*, 156–157. [CrossRef]
51. Ireson, C.R.; Alavijeh, M.S.; Palmer, A.M.; Fowler, E.R.; Jones, H.J. The role of mouse tumour models in the discovery and development of anticancer drugs. *Br. J. Cancer* **2019**, *121*, 101–108. [CrossRef]

52. Ciociola, A.A.; Cohen, L.B.; Kulkarni, P.; Gastroenterology, F.D.-R.M.C.o.t.A.C.o. How drugs are developed and approved by the FDA: Current process and future directions. *Am. J. Gastroenterol.* **2014**, *109*, 620–623. [CrossRef]
53. Schiffman, J.D.; Breen, M. Comparative oncology: What dogs and other species can teach us about humans with cancer. *Philos. Trans. R Soc. Lond. B Biol. Sci.* **2015**, *370*. [CrossRef]
54. Hahn, K.A.; Bravo, L.; Adams, W.H.; Frazier, D.L. Naturally occurring tumors in dogs as comparative models for cancer therapy research. *In Vivo* **1994**, *8*, 133–143. [PubMed]
55. Reif, J.S.; Bruns, C.; Lower, K.S. Cancer of the nasal cavity and paranasal sinuses and exposure to environmental tobacco smoke in pet dogs. *Am. J. Epidemiol.* **1998**, *147*, 488–492. [CrossRef] [PubMed]
56. Reif, J.S.; Dunn, K.; Ogilvie, G.K.; Harris, C.K. Passive smoking and canine lung cancer risk. *Am. J. Epidemiol.* **1992**, *135*, 234–239. [CrossRef] [PubMed]
57. Laflamme, D.; Izquierdo, O.; Eirmann, L.; Binder, S. Myths and misperceptions about ingredients used in commercial pet foods. *Vet. Clin. North. Am. Small Anim. Pract.* **2014**, *44*, 689–698. [CrossRef]
58. Song, S.J.; Lauber, C.; Costello, E.K.; Lozupone, C.A.; Humphrey, G.; Berg-Lyons, D.; Caporaso, J.G.; Knights, D.; Clemente, J.C.; Nakielny, S.; et al. Cohabiting family members share microbiota with one another and with their dogs. *Elife* **2013**, *2*, e00458. [CrossRef]
59. Lindblad-Toh, K.; Wade, C.M.; Mikkelsen, T.S.; Karlsson, E.K.; Jaffe, D.B.; Kamal, M.; Clamp, M.; Chang, J.L.; Kulbokas, E.J., 3rd; Zody, M.C.; et al. Genome sequence, comparative analysis and haplotype structure of the domestic dog. *Nature* **2005**, *438*, 803–819. [CrossRef]
60. Vonholdt, B.M.; Pollinger, J.P.; Lohmueller, K.E.; Han, E.; Parker, H.G.; Quignon, P.; Degenhardt, J.D.; Boyko, A.R.; Earl, D.A.; Auton, A.; et al. Genome-wide SNP and haplotype analyses reveal a rich history underlying dog domestication. *Nature* **2010**, *464*, 898–902. [CrossRef]
61. Hoffman, M.M.; Birney, E. Estimating the neutral rate of nucleotide substitution using introns. *Mol. Biol Evol* **2007**, *24*, 522–531. [CrossRef]
62. Paoloni, M.C.; Khanna, C. Comparative oncology today. *Vet. Clin. North. Am. Small Anim Pract* **2007**, *37*, 1023–1032. [CrossRef]
63. Khanna, C.; Lindblad-Toh, K.; Vail, D.; London, C.; Bergman, P.; Barber, L.; Breen, M.; Kitchell, B.; McNeil, E.; Modiano, J.F.; et al. The dog as a cancer model. *Nat. Biotechnol.* **2006**, *24*, 1065–1066. [CrossRef]
64. Di Cerbo, A.; Palmieri, B.; De Vico, G.; Iannitti, T. Onco-epidemiology of domestic animals and targeted therapeutic attempts: Perspectives on human oncology. *J. Cancer Res. Clin. Oncol* **2014**, *140*, 1807–1814. [CrossRef] [PubMed]
65. Modiano, J.F.; Breen, M.; Burnett, R.C.; Parker, H.G.; Inusah, S.; Thomas, R.; Avery, P.R.; Lindblad-Toh, K.; Ostrander, E.A.; Cutter, G.C.; et al. Distinct B-cell and T-cell lymphoproliferative disease prevalence among dog breeds indicates heritable risk. *Cancer Res.* **2005**, *65*, 5654–5661. [CrossRef] [PubMed]
66. Nishiya, A.T.; Massoco, C.O.; Felizzola, C.R.; Perlmann, E.; Batschinski, K.; Tedardi, M.V.; Garcia, J.S.; Mendonca, P.P.; Teixeira, T.F.; Zaidan Dagli, M.L. Comparative aspects of canine melanoma. *Vet. Sci.* **2016**, *3*, 7. [CrossRef] [PubMed]
67. Dobson, J.M. Breed-predispositions to cancer in pedigree dogs. *ISRN Vet. Sci* **2013**, *2013*, 941275. [CrossRef] [PubMed]
68. Klopfenstein, M.; Howard, J.; Rossetti, M.; Geissbuhler, U. Life expectancy and causes of death in Bernese mountain dogs in Switzerland. *BMC Vet. Res.* **2016**, *12*, 153. [CrossRef] [PubMed]
69. Lewis, T.W.; Wiles, B.M.; Llewellyn-Zaidi, A.M.; Evans, K.M.; O'Neill, D.G. Longevity and mortality in Kennel Club registered dog breeds in the UK in 2014. *Canine Genet. Epidemiol.* **2018**, *5*, 10. [CrossRef] [PubMed]
70. Dobson, J.; Hoather, T.; McKinley, T.J.; Wood, J.L. Mortality in a cohort of flat-coated retrievers in the UK. *Vet. Comp. Oncol.* **2009**, *7*, 115–121. [CrossRef]
71. Zink, M.C.; Farhoody, P.; Elser, S.E.; Ruffini, L.D.; Gibbons, T.A.; Rieger, R.H. Evaluation of the risk and age of onset of cancer and behavioral disorders in gonadectomized Vizslas. *J. Am. Vet. Med. Assoc.* **2014**, *244*, 309–319. [CrossRef]
72. Henry, C.J.; Higginbotham, H.M. *Cancer Management in Small Animal Practice*; Saunders Elsevier: Philadelphia, PA, USA, 2009.
73. Fan, T.M.; Khanna, C. Comparative Aspects of Osteosarcoma Pathogenesis in Humans and Dogs. *Vet. Sci* **2015**, *2*, 210–230. [CrossRef]

74. Urfer, S.R.; Gaillard, C.; Steiger, A. Lifespan and disease predispositions in the Irish Wolfhound: A review. *Vet. Q.* **2007**, *29*, 102–111. [CrossRef]
75. Zandvliet, M. Canine lymphoma: A review. *Vet. Q.* **2016**, *36*, 76–104. [CrossRef] [PubMed]
76. Gajewski, T.F.; Schreiber, H.; Fu, Y.X. Innate and adaptive immune cells in the tumor microenvironment. *Nat. Immunol.* **2013**, *14*, 1014–1022. [CrossRef] [PubMed]
77. Ribas, A. Adaptive immune resistance: How cancer protects from immune attack. *Cancer Discov.* **2015**, *5*, 915–919. [CrossRef] [PubMed]
78. Goulart, M.R.; Pluhar, G.E.; Ohlfest, J.R. Identification of myeloid derived suppressor cells in dogs with naturally occurring cancer. *PLoS ONE* **2012**, *7*, e33274. [CrossRef]
79. Pinheiro, D.; Chang, Y.M.; Bryant, H.; Szladovits, B.; Dalessandri, T.; Davison, L.J.; Yallop, E.; Mills, E.; Leo, C.; Lara, A.; et al. Dissecting the regulatory microenvironment of a large animal model of non-Hodgkin lymphoma: Evidence of a negative prognostic impact of FOXP3+ T cells in canine B cell lymphoma. *PLoS ONE* **2014**, *9*, e105027. [CrossRef]
80. Coy, J.; Caldwell, A.; Chow, L.; Guth, A.; Dow, S. PD-1 expression by canine T cells and functional effects of PD-1 blockade. *Vet. Comp. Oncol.* **2017**, *15*, 1487–1502. [CrossRef]
81. Zerdes, I.; Wallerius, M.; Sifakis, E.G.; Wallmann, T.; Betts, S.; Bartish, M.; Tsesmetzis, N.; Tobin, N.P.; Coucoravas, C.; Bergh, J.; et al. STAT3 activity promotes programmed-death ligand 1 expression and suppresses immune responses in breast cancer. *Cancers* **2019**, *11*, 1479. [CrossRef]
82. Reardon, S. CRISPR gene-editing creates wave of exotic model organisms. *Nature* **2019**, *568*, 441–442. [CrossRef]
83. Simoff, I.; Karlgren, M.; Backlund, M.; Lindstrom, A.C.; Gaugaz, F.Z.; Matsson, P.; Artursson, P. Complete knockout of endogenous Mdr1 (Abcb1) in MDCK cells by CRISPR-Cas9. *J. Pharm. Sci.* **2016**, *105*, 1017–1021. [CrossRef]
84. Amoasii, L.; Hildyard, J.C.W.; Li, H.; Sanchez-Ortiz, E.; Mireault, A.; Caballero, D.; Harron, R.; Stathopoulou, T.R.; Massey, C.; Shelton, J.M.; et al. Gene editing restores dystrophin expression in a canine model of Duchenne muscular dystrophy. *Science* **2018**, *362*, 86–91. [CrossRef]
85. Liongue, C.; Ward, A.C. Evolution of the JAK-STAT pathway. *JAKSTAT* **2013**, *2*, e22756. [CrossRef] [PubMed]
86. Morris, R.; Kershaw, N.J.; Babon, J.J. The molecular details of cytokine signaling via the JAK/STAT pathway. *Protein Sci.* **2018**, *27*, 1984–2009. [CrossRef] [PubMed]
87. Polak, K.L.; Chernosky, N.M.; Smigiel, J.M.; Tamagno, I.; Jackson, M.W. Balancing STAT Activity as a Therapeutic Strategy. *Cancers* **2019**, *11*, 1716. [CrossRef] [PubMed]
88. Yue-Ting, K.L.; Ramaiyer, M.; E. Johnson, D.; R. Grandis, J. Targeting STAT3 in cancer with nucleotide therapeutics. *Cancers* **2019**, *11*, 1681. [CrossRef]
89. Gough, D.J.; Corlett, A.; Schlessinger, K.; Wegrzyn, J.; Larner, A.C.; Levy, D.E. Mitochondrial STAT3 supports Ras-dependent oncogenic transformation. *Science* **2009**, *324*, 1713–1716. [CrossRef]
90. Meier, J.A.; Larner, A.C. Toward a new STATe: The role of STATs in mitochondrial function. *Semin. Immunol.* **2014**, *26*, 20–28. [CrossRef]
91. Frank, D.A. STAT3 as a central mediator of neoplastic cellular transformation. *Cancer Lett.* **2007**, *251*, 199–210. [CrossRef]
92. Bar-Natan, M.; Nelson, E.A.; Xiang, M.; Frank, D.A. STAT signaling in the pathogenesis and treatment of myeloid malignancies. *JAKSTAT* **2012**, *1*, 55–64. [CrossRef]
93. Roeser, J.C.; Leach, S.D.; McAllister, F. Emerging strategies for cancer immunoprevention. *Oncogene* **2015**, *34*, 6029–6039. [CrossRef]
94. Yu, H.; Jove, R. The STATs of cancer–new molecular targets come of age. *Nat. Rev. Cancer* **2004**, *4*, 97–105. [CrossRef]
95. Sen, M.; Thomas, S.M.; Kim, S.; Yeh, J.I.; Ferris, R.L.; Johnson, J.T.; Duvvuri, U.; Lee, J.; Sahu, N.; Joyce, S.; et al. First-in-human trial of a STAT3 decoy oligonucleotide in head and neck tumors: Implications for cancer therapy. *Cancer Discov.* **2012**, *2*, 694–705. [CrossRef] [PubMed]
96. Park, J.S.; Kwok, S.K.; Lim, M.A.; Kim, E.K.; Ryu, J.G.; Kim, S.M.; Oh, H.J.; Ju, J.H.; Park, S.H.; Kim, H.Y.; et al. STA-21, a promising STAT-3 inhibitor that reciprocally regulates Th17 and Treg cells, inhibits osteoclastogenesis in mice and humans and alleviates autoimmune inflammation in an experimental model of rheumatoid arthritis. *Arthritis Rheumatol.* **2014**, *66*, 918–929. [CrossRef] [PubMed]

97. Kunigal, S.; Lakka, S.S.; Sodadasu, P.K.; Estes, N.; Rao, J.S. Stat3-siRNA induces Fas-mediated apoptosis in vitro and in vivo in breast cancer. *Int. J. Oncol.* **2009**, *34*, 1209–1220. [PubMed]
98. Darnell, J.E., Jr. Transcription factors as targets for cancer therapy. *Nat. Rev. Cancer* **2002**, *2*, 740–749. [CrossRef] [PubMed]
99. Ferbeyre, G.; Moriggl, R. The role of Stat5 transcription factors as tumor suppressors or oncogenes. *Biochim. Biophys. Acta* **2011**, *1815*, 104–114. [CrossRef]
100. Zhang, H.F.; Lai, R. STAT3 in Cancer-Friend or Foe? *Cancers* **2014**, *6*, 1408–1440. [CrossRef]
101. Starr, R.; Hilton, D.J. Negative regulation of the JAK/STAT pathway. *Bioessays* **1999**, *21*, 47–52. [CrossRef]
102. Wake, M.S.; Watson, C.J. STAT3 the oncogene - still eluding therapy? *FEBS J.* **2015**, *282*, 2600–2611. [CrossRef]
103. Carmi-Levy, I.; Homey, B.; Soumelis, V. A modular view of cytokine networks in atopic dermatitis. *Clin. Rev. Allergy Immunol.* **2011**, *41*, 245–253. [CrossRef]
104. Brandt, E.B.; Sivaprasad, U. Th2 Cytokines and Atopic Dermatitis. *J. Clin. Cell Immunol* **2011**, 2. [CrossRef]
105. Ong, P.Y.; Leung, D.Y. Immune dysregulation in atopic dermatitis. *Curr. Allergy Asthma Rep.* **2006**, *6*, 384–389. [CrossRef] [PubMed]
106. Hill, P.B.; Lo, A.; Eden, C.A.; Huntley, S.; Morey, V.; Ramsey, S.; Richardson, C.; Smith, D.J.; Sutton, C.; Taylor, M.D.; et al. Survey of the prevalence, diagnosis and treatment of dermatological conditions in small animals in general practice. *Vet. Rec.* **2006**, *158*, 533–539. [CrossRef] [PubMed]
107. Olivry, T.; Bizikova, P. A systematic review of randomized controlled trials for prevention or treatment of atopic dermatitis in dogs: 2008-2011 update. *Vet. Dermatol.* **2013**, *24*, 97–117, e25–e26. [CrossRef] [PubMed]
108. Marsella, R.; Olivry, T.; Maeda, S. Cellular and cytokine kinetics after epicutaneous allergen challenge (atopy patch testing) with house dust mites in high-IgE beagles. *Vet. Dermatol.* **2006**, *17*, 111–120. [CrossRef] [PubMed]
109. Nuttall, T.J.; Knight, P.A.; McAleese, S.M.; Lamb, J.R.; Hill, P.B. Expression of Th1, Th2 and immunosuppressive cytokine gene transcripts in canine atopic dermatitis. *Clin. Exp. Allergy* **2002**, *32*, 789–795. [CrossRef] [PubMed]
110. Schlotter, Y.M.; Rutten, V.P.; Riemers, F.M.; Knol, E.F.; Willemse, T. Lesional skin in atopic dogs shows a mixed Type-1 and Type-2 immune responsiveness. *Vet. Immunol. Immunopathol.* **2011**, *143*, 20–26. [CrossRef]
111. Gonzales, A.J.; Bowman, J.W.; Fici, G.J.; Zhang, M.; Mann, D.W.; Mitton-Fry, M. Oclacitinib (APOQUEL((R))) is a novel Janus kinase inhibitor with activity against cytokines involved in allergy. *J. Vet. Pharmacol. Ther.* **2014**, *37*, 317–324. [CrossRef]
112. Cosgrove, S.B.; Wren, J.A.; Cleaver, D.M.; Walsh, K.F.; Follis, S.I.; King, V.I.; Tena, J.K.; Stegemann, M.R. A blinded, randomized, placebo-controlled trial of the efficacy and safety of the Janus kinase inhibitor oclacitinib (Apoquel(R)) in client-owned dogs with atopic dermatitis. *Vet. Dermatol.* **2013**, *24*, 587–597, e141–e142. [CrossRef]
113. Cosgrove, S.B.; Wren, J.A.; Cleaver, D.M.; Martin, D.D.; Walsh, K.F.; Harfst, J.A.; Follis, S.L.; King, V.L.; Boucher, J.F.; Stegemann, M.R. Efficacy and safety of oclacitinib for the control of pruritus and associated skin lesions in dogs with canine allergic dermatitis. *Vet. Dermatol.* **2013**, *24*, 479-e114. [CrossRef]
114. Collard, W.T.; Hummel, B.D.; Fielder, A.F.; King, V.L.; Boucher, J.F.; Mullins, M.A.; Malpas, P.B.; Stegemann, M.R. The pharmacokinetics of oclacitinib maleate, a Janus kinase inhibitor, in the dog. *J. Vet. Pharmacol Ther* **2014**, *37*, 279–285. [CrossRef]
115. Lu, Z.; Hong, C.C.; Jark, P.C.; Assumpcao, A.; Bollig, N.; Kong, G.; Pan, X. JAK1/2 Inhibitors AZD1480 and CYT387 inhibit canine B-Cell lymphoma growth by increasing apoptosis and disrupting cell proliferation. *J. Vet. Intern. Med.* **2017**, *31*, 1804–1815. [CrossRef] [PubMed]
116. Assumpcao, A.; Jark, P.C.; Hong, C.C.; Lu, Z.; Ruetten, H.M.; Heaton, C.M.; Pinkerton, M.E.; Pan, X. STAT3 Expression and Activity are Up-Regulated in Diffuse Large B Cell Lymphoma of Dogs. *J. Vet. Intern. Med.* **2018**, *32*, 361–369. [CrossRef] [PubMed]
117. Krol, M.; Pawlowski, K.M.; Dolka, I.; Musielak, O.; Majchrzak, K.; Mucha, J.; Motyl, T. Density of Gr1-positive myeloid precursor cells, p-STAT3 expression and gene expression pattern in canine mammary cancer metastasis. *Vet. Res. Commun.* **2011**, *35*, 409–423. [CrossRef] [PubMed]
118. Van Garderen, E.; Swennenhuis, J.F.; Hellmen, E.; Schalken, J.A. Growth hormone induces tyrosyl phosphorylation of the transcription factors Stat5a and Stat5b in CMT-U335 canine mammary tumor cells. *Domest. Anim. Endocrinol.* **2001**, *20*, 123–135. [CrossRef]

119. Brown, M.E.; Bear, M.D.; Rosol, T.J.; Premanandan, C.; Kisseberth, W.C.; London, C.A. Characterization of STAT3 expression, signaling and inhibition in feline oral squamous cell carcinoma. *BMC Vet. Res.* **2015**, *11*, 206. [CrossRef]
120. Petterino, C.; Podesta, G.; Ratto, A.; Drigo, M.; Pellegrino, C. Immunohistochemical study of phospho-Stat3-ser727 expression in feline mammary gland tumours. *Vet. Res. Commun.* **2007**, *31*, 173–184. [CrossRef]
121. Petterino, C.; Ratto, A.; Podesta, G.; Drigo, M.; Pellegrino, C. Immunohistochemical evaluation of STAT3-p-tyr705 expression in feline mammary gland tumours and correlation with histologic grade. *Res. Vet. Sci.* **2007**, *82*, 218–224. [CrossRef]
122. Willmann, M.; Hadzijusufovic, E.; Hermine, O.; Dacasto, M.; Marconato, L.; Bauer, K.; Peter, B.; Gamperl, S.; Eisenwort, G.; Jensen-Jarolim, E.; et al. Comparative oncology: The paradigmatic example of canine and human mast cell neoplasms. *Vet. Comp. Oncol.* **2019**, *17*, 1–10. [CrossRef]
123. Makielski, K.M.; Mills, L.J.; Sarver, A.L.; Henson, M.S.; Spector, L.G.; Naik, S.; Modiano, J.F. Risk factors for development of canine and human osteosarcoma: A comparative review. *Vet. Sci.* **2019**, *6*, 48. [CrossRef]
124. Rowell, J.L.; McCarthy, D.O.; Alvarez, C.E. Dog models of naturally occurring cancer. *Trends Mol. Med.* **2011**, *17*, 380–388. [CrossRef]
125. Mirabello, L.; Troisi, R.J.; Savage, S.A. Osteosarcoma incidence and survival rates from 1973 to 2004: Data from the surveillance, epidemiology, and end results program. *Cancer* **2009**, *115*, 1531–1543. [CrossRef] [PubMed]
126. Fenger, J.M.; London, C.A.; Kisseberth, W.C. Canine osteosarcoma: A naturally occurring disease to inform pediatric oncology. *ILAR J.* **2014**, *55*, 69–85. [CrossRef] [PubMed]
127. Fossey, S.L.; Liao, A.T.; McCleese, J.K.; Bear, M.D.; Lin, J.; Li, P.K.; Kisseberth, W.C.; London, C.A. Characterization of STAT3 activation and expression in canine and human osteosarcoma. *BMC Cancer* **2009**, *9*, 81. [CrossRef] [PubMed]
128. Fossey, S.L.; Bear, M.D.; Lin, J.; Li, C.; Schwartz, E.B.; Li, P.K.; Fuchs, J.R.; Fenger, J.; Kisseberth, W.C.; London, C.A. The novel curcumin analog FLLL32 decreases STAT3 DNA binding activity and expression, and induces apoptosis in osteosarcoma cell lines. *BMC Cancer* **2011**, *11*, 112. [CrossRef] [PubMed]
129. Couto, J.I.; Bear, M.D.; Lin, J.; Pennel, M.; Kulp, S.K.; Kisseberth, W.C.; London, C.A. Biologic activity of the novel small molecule STAT3 inhibitor LLL12 against canine osteosarcoma cell lines. *BMC Vet. Res.* **2012**, *8*, 244. [CrossRef] [PubMed]
130. Onimoe, G.I.; Liu, A.; Lin, L.; Wei, C.C.; Schwartz, E.B.; Bhasin, D.; Li, C.; Fuchs, J.R.; Li, P.K.; Houghton, P.; et al. Small molecules, LLL12 and FLLL32, inhibit STAT3 and exhibit potent growth suppressive activity in osteosarcoma cells and tumor growth in mice. *Invest. New Drugs* **2012**, *30*, 916–926. [CrossRef]
131. Paoloni, M.; Davis, S.; Lana, S.; Withrow, S.; Sangiorgi, L.; Picci, P.; Hewitt, S.; Triche, T.; Meltzer, P.; Khanna, C. Canine tumor cross-species genomics uncovers targets linked to osteosarcoma progression. *BMC Genomics* **2009**, *10*, 625. [CrossRef]
132. Rodriguez, C.O., Jr. Using canine osteosarcoma as a model to assess efficacy of novel therapies: Can old dogs teach us new tricks? *Adv. Exp. Med. Biol.* **2014**, *804*, 237–256. [CrossRef]
133. London, C.A.; Seguin, B. Mast cell tumors in the dog. *Vet. Clin. North. Am. Small Anim. Pract.* **2003**, *33*, 473–489. [CrossRef]
134. Misdorp, W. Mast cells and canine mast cell tumours. A review. *Vet. Q.* **2004**, *26*, 156–169. [CrossRef]
135. Shoop, S.J.; Marlow, S.; Church, D.B.; English, K.; McGreevy, P.D.; Stell, A.J.; Thomson, P.C.; O'Neill, D.G.; Brodbelt, D.C. Prevalence and risk factors for mast cell tumours in dogs in England. *Canine Genet. Epidemiol.* **2015**, *2*, 1. [CrossRef] [PubMed]
136. Lim, K.H.; Tefferi, A.; Lasho, T.L.; Finke, C.; Patnaik, M.; Butterfield, J.H.; McClure, R.F.; Li, C.Y.; Pardanani, A. Systemic mastocytosis in 342 consecutive adults: Survival studies and prognostic factors. *Blood* **2009**, *113*, 5727–5736. [CrossRef] [PubMed]
137. Letard, S.; Yang, Y.; Hanssens, K.; Palmerini, F.; Leventhal, P.S.; Guery, S.; Moussy, A.; Kinet, J.P.; Hermine, O.; Dubreuil, P. Gain-of-function mutations in the extracellular domain of KIT are common in canine mast cell tumors. *Mol. Cancer Res.* **2008**, *6*, 1137–1145. [CrossRef] [PubMed]
138. London, C.A.; Galli, S.J.; Yuuki, T.; Hu, Z.Q.; Helfand, S.C.; Geissler, E.N. Spontaneous canine mast cell tumors express tandem duplications in the proto-oncogene c-kit. *Exp. Hematol.* **1999**, *27*, 689–697. [CrossRef]

139. Nagata, H.; Worobec, A.S.; Oh, C.K.; Chowdhury, B.A.; Tannenbaum, S.; Suzuki, Y.; Metcalfe, D.D. Identification of a point mutation in the catalytic domain of the protooncogene c-kit in peripheral blood mononuclear cells of patients who have mastocytosis with an associated hematologic disorder. *Proc. Natl. Acad. Sci. USA* **1995**, *92*, 10560–10564. [CrossRef]
140. Metcalfe, D.D. Mast cells and mastocytosis. *Blood* **2008**, *112*, 946–956. [CrossRef]
141. Iemura, A.; Tsai, M.; Ando, A.; Wershil, B.K.; Galli, S.J. The c-kit ligand, stem cell factor, promotes mast cell survival by suppressing apoptosis. *Am. J. Pathol.* **1994**, *144*, 321–328.
142. Moller, C.; Alfredsson, J.; Engstrom, M.; Wootz, H.; Xiang, Z.; Lennartsson, J.; Jonsson, J.I.; Nilsson, G. Stem cell factor promotes mast cell survival via inactivation of FOXO3a-mediated transcriptional induction and MEK-regulated phosphorylation of the proapoptotic protein Bim. *Blood* **2005**, *106*, 1330–1336. [CrossRef]
143. Hahn, K.A.; Ogilvie, G.; Rusk, T.; Devauchelle, P.; Leblanc, A.; Legendre, A.; Powers, B.; Leventhal, P.S.; Kinet, J.P.; Palmerini, F.; et al. Masitinib is safe and effective for the treatment of canine mast cell tumors. *J. Vet. Intern. Med.* **2008**, *22*, 1301–1309. [CrossRef]
144. London, C.A.; Malpas, P.B.; Wood-Follis, S.L.; Boucher, J.F.; Rusk, A.W.; Rosenberg, M.P.; Henry, C.J.; Mitchener, K.L.; Klein, M.K.; Hintermeister, J.G.; et al. Multi-center, placebo-controlled, double-blind, randomized study of oral toceranib phosphate (SU11654), a receptor tyrosine kinase inhibitor, for the treatment of dogs with recurrent (either local or distant) mast cell tumor following surgical excision. *Clin. Cancer Res.* **2009**, *15*, 3856–3865. [CrossRef]
145. Chaix, A.; Lopez, S.; Voisset, E.; Gros, L.; Dubreuil, P.; De Sepulveda, P. Mechanisms of STAT protein activation by oncogenic KIT mutants in neoplastic mast cells. *J. Biol. Chem.* **2011**, *286*, 5956–5966. [CrossRef] [PubMed]
146. Baumgartner, C.; Cerny-Reiterer, S.; Sonneck, K.; Mayerhofer, M.; Gleixner, K.V.; Fritz, R.; Kerenyi, M.; Boudot, C.; Gouilleux, F.; Kornfeld, J.W.; et al. Expression of activated STAT5 in neoplastic mast cells in systemic mastocytosis: Subcellular distribution and role of the transforming oncoprotein KIT D816V. *Am. J. Pathol.* **2009**, *175*, 2416–2429. [CrossRef] [PubMed]
147. Keller, A.; Wingelhofer, B.; Peter, B.; Bauer, K.; Berger, D.; Gamperl, S.; Reifinger, M.; Cerny-Reiterer, S.; Moriggl, R.; Willmann, M.; et al. The JAK2/STAT5 signaling pathway as a potential therapeutic target in canine mastocytoma. *Vet. Comp. Oncol.* **2018**, *16*, 55–68. [CrossRef] [PubMed]
148. Peter, B.; Bibi, S.; Eisenwort, G.; Wingelhofer, B.; Berger, D.; Stefanzl, G.; Blatt, K.; Herrmann, H.; Hadzijusufovic, E.; Hoermann, G.; et al. Drug-induced inhibition of phosphorylation of STAT5 overrides drug resistance in neoplastic mast cells. *Leukemia* **2018**, *32*, 1016–1022. [CrossRef] [PubMed]
149. Page, B.D.; Khoury, H.; Laister, R.C.; Fletcher, S.; Vellozo, M.; Manzoli, A.; Yue, P.; Turkson, J.; Minden, M.D.; Gunning, P.T. Small molecule STAT5-SH2 domain inhibitors exhibit potent antileukemia activity. *J. Med. Chem.* **2012**, *55*, 1047–1055. [CrossRef]
150. Park, J.S.; Withers, S.S.; Modiano, J.F.; Kent, M.S.; Chen, M.; Luna, J.I.; Culp, W.T.N.; Sparger, E.E.; Rebhun, R.B.; Monjazeb, A.M.; et al. Canine cancer immunotherapy studies: Linking mouse and human. *J. Immunother. Cancer* **2016**, *4*, 97. [CrossRef]
151. Alvarez, C.E. Naturally occurring cancers in dogs: Insights for translational genetics and medicine. *ILAR J.* **2014**, *55*, 16–45. [CrossRef]
152. Furdos, I.; Fazekas, J.; Singer, J.; Jensen-Jarolim, E. Translating clinical trials from human to veterinary oncology and back. *J. Transl. Med.* **2015**, *13*, 265. [CrossRef]
153. Riccardo, F.; Aurisicchio, L.; Impellizeri, J.A.; Cavallo, F. The importance of comparative oncology in translational medicine. *Cancer Immunol. Immunother.* **2015**, *64*, 137–148. [CrossRef]

Permissions

All chapters in this book were first published by MDPI; hereby published with permission under the Creative Commons Attribution License or equivalent. Every chapter published in this book has been scrutinized by our experts. Their significance has been extensively debated. The topics covered herein carry significant findings which will fuel the growth of the discipline. They may even be implemented as practical applications or may be referred to as a beginning point for another development.

The contributors of this book come from diverse backgrounds, making this book a truly international effort. This book will bring forth new frontiers with its revolutionizing research information and detailed analysis of the nascent developments around the world.

We would like to thank all the contributing authors for lending their expertise to make the book truly unique. They have played a crucial role in the development of this book. Without their invaluable contributions this book wouldn't have been possible. They have made vital efforts to compile up to date information on the varied aspects of this subject to make this book a valuable addition to the collection of many professionals and students.

This book was conceptualized with the vision of imparting up-to-date information and advanced data in this field. To ensure the same, a matchless editorial board was set up. Every individual on the board went through rigorous rounds of assessment to prove their worth. After which they invested a large part of their time researching and compiling the most relevant data for our readers.

The editorial board has been involved in producing this book since its inception. They have spent rigorous hours researching and exploring the diverse topics which have resulted in the successful publishing of this book. They have passed on their knowledge of decades through this book. To expedite this challenging task, the publisher supported the team at every step. A small team of assistant editors was also appointed to further simplify the editing procedure and attain best results for the readers.

Apart from the editorial board, the designing team has also invested a significant amount of their time in understanding the subject and creating the most relevant covers. They scrutinized every image to scout for the most suitable representation of the subject and create an appropriate cover for the book.

The publishing team has been an ardent support to the editorial, designing and production team. Their endless efforts to recruit the best for this project, has resulted in the accomplishment of this book. They are a veteran in the field of academics and their pool of knowledge is as vast as their experience in printing. Their expertise and guidance has proved useful at every step. Their uncompromising quality standards have made this book an exceptional effort. Their encouragement from time to time has been an inspiration for everyone.

The publisher and the editorial board hope that this book will prove to be a valuable piece of knowledge for researchers, students, practitioners and scholars across the globe.

List of Contributors

Benedikt Linder and Donat Kögel
Experimental Neurosurgery, Department of Neurosurgery, Neuroscience Center, Goethe University Hospital, 60528 Frankfurt am Main, Germany

Ulrike Weirauch, Alexander Ewe and Achim Aigner
Rudolf-Boehm-Institute for Pharmacology and Toxicology, Clinical Pharmacology, University of Leipzig, 04107 Leipzig, Germany

Anja Uhmann
Institute of Human Genetics, Tumor Genetics Group, University of Göttingen, 37073 Göttingen, Germany

Volker Seifert
Department of Neurosurgery, Goethe University Hospital, 60528 Frankfurt am Main, Germany

Michel Mittelbronn
Institute of Neurology (Edinger-Institute), University Hospital Frankfurt, Goethe University, 60528 Frankfurt am Main, Germany
Luxembourg Centre for Systems Biomedicine (LCSB), University of Luxembourg, 4362 Esch-sur-Alzette, Luxembourg
Laboratoire national de santé (LNS), 3555 Dudelange, Luxembourg
Luxembourg Centre of Neuropathology (LCNP), 3555 Dudelange, Luxembourg
NORLUX Neuro-Oncology Laboratory, Luxembourg Institute of Health (LIH), 1526 Luxembourg, Luxembourg

Patrick N. Harter
Institute of Neurology (Edinger-Institute), University Hospital Frankfurt, Goethe University, 60528 Frankfurt am Main, Germany
German Cancer Consortium DKTK Partner site Frankfurt/Main, Frankfurt am Main, Germany and German Cancer Research Center DKFZ, 69120 Heidelberg, Germany

Katharina Wöss, Natalija Simonović, Birgit Strobl, Sabine Macho-Maschler and Mathias Müller
Institute of Animal Breeding and Genetics, University of Veterinary Medicine Vienna, A-1210 Vienna, Austria

Yung-Hsing Huang, Yuen Morrissey, Chengsheng Wu and Raymond Lai
Department of Laboratory Medicine and Pathology, University of Alberta, Edmonton, AB T6G 2R3, Canada

Mohammad Reza Vakili, Igor Paiva, Amir Hasan Soleimani, Forugh Sanaee and Afsaneh Lavasanifar
Faculty of Pharmacy and Pharmaceutical Sciences, University of Alberta, Edmonton, AB T6G 2H7, Canada

Ommoleila Molavi
Department of Laboratory Medicine and Pathology, University of Alberta, Edmonton, AB T6G 2R3, Canada
Faculty of Pharmacy, Tabriz University of Medical Science, Tabriz East Azerbaijan Province, Iran

Florence Boutillon, Natascha Pigat, Lucila Sackmann Sala, Jacques-Emmanuel Guidotti and Vincent Goffin
Institut Necker Enfants Malades, Inserm U1151, 75014 Paris, France
Faculté de Médecine, Université Paris Descartes, 75014 Paris, France

Edouard Reyes-Gomez
Unité d'Histologie et d'Anatomie Pathologique, Ecole Nationale Vétérinaire d'Alfort, 94704 Maisons-Alfort, France
Laboratoire d'Anatomo-Cytopathologie, BioPôle Alfort, Ecole Nationale Vétérinaire d'Alfort, 94704 Maisons-Alfort, France
U955—IMRB, Inserm, Ecole Nationale Vétérinaire d'Alfort, UPEC, 94704 Maisons-Alfort, France

Richard Moriggl
Institute of Animal Breeding and Genetics, University of Veterinary Medicine Vienna, 1210 Vienna, Austria
Medical University of Vienna, 1090 Vienna, Austria

Gaëtan Jego and François Hermetet
INSERM, LNC UMR1231, team HSP-Pathies, University of Bourgogne Franche-Comté, F-21000 Dijon, France
UFR des Sciences de Santé, University of Burgundy and Franche-Comté, F-21000 Dijon, France

François Girodon
INSERM, LNC UMR1231, team HSP-Pathies, University of Bourgogne Franche-Comté, F-21000 Dijon, France
UFR des Sciences de Santé, University of Burgundy and Franche-Comté, F-21000 Dijon, France
Haematology laboratory, Dijon University Hospital, F-21000 Dijon, France

Carmen Garrido
INSERM, LNC UMR1231, team HSP-Pathies, University of Bourgogne Franche-Comté, F-21000 Dijon, France
UFR des Sciences de Santé, University of Burgundy and Franche-Comté, F-21000 Dijon, France
Centre Georges François Leclerc, 21000 Dijon, France

Sabrina Ernst
Institute of Biochemistry and Molecular Biology, RWTH Aachen University, 52074 Aachen, Germany
Confocal Microscopy Facility, Interdisciplinary Center for Clinical Research IZKF, RWTH Aachen University, 52074 Aachen, Germany

Gerhard Müller-Newen
Institute of Biochemistry and Molecular Biology, RWTH Aachen University, 52074 Aachen, Germany

Yue-Ting K. Lau, Malini Ramaiyer, Daniel E. Johnson and Jennifer R. Grandis
Department Otolaryngology – Head and Neck Surgery, University of California at San Francisco, 1450 3rd Street, Room HD268, San Francisco, CA 94143, USA

Srikanta Dash
Department of Pathology and Laboratory Medicine, Tulane University Health Sciences Center, New Orleans, LA 70112, USA
Section of Gastroenterology and Hepatology, Tulane University Health Sciences Center, New Orleans, LA 70112, USA

Yucel Aydin, Kyoungsub Song, Dong Lin, Hanadi Osman, Brady Youngquist and John W. Scott
Department of Pathology and Laboratory Medicine, Tulane University Health Sciences Center, New Orleans, LA 70112, USA

Nathan J. Shores and Ramazan Kurt
Section of Gastroenterology and Hepatology, Tulane University Health Sciences Center, New Orleans, LA 70112, USA

Paul Thevenot and Ari Cohen
Liver Transplant Surgery Section, Ochsner Medical Center, New Orleans, LA 70121, USA

Antonio Vella, Alice Giacomazzi, Tiziana Cestari and Vincenzo Bronte
Unit of Immunology, Azienda Ospedaliera Universitaria Integrata, 37134 Verona, Italy

Elisabetta D'Aversa and Giulia Breveglieri
Department of Life Sciences and Biotechnology, University of Ferrara, 44100 Ferrara, Italy

Martina Api, Marisole Allegri, Benedetta Fabrizzi and Valentino Bezzerri
Cystic Fibrosis Center, Azienda Ospedaliero Universitaria Ospedali Riuniti, 60126 Ancona, Italy

Elena Marinelli Busilacchi and Antonella Poloni
Hematology Clinic, Università Politecnica delle Marche -AOU Ospedali Riuniti, 60126 Ancona, Italy

Claudio Sorio
Department of Medicine, University of Verona, 37134 Verona, Italy

Gloria Bedini and Giovanna D'Amico
Immunology and Cell Therapy Unit, Tettamanti Research Center, University of Milano-Bicocca, 20900 Monza, Italy

Antonio Benedetti
Department of Gastroenterology and Hepatology, Università Politecnica delle Marche, 60126 Ancona, Italy

Chiara Bovo
Hospital Health Direction, Azienda Ospedaliera Universitaria Integrata, 37126 Verona, Italy

Seth J. Corey
Department of Pediatric Hematology/Oncology and Stem Cell Transplantation, Cleveland Clinic, Cleveland, OH 44195, USA

Monica Borgatti
Department of Life Sciences and Biotechnology, University of Ferrara, 44100 Ferrara, Italy
Biotechnology Center, University of Ferrara, 44100 Ferrara, Italy

Marco Cipolli
Cystic Fibrosis Center, Azienda Ospedaliera Universitaria Integrata, 37126 Verona, Italy

Marie Brachet-Botineau and Fabrice Gouilleux
Leukemic Niche and Oxidative metabolism (LNOx), CNRS ERL 7001, University of Tours, 37000 Tours, France

Marion Polomski, Ludovic Juen, Damien Hédou, Marie-Claude Viaud-Massuard and Gildas Prié
Innovation Moléculaire et Thérapeutique (IMT), EA 7501, University of Tours, 37000 Tours, France

Heidi A. Neubauer
Institute of Animal Breeding and Genetics, University of Veterinary Medicine Vienna, A-1210 Vienna, Austria

Matthias Kieslinger, Alexander Swoboda, Nina Kramer, Barbara Pratscher, Birgitt Wolfesberger and Iwan A. Burgener
Division of Small Animal Internal Medicine, Department of Companion Animals and Horses, University of Veterinary Medicine Vienna, 1210 Vienna, Austria

Index

O

P

R

S

T

X

Printed in the USA
CPSIA information can be obtained
at www.ICGtesting.com
JSHW051414091023
49903JS00006B/412

9 781646 465552